Lecture Notes in Computer Science   12080

More information about this series at http://www.springer.com/series/7409

Michael R. Berthold · Ad Feelders ·
Georg Krempl (Eds.)

Advances in Intelligent Data Analysis XVIII

18th International Symposium on Intelligent Data Analysis, IDA 2020
Konstanz, Germany, April 27–29, 2020
Proceedings

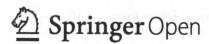
Springer Open

Editors
Michael R. Berthold
University of Konstanz
Konstanz, Germany

Ad Feelders
Utrecht University
Utrecht, The Netherlands

Georg Krempl
Utrecht University
Utrecht, The Netherlands

ISSN 0302-9743 ISSN 1611-3349 (electronic)
Lecture Notes in Computer Science
ISBN 978-3-030-44583-6 ISBN 978-3-030-44584-3S (eBook)
https://doi.org/10.1007/978-3-030-44584-3

LNCS Sublibrary: SL3 – Information Systems and Applications, incl. Internet/Web, and HCI

Preface

We are proud to present the proceedings of the 18th International Symposium on Intelligent Data Analysis (IDA 2020), which was held during April 27–29, 2020, in Konstanz, Germany. The first symposium of this series was organized in 1995 and held biannually until 2009, when the conference switched to being held annually. Following demand expressed by the IDA community in a survey held in 2018, IDA 2020 was the first of the series to take place in spring rather than fall, as was common before.

The switch to April, and a more organized outreach to the community, coincided with an increase in the number of submissions from 65 in 2018, to 114 in 2020. After a rigorous review process, 45 of these 114 submissions were accepted for presentation. Almost all submissions were reviewed by at least three Program Committee (PC) members (only two papers had two reviews) and a substantial number of submissions received more than three reviews. In addition to the PC, the review process also involved program chair advisors – a select set of senior researchers with a multi-year involvement in the IDA symposium series. Whenever a program chair advisor flagged a paper with an informed, thoughtful, positive review due to the paper presenting a particularly interesting and novel idea, the paper was accepted irrespective of the other reviews. Each accepted paper was offered a slot for either oral presentation (15 papers) or poster presentation (30 papers).

We wish to express our gratitude to the authors of all submitted papers for their high-quality contributions; to the PC members and additional reviewers for their efforts in reviewing, discussing, and commenting on all submitted papers; to the program chair advisors for their active involvement; and to the IDA council for their ongoing guidance and support. Many people have helped behind the scenes to make IDA 2020 possible, but this year we are particularly grateful to our publicity chairs who helped spread the word: Daniela Gawehns and Hugo Manuel Proença!

February 2020

Georg Krempl
Ad Feelders
Michael R. Berthold

Organization

Program Chairs

Georg Krempl   Utrecht University, The Netherlands
Ad Feelders   Utrecht University, The Netherlands

Program Chair Advisors

Niall Adams	Imperial College London, UK
Michael R. Berthold	University of Konstanz, Germany
Hendrik Blockeel	Katholieke Universiteit Leuven, Belgium
Elizabeth Bradley	University of Colorado Boulder, USA
Tijl De Bie	Ghent University, Belgium
Wouter Duivesteijn	Eindhoven University of Technology, The Netherlands
Elisa Fromont	Université de Rennes 1, France
Johannes Fürnkranz	Johannes Kepler University Linz, Austria
Jaakko Hollmén	Aalto University, Finland
Frank Höppner	Ostfalia University of Applied Sciences, Germany
Frank Klawonn	Ostfalia University of Applied Sciences, Germany
Arno Knobbe	Leiden University, The Netherlands
Rudolf Kruse	University of Magdeburg, Germany
Nada Lavrač	Jozef Stefan Institute, Slovenia
Matthijs van Leeuwen	Leiden University, The Netherlands
Xiaohui Liu	Brunel University, UK
Panagiotis Papapetrou	Stockholm University, Sweden
Arno Siebes	Utrecht University, The Netherlands
Stephen Swift	Brunel University, UK
Hannu Toivonen	University of Helsinki, Finland
Allan Tucker	Brunel University, UK
Albrecht Zimmermann	Université Caen Normandie, France

Program Committee

Fabrizio Angiulli	DEIS, University of Calabria, Italy
Martin Atzmueller	Tilburg University, The Netherlands
José Luis Balcázar	Universitat Politècnica de Catalunya, Spain
Giacomo Boracchi	Politecnico di Milano, Italy
Christian Borgelt	Universität Salzburg, Austria
Henrik Boström	KTH Royal Institute of Technology, Sweden
Paula Brito	University of Porto, Portugal
Dariusz Brzezinski	Poznań University of Technology, Poland
José Del Campo-Ávila	Universidad de Málaga, Spain

Cassio de Campos	Eindhoven University of Technology, The Netherlands
Andre de Carvalho	University of São Paulo, Brazil
Paulo Cortez	University of Minho, Portugal
Bruno Cremilleux	Université de Caen Normandie, France
Brett Drury	LIAAD-INESC-TEC, Portugal
Saso Dzeroski	Jozef Stefan Institute, Slovenia
Nuno Escudeiro	Instituto Superior de Engenharia do Porto, Portugal
Douglas Fisher	Vanderbilt University, USA
Joao Gama	University of Porto, Portugal
Lawrence Hall	University of South Florida, USA
Barbara Hammer	Bielefeld University, Germany
Martin Holena	Institute of Computer Science, Czech Republic
Tomas Horvath	Eötvös Loránd University, Hungary
Francois Jacquenet	Laboratoire Hubert Curien, France
Baptiste Jeudy	Laboratoire Hubert Curien, France
Ulf Johansson	Jönköping University, Sweden
Alipio M. Jorge	University of Porto, Portugal
Irena Koprinska	The University of Sydney, Australia
Daniel Kottke	University of Kassel, Germany
Petra Kralj Novak	Jozef Stefan Institute, Slovenia
Mark Last	Ben-Gurion University of the Negev, Israel
Niklas Lavesson	Jönköping University, Sweden
Daniel Lawson	University of Bristol, UK
Jefrey Lijffijt	Ghent University, Belgium
Ling Luo	The University of Melbourne, Australia
George Magoulas	Birkbeck University of London, UK
Vlado Menkovski	Eindhoven University of Technology, The Netherlands
Vera Migueis	University of Porto, Portugal
Decebal Constantin Mocanu	Eindhoven University of Technology, The Netherlands
Emilie Morvant	University of Saint-Etienne, LaHC, France
Mohamed Nadif	Paris Descartes University, France
Siegfried Nijssen	Université Catholique de Louvain, Belgium
Andreas Nuernberger	Otto-von-Guericke University of Magdeburg, Germany
Kaustubh Raosaheb Patil	Massachusetts Institute of Technology, USA
Mykola Pechenizkiy	Eindhoven University of Technology, The Netherlands
Jose-Maria Pena	Universidad Politécnica de Madrid, Spain
Ruggero G. Pensa	University of Torino, Italy
Marc Plantevit	LIRIS, Université Claude Bernard Lyon 1, France
Lubos Popelinsky	Masaryk University, Czech Republic
Eric Postma	Tilburg University, The Netherlands
Miguel A. Prada	Universidad de Leon, Spain
Ronaldo Prati	Universidade Federal do ABC, UFABC, Brazil
Peter van der Putten	Leiden University and Pegasystems, The Netherlands
Jesse Read	École Polytechnique, France
Antonio Salmeron	University of Almería, Spain
Vítor Santos Costa	University of Porto, Portugal

Contents

Multivariate Time Series as Images: Imputation Using Convolutional Denoising Autoencoder

Abdullah Al Safi, Christian Beyer[(✉)], Vishnu Unnikrishnan,
and Myra Spiliopoulou

Fakultät für Informatik, Otto-von-Guericke-Universität,
Postfach 4120, 39106 Magdeburg, Germany
abdullah.safi@st.ovgu.de,
{christian.beyer,vishnu.unnikrishnan,myra}@ovgu.de

Abstract. Missing data is a common occurrence in the time series domain, for instance due to faulty sensors, server downtime or patients not attending their scheduled appointments. One of the best methods to impute these missing values is *Multiple Imputations by Chained Equations (MICE)* which has the drawback that it can only model linear relationships among the variables in a multivariate time series. The advancement of deep learning and its ability to model non-linear relationships among variables make it a promising candidate for time series imputation. This work proposes a modified Convolutional Denoising Autoencoder (CDA) based approach to impute multivariate time series data in combination with a preprocessing step that encodes time series data into 2D images using Gramian Angular Summation Field (GASF). We compare our approach against a standard feed-forward Multi Layer Perceptron (MLP) and MICE. All our experiments were performed on 5 UEA MTSC multivariate time series datasets, where 20 to 50% of the data was simulated to be missing completely at random. The CDA model outperforms all the other models in 4 out of 5 datasets and is tied for the best algorithm in the remaining case.

Keywords: Convolutional Denoising Autoencoder · Gramian Angular Summation Field · MICE · MLP. · Imputation · Time series

1   Introduction

Time series data resides in various domains of industries and research fields and is often corrupted with missing data. For further use or analysis, the data often needs to be complete, which gives the rise to the need for imputation techniques with enhanced capabilities of introducing least possible error into the data. One of the most prominent imputation methods is MICE which uses iterative regression and value replacement to achieve state-of-the-art imputation quality but has the drawback that it can only model linear relationships among variables (dimensions).

© The Author(s) 2020
M. R. Berthold et al. (Eds.): IDA 2020, LNCS 12080, pp. 1–13, 2020.
https://doi.org/10.1007/978-3-030-44584-3_1

In past few years, different deep learning architectures were able to break into different problem domains, often exceeding previously achieved performances by other algorithms [7]. Areas like speech recognition, natural language processing, computer vision, etc. were greatly impacted and improved by deep learning architectures. Deep learning models have a robust capability of modelling latent representation of the data and non-linear patterns, given enough training data. Hence, this work presents a deep learning based imputation model called Convolutional Denoising Autoencoder (CDA) with altered convolution and pooling operations in Encoder and Decoder segments. Instead of using the traditional steps of convolution and pooling, we use deconvolution and upsampling which was inspired by [5]. The time series to image transformation mechanisms proposed in [12] and [13] were inherited as a preprocessing step as CDA models are typically designed for images. As rival imputation models, Multiple Imputation by Chained Equations (MICE) and a Multi Layer Perceptron (MLP) based imputation were incorporated.

2   Related Work

Three distinct types of missingness in data were identified in [8]. The first one is *Missing Completely At Random (MCAR)*, where the missingness of the data does not depend on itself or any other variables. In *Missing At Random (MAR)* the missing value depends on other variables but not on the variable where the data is actually missing and in *Missing Not At Random (MNAR)* the missingness of an observation depends on the concerned variable itself. All the experiments in this study were carried out on MCAR missingness as reproducing MAR and MNAR missingness can be challenging and hard to distinguish [5].

Multiple Imputation by Chained Equations (MICE) has secured its place as a principal method for imputing missing data [1]. Costa et al. in [3] experimented and showed that MICE offered the better imputation quality than a Denoising Autoencoder based model for several missing percentages and missing types.

A novel approach was proposed in [14], incorporating General Adversarial Networks (GAN) to perform imputations, thus authors named it Generative Adversarial Imputation Nets (GAIN). The approach imputed significantly well against some state-of-the-art imputation methods including MICE. An Autoencoder based approach was proposed in [4], which was compared against an Artificial Neural Network (NN) model on MCAR missing type and several missing percentages. The proposed model performed well against NN. A novel Denoising Autoencoder based imputation using partial loss (DAPL) approach was presented in [9], where different missing data percentages and MCAR missing type were simulated in a breast cancer dataset. The comparisons incorporated statistical, machine learning based approaches and standard Denoising Autoencoder (DAE) model where DAPL outperformed DAE and all the other models. An MLP based imputation approach was presented for MCAR missingness in [10] and also outperformed other statistical models. A Convolutional Denoising Autoencoder model which did not impute missing data but denoised audio signals was presented in [15]. A Denoising Autoencoder with more units in the encoder layer than input layer was presented in [5] and achieved good imputation results against MICE. Our work was inspired from both of these works which is why we combined the two approaches into a Convolutional Denoising Autoencoder which maps input data into a higher subspace in the Encoder.

3 Methodology

In this section we first describe how we introduce missing data in our datasets, then we show the process used to turn multivariate time series into images which is required by one of our imputation methods and finally we introduce the imputation methods which were compared in this study.

3.1 Simulating Missing Data

Simulating missing data is a mechanism of artificially introducing unobserved data into a complete time series dataset. Our experiment incorporated 20%, 30%, 40% and 50% of missing data and the missing type was MCAR. Introducing MCAR missingness is quite a simple approach as it does not depend on observed or unobserved data. Many studies assume MCAR missing type quite often when there is no concrete evidence of missingness type [6]. In this experimental framework, values at randomly selected indices were erased from randomly selected variables which simulated MCAR missingness of different percentages.

3.2 Translating Time Series into Images

A novel approach of encoding time series data into various types of images using Gramian Angular Field (GAF) was presented in [12] to improve classification and imputation. One of the variants of GAF was Gramian Angular Summation Field (GASF), which comprised of multiple steps to perform the encoding. First, the time series is scaled within $[-1, 1]$ range.

$$x_i' = \frac{(x_i - Max(X)) + (x_i - Min(X))}{Max(X) - Min(X)} \tag{1}$$

Here, x_i is a specific value at timepoint i where x_i' is derived by scaling and X is the time series. The time series is scaled within $[-1, 1]$ range in order to be represented as polar coordinates achieved by applying angular cosine.

$$\theta_i = arccos(x_i')\{-1 <= x_i' <= 1, x_i' \in X\} \tag{2}$$

The polar encoded time series vector is then transformed into a matrix. If the length of the time series vector is n, then the transformed matrix is of shape $(n \times n)$.

$$GASF_{i,j} = cos(\theta_i + \theta_j) \tag{3}$$

The GASF represents the temporal features in the form of an image where the timestamps move along top-left to bottom-right, thereby preserving the time factor in the data. Figure 1 shows the different steps of time series to image transformation.

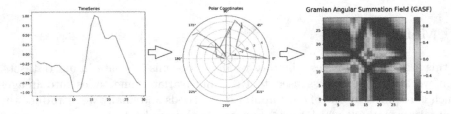

Fig. 1. Time series to image transformation

The methods of encoding time series into images described in [12] were only applicable for univariate time series. The GASF transformation generates one image for one time series dimension and thus it is possible to generate multiple images for multivariate time series. An approach which vertically stacked images transformed from different variables was presented in [13], see Fig. 2. The images were grayscaled and the different orders of vertical stacking (ascending, descending and random) were examined by performing a statistical test. The stacking order did not impact classification accuracy.

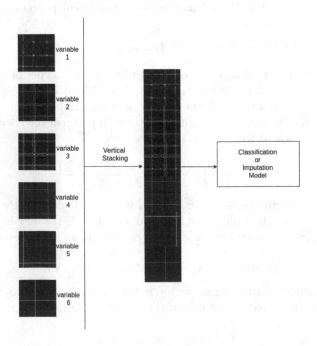

Fig. 2. Vertical stacking of images transformed from different variables

3.3   Convolutional Denoising Autoencoder

Autoencoder is a very popular unsupervised deep learning model frequently found in different application areas. Autoencoder is unsupervised in fashion and reconstructs the original input by discovering robust features in the hidden layer representation. The latent representation of high dimensional data in the hidden layer contributes in reconstructing the original data. The architecture of Autoencoder consists of two principal segments named Encoder and Decoder. The Encoder usually compresses the original representation of the data into lower dimension. The Decoder decodes the low dimensional representation of the input back into its original dimensional representation.

$$Encoder(x^n) = s(x^n W_E + b_E) = x^d \tag{4}$$

$$Decoder(x^d) = s(x^d W_D + b_D) = x^n \tag{5}$$

Here, x^n is the original input with n dimensions. s is any non-linear activation function, W is weight and b is bias.

Denoising Autoencoder model is an extension of Autoencoder where the input is reconstructed from a corrupted version of it. There are different ways of adding corruption, such as Gaussian noise, setting some values to zero etc. The noisy input is fed as input and the model minimizes the loss between the clean input and corrupted reconstructed input. The objective function looks as follows

$$RMSE(X, X') \frac{1}{n} \sqrt{|X_{clean} - X'_{reconstructed}|^2} \tag{6}$$

Convolutional Denoising Autoencoder (CDA) incorporates convolution operation which is ideally performed in Convolutional Neural Networks (CNN). CNN is a methodology, where the layers of perceptrons are replaced by convolution layers and convolution operation is performed on the data. Convolution is defined as multiplication of two function within a finite or infinite range, where two functions refer to input data (e.g. Image) and a fixed size kernel consecutively. The kernel traverses through the input space to generate feature maps. The feature maps consist of important features of the data. The multiple features are pooled, preserving important features.

The combination of convoluted feature maps generation and pooling is performed in the Encoder layer of CDA where the corrupted version of the input is fed into the input layer of the network. The Decoder layer performs Deconvolutiont and Upsampling which decompresses the output coming from Encoder layer back into the shape of input data. The loss between reconstructed data and clean data is minimized. In this work, the default architecture of CDA is tweaked in the favor of imputing multivariate time series data. Deconvolution and Upsampling were performed in the Encoder layer and Convolution and Maxpooling was performed in Decoder layer. The motivation behind this specific tweaking came from [5], where a Denoising Autoencoder was designed with more hidden units in the Encoder layer than input layer. The high dimensional representation in Encoder layer created additional feature which was the contributor of data recovery.

3.4 Competitor Models

Multiple Imputation by Chained Equations (MICE): MICE, which is sometimes addressed as fully conditional specification or sequential regression multiple imputation, has emerged in the statistical literature as the principal method of addressing missing data [1]. MICE creates multiple versions of the imputed datasets through multiple imputation technique.

The steps for performing MICE are the following:

- A simple imputation method is performed across the time series (mean, mode or median). The missing time points are referred as "placeholders".
- If there are total m variables having missing points, then one of the variables are set back to missing state. The variable with "missing state" label is considered as dependent variable and other variables are considered as predictors.
- A regression is performed over these settings and "missing state" variable is imputed. Different regressions are supported in this architecture but since the dataset only contains continuous values, linear, ridge or lasso regression are chosen.
- The remaining $m - 1$ "missing state" are regressed and imputed by the same way. Once all the m variables are imputed, one iteration is completed. More iterations are performed and the imputations are placed in the time series in each iteration.
- The number of iterations can be determined by observing whether coefficients of the regression model are converged or not.

According to the experimental setup of our work, MICE had three different regression supports, namely Linear, Ridge and Lasso regression.

Multi Layer Perceptron (MLP) Based Imputation: The imputation mechanism of MLP is inspired by the MICE algorithm. Nevertheless, MLP based imputation models do not perform the chained or multiple imputations like MICE but improve the quality of imputation over several epochs as stochastic gradient descent optimizes the weights and biases per epoch. A concrete MLP architecture was described in literature [10] which was a three layered MLP with the hyperbolic tangent activation function in the hidden layer and the identity function (linear) as the activation function for the output layer. The train and test split were slightly different, where training set and test set consisted of both observed and unobserved data.

The imputation process of MLP model in our work is similar to MICE but the non-linear activation function of MLP facilitates finding complex non-linear patterns. However, the imputation of a variable is performed only once, in contrast to the multiple iterations in MICE.

4 Experiments

In this section we present the used datasets, the preprocessing steps that were conducted before training, the chosen hyperparameters and our evaluation method. Our complete imputation process for the CDA model is depicted in Fig. 3. The process for the competitors is the same except that corrupting the training data and turning the time series into images is not being done.

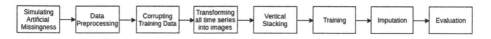

Fig. 3. Experiment steps for the CDA model

4.1 Datasets and Data Preprocessing

Our experiments were conducted on 5 time series datasets from the UEA MTSC repository [2]. Each dataset in UEA time series archive has training and test splits and specific number of dimensions. Each training or test split represents a time series. The table below presents all the relevant structural details (Table 1).

Table 1. A structural summary of the 5 UEA MTSC dataset

Dataset name	Number of series	Dimensions	Length	Classes
ArticularyWordRecognition	275	9	144	25
Cricket	108	6	1197	12
Handwriting	150	3	152	26
StandWalkJump	12	4	2500	3
UWaveGestureLibrary	120	3	315	8

The Length column of the table denotes the length of each time series. In our framework, each time series was transformed into images. The number of time series for any of the datasets was not very high in number. As we had selected a deep learning model for imputation, such low number of samples could cause overfitting. Experiments showed us that the default number of time series could not perform well. Therefore, the main idea was to increase the number of time series by splitting them into multiple parts and reducing their corresponding lengths. This modification facilitated us by introducing more patterns for learning which aided in imputation. The final lengths chosen were those that yielded the best results. The table below presents the modified number of time series and lengths for each dataset (Table 2).

Table 2. Modified number of time series and lengths

Dataset name	Number of series	Dimension	Length
ArticularyWordRecognition	6600	9	6
Cricket	6804	6	19
Handwriting	1200	3	19
StandWalkJump	3000	4	10
UWaveGestureLibrary	1800	3	21

The evaluation of the imputation models require a complete dataset and the corresponding incomplete dataset. Therefore, artificial missingness was introduced at different percentages (20%, 30%, 40% and 50%) into all the datasets. After simulating artificial missingness, each dataset has an observed part, which contains all the time series segments where no variables are missing and an unobserved part, where at least one variable is missing. After simulating artificial missingness, each dataset had an observed and unobserved split and the observed data was further processed for training. As CDA models learn denoising from a corrupted version of the input, we introduced noise by discarding a certain amount of values for each observed case from specific variables and replacing them by the mean of the corresponding variables. A higher amount of noise has seen to be contributing more in learning dependencies of different variables, which leads to denoising of good quality [11]. The variables selected for adding noise were the same variables having missing data in unobserved data. Different amount of noise was examined but 90% noise lead to good results. Unobserved data was also mean imputed as the CDA model would apply the denoising technique on the "mean-noise" for imputation. So the CDA learns to deal with "mean-noise" on the observed part and is then applied on mean imputed unobserved part to create the final imputation.

The next step was to perform time series to image transformation where, all the observed and unobserved chunks were rescaled between −1 to 1 using min-max scaling. Rescaled data was further transformed into polar coordinates and then GASF encoded image was achieved for each dimension. Multiple images referring to multiple variables were vertically aggregated. Finally, both observed and unobserved splits consisted their own set of images.

Note that, the following data preprocessing was performed only for CDA based imputation models. The competitor models imputed using the raw format of the data.

4.2   Model Architecture and Hyperparameters

Our Model architecture was different from a general CDA, where the Encoder layer incorporates Deconvolution and Upsampling operations and the Decoder layer incorporates Convolution and Maxpooling operations. The Encoder and Decoder both have 3 layers. The table below demonstrates the structure of the imputation model (Table 3).

Table 3. The architecture of CDA based imputation model

	Operation	Layer name	Kernel size	Number of feature maps
Encoder	Upsampling	up_0	(2, 2)	–
	Deconvolution	deconv_0	(5, 5)	64
	Upsampling	up_1	(2, 2)	–
	Deconvolution	deconv_1	(7, 7)	64
	Upsampling	up_2	(2, 2)	–
	Deconvolution	deconv_2	(5, 6)	128
Decoder	Convolution	conv_0	(5, 6)	128
	Maxpool	pool_0	(2, 2)	–
	Convolution	conv_1	(7, 7)	64
	Maxpool	pool_1	(2, 2)	–
	Convolution	conv_2	(5, 5)	64
	Maxpool	pool_2	(2, 2)	–

Hyperparameter specification was achieved by performing random search on different random combinations of hyperparameter values and the root mean square error (RMSE) was used to decide on the best combination. The random search allowed us to avoid the exhaustive searching unlike grid search. Applying random search, we selected stochastic gradient descent (SGD) as optimizer, which backpropagates the error to optimize the weights and biases. The number of epochs was 100 and the batch size was 16.

4.3   Competitor Model's Architecture and Hyperparameters

As competitor models, MICE and MLP based imputation models were selected. MLP based model had 3 hidden layers and number of hidden units were 2/3 of the number of input units in each layer. The hyperparameters for both of the models were tuned by using random search.

Hyperbolic Tangent Function was selected as activation function with a dropout of 0.3. Stochastic Gradient Descent operated as optimizer for 150 epochs and with a batch size of 20.

MICE based imputation was demonstrated using Linear, Ridge and Lasso regression and 10 iterations were performed for each of them.

4.4   Training

Based on the preprocessed data and model architecture described above, the training is started. L2 regularization was used with weight of 0.01 and stochastic gradient descent was used as the optimizer which outperformed Adam and Adagrad optimizers. The whole training process was about learning to minimize loss between the clean and corrupted data so that it can be applied on the unobserved data (noisy data after mean imputation) to perform imputation. The training and validation split was 70% and 30%. Experiments show that, the training and validation loss was saturated approximately after 10–15 epochs, which was observed for most of the cases.

The training was conducted on a machine with Nvidia RTX 2060 with RAM memory of 16 GB. The programming language for the training and all the steps above was Python 3.7 and the operating system was Ubuntu 16.04 LTS.

4.5   Evaluation Criteria

As all the time series dataset contain continuous numeric values, Root Mean Square Error (RMSE) was selected for evaluation. In out experimental setup, RMSE is not calculated on overall time series but only missing data points are taken into account to be compared with ground truth while calculating RMSE $RMSE = \sqrt{\frac{1}{m}\Sigma_{i=1}^{m}(x_i - x_i')^2}$. Where m is the total number of missing time points and I represents all the indices of missing values across the time series.

5   Results

Our proposed CDA based imputation model was compared with MLP and three different versions of MICE, each using a different type of regression. Figure 4 presents the RMSE values for 20%, 30% 40% and 50% missingness.

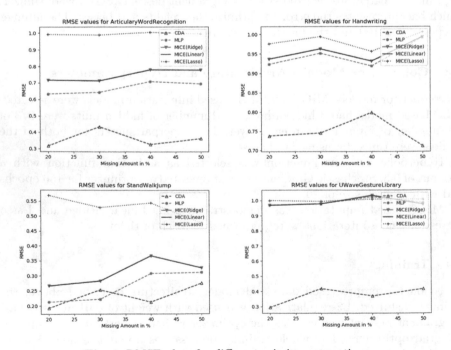

Fig. 4. RMSE plots for different missing proportions

The RMSE values for the CDA based model are the lowest at every percentage of missingness on the *Handwriting, ArticularyWordRecognition, UWaveGestureLibrary and Cricket* dataset. The depiction of the results on the *Cricket* dataset is omitted due to space limitations. Unexpectedly, in *StandWalkJump* dataset the performance of MLP and CDA model are very similar, and MLP is even better at 30% missingness. MICE (Linear) and MICE (Ridge) are identical in imputation for all the datasets. MICE (Lasso) performed worst of all the models, which implies that changing the regression type could potentially cause an impact on the imputation quality. The MLP model beat all the MICE models but was outperformed by the CDA model in at least for 80% of the cases.

6   Conclusion

In this work, we introduce an architecture of a Convolutional Denoising Autoencoder (CDA) adapted for multivariate time series imputation which inflates the size of the hidden layers in the Encoder instead of reducing them. We also employ a preprocessing step that turns the time series into 2D images based on Gramian Angular Summation Fields in order to make the data more suitable for our CDA. We compare our method against a standard Multi Layer Perceptron (MLP) and the state-of-the-art imputation method Multiple Imputations by Chained Equations (MICE) with three different types of regression (Linear, Ridge and Lasso). Our experiments were conducted on five different multivariate time series datasets, for which we simulated 20%, 30%, 40% and 50% missingness with data missing completely at random. Our results show that the CDA based imputation outperforms MICE on all five datasets and also beats the MLP on four datasets. On the fifth dataset CDA and MLP perform very similarly, but CDA is still better on four out of the five degrees of missingness. Additionally we present a preprocessing step on the datasets which manipulates the time series lengths to generate more training samples for our model which led to a better performance. The results show that the CDA model performs strongly against both linear and non-linear regression based imputation models. Deep Learning Networks are usually computationally more intensive than MICE but the imputation quality of CDA was convincing enough to be chosen over MICE or MLP based imputation.

In the future we plan to investigate also other types of missing data apart from *Missing Completely At Random (MCAR)* and want to incorporate more datasets as well as other deep learning based approaches for imputation.

Acknowledgments. This work is partially funded by the German Research Foundation, project OSCAR "Opinion Stream Classification with Ensembles and Active Learners". The principal investigators of OSCAR are Myra Spiliopoulou and Eirini Ntoutsi. Additionally, Christian Beyer is also partially funded by a PhD grant from the federal state of Saxony-Anhalt.

References

1. Azur, M.J., Stuart, E.A., Frangakis, C., Leaf, P.J.: Multiple imputation by chained equations: what is it and how does it work? Int. J. Methods Psychiatr. Res. **20**(1), 40–49 (2011)
2. Bagnall, A., et al.: The UEA multivariate time series classification archive, 2018. arXiv preprint arXiv:1811.00075 (2018)
3. Costa, A.F., Santos, M.S., Soares, J.P., Abreu, P.H.: Missing data imputation via denoising autoencoders: the untold story. In: Duivesteijn, W., Siebes, A., Ukkonen, A. (eds.) IDA 2018. LNCS, vol. 11191, pp. 87–98. Springer, Cham (2018). https://doi.org/10.1007/978-3-030-01768-2_8
4. Duan, Y., Lv, Y., Kang, W., Zhao, Y.: A deep learning based approach for traffic data imputation. In: 17th International IEEE Conference on Intelligent Transportation Systems (ITSC), pp. 912–917. IEEE (2014)
5. Gondara, L., Wang, K.: MIDA: multiple imputation using denoising autoencoders. In: Phung, D., Tseng, V.S., Webb, G.I., Ho, B., Ganji, M., Rashidi, L. (eds.) PAKDD 2018, Part III. LNCS (LNAI), vol. 10939, pp. 260–272. Springer, Cham (2018). https://doi.org/10.1007/978-3-319-93040-4_21
6. Kang, H.: The prevention and handling of the missing data. Korean J. Anesthesiol. **64**(5), 402–406 (2013)
7. LeCun, Y., Bengio, Y., Hinton, G.: Deep learning. Nature **521**(7553), 436–444 (2015)
8. Little, R.J., Rubin, D.B.: Statistical Analysis with Missing Data, vol. 793. Wiley, Hoboken (2019)
9. Qiu, Y.L., Zheng, H., Gevaert, O.: A deep learning framework for imputing missing values in genomic data. bioRxiv, p. 406066 (2018)
10. Silva-Ramírez, E.L., Pino-Mejías, R., López-Coello, M., Cubiles-de-la Vega, M.D.: Missing value imputation on missing completely at random data using multilayer perceptrons. Neural Netw. **24**(1), 121–129 (2011)
11. Vincent, P., Larochelle, H., Bengio, Y., Manzagol, P.A.: Extracting and composing robust features with denoising autoencoders. In: Proceedings of the 25th International Conference on Machine Learning, pp. 1096–1103 (2008)
12. Wang, Z., Oates, T.: Imaging time-series to improve classification and imputation. In: Twenty-Fourth International Joint Conference on Artificial Intelligence (2015)
13. Yang, C.L., Yang, C.Y., Chen, Z.X., Lo, N.W.: Multivariate time series data transformation for convolutional neural network. In: 2019 IEEE/SICE International Symposium on System Integration (SII), pp. 188–192. IEEE (2019)
14. Yoon, J., Jordon, J., Van Der Schaar, M.: Gain: missing data imputation using generative adversarial nets. arXiv preprint arXiv:1806.02920 (2018)
15. Zhao, M., Wang, D., Zhang, Z., Zhang, X.: Music removal by convolutional denoising autoencoder in speech recognition. In: 2015 Asia-Pacific Signal and Information Processing Association Annual Summit and Conference (APSIPA), pp. 338–341. IEEE (2015)

Dual Sequential Variational Autoencoders for Fraud Detection

Ayman Alazizi[1,2]([✉]), Amaury Habrard[1], François Jacquenet[1],
Liyun He-Guelton[2], and Frédéric Oblé[2]

[1] Univ. Lyon, Univ. St-Etienne, UMR CNRS 5516, Laboratoire Hubert-Curien,
42000 Saint-Etienne, France
{ayman.alazizi,amaury.habrard,francois.jacquenet}@univ-st-etienne.fr
[2] Worldline, 95870 Bezons, France
{ayman.alazizi,liyun.he-guelton,frederic.oble}@worldline.com

Abstract. Fraud detection is an important research area where machine learning has a significant role to play. An important task in that context, on which the quality of the results obtained depends, is feature engineering. Unfortunately, this is very time and human consuming. Thus, in this article, we present the DuSVAE model that consists of a generative model that takes into account the sequential nature of the data. It combines two variational autoencoders that can generate a condensed representation of the input sequential data that can then be processed by a classifier to label each new sequence as fraudulent or genuine. The experiments we carried out on a large real-word dataset, from the Worldline company, demonstrate the ability of our system to better detect frauds in credit card transactions without any feature engineering effort.

Keywords: Anomaly detection · Fraud detection · Sequential data · Variational autoencoder

1 Introduction

An anomaly (also called outlier, change, deviation, surprise, peculiarity, intrusion, etc.) is a pattern, in a dataset, that does not conform to an expected behavior. Thus, anomaly detection is the process of finding anomalies in a dataset [4]. Fraud detection, a subdomain of anomaly detection, is a research area where the use of machine learning can have a significant financial impact for companies suffering from large frauds and it is not surprising that a very large amount of research has been conducted over many years in that field [1].

At the Wordline company, we process billions of electronic transactions per year in our highly secured data centers. It is obvious that detecting frauds in that context is a very difficult task. For many years, the detection of credit card frauds within Wordline has been based on a set of rules manually designed by experts. Nevertheless such rules are difficult to maintain, difficult to transfer to other business lines, and dependent on experts who need a very long training

M. R. Berthold et al. (Eds.): IDA 2020, LNCS 12080, pp. 14–26, 2020.
https://doi.org/10.1007/978-3-030-44584-3_2

period. The contribution of machine learning in this context seems obvious and Wordline has decided for several years to develop research in this field.

Firstly, Worldline has put a lot of effort in feature engineering [3,9,12] to develop discriminative handcrafted features. This improved drastically supervised learning of classifiers that aim to label card transactions as genuine or fraudulent. Nevertheless, designing such features requires a huge amount of time and human resources which is very costly. Thus developing automatic feature engineering methods becomes a critical issue to improve the efficiency of our models. However, in our industrial setting, we have to face with many issues among which the presence of highly imbalanced data where the fraud ratio is about 0.3%. For this reason, we first focused on classic unsupervised approaches in anomaly detection where the objective is to learn a model from normal data and then isolate non-compliant samples and consider them as anomalies [5,17,19,21,22].

In this context, Deep autoencoder [7] is considered as a powerful data modeling tool in the unsupervised setting. An autoencoder (AE) is made up of two parts: an encoder designed to generate a compressed coding from the training input data and a decoder that reconstructs the original input from the compressed coding. In the context of anomaly detection [6,20,22], an autoencoder is generally trained by minimizing the reconstruction error only on normal data. Afterwards, the reconstruction error is applied as an anomaly score. This assumes that the reconstruction error for a normal data should be small as it is close to the learning data, while the reconstruction error for an abnormal data should be high.

However, this assumption is not always valid. Indeed, it has been observed that sometimes the autoencoder generalizes so well that it can also reconstruct anomalies, which leads to view some anomalies as normal data. This can also be the case when some abnormal data share some characteristics of normal data in the training set or when the decoder is "too powerful" to properly decode abnormal codings. To solve the shortcomings of autoencoders, [13,18] proposed the *negative learning technique* that aims to control the compressing capacity of an autoencoder by optimizing conflicting objectives of normal and abnormal data. Thus, this approach looks for a solution in the gradient direction for the desired normal input and in the opposite direction for the undesired input.

This approach could be very appealing to deal with fraud detection problems but we found that it is sometimes not sufficient in the context of our data. Indeed, it is generally almost impossible to obtain in advance a dataset containing all representative frauds, especially in the context where unknown fraudulent transactions occur on new terminals or via new fraudulent behaviors. This has led us to consider more complex models with variational autoencoders (VAE), a probabilistic generative extension of AE, able to model complex generative distributions that we found more adapted to efficiently model new possible frauds.

Another important point for credit card fraud detection is the sequential aspect of the data. Indeed, to test a card for example, a fraudster may try to make several (small) transactions in a short time interval, or directly perform an abnormally high transaction with respect to existing transactions of the true card holder. In fact this sequential aspect has been addressed either indirectly via aggregated features [3], that we would like to avoid designing, or directly by sequential models such as LSTM, but [9] report nevertheless that the LSTM did not improve much the detection performance for e-commerce transactions. One of the main contribution of this paper is to propose a method to identify fraudulent sequences of credit transactions in the context of highly imbalanced data. For this purpose, we propose a model called DuSVAE, for Dual Sequential Variational Autoencoders, that consists of a combination of two variational autoencoders. The first one is trained from fraudulent sequences of transactions in order to be able to project the input data into another feature space and to assign a fraud score to each sequence thanks to the reconstruction error information. Once this model is trained, we plug a second VAE at the output of the first one. This second VAE is then trained with a negative learning approach with the objective to maximize the reconstruction error of the fraudulent sequences and minimize the reconstruction error of the genuine ones.

Our method has been evaluated on a Wordline dataset for credit card fraud detection. The obtained results show that DuSVAE can extract hidden representations able to provide results close to those obtained after a significant work of feature engineering, therefore saving time and human effort. It is even possible to improve the results when combining engineered features with DuSVAE.

The article is organized as follows: some preliminaries about the techniques used in this work are given in Sect. 2. Then we describe the architecture and the training strategy of the DusVAE method in Sect. 3. Experiments are presented in Sect. 4 after a presentation of the dataset and useful metrics. Finally Sect. 5 concludes this article.

2   Preliminaries

In this section, we briefly describe the main techniques that are used in DuSVAE: vanilla and variational autoencoders, negative learning and mixture of experts.

2.1   Autoencoder (AE)

An AE is a neural network [7], which is optimized in an unsupervised manner, usually used to reduce the dimensionality of the input data. It is made up of two parts linked together: an encoder $E(x)$ and a decoder $\mathcal{D}(z)$. Given an input sample x, the encoder generates z, a condensed representation of x. The decoder is then tuned to reconstruct the original input x from the encoded representation z. The objective function used during the training of the AE is given by:

$$\mathcal{L}_{AE}(x) = \|x - \mathcal{D}(E(x))\| \tag{1}$$

where $\| \cdot \|$ denotes an arbitrary distance function. The ℓ_2 norm is typically applied here. The AE can be optimized for example using stochastic gradient descent (SGD) [10].

2.2   Variational Autoencoder (VAE)

A VAE [11,16] is an attractive probabilistic generative version of the standard autoencoder. It can learn a complex distribution and then use it as a generative model defined by a prior $p(z)$ and conditional distribution $p_\theta(x|z)$. Due to the fact that the true likelihood of the data is generally intractable, a VAE is trained through maximizing the evidence lower bound (ELBO):

$$\mathcal{L}(x; \theta, \phi) = \mathbb{E}_{q_\phi(z|x)} \left[\log p_\theta(x|z)\right] - D_{\text{KL}}\left(q_\phi(z|x)\|p(z)\right) \qquad (2)$$

where the first term $\mathbb{E}_{q_\phi(z|x)}\left[\log p_\theta(x|z)\right]$ is a negative reconstruction loss that enforces $q_\phi(z|x)$ (the encoder) to generate a meaningful latent vector z, so that $p_\theta(x|z)$ (the decoder) can reconstruct the input x from z. The second term $D_{\text{KL}}\left(q_\phi(z|x)\|p(z)\right)$ is a KL regularization loss that minimizes the KL divergence between the approximate posterior $q_\phi(z|x)$ and the prior $p(z) = \mathcal{N}(\mathbf{0}, \mathbf{I})$.

2.3   Negative Learning

Negative learning is a technique used for regularizing the training of the AE in the presence of labelled data by limiting reconstruction capability (LRC) [13]. The basic idea is to maximize the reconstruction error for abnormal instances, while minimizing the reconstruction error for normal ones in order to improve the discriminative ability of the AE. Given an input instance $x \in \mathbb{R}^n$ and $y \in \{0, 1\}$ denotes its associated label where $y = 1$ stands for a fraudulent instance and $y = 0$ for a genuine one. The objective function of LRC to be minimized is:

$$(1 - y)\mathcal{L}_{AE}(x) - (y)\mathcal{L}_{AE}(x) \qquad (3)$$

Training LRC-based models has the major disadvantage to be generally unstable due to the fact that the anomaly reconstruction error is not upper bounded. The LRC approach tends then to maximize the reconstruction error for known anomalies rather than minimizing the reconstruction error for normal points leading to a bad reconstruction of normal data points. To overcome this problem, [18] has proposed Autoencoding Binary Classifiers (ABC) for supervised anomaly detection that improves LRC by using an objective function based on a bounded reconstruction loss for anomalies, leading to a better training stability. The objective function of the ABC to be minimized is:

$$(1 - y)\mathcal{L}_{AE}(x) - y \log_2(1 - e^{-\mathcal{L}_{AE}(x)}) \qquad (4)$$

2.4   Mixture-of-Experts Layer (MoE)

In addition to the previous methods, we now present the notion of MoE layer [8] that will be used in our model.

The MoE layer aims to combine the outputs of a group of n neural networks called experts $EX_1, EX_2, ..., EX_n$. The experts have their specific parameters but work on the same input, their n output are combined linearly with the

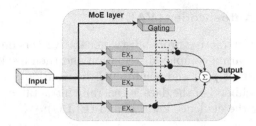

Fig. 1. An illustration of the MoE layer architecture

outputs of the gating network G which weights the experts according to the input x. See Fig. 1 for an illustration. Let $E_i(x)$ be the output of expert EX_i, and $G(x)_i$ be the ith attribute of $G(x)$, then the output y of the MoE is defined as follows:

$$y = \sum_{i=1}^{n} G(x)_i EX_i(x). \tag{5}$$

The intuition behind MoE layers is to train different network experts that can focus on specific peculiarities of the data and then choose an appropriate combination of experts with respect to the input x. In our industrial context, such a layer would help us to take into account different behaviors from millions of cardholders, which results in a variety of data distributions. The different expert networks can thus model various behaviors observed in the dataset and be combined adequately in function of the input data.

3   The DuSVAE Model

In this section, we present our approach to extract a hidden representation of input sequences to be used for anomaly/fraud detection. We first introduce the model architecture with the loss functions used, then we describe the learning procedure used to train the model.

3.1   Model Architecture

We assume in the following that we are given as input a set of sequences $\mathcal{X} = \{x \mid x = (t^1, t^2, ..., t^m) \text{ with } t^i \in \mathbb{R}^d\}$, every sequence being composed of m transactions encoded by numerical vectors. Each sequence is associated to a label $y \in \{0, 1\}$ such that $y = 1$ indicates a fraudulent sequence and $y = 0$ a genuine one. We label a sequence as fraudulent if its last transaction is a fraud.

As illustrated in Fig. 2, our approach consists of two sequential variational autoencoders. The first one is trained only on fraudulent sequences of the training data. We use the generative capacity of this autoencoder to generate diverse and representative instances of fraudulent instances with respect to the sequences given as input. This autoencoder has the objective to prepare the data for the

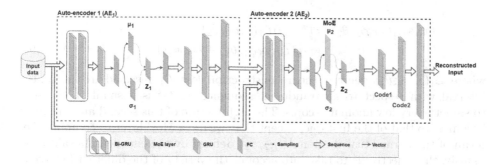

Fig. 2. The DuSVAE model architecture

second autoencoder and to provide also a first anomaly/fraud score with the reconstruction error.

The first layers of the autoencoders are bi-directional GRU layers allowing us to handle sequential data. The remaining parts of the encoder and the decoder contain GRU and fully connected (FC) layers, as shown in Fig. 2. The loss function used to optimize the reconstruction error of the first autoencoder is defined as follows:

$$\mathcal{L}_{rec}(x, \phi_1, \theta_1) = mse(x, \mathcal{D}_{\theta_1}(E_{\phi_1}(x))) + D_{\mathrm{KL}}(q_{\phi_1}(z|x)\|p(z)), \qquad (6)$$

where mse is the mean square error function and $p(z) = \mathcal{N}(\mathbf{0}, \mathbf{I})$. The encoder $E_{\phi_1}(x)$ generates a latent representation z according to $q_{\phi_1}(z|x) = \mathcal{N}(\mu_1, \sigma_1)$. The decoder $\mathcal{D}_{\theta_1}$ tries to reconstruct the input sequence from z. In order to avoid mode collapse between the reconstructed transactions of the sequence, we add the following loss function to control the reconstruction of individual transactions with respect to relative distances from an input sequence x:

$$\mathcal{L}_{trxAE}(x, \phi_1, \theta_1) = \sum_{i=1}^{m} \sum_{j=i+1}^{m} \frac{1}{d}\|abs(t^i - t^j) - abs(\bar{t}^i - \bar{t}^j)\|_1 \qquad (7)$$

where $\bar{t}^i$ is the reconstruction obtained by the AE for the ith transaction of the sequence and $abs(t)$ returns a vector where the features are the absolute values of the original input vector t.

So, we train the parameters (ϕ_1, θ_1) of the first autoencoder by minimizing the following loss function over all the fraudulent sequences of the training samples:

$$\mathcal{L}_1(x, \phi_1, \theta_1) = \mathcal{L}_{rec}(x, \phi_1, \theta_1) + \lambda \mathcal{L}_{trx}(x, \phi_1, \theta_1), \qquad (8)$$

where λ is a tradeoff parameter.

The second autoencoder is then trained over all the training sequences by negative learning. It takes as input both a sequence x and its reconstructed version from the first autoencoder $AE_1(x)$ that corresponds to the output of its last layer. The loss function considered to optimize the parameters (ϕ_2, θ_2) of the second autoencoder is then defined as follows:

$$\mathcal{L}_2(x, AE_1(x), \phi_2, \theta_2) = (1 - y)\mathcal{L}_1(x, \phi_2, \theta_2)$$
$$-y(\overline{\mathcal{L}_1}(x, \phi_1, \theta_1) + \epsilon) \log_2(1 - e^{-\mathcal{L}_1(x,\phi_2,\theta_2)}), \quad (9)$$

where $\overline{\mathcal{L}_1}(x, \phi_1, \theta_1)$ denotes the reconstruction loss $\mathcal{L}_1$ rescaled in the $[0, 1]$-interval with respect to all fraudulent sequences and ϵ is a small value used to smooth very low anomaly scores. The architecture of this second autoencoder is similar to that of the first one, except that we use a MoE layer to compute the mean of the normal distribution $\mathcal{N}(\mu_2, \sigma_2)$ defined by the encoder. As said previously, the objective is to take into account the variety of the different behavior patterns found in our genuine data. The experts used in that layer are simple one-layer feed-forward neural networks.

3.2 The Training Strategy

The global learning algorithm is presented in Algorithm 1. We have two training phases, the first one focuses on training the first autoencoder AE_1 as a backing model for the second phase. It is trained only on fraudulent sequences by minimizing Eq. 8. Once the model has converged, we freeze its weights and start the second phase. For training the second autoencoder AE_2, we use both genuine and fraudulent sequences and their reconstructed versions given by AE_1. We then optimize the weights of AE_2 by minimizing Eq. 9. To control the imbalance ratio, the training is done at each iteration by sampling n examples from fraudulent sequences and n from genuine sequences. We repeat this step iteratively by increasing the number n of sampled transactions for each novel iteration until the model converges.

Algorithm 1. Dual sequential variational autoencoder (DuSVAE)

1: **Input:** $\mathcal{X}_g$ genuine data, $\mathcal{X}_f$ fraudulent data.
2: **Parameters:** n number of sampled examples; h increment step.
3: **Output:** AE_1 Autoencoder, AE_2 Autoencoder.
4: **repeat**
5:     Train AE_1 on $\mathcal{X}_f$ by minimizing Equation 8
6: **until** convergence
7: Freeze the weights of AE_1
8: **repeat**
9:     $\mathcal{X}_1 \leftarrow Sample(\mathcal{X}_f, n) \cup Sample(\mathcal{X}_g, n)$
10:    $\mathcal{X}_2 \leftarrow AE_1(\mathcal{X}_1)$
11:    Train AE_2 on $(\mathcal{X}_1, \mathcal{X}_2)$ by minimizing Equation 9
12:    **if** $n \leq |\mathcal{X}_f|$ **then**
13:        $n \leftarrow n + h$
14:    **end if**
15: **until** convergence

Table 1. Properties of the Worldline dataset used in the experiments.

	Train (01/01-21/03)	Validation (22/03-31/03)	Test (01/04-30/04)
# of genuine	25,120,194	3,019,078	9,287,673
# of fraud	88,878	9,631	29,614
Total	25,209,072	3,028,709	9,317,287
Imbalance ratio	0.003526	0.00318	0.003178

4   Experiments

In this section, we provide an experimental evaluation of our approach on a real-world dataset of credit card e-payment transactions provided by Worldline. First, we present the dataset, then we present the metrics used to evaluate the models learned by our system and finally, we compare DuSVAE with other state-of-the-art approaches.

4.1   Dataset

The dataset provided by Wordline covers 4 months of credit card e-payment transactions made by European cardholders in e-commerce mode that has been splitted into **Train**, **Validation** and **Test** sets used respectively to train, tune and test the learned models. Its main challenges have been studied in [2], one of them being the imbalance ratio as we can see on Table 1 that presents the main characteristics of this dataset.

Each transaction is described by 12 features. A Boolean value is assigned to each transaction to specify whether it corresponds to a fraud or not. This labeling is handled by a team of human experts.

Since most features have a large number of values, using brute one-hot encoding would generate a huge number of features. For example the "Merchant Category Code" feature has 283 possible values and one-hot encoding would produce 283 new features. That would make our approach inefficient. Thus, before using one-hot encoding, we transform each categorical value of each feature by a score which is its risk to be associated with a fraudulent transaction. Let's consider for example a categorical feature f. We can compute the probability of the jth value of feature f to be associated with a fraudulent transaction, denoted as β_j, as follows: $\beta_j = \frac{N^+_{f=j}}{N_{f=j}}$, where $N^+_{f=j}$ is the number of fraudulent transactions where the value of feature f is equal to j and $N_{f=j}$ is the total number of transactions where the value of feature f is equal to j. In order to take into account the number of transactions related to a particular value of a given feature, we follow [14]. For each value j of a given feature, the fraud score S_j for this value is defined as follows:

$$S_j = \alpha'_j \beta_j + \left(1 - \alpha'_j\right) \text{AFP} \tag{10}$$

This score computes a weighted value of β_j and the probability of having a fraud in a day (Average Fraud Probability: AFP). The weight α'_j is a normalized value of α_j in the range $[0, 1]$, where α_j is defined as the proportion of the number of transactions for that value on the total number N of transactions: $\alpha_j = \frac{N_{f=j}}{N}$.

Having replaced each value for each feature by its score, we can then run one-hot encoding and thus significantly reduce the number of features generated. For example, the "Merchant Category Code" feature has 283 possible values and instead of generating 283 features, this technique produces only 29 features.

Finally, to generate sequences from transactions, we grouped all the transactions by cardholder ID and we ordered each cardholder's transactions by time. Then, with a sliding window over the transactions we obtained a time-ordered sequence of transactions for each cardholder. For each sequence, we have assigned the label *fraudulent* or *genuine* of its last transaction.

4.2 Metrics

In the context of fraud detection, fortunately, the number of fraudulent transactions is significantly lower than the number of normal transactions. This leads to a very imbalanced dataset. In this situation, the traditional performance measures are not appropriate. Indeed, with an overall fraud rate of 0.3%, classifying each transaction as normal leads to an accuracy of 99.7%, despite the fact that the model is absolutely naive. That means we have to choose appropriate performance measures that are robust in the case of imbalanced data. In this work we rely on the area under the precision-recall curve (AUC-PR) as a robust and clear measure of the accuracy of the classifier in an imbalanced setting. Each point of the precision-recall curve corresponds to the precision of the classifier at a specific recall level.

Once an alert is raised after a fraud has been detected, fraud experts can contact the card-holder to check the validity of suspicious transactions. So, within a single day, the fraud experts have to check a large number of transactions provided by the fraud detection system. Consequently, the precision of the transactions highlighted as fraud is an important metric because that can help human experts at Worldline to focus on the most important frauds and leave aside minor frauds due to lack of time to process them. For this purpose, we rely on the $P_{@K}$ as a global metric to compare models. It is the average of the precision of the first K transactions which are calculated according to the following equation.

$$Average P_{@K} = \frac{1}{K} \sum_{i=1}^{K} P_{@i} \tag{11}$$

4.3 Comparison with the State of the Art

We compare our approach with the following methods: variational autoencoder [11, 16] trained on fraudulent or genuine data only (VAE(F) or VAE(G) respectively); limiting reconstruction capability (LRC) [13] and autoencoding binary classifiers for supervised anomaly detection (ABC) [18]. It is important to note

Table 2. AUC-PR achieved by CatBoost using various autoencoder models

Models	Raw		Reconstructed		Reconstruction error	Code1	Code2
	Trx	Seq	Trx	Seq			
VAE (F)	0.19	0.40	0.36	0.38	0.29	0.30	0.27
VAE (G)			0.42	0.43	0.31	0.32	0.33
LRC			0.46	0.46	0.17	0.28	0.13
ABC			0.48	0.50	**0.37**	0.32	0.3
DuSVAE			**0.51**	**0.53**	0.36	**0.50**	**0.49**

that ABC and LRC are not sequential models by nature. So, to make our comparison more fair, we adapted their implementation to allow them to process sequential data. As a classifier, we used CatBoost [15] which is robust in the context of imbalanced data and efficient on GPUs.

First, as we can observe in Table 2, the AUC-PR values obtained by running CatBoost directly on transactions and sequences of transactions are respectively equal to 0.19 and 0.40. If we look at the AUC-PR values obtained by running CatBoost on the reconstructed transactions and sequences of transactions, we can observe that the results are always greater than those obtained by running CatBoost on raw data. Moreover it is interesting to note that DuSVAE achieved the best results (0.51 and 0.53) compared to other state-of-the-art systems.

Now, if we look at the performance obtained by CatBoost on the hidden representation vectors Code1 and Code2, we observe that DuSVAE outperforms the results obtained by other state-of-the-art systems and those results are quite similar to the ones obtained on the reconstructed sequences of transactions. This is interesting because it means that using DuSVAE a condensed representation of the input data can be obtained, which still gives approximately the same results as on the reconstructed sequences of transactions but that are of higher dimensionality (about 10 times more) and can be less efficiently processed by the classifier. Finally, when using the reconstruction error as a score to classify fraudulent data, as done usually in anomaly detection, we can observe that DuSVAE is competitive with the best method. However, the performance level of Code1 and Code2 with CatBoost being significantly better makes the use of the hidden representations a better strategy than using the reconstruction error.

We then evaluated the impact of handcrafted features built by Worldline on the classifier performance. As we can see on the first two lines of Table 3, adding handcrafted features to the original sequential raw dataset leads to much better results both from the point of view of AUC-PR measure and P@K measure.

Now if we consider using DuSVAE (rows 3 and 4 of Table 3), we can also notice a significant improvement of the results obtained on the raw dataset of sequences augmented by handcrafted features compared to the results obtained on the original one without these additional features. This is observed for both the AUC-PR measure and the P@K measure. We see that, for the moment, by using a classifier on the sequences reconstructed by DuSVAE on just the

Table 3. AUC-PR and P@K achieved by CatBoost for sequence classification.

Input	AUC-PR	P@100	P@500
Raw data	0,40	0.43	0.11
Raw data + Handcrafted features	0,60	0.62	0.938
DuSVAE (The input:raw data)	0,53	0.88	0.72
DuSVAE (The input: raw data + Handcrafted features)	0,65	0.85	0.941

raw dataset (AUC-PR = 0.53), we cannot reach the results obtained when we use this classifier on the raw dataset augmented by handcrafted features (AUC-PR = 0.60). This can be explained by the fact that those features are based on history and profiling techniques that embed information covering a period of time larger than the one used for our dataset. Nevertheless we are not so far and the fact that using DuSVAE on the dataset augmented by handcrafted features (AUC-PR = 0.65) leads to better results than using the classifier without DuSVAE (AUC-PR = 0.60) is promising.

Table 3 also shows that the very good P@K values obtained when running the classifier on the sequences of transactions reconstructed by DuSVAE mean that DuSVAE can be a very significant help for experts to focus on real fraudulent transactions and not waste time on fake ones.

5   Conclusion

In this paper, we presented the DuSVAE model which is a new fraud detection technique. Our model combines two sequential variational autoencoders to produce a condensed representation vector of the input sequential data that can then be used by a classifier to label new sequences of transactions as genuine or fraudulent. Our experiments have shown that the DuSVAE model produces much better results, in terms of AUC-PR and $P_{@K}$ measures, than state-of-the-art systems. Moreover, the DuSVAE model produces a condensed representation of the input data which can replace very favorably the handcrafted features. Indeed, running a classifier on the condensed representation of the input data built by the DuSVAE model leads to outperform the results obtained on the raw data, with or without handcrafted features.

We believe that a first interesting way to further improve our results will be to focus on attention mechanisms to better take into account the history of past transactions in the detection of present frauds. A second approach will be to better take into account the temporal aspects in the sequential representation of our data and to reflect it in the core algorithm.

References

1. Abdallah, A., Maarof, M.A., Zainal, A.: Fraud detection system: a survey. J. Netw. Comput. Appl. **68**, 90–113 (2016)
2. Alazizi, A., Habrard, A., Jacquenet, F., He-Guelton, L., Oblé, F., Siblini, W.: Anomaly detection, consider your dataset first, an illustration on fraud detection. In: Proceedings of ICTAI 2019. IEEE (2019)
3. Bahnsen, A.C., Aouada, D., Stojanovic, A., Ottersten, B.: Feature engineering strategies for credit card fraud detection. Expert Syst. Appl. **51**, 134–142 (2016)
4. Chandola, V., Banerjee, A., Kumar, V.: Anomaly detection: a survey. ACM Comput. Surv. **41**(3), 15:1–15:58 (2009)
5. Golan, I., El-Yaniv, R.: Deep anomaly detection using geometric transformations. In: Proceedings of NIPS, pp. 9758–9769 (2018)
6. Hasan, M., Choi, J., Neumann, J., Roy-Chowdhury, A.K., Davis, L.S.: Learning temporal regularity in video sequences. In: Proceedings of CVPR, pp. 733–742 (2016)
7. Hinton, G.E.: Connectionist learning procedures. Artif. Intell. **40**(1–3), 185–234 (1989)
8. Jacobs, R.A., Jordan, M.I., Nowlan, S.J., Hinton, G.E., et al.: Adaptive mixtures of local experts. Neural Comput. **3**(1), 79–87 (1991)
9. Jurgovsky, J., et al.: Sequence classification for credit-card fraud detection. Expert Syst. Appl. **100**, 234–245 (2018)
10. Kingma, D.P., Ba, J.: Adam: a method for stochastic optimization. arXiv:1412.6980 (2014)
11. Kingma, D.P., Welling, M.: Auto-encoding variational bayes. In: Proceedings of ICLR (2014)
12. Lucas, Y., et al.: Towards automated feature engineering for credit card fraud detection using multi-perspective HMMs. Future Gener. Comput. Syst. **102**, 393–402 (2020)
13. Munawar, A., Vinayavekhin, P., De Magistris, G.: Limiting the reconstruction capability of generative neural network using negative learning. In: Proceedings of the International Workshop on Machine Learning for Signal Processing, pp. 1–6 (2017)
14. Pozzolo, A.D.: Adaptive machine learning for credit card fraud detection. Ph.D. thesis, Université libre de Bruxelles (2015)
15. Prokhorenkova, L., Gusev, G., Vorobev, A., Dorogush, A.V., Gulin, A.: Catboost: unbiased boosting with categorical features. In: Proceedings of NIPS, pp. 6638–6648 (2018)
16. Rezende, D.J., Mohamed, S., Wierstra, D.: Stochastic backpropagation and approximate inference in deep generative models. arXiv:1401.4082 (2014)
17. Sabokrou, M., Khalooei, M., Fathy, M., Adeli, E.: Adversarially learned one-class classifier for novelty detection. In: Proceedings of CVPR, pp. 3379–3388 (2018)
18. Yamanaka, Y., Iwata, T., Takahashi, H., Yamada, M., Kanai, S.: Autoencoding binary classifiers for supervised anomaly detection. arXiv:1903.10709 (2019)
19. Zhai, S., Cheng, Y., Lu, W., Zhang, Z.: Deep structured energy based models for anomaly detection. arXiv:1605.07717 (2016)
20. Zhao, Y., Deng, B., Shen, C., Liu, Y., Lu, H., Hua, X.S.: Spatio-temporal autoencoder for video anomaly detection. In: Proceedings of the ACM International Conference on Multimedia, pp. 1933–1941 (2017)

21. Zimek, A., Schubert, E., Kriegel, H.P.: A survey on unsupervised outlier detection in high-dimensional numerical data. Stat. Anal. Data Mining: ASA Data Sci. J. **5**(5), 363–387 (2012)
22. Zong, B., et al.: Deep autoencoding gaussian mixture model for unsupervised anomaly detection. In: Proceedings of ICLR (2018)

A Principled Approach to Analyze Expressiveness and Accuracy of Graph Neural Networks

Asma Atamna[1,3]($\boxtimes$), Nataliya Sokolovska[2], and Jean-Claude Crivello[3]

[1] LTCI, Télécom Paris, Institut Polytechnique de Paris, Palaiseau, France
`asma.atamna@telecom-paris.fr`
[2] NutriOmics, INSERM, Sorbonne University, Paris, France
`nataliya.sokolovska@sorbonne-universite.fr`
[3] ICMPE (UMR 7182), CNRS, University of Paris-Est, Thiais, France
`crivello@icmpe.cnrs.fr`

Abstract. Graph neural networks (GNNs) have known an increasing success recently, with many GNN variants achieving state-of-the-art results on node and graph classification tasks. The proposed GNNs, however, often implement complex node and graph embedding schemes, which makes it challenging to explain their performance. In this paper, we investigate the link between a GNN's *expressiveness*, that is, its ability to map different graphs to different representations, and its generalization performance in a graph classification setting. In particular, we propose a principled experimental procedure where we (i) define a practical measure for expressiveness, (ii) introduce an expressiveness-based loss function that we use to train a simple yet practical GNN that is permutation-invariant, (iii) illustrate our procedure on benchmark graph classification problems and on an original real-world application. Our results reveal that expressiveness alone does not guarantee a better performance, and that a powerful GNN should be able to produce graph representations that are *well separated* with respect to the class of the corresponding graphs.

Keywords: Graph neural networks · Classification · Expressiveness

1 Introduction

Many real-world data present an inherent structure and can be modelled as sequences, graphs, or hypergraphs [2,5,9,15]. Graph-structured data, in particular, are very common in practice and are at the heart of this work.

We consider the problem of graph classification. That is, given a set $\mathcal{G} = \{G_i\}_{i=1}^m$ of arbitrary graphs and their respective labels $\{y_i\}_{i=1}^m$, where $y_i \in \{1, \ldots, C\}$ and C is the number of classes, we aim at finding a mapping

Supported by the Emergence@INC-2018 program of the French National Center for Scientific Research (CNRS) and the *DiagnoLearn* ANR JCJC project.

M. R. Berthold et al. (Eds.): IDA 2020, LNCS 12080, pp. 27–39, 2020.
https://doi.org/10.1007/978-3-030-44584-3_3

$f_\theta : \mathcal{G} \to \{1, \ldots, C\}$ that minimizes the classification error, where θ denotes the parameters to optimize.

Graph neural networks (GNNs) and their deep learning variants, the graph convolutional networks (GCNs) [1,7,9,10,13,17,20,27], have gained considerable interest recently. GNNs learn latent node representations by recursively aggregating the neighboring node features for each node, thereby capturing the structural information of a node's neighborhood.

Despite the profusion of GNN variants, some of which achieve state-of-the-art results on tasks like node classification, graph classification, and link prediction, GNNs remain very little studied. In particular, it is often unclear what a GNN learns and how the learned graph (or node) mapping influences its generalization performance. In a recent work, [25] present a theoretical framework to analyze the expressive power of GNNs, where a GNN's *expressiveness* is defined as its ability to compute different graph representations for different graphs. Theoretical conditions under which a GNN is maximally expressive are derived. Although it is reasonable to assume that a higher expressiveness would result in a higher accuracy on classification tasks, this link has not been explicitly studied so far.

In this paper, we design a principled experimental procedure to analyze the link between expressiveness and the test accuracy of GNNs. In particular:

- We define a practical measure to estimate the expressiveness of GNNs;
- We use this measure to define a new penalized loss function that allows training GNNs with varying expressive power.

To illustrate our experimental framework, we introduce a simple yet practical architecture, the Simple Permutation-Invariant Graph Convolutional Network (SPI-GCN). We also present an original graph data set of metal hydrides that we use along with benchmark graph data sets to evaluate SPI-GCN.

This paper is organized as follows. Section 2 discusses the related work. Section 3 introduces preliminary notations and concepts related to graphs and GNNs. In Sect. 4, we introduce our graph neural network, SPI-GCN. In Sect. 5, we present a practical expressiveness estimator and a new expressiveness-based loss function as part of our experimental framework. Section 6 presents our results and Sect. 7 concludes the paper.

2   Related Work

Graph neural networks (GNNs) were first introduced in [11,19]. They learn latent node representations by iteratively aggregating neighborhood information for each node. Their more recent deep learning variants, the graph convolutional networks (GCNs), generalize conventional convolutional neural networks to irregular graph domains. In [13], the authors present a GCN for node classification where the computed node representations can be interpreted as the graph coloring returned by the 1-dimensional Weisfeiler-Lehman (WL) algorithm [24]. A related GCN that is invariant to node permutation is presented in [27]. The graph convolution operator is closely related to the one in [13], and the authors introduce a permutation-invariant pooling operator that sorts the convolved nodes before feeding them to a 1-dimensional classical convolution layer for graph-level classification. A popular GCN is PATCHY-SAN [17]. Its graph convolution operator extracts normalized local "patches" (neighborhood representations) of the graph which are then sorted and fed to a 1-dimensional traditional convolution layer for graph-level classification. The method, however, requires the definition of a node ordering and running the WL algorithm in a preprocessing step. On the other hand, the normalization of the extracted patches implies sorting the nodes again and using the external graph software NAUTY [14].

Despite the success of GNNs, there are relatively few papers that analyze their properties, either mathematically or empirically. A notable exception is the recent work by [25] that studies the expressive power of GNNs. The authors prove that (i) GNNs are at most as powerful as the WL test in distinguishing graph structures and that (ii) if the graph function of a GNN—i.e. its graph embedding scheme—is injective, then the GNN is as powerful as the WL test. The authors also present the Graph Isomorphism Network (GIN), which approximates the theoretical maximally expressive GNN. In another study [4], the authors present a simple neural network defined on a set of graph augmented features and show that their architecture can be obtained by linearizing graph convolutions in GNNs.

Our work is related to [25] in that we adopt the same definition of expressiveness, that is, the ability of a GNN to compute distinct graph representations for distinct input graphs. However, we go one step further and investigate how the graph function learned by GNNs affects their generalization performance. On the other hand, our SPI-GCN extends the GCN in [13] to graph-level classification. Our SPI-GCN is also related to [27] in that we use a similar graph convolution operator inspired by [13]. Unlike [27], however, our architecture does not require any node ordering, and we only use a simple multilayer perceptron (MLP) to perform classification.

3   Some Graph Concepts

A graph G is a pair (V, E) of a set $V = \{v_1, \ldots, v_n\}$ of vertices (or nodes) v_i, and a set $E \subseteq V \times V$ of edges (v_i, v_j). In this work, we represent a graph G by two matrices: (i) an *adjacency matrix* $A \in \mathbb{R}^{n \times n}$ such that $a_{ij} = 1$ if there is an edge between nodes v_i and v_j and $a_{ij} = 0$ otherwise,[1] and (ii) a *node feature matrix* $X \in \mathbb{R}^{n \times d}$, with d being the number of node features. Each row $x_i \in \mathbb{R}^d$ of X contains the feature representation of a node v_i, where d is the dimension of the feature space. Since we only consider node features in this paper (as opposed to *edge features* for instance), we will refer to the node feature matrix X simply as the feature matrix in the rest of this paper.

---

[1] Given a matrix M, m_i denotes its ith row and m_{ij} denotes the entry at its ith row and jth column. More generally, we denote matrices by capital letters and vectors by small letters. Scalars, on the other hand, are denoted by small italic letters.

An important notion in graph theory is *graph isomorphism*. Two graphs $G_1 = (V_1, E_1)$ and $G_2 = (V_2, E_2)$ are isomorphic if there exists a bijection $g : V_1 \rightarrow V_2$ such that every edge (u, v) is in E_1 if and only if the edge $(g(u), g(v))$ is in E_2. Informally, this definition states that two graphs are isomorphic if there exists a vertex permutation such that when applied to one graph, we recover the vertex and edge sets of the other graph.

3.1   Graph Neural Networks

Consider a graph G with adjacency matrix A and feature matrix X. GNNs use the graph structure (A) and the node features (X) to learn a node-level or a graph-level representation—or *embedding*—of G. GNNs iteratively update a node representation by aggregating its neighbors' representations. At iteration l, a node representation captures its l-hop neighborhood's structural information. Formally, the lth layer of a general GNN can be defined as follows:

$$a_i^{l+1} = \text{AGGREGATE}^l(\{z_j^l : j \in N(i)\}) \tag{1}$$

$$z_i^{l+1} = \text{COMBINE}^l(z_i^l, a_i^{l+1}) \ , \tag{2}$$

where z_i^{l+1} is the feature vector of node v_i at layer l and where $z_i^0 = x_i$. While COMBINE usually consists in concatenating node representations from different layers, different—and often complex—architectures for AGGREGATE have been proposed. In [13], the presented GCN merges the AGGREGATE and COMBINE functions as follows:

$$z_i^{l+1} = \text{ReLU}\left(\text{mean}(\{z_j^l : j \in N(i) \cup \{i\}\}) \cdot W^l\right) \ , \tag{3}$$

where ReLU is a rectified linear unit and W^l is a trainable weight matrix. GNNs for graph classification have an additional module that aggregates the node-level representations to produce a graph-level one as follows:

$$z_G = \text{READOUT}(\{z_i^L : v_i \in V\}) \ , \tag{4}$$

for a GNN with L layers. In [25], the authors discuss the impact that the choice of AGGREGATE^l, COMBINE^l, and READOUT has on the so-called *expressiveness* of the GNN, that is, its ability to map different graphs to different embeddings. They present theoretical conditions under which a GNN is maximally expressive.

We now present a simple yet practical GNN architecture on which we illustrate our experimental framework.

4   Simple Permutation-Invariant Graph Convolutional Network (SPI-GCN)

Our Simple Permutation-Invariant Graph Convolutional Network (SPI-GCN) consists of the following sequential modules: (1) a *graph convolution module* that encodes local graph structure and node features in a substructure feature matrix whose rows represent the nodes of the graph, (2) a *sum-pooling layer* as a READOUT function to produce a single-vector representation of the input graph, and (3) a *prediction module* consisting of dense layers that reads the vector representation of the graph and outputs predictions.

Let G be a graph represented by the adjacency matrix $A \in \mathbb{R}^{n \times n}$ and the feature matrix $X \in \mathbb{R}^{n \times d}$, where n and d represent the number of nodes and the dimension of the feature space respectively. Without loss of generality, we consider graphs without self-loops.

4.1   Graph Convolution Module

Given a graph G with its adjacency and feature matrices, A and X, we define the first convolution layer as follows:

$$Z = f(\hat{D}^{-1}\hat{A}XW) \, , \tag{5}$$

where $\hat{A} = A + I_n$ is the adjacency matrix of G with added self-loops, $\hat{D}$ is the diagonal node degree matrix of $\hat{A}$,[2] $W \in \mathbb{R}^{d \times d'}$ is a trainable weight matrix, f is a nonlinear activation function, and $Z \in \mathbb{R}^{n \times d'}$ is the convolved graph. To stack multiple convolution layers, we generalize the propagation rule in (5) as follows:

$$Z^{l+1} = f^l(\hat{D}^{-1}\hat{A}Z^l W^l) \, , \tag{6}$$

where $Z^0 = X$, Z^l is the output of the lth convolution layer, W^l is a trainable weight matrix, and f^l is the nonlinear activation function applied at layer l. Similarly to the GCN presented in [13] from which we draw inspiration, our graph convolution module merges the AGGREGATE and COMBINE functions (see (1) and (2)), and we can rewrite (6) as:

$$z_i^{l+1} = f^l \left(\text{mean}(\{z_j^l : j \in N(i) \cup \{i\}\}) \cdot W^l \right) \, , \tag{7}$$

where z_i^{t+1} is the ith row of Z^{l+1}.

We return the result of the last convolution layer, that is, for a network with L convolution layers, the result of the convolution is the last substructure feature matrix Z^L. Note that (6) is able to process graphs with varying node numbers.

4.2   Sum-Pooling Layer

The *sum-pooling* layer produces a graph-level representation z_G by summing the rows of Z^L, previously returned by the convolution module. Formally:

$$z_G = \sum_{i=1}^{n} z_i^L \, . \tag{8}$$

---

[2] If G is a directed graph, $\hat{D}$ corresponds to the *outdegree* diagonal matrix of $\hat{A}$.

The resulting vector $z_G \in \mathbb{R}^{d_L}$ contains the final vector representation (or *embedding*) of the input graph G in a d_L-dimensional space. This vector representation is then used for prediction—graph classification in our case.

Using a sum pooling operator is a simple idea that has been used in GNNs such as [1,21]. Additionally, it results in the invariance of our architecture to node permutation, as stated in Theorem 1.

Theorem 1. *Let G and G_ς be two arbitrary isomorphic graphs. The sum-pooling layer of SPI-GCN produces the same vector representation for G and G_ς.*

This invariance property is crucial for GNNs as it ensures that two isomorphic—and hence equivalent—graphs will result in the same output. The proof of Theorem 1 is straightforward and omitted for space limitations.

4.3 Prediction Module

The prediction module of SPI-GCN is a simple MLP that takes as input the graph-level representation z_G returned by the sum-pooling layer and returns either: (i) a probability p in case of binary classification or (ii) a vector p of probabilities such that $\sum_i p_i = 1$ in case of multi-class classification.

Note that SPI-GCN can be trained in an end-to-end fashion through backpropagation. Additionally, since only one graph is treated in a forward pass, the training complexity of SPI-GCN is linear in the number of graphs.

In the next section, we describe a practical methodology for studying the expressiveness of SPI-GCN and its connection to the generalization performance of the algorithm.

5   Investigating Expressiveness of SPI-GCN

We start here by introducing a practical definition of expressiveness. We then show how the defined measure can be used to train SPI-GCN and help understand the impact expressiveness has on its generalization performance.

5.1 Practical Measure of Expressiveness

The expressiveness of a GNN, as defined in [25], is its ability to map different graph structures to different embeddings and, therefore, reflects the injectivity of its graph embedding function. Since studying injectivity can be tedious, we characterize expressiveness—and hence injectivity—as a function of the pairwise distance between graph embeddings.

Let $\{z_{G_i}\}_{i=1}^m$ be the set of graph embeddings computed by a GNN $\mathcal{A}$ for a given input graph data set $\{G_i\}_{i=1}^m$. We define $\mathcal{A}$'s expressiveness, $\mathcal{E}(\mathcal{A})$, as follows:

$$\mathcal{E}(\mathcal{A}) = \text{mean}(\{\|z_{G_i} - z_{G_j}\|_2 : i, j = 1, \ldots, m, \ i \neq j\}) , \qquad (9)$$

that is, $\mathcal{E}(\mathcal{A})$ is the average pairwise Euclidean distance between graph embeddings produced by $\mathcal{A}$. While not strictly equivalent to injectivity, $\mathcal{E}$ is a reasonable indicator thereof, as the average pairwise distance reflects the *diversity* within graph representations which, in turn, is expected to be higher for more diverse input graph data sets. For permutation-invariant GNNs like SPI-GCN,[3] $\mathcal{E}$ is zero when all graphs $\{G_i\}_{i=1}^m$ are isomorphic.

5.2 Penalized Cross Entropy Loss

We train SPI-GCN using a *penalized cross entropy loss*, $\mathcal{L}_p$, that consists of a classical cross entropy augmented with a penalty term defined as a function of the expressiveness of SPI-GCN. Formally:

$$\mathcal{L}_p = \text{cross-entropy}(\{y_i\}_{i=1}^m, \{\hat{y}_i\}_{i=1}^m) - \alpha \cdot \mathcal{E}(\text{SPI-GCN}) \ , \qquad (10)$$

where $\{y_i\}_{i=1}^m$ (resp. $\{\hat{y}_i\}_{i=1}^m$) is the set of real (resp. predicted) graph labels, α is a non-negative penalty factor, and $\mathcal{E}$ is defined in (9) with $\{z_{G_i}\}_{i=1}^m$ being the graph embeddings computed by SPI-GCN.

By adding the penalty term $-\alpha \cdot \mathcal{E}(\text{SPI-GCN})$ in $\mathcal{L}_p$, the expressiveness is maximized while the cross entropy is minimized during the training process. The penalty factor α controls the importance attributed to $\mathcal{E}(\text{SPI-GCN})$ when $\mathcal{L}_p$ is minimized. Consequently, higher values of α allow to train more expressive variants of SPI-GCN whereas for $\alpha = 0$, only the cross entropy is minimized.

In the next section, we assess the performance of SPI-GCN for different values of α. We also compare SPI-GCN with other more complex GNN architectures, including the state-of-the-art method.

6   Experiments

We carry out a first set of experiments where we compare our approach, SPI-GCN, with two recent GCNs. In a second set of experiments, we train different instances of SPI-GCN with increasing values of the penalty factor α (see (10)) in an attempt to understand how the expressiveness of SPI-GCN affects its test accuracy, and whether it is the determining factor of its generalization performance, as implicitly suggested in [25]. Our code and data are available at https://github.com/asmaatamna/SPI-GCN.

6.1   Data Sets

We use nine public benchmark data sets including five bioinformatics data sets (MUTAG [6], PTC [22], ENZYMES [3], NCI1 [23], PROTEINS [8]), two social network data sets (IMDB-BINARY, IMDB-MULTI [26]), one image data set where images are represented as region adjacency graphs (COIL-RAG [18]), and one synthetic data set (SYNTHIE [16]). We also evaluate SPI-GCN on an original real-world data set collected at the ICMPE,[4] HYDRIDES, that contains metal hydrides in graph format, labelled as *stable* or *unstable* according to specific energetic properties that determine their ability to store hydrogen efficiently.

---

[3] As mentioned previously, we state that permutation-invariance is a minimal requirement for any practical GNN.

[4] East Paris Institute of Chemistry and Materials Science, France.

6.2   Architecture of SPI-GCN

The instance of SPI-GCN that we use for experiments has two graph convolution layers of 128 and 32 hidden units respectively, followed by a hyperbolic tangent function and a softmax function (per node) respectively. The sum-pooling layer is a classical sum applied row-wise; it is followed by a prediction module consisting of a MLP with one hidden layer of 256 hidden units followed by a batch normalization layer and a ReLU. We choose this architecture by trial and error and keep it unchanged throughout the experiments.

6.3   Comparison with Other Methods

In these experiments, we consider the simplest variant of SPI-GCN where the penalty term in (10) is discarded by setting $\alpha = 0$. That is, the algorithm is trained using only the cross entropy loss.

Baselines. We compare SPI-GCN with the well-known GCN, PATCHY-SAN (PSCN) [17], the Deep Graph Convolutional Neural Network (DGCNN) [27] that uses a similar convolution module to ours, and the recent state-of-the-art Graph Isomorphism Network (GIN) [25].

Experimental Procedure. We train SPI-GCN using full batch ADAM optimizer [12], with cross entropy as the loss function to minimize ($\alpha = 0$ in (10)). Upon experimentation, we set ADAM's hyperparameters as follows. The algorithm is trained for 200 epochs on all data sets and the learning rate is set to 10^{-3}. To estimate the accuracy, we perform 10-fold cross validation using 9 folds for training and one fold for testing each time. We report the average (test) accuracy and the corresponding standard deviation in Table 1. Note that we only use node attributes in our experiments. In particular, SPI-GCN does not exploit node or edge labels of the data sets. When node attributes are not available, we use the identity matrix as the feature matrix for each graph.

We follow the same procedure for DGCNN. We use the authors' implementation[5] and perform 10-fold cross validation with the recommended values for training epochs, learning rate, and SortPooling parameter k, for each data set.

For PSCN, we report the results from the original paper [17] (for receptive field size $k = 10$) as we could not find an authors' public implementation of the algorithm. The experiments were conducted using a similar procedure as ours.

For GIN, we also report the published results [25] (GIN-0 in the paper), as it was not straightforward to use the authors' implementation.

Results. Table 1 shows the results for our algorithm (SPI-GCN), DGCNN [27], PSCN [17], and the state-of-the-art GIN [25]. We observe that SPI-GCN is highly competitive with other algorithms despite using the same architecture for all data sets. The only noticeable exceptions are on the NCI1 and IMDB-BINARY data sets, where the best approach (GIN) is up to 1.28 times better. On the other hand, SPI-GCN appears to be highly competitive on classification tasks

---

[5] https://github.com/muhanzhang/pytorch_DGCNN.

with more than 3 classes (ENZYMES, COIL-RAG, SYNTHIE). The difference in accuracy is particularly significant on COIL-RAG (100 classes), where SPI-GCN is around 34 times better than DGCNN, suggesting that the features extracted by SPI-GCN are more suitable to characterize the graphs at hand. SPI-GCN also achieves a very reasonable accuracy on the HYDRIDES data set and is 1.06 times better than DGCNN on ENZYMES.

The results in Table 1 show that despite its simplicity, SPI-GCN is competitive with other practical graph algorithms and, hence, it is a reasonable architecture to consider for our next set of experiments involving expressiveness.

Table 1. Accuracy results for SPI-GCN and three other deep learning methods (DGCNN, PSCN, GIN).

Algorithm	SPI-GCN	DGCNN	PSCN	GIN
MUTAG	84.40 ± 8.14	86.11 ± 7.14	88.95 ± 4.37	$\mathbf{89.4 \pm 5.6}$
PTC	56.41 ± 5.71	55.00 ± 5.10	62.29 ± 5.68	$\mathbf{64.6 \pm 7.0}$
NCI1	64.11 ± 2.37	72.73 ± 1.56	76.34 ± 1.68	$\mathbf{82.7 \pm 1.7}$
PROTEINS	72.06 ± 3.18	72.79 ± 3.58	75.00 ± 2.51	$\mathbf{76.2 \pm 2.8}$
ENZYMES	$\mathbf{50.17 \pm 5.60}$	47.00 ± 8.36	–	–
IMDB-BINARY	60.40 ± 4.15	68.60 ± 5.66	71.00 ± 2.29	$\mathbf{75.1 \pm 5.1}$
IMDB-MULTI	44.13 ± 4.61	45.20 ± 3.75	45.23 ± 2.84	$\mathbf{52.3 \pm 2.8}$
COIL-RAG	$\mathbf{74.38 \pm 2.42}$	2.21 ± 0.33	–	–
SYNTHIE	$\mathbf{71.00 \pm 6.44}$	54.25 ± 4.34	–	–
HYDRIDES	82.75 ± 2.67	–	–	–

6.4 Expressiveness Experiments

Through these experiments, we try to answer the following questions:

- Do more expressive GNNs perform better on graph classification tasks? That is, is the injectivity of a GNN's graph function the determining factor of its performance?
- Can the performance be explained by another factor? If yes, what is it?

To this end, we train increasingly injective instances of SPI-GCN on the penalized cross entropy loss $\mathcal{L}_p$ (10) by setting the penalty factor α to increasingly large values. Then, for each trained instance, we investigate (i) its test accuracy, (ii) its expressiveness $\mathcal{E}(\text{SPI-GCN})$ (9), and (iii) the *average normalized Inter-class Graph Embedding Distance* (IGED), defined as the average pairwise Euclidean distance between mean graph embeddings taken class-wise divided by $\mathcal{E}(\text{SPI-GCN})$. Formally:

$$\text{IGED} = \frac{\text{mean}(\{||z_c^* - z_{c'}^*||_2 : c, c' = 1, \dots, C, \ c \neq c'\})}{\mathcal{E}(\text{SPI-GCN})}, \tag{11}$$

Table 2. Expressiveness experiments results. SPI-GCN is trained on the penalized cross entropy loss, $\mathcal{L}_p$, with increasing values of the penalty factor α. For each data set, and for each value of α, we report the test accuracy (a), the expressiveness $\mathcal{E}$(SPI-GCN) (b), and the IGED (c). Highlighted are the maximal values for each quantity.

α	0	10^{-3}	10^{-1}	1	10	
MUTAG	84.40 ± 8.14	84.40 ± 8.14	$\mathbf{86.07 \pm 9.03}$	82.56 ± 7.33	81.45 ± 6.68	(a)
	0.09 ± 0.01	0.09 ± 0.01	0.12 ± 0.01	5.96 ± 1.08	$\mathbf{6.32 \pm 0.76}$	(b)
	0.68 ± 0.16	0.68 ± 0.16	0.82 ± 0.18	$\mathbf{1.21 \pm 0.23}$	1.20 ± 0.22	(c)
PTC	56.41 ± 5.71	54.97 ± 6.05	54.64 ± 6.33	57.88 ± 8.65	$\mathbf{58.70 \pm 7.40}$	(a)
	0.09 ± 0.01	0.09 ± 0.01	0.11 ± 0.01	8.41 ± 3.13	$\mathbf{9.03 \pm 2.94}$	(b)
	0.26 ± 0.05	0.26 ± 0.05	0.26 ± 0.06	0.41 ± 0.22	$\mathbf{0.42 \pm 0.22}$	(c)
NCI1	64.11 ± 2.37	$\mathbf{64.21 \pm 2.36}$	64.01 ± 2.87	63.48 ± 1.36	63.19 ± 1.72	(a)
	0.09 ± 0.004	0.09 ± 0.005	1.07 ± 0.19	16.83 ± 0.49	$\mathbf{16.91 \pm 0.52}$	(b)
	0.18 ± 0.02	0.19 ± 0.03	0.59 ± 0.05	$\mathbf{0.62 \pm 0.05}$	$\mathbf{0.62 \pm 0.05}$	(c)
PROTEINS	$\mathbf{72.06 \pm 3.18}$	71.78 ± 3.55	71.51 ± 3.26	70.97 ± 3.49	71.42 ± 3.23	(a)
	5.89 ± 1.34	13.07 ± 3.21	$\mathbf{35.88 \pm 4.89}$	$\mathbf{35.88 \pm 4.89}$	$\mathbf{35.88 \pm 4.89}$	(b)
	$\mathbf{0.74 \pm 0.09}$	$\mathbf{0.74 \pm 0.09}$	$\mathbf{0.74 \pm 0.09}$	$\mathbf{0.74 \pm 0.09}$	$\mathbf{0.74 \pm 0.09}$	(c)
ENZYMES	50.17 ± 5.60	50.17 ± 5.60	29.33 ± 5.93	29.33 ± 5.54	29.33 ± 5.88	(a)
	0.79 ± 0.21	1.85 ± 0.64	23.22 ± 2.99	23.33 ± 3.02	$\mathbf{23.35 \pm 3.01}$	(b)
	$\mathbf{0.44 \pm 0.06}$	0.42 ± 0.10	0.42 ± 0.10	0.42 ± 0.10	0.42 ± 0.10	(c)
IMDB-BIN.	60.40 ± 4.15	$\mathbf{61.70 \pm 4.96}$	61.10 ± 3.75	54.40 ± 3.10	54.20 ± 5.15	(a)
	0.12 ± 0.01	0.12 ± 0.01	0.16 ± 0.01	$\mathbf{12.43 \pm 2.37}$	11.70 ± 2.89	(b)
	$\mathbf{0.15 \pm 0.03}$	$\mathbf{0.15 \pm 0.03}$	$\mathbf{0.15 \pm 0.03}$	0.12 ± 0.08	0.12 ± 0.08	(c)
IMDB-MUL.	44.13 ± 4.61	44.60 ± 5.41	$\mathbf{44.80 \pm 4.51}$	39.73 ± 4.34	38.87 ± 4.42	(a)
	0.08 ± 0.01	0.08 ± 0.01	0.64 ± 0.14	$\mathbf{10.38 \pm 1.05}$	9.91 ± 1.15	(b)
	$\mathbf{0.16 \pm 0.02}$	$\mathbf{0.16 \pm 0.02}$	$\mathbf{0.16 \pm 0.09}$	0.15 ± 0.09	0.15 ± 0.09	(c)
COIL-RAG	$\mathbf{74.38 \pm 2.42}$	74.38 ± 2.45	72.49 ± 3.21	52.08 ± 4.89	28.72 ± 3.62	(a)
	0.08 ± 0.002	0.081 ± 0.002	0.13 ± 0.01	2.00 ± 0.18	$\mathbf{2.33 \pm 0.14}$	(b)
	0.95 ± 0.01	0.95 ± 0.01	0.96 ± 0.01	$\mathbf{0.98 \pm 0.02}$	$\mathbf{0.98 \pm 0.02}$	(c)
SYNTHIE	71.00 ± 6.44	71.00 ± 6.04	$\mathbf{74.00 \pm 6.44}$	73.00 ± 7.57	73.75 ± 7.52	(a)
	1.60 ± 0.20	1.86 ± 0.24	$\mathbf{29.97 \pm 2.16}$	29.50 ± 2.18	29.37 ± 2.18	(b)
	$\mathbf{0.73 \pm 0.07}$	0.72 ± 0.08	0.61 ± 0.11	0.59 ± 0.12	0.58 ± 0.12	(c)
HYDRIDES	82.75 ± 2.67	82.65 ± 2.44	$\mathbf{83.92 \pm 4.30}$	77.45 ± 3.25	76.37 ± 2.57	(a)
	0.13 ± 0.01	0.13 ± 0.01	1.68 ± 0.87	4.75 ± 0.41	$\mathbf{5.03 \pm 0.75}$	(b)
	0.50 ± 0.11	0.50 ± 0.11	0.8 ± 0.19	$\mathbf{0.85 \pm 0.21}$	0.72 ± 0.22	(c)

where z_k^* is the mean graph embedding for class k. The IGED can be interpreted as an estimate of how well the graph embeddings computed by SPI-GCN are *separated* with respect to their respective class.

Experimental Procedure. We train SPI-GCN on the penalized cross entropy loss $\mathcal{L}_p$ (10) where we sequentially choose α from $\{0, 10^{-3}, 10^{-1}, 1, 10\}$. We do so using full batch ADAM optimizer that we run for 200 epochs with a learning rate of 10^{-3}, on all the graph data sets introduced previously. For each data set and for each value of α, we perform 10-fold cross validation using 9 folds for training and one fold for testing. We report in Table 2 the average and standard deviation of: (a) the test accuracy, (b) the expressiveness $\mathcal{E}(\text{SPI-GCN})$, and (c) the IGED (11), for each value of α and for each data set.

Results. We observe from Table 2 that using a penalty term in $\mathcal{L}_p$ to maximize the expressiveness—or injectivity—of SPI-GCN helps to improve the test accuracy on some data sets, notably on MUTAG, PTC, and SYNTHIE. However, larger values of $\mathcal{E}(\text{SPI-GCN})$ do not correspond to a higher test accuracy except for two cases (PTC, SYNTHIE). Overall, $\mathcal{E}(\text{SPI-GCN})$ increases when α increases, as expected, since the expressiveness is maximized during training when $\alpha > 0$. The IGED, on the other hand, is correlated to the best performance in four out of ten cases (ENZYMES, IMDB-BINARY, and IMDB-MULTI), where the test accuracy is maximal when the IGED is maximal. On HYDRIDES, the difference in IGED for $\alpha = 10^{-1}$ (highest accuracy) and $\alpha = 1$ (highest IGED value) is negligible.

Our empirical results indicate that while optimizing the expressiveness of SPI-GCN may result in a higher test accuracy in some cases, more expressive GNNs do not systematically perform better in practice. The IGED, however, which reflects a GNN's ability to compute graph representations that are correctly clustered according to their effective class, better explains the generalization performance of the GNN.

7 Conclusion

In this paper, we challenged the common belief that more expressive GNNs achieve a better performance. We introduced a principled experimental procedure to analyze the link between the expressiveness of a GNN and its test accuracy in a graph classification setting. To the best of our knowledge, our work is the first that explicitly studies the generalization performance of GNNs by trying to uncover the factors that control it, and paves the way for more theoretical analyses. Interesting directions for future work include the design of better expressiveness estimators, as well as different (possibly more complex) penalized loss functions.

References

1. Atwood, J., Towsley, D.: Diffusion-convolutional neural networks. In: NIPS (2016)
2. Bojchevski, A., Shchur, O., Zügner, D., Günnemann, S.: NetGAN: generating graphs via random walks. In: ICML (2018)
3. Borgwardt, K.M., Ong, C.S., Schönauer, S., Vishwanathan, S.V.N., Smola, A.J., Kriegel, H.P.: Protein function prediction via graph kernels. Bioinformatics **21**(1), 47–56 (2005)
4. Chen, T., Bian, S., Sun, Y.: Are powerful graph neural nets necessary? A dissection on graph classification (2019). arXiv:1905.04579v2

5. Collobert, R., Weston, J., Bottou, L., Karlen, M., Kavukcuoglu, K., Kuksa, P.: Natural language processing (almost) from scratch. J. Mach. Learn. Res. **12**, 2493–2537 (2011)
6. Debnath, A.K., Lopez de Compadre, R.L., Debnath, G., Shusterman, A.J., Hansch, C.: Structure-activity relationship of mutagenic aromatic and heteroaromatic nitro compounds. Correlation with molecular orbital energies and hydrophobicity. J. Med. Chem. **34**(2), 786–797 (1991)
7. Defferrard, M., Bresson, X., Vandergheynst, P.: Convolutional neural networks on graphs with fast localized spectral filtering. In: NIPS (2016)
8. Dobson, P.D., Doig, A.J.: Distinguishing enzyme structures from non-enzymes without alignments. J. Mol. Biol. **330**(4), 771–783 (2003)
9. Duvenaud, D.K., et al.: Convolutional networks on graphs for learning molecular fingerprints. In: NIPS (2015)
10. Gilmer, J., Schoenholz, S.S., Riley, P.F., Vinyals, O., Dahl, G.E.: Neural message passing for quantum chemistry. In: ICML (2017)
11. Gori, M., Monfardini, G., Scarselli, F.: A new model for learning in graph domains. In: IJCNN (2005)
12. Kingma, D.P., Ba, J.: Adam: a method for stochastic optimization. In: ICLR (2015)
13. Kipf, T.N., Welling, M.: Semi-supervised classification with graph convolutional networks. In: ICLR (2017)
14. Mckay, B.D., Piperno, A.: Practical graph isomorphism, II. J. Symb. Comput. **60**, 94–112 (2014)
15. Min, S., Lee, B., Yoon, S.: Deep learning in bioinformatics. Brief. Bioinform. **18**(5), 851–869 (2017)
16. Morris, C., Kriege, N.M., Kersting, K., Mutzel, P.: Faster kernels for graphs with continuous attributes via hashing. In: ICDM (2016)
17. Niepert, M., Ahmed, M., Kutzkov, K.: Learning convolutional neural networks for graphs. In: ICML (2016)
18. Riesen, K., Bunke, H.: IAM graph database repository for graph based pattern recognition and machine learning. In: da Vitoria Lobo, N., et al. (eds.) SSPR /SPR 2008. LNCS, vol. 5342, pp. 287–297. Springer, Heidelberg (2008). https://doi.org/10.1007/978-3-540-89689-0_33
19. Scarselli, F., Gori, M., Tsoi, A.C., Hagenbuchner, M., Monfardini, G.: The graph neural network model. Trans. Neural Netw. **20**(1), 61–80 (2009)
20. Schlichtkrull, M., Kipf, T.N., Bloem, P., van den Berg, R., Titov, I., Welling, M.: Modeling relational data with graph convolutional networks. In: Gangemi, A., et al. (eds.) ESWC 2018. LNCS, vol. 10843, pp. 593–607. Springer, Cham (2018). https://doi.org/10.1007/978-3-319-93417-4_38
21. Simonovsky, M., Komodakis, N.: Dynamic edge-conditioned filters in convolutional neural networks on graphs. In: CVPR 2017 (2017)
22. Srinivasan, A., Helma, C., King, R.D., Kramer, S.: The predictive toxicology challenge 2000–2001. Bioinformatics **17**(1), 107–108 (2001)
23. Wale, N., Watson, I.A., Karypis, G.: Comparison of descriptor spaces for chemical compound retrieval and classification. Knowl. Inf. Syst. **14**(3), 347–375 (2008)
24. Weisfeiler, B., Lehman, A.A.: A reduction of a graph to a canonical form and an algebra arising during this reduction. Nauchno-Technicheskaya Informatsia **9**(2), 12–16 (1968)

25. Xu, K., Hu, W., Leskovec, J., Jegelka, S.: How powerful are graph neural networks? In: ICLR (2019)
26. Yanardag, P., Vishwanathan, S.: Deep graph kernels. In: SIGKDD (2015)
27. Zhang, M., Cui, Z., Neumann, M., Chen, Y.: An end-to-end deep learning architecture for graph classification. In: AAAI (2018)

Efficient Batch-Incremental Classification
Using UMAP for Evolving Data Streams

Maroua Bahri[1,3(✉)], Bernhard Pfahringer[2], Albert Bifet[1,2],
and Silviu Maniu[3,4,5]

[1] LTCI, Télécom Paris, IP-Paris, Paris, France
{maroua.bahri,albert.bifet}@telecom-paris.fr
[2] University of Waikato, Hamilton, New Zealand
bernhard@waikato.ac.nz
[3] Université Paris-Saclay, LRI, CNRS, Orsay, France
silviu.maniu@lri.fr
[4] Inria, Paris, France
[5] ENS DI, CNRS, École Normale Supérieure, Université PSL, Paris, France

Abstract. Learning from potentially infinite and high-dimensional data streams poses significant challenges in the classification task. For instance, k-Nearest Neighbors (kNN) is one of the most often used algorithms in the data stream mining area that proved to be very resource-intensive when dealing with high-dimensional spaces. Uniform Manifold Approximation and Projection (UMAP) is a novel manifold technique and one of the most promising dimension reduction and visualization techniques in the non-streaming setting because of its high performance in comparison with competitors. However, there is no version of UMAP that copes with the challenging context of streams. To overcome these restrictions, we propose a batch-incremental approach that pre-processes data streams using UMAP, by producing successive embeddings on a stream of disjoint batches in order to support an incremental kNN classification. Experiments conducted on publicly available synthetic and real-world datasets demonstrate the substantial gains that can be achieved with our proposal compared to state-of-the-art techniques.

Keywords: Data stream · k-Nearest Neighbors · Dimension reduction · UMAP

1 Introduction

With the evolution of technology, several kinds of devices and applications are continuously generating large amounts of data in a fast-paced way as *streams*. Hence, the data stream mining area has become indispensable and ubiquitous in many real-world applications that require real-time – or near real-time –

This work was done in the context of IoTA AAP Emergence DigiCosme Project and was funded by Labex DigiCosme.

M. R. Berthold et al. (Eds.): IDA 2020, LNCS 12080, pp. 40–53, 2020.
https://doi.org/10.1007/978-3-030-44584-3_4

processing, e.g., social networks, weather forecast, spam filters, and more. Unlike traditional datasets, the dynamic environment and the tremendous volume of data streams make them impossible to store or to scan multiple times [12].

Classification is an active area of research in data stream mining field where several researchers are paying attention to develop new – or improve existing – algorithms [14]. However, the dynamic nature of data streams has outpaced the capability of traditional classification algorithms to process data streams. In this context, a multitude of supervised algorithms for static datasets that have been widely studied in the offline processing, and proved to be of limited effectiveness on large data, have been extended to work within a streaming framework [3,5,11,18]. Data stream mining approaches can be divided into two main types [23]: (i) *instance-incremental* approaches which update the model with each instance as soon as it arrives, such as Self-Adjusting Memory kNN (sam$k$NN) [18], and Hoeffding Adaptive Tree (HAT) [4]; and (ii) *batch-incremental* approaches which make no change/increment to their model until a batch is completed, e.g., support vector machines [10], and batch-incremental ensemble of decision trees [15]. Nevertheless, the high dimensionality of data complicates the classification task for some algorithms and increases their computational resources, most notably the k-Nearest Neighbors (kNN) because it needs the entire dataset to predict the labels for test instances [23]. To cope with this issue, a promising approach is *feature transformation* which transforms the input features into a new set of features, containing the most relevant components, in some lower-dimensional space.

In attempt to improve the performance of kNN, we incorporate a batch-incremental feature transformation strategy to tackle potentially high-dimensional and possibly infinite batches of evolving data streams while ensuring effectiveness and quality of learning (e.g. *accuracy*). This is achieved via a new manifold technique that has attracted a lot of attention recently: Uniform Manifold Approximation and Projection (UMAP) [21], built upon rigorous mathematical foundations, namely Riemannian geometry. To the best of our knowledge, no incremental version of UMAP exists which makes it not applicable on large datasets. The main contributions are summarized as follows:

- *Batch-Incremental UMAP*: a new batch-incremental novel manifold learning technique, based on extending the UMAP algorithm to data streams.
- *UMAP-kNearest Neighbors (UMAP-kNN)*: a new batch-incremental kNN algorithm for data streams classification using UMAP.
- *Empirical experiments*: we provide an experimental study, on various datasets, that discusses the implication of parameters on the algorithms performance;

The paper is organized as follows. Section 2 reviews the prominent related work. Section 3 provides the background of UMAP, followed by the description of our approach. In Sect. 4 we present and discuss the results of experiments on diverse datasets. Finally, we draw our conclusions and present future directions.

2 Related Work

Dimensionality reduction (DR) is a powerful tool in data science to look for hidden structure in data and reduce the resources usage of learning algorithms. The problem of dimensionality has been widely studied [25] and used throughout different domains, such as image processing and face recognition. Dimensionality reduction techniques facilitate the classification task, by removing redundancies and extracting the most relevant features in the data, and permits a better data visualization. A common taxonomy divides these approaches into two major groups: *matrix factorization* and *graph-based* approaches.

Matrix factorization algorithms require matrix computation tools, such as Principal Components Analysis (PCA) [16]. It is a well-known linear technique that uses singular value decomposition and aims to find a lower-dimensional basis by converting the data into features called principal components by computing the eigenvalues and eigenvectors of a covariance matrix. This straightforward technique is computationally cheap but ineffective with data streams since it relies on the whole dataset. Therefore, some incremental versions of PCA have been developed to handle streams of data [13,24,26].

Graph/Neighborhood-based techniques are leveraged in the context of dimension reduction and visualization by using the insight that similar instances in a large space should be represented by close instances in a low-dimensional space, whereas dissimilar instances should be well separated. t-distributed Stochastic Neighbor Embedding (tSNE) [20] is one of the most prominent DR techniques in the literature. It has been proposed to visualize high-dimensional data embedded in a lower space (typically 2 or 3 dimensions). In addition to the fact that it is computationally expensive, tSNE does not preserve distances between all instances and can affect any density–or distance–based algorithm and hence conserves more of the local structure than the global structure.

3 Batch-Incremental Classification

In the following, we assume a data stream S is a *sequence* of instances $X_1, \ldots, X_N$, where N denotes the number of available observations so far. Each instance X_i is composed of a vector of d attributes $X_i = (x_i^1, \ldots, x_i^d)$. The dimensionality reduction of S comprises the process of finding a low-dimensional presentation $S' = Y_1, \ldots, Y_N$, where $Y_i = (y_i^1, \ldots, y_i^p)$ and $p \ll d$.

3.1 Prior Work

Unlike tSNE [20], UMAP has no restriction on the projected space size making it useful not only for visualization but also as a general dimension reduction technique for machine learning algorithms. It starts by constructing open balls over all instances and building simplicial complexes. Dimension reduction is obtained by finding a representation, in a lower space, that closely resembles the topological structure in the original space. Given the new dimension, an equivalent

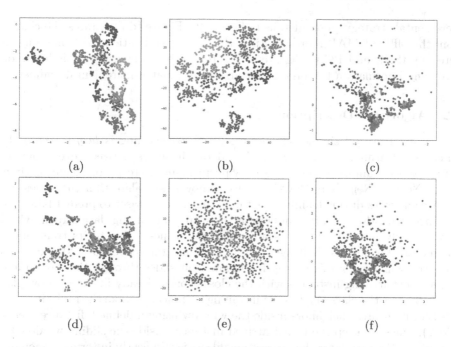

(a)   (b)   (c)

(d)   (e)   (f)

Fig. 1. Projection of CNAE dataset in 2-dimensional space. Offline: (a) UMAP, (b) tSNE, and (c) PCA. Batch-incremental: (d) UMAP, (e) tSNE, and (f) PCA. (Color figure online)

fuzzy topological representation is then constructed [21]. Then, UMAP optimizes it by minimizing the cross-entropy between these two fuzzy topological representations. UMAP offers better visualization quality than tSNE by preserving more of the global structure in a shorter running time. To the best of our knowledge, none of these techniques has a streaming version. Ultimately, both techniques are essentially transductive[1] and do not learn a mapping function from the input space. Hence, they need to process all the data for each new unseen instance, which prevents them from being usable in data streams classification models.

Figure 1 shows the projection of CNAE dataset (see Table 1) into 2-dimensions in an offline/online fashions where each color represents a label. In Fig. 1a, we note that UMAP offers the most interesting visualization while separating classes (9 classes). The overlap in the new space, for instance with tSNE in Fig. 1b, can potentially affect later classification task, notably distance-based algorithms, since properties like global distances and density may be lost. On the other hand, linear transformation, such as PCA, cannot discriminate between instances which prevents them from being represented in the form of clusters (Fig. 1c). To motivate our choice, we project the same dataset using our batch-

---

[1] Transductive learning consists on learning on a full given dataset (including unknown label), but prediction is only made on the known set of unlabeled instances from the same dataset. This is achieved by clustering data instances.

incremental strategy (more details in Sect. 3.2). Figure 1d illustrates the change from the offline UMAP representation; this is not as drastic as the ones engendered by tSNE and PCA (Figs. 1e and f, respectively) showing their limits on capturing information from data that arrives in a batch-incremental manner.

3.2   Algorithm Description

A very efficient and simple scheme in supervised learning is *lazy learning* [1]. Since lazy learning approaches are based on distances between every pair of instances, they unfortunately have a low performance in terms of execution time. The k-Nearest Neighbors (kNN) is a well-known lazy algorithm that does not require any work during training, so it uses the entire dataset to predict labels for test instances. However, it is impossible to store an evolving data stream which is potentially infinite – nor to scan it multiple times – due to its tremendous volume. To tackle this challenge, a basic incremental version of kNN has been proposed which uses a fixed-length window that slides through the stream and merges new arriving instances with the closest ones already in the window [23].

To predict the class label for an incoming instance, we take the majority class labels of its nearest neighbors inside the window using a defined distance metric (Eq. 2). Since we keep the recent arrived instances inside the sliding window for prediction, the search for the nearest neighbors is still costly in terms of memory and time [3,7] and high-dimensional streams require further resources.

Given a window w, the distance between X_i and X_j is defined as follows:

$$D_{X_j}(X_i) = \sqrt{\|X_i - X_j\|^2}. \tag{1}$$

Similarly, the k-Nearest Neighbors distance is defined as follows:

$$D_{w,k}(X_i) = \min_{\binom{w}{k}, X_j \in w} \sum_{j=1}^{k} D_{X_j}(X_i), \tag{2}$$

where $\binom{w}{k}$ denotes the subset of the kNN to X_i in w.

When dealing with high-dimensional data, a pre-processing phase before applying a learning algorithm is imperative to avoid the curse of dimensionality from a *computational* point of view. The latter may increase the resources usage and decrease the performance of some algorithms, such as kNN. The main idea to mitigate this curse consists of using an efficient strategy with consistent and promising results such as UMAP.

Since UMAP is a transductive technique, an instance-incremental learning approach that includes UMAP does not work because the entire stream needs to be processed for each new incoming instance. By doing it this way, the process will be costly and will not respond to the streaming requirements. To alleviate the processing cost considering the framework within which several challenges shall be respected, including the memory constraint and the incremental behavior of data, we adopt a batch-incremental strategy. In the following, we introduce the procedure of our novel approach, batch-incremental UMAP-kNN.

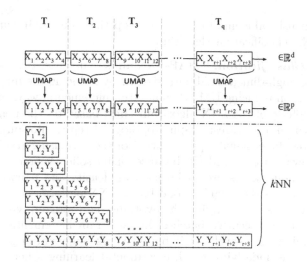

Fig. 2. Batch-incremental UMAP-kNN scheme

Step 1: ***Partition of the Stream.*** During this step, we assume that data arrive in batches – or chunks – by dividing the stream into disjoint partitions $S_1, S_2, \ldots$ of size s. The first part of Fig. 2 shows a stream of instances divided into batches, so instead of having instances available one at a time, they will arrive as a group of instances simultaneously, $S_1, \ldots, S_q$, where S_q is the qth chunk. A simple example of data stream is a video sequence where at each instant we have a succession of images.

Step 2: ***Data Pre-processing.*** We aim to construct a low-dimensional $Y_i \in p$, from an infinite stream of high-dimensional data $X_i \in d$, where $p \ll d$. As mentioned before, UMAP is unable to compress data incrementally and needs to transform more than one observation at a time because it builds a neighborhood-graph on a set of instances and then lays it out in a lower dimensional space [21]. Thus, our proposed approach operates on batches of the stream where a single batch S_i of data is processed at a time T_i. The two first steps in Fig. 2 depict the application of UMAP on the disjoint batches. Once a batch is complete, throughout the second step, we apply UMAP on it independently from the chunks that have been already processed, so each $S_i \in \mathbb{R}^d$ will be transformed and represented by $S_i' \in \mathbb{R}^p$. This new representation is very likely devoid of redundancies, irrelevant attributes, and is obtained by finding potentially useful non-linear combinations of existing attributes, i.e. by repacking relevant information of the larger feature space and encoding it more compactly.

For UMAP to learn when moving from a batch to another, we seed each chunk's embedding with the outcome of the previous one, i.e., match the prior initial coordinates for instances in the current embedding to the final coordinates in the preceding one. This will help to avoid losing the topological information of the stream and to keep stability in successive embeddings as we transition from one batch to its successor. Afterwards, we use the compressed representation of the high-dimensional chunk for the next step that consists in supporting the incremental kNN classification algorithm.

Step 3: kNN Classification. UMAP-kNN aims to decrease the computational costs of kNN on high-dimensional data stream by reducing the input space size using the dimension reducing UMAP, in a batch-incremental way. In addition to the prediction phase of the kNN algorithm that, based on the neighborhood[2], UMAP operates on a k-nearest graph (topological representation) as well and optimizes the low-dimensional representation of the data using gradient descent. One nice takeaway is that UMAP, because of its solid theoretical backing as a manifold technique, keeps properties such as density and pairwise distances. Thus, it does not bias the neighborhood-based kNN performance.

This step consists of classifying the evolving data stream, where the learning task occurs on consecutive batches, i.e. we train incrementally kNN with instances becoming successively available in chunk buffers after pre-processing. Figure 2 shows the underlying batch-incremental learning scheme used which is built upon the divide-and-conquer strategy. Since UMAP is independently applied to batches, so once a chunk is complete and has been transformed in $\mathbb{R}^p$, we feed the half of the batch to the sliding window and we predict incrementally the class label for the second half (the rest of instances).

Given that kNN is adaptive, the main novelty of UMAP-kNN is in how it merges the current batch to previous ones. This is done by adding it to the instances from previous chunks inside the kNN window. Even if past chunks have been discarded, only some of them have been stored and maintained while the adaptive window scrolls. Thereafter, instances kept temporarily inside the window are going to be used to define the neighborhood and predict the class labels for later incoming instances. As presented in Fig. 2, the intuitive idea to combine results from different batches is to use the half of each batch for training and the second half for prediction. In general, due to the possibility of having often very different successive embeddings, one would expect that this may affect the global performance of our approach. Thus, we adopt this scheme to maintain a stability over an adaptive batch-incremental manifold classification approach.

4   Experiments

In this section, we present a series of experiments carried out on various datasets based on three main results: the accuracy, the memory (MB), and the time (Sec).

4.1   Datasets

We use 3 synthetic and 6 real-world datasets from different scenarios that have been thoroughly used in the literature to evaluate the performance of data

---

[2] The distances between the new incoming instance and the instances already available inside the adaptive window are computed in order to assign it to a particular class.

streams classifiers. Table 1 presents a short description of each dataset, and further details are provided in what follows.

Tweets. The dataset was created using the tweets text data generator provided by MOA [6] that simulates sentiment analysis on tweets, where messages can be classified depending on whether they convey positive or negative feelings. $Tweets_{1,2,3}$ produce instances of 500, 1,000 and 1,500 attributes respectively.

Har. Human Activity Recognition dataset [2] built from several subjects performing daily living activities, such as walking, sitting, standing and laying, while wearing a waist-mounted smartphone equipped with sensors. The sensor signals were pre-processed using noise filters and attributes were normalized.

CNAE. CNAE is the national classification of economic activities dataset [9]. Instances represent descriptions of Brazilian companies categorized into 9 classes. The original texts were pre-processed to obtain the current highly sparse data.

Enron. The Enron corpus dataset is a large set of email messages that was made public during the legal investigation concerning the Enron corporation [17]. This cleaned version of Enron consists of $1,702$ instances and $1,000$ attributes.

Table 1. Overview of the datasets

Dataset	#Instances	#Attributes	#Classes	Type
$Tweets_1$	1,000,000	500	2	Synthetic
$Tweets_2$	1,000,000	1,000	2	Synthetic
$Tweets_3$	1,000,000	1,500	2	Synthetic
Har	10,299	561	6	Real
CNAE	1,080	856	9	Real
Enron	1,702	1,000	2	Real
IMDB	120,919	1,001	2	Real
Nomao	34,465	119	2	Real
Covt	581,012	54	7	Real

IMDB. IMDB movie reviews dataset was proposed for sentiment analysis [19], where each review is encoded as a sequence of word indexes (integers).

Nomao. Nomao dataset [8] was provided by Nomao Labs where data come from several sources on the web about places (name, address, localization, etc.).

Covt. The forest covertype dataset obtained from US forest service resource information system data where each class label presents a different cover type.

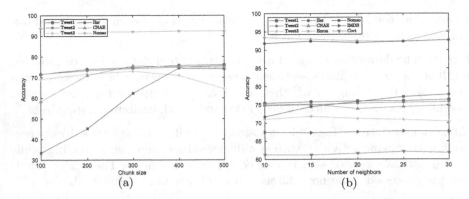

Fig. 3. (a) Varying the chunk size. (b) Varying the neighborhood size for UMAP.

4.2 Results and Discussions

We compare our proposed classifier, UMAP-kNN, to various commonly-used baseline methods in dimensionality reduction and machine learning areas. PCA [24], tSNE (fixing the perplexity to 30, which is the best value as reported in [20]), SAM-kNN (SkNN) [18]. We use HAT, a classifier with a different structure based on trees [4], to assess its performance with the neighborhood-based UMAP. For fair comparison, we compare the performance of UMAP-kNN approach with a competitor using UMAP as well in the same batch-incremental manner. Actually, incremental kNN has two crucial parameters: (i) the number of neighbors k fixed to 5; and (ii) the window size w, that maintains the low-dimensional data, fixed to 1000. According to previous studies such as [7], a bigger window will increase the resources usage and smaller size will impact the accuracy.

The experiments were conducted on a machine equipped with an Intel Core i5 CPU and 4 GB of RAM. All experiments were implemented and evaluated in Python by extending the Scikit-multiflow framework[3] [22].

Figure 3a depicts the influence of the chunk size on the accuracy using UMAP-kNN with some datasets. Generally, fixing the chunk size imposes the following dilemma: choosing a small size so that we obtain an accurate reflection of the current data or choosing a large size that may increase the accuracy since more data are available. The ideal would be to use a batch with the maximum of instances to represent as possible the whole stream. In practice, the chunk size needs to be small enough to fit in the main memory otherwise the running time of the approach will increase. Since UMAP is relatively slow, we choose small chunk sizes to overcome this issue with UMAP-kNN. Based on the obtained results, we fix the chunk size to 400 for the best trade-off accuracy-memory.

We investigate the behavior of a crucial parameter that controls UMAP, number of neighbors, via the classification performance of our approach. Based

---

[3] https://scikit-multiflow.github.io/.

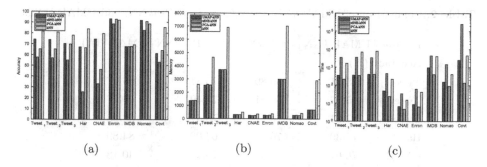

Fig. 4. Comparison of UMAP-kNN, tSNE-kNN, PCA-kNN, and kNN (with the entire datasets) while projecting into 3-dimensions: (a) Accuracy. (b) Memory.

on the size of the neighborhood, UMAP constructs the manifold and focuses on preserving local and global structures. Figure 3b shows the accuracy when the number of neighbors is varied on diverse datasets. We notice that for all datasets, the accuracy is consistently the same with no large differences, e.g. Har. Since a large neighborhood leads to a slower learning process, in the following we fix the neighborhood size to 15.

tSNE is a visualization technique, so we are limited to project high-dimensional data into 2 or 3 dimensions. In order to evaluate the performance of our proposal in a fair comparison against each of tSNE-kNN and PCA-kNN, we project data into 3-dimensional space. We illustrate in Fig. 4a that UMAP-kNN makes significantly more accurate predictions beating consistently the best performing baselines (tSNE-kNN and PCA-kNN) notably with CNAE and the tweets datasets. Figure 4b depicts the quantity of memory needed by the three algorithms which is practically the same for some datasets. Compared to kNN that uses the whole data without projection, we notice that UMAP-kNN consumes much less memory whilst sacrificing a bit in accuracy because we are removing many attributes. Figure 4c shows that our approach is consistently faster than tSNE-kNN because tSNE computes the distances between every pair of instances to project. But PCA-kNN is a bit faster thanks to the simplicity of PCA. But with this trade-off our approach performs good on almost all datasets.

In addition to its good classification performance in comparison with competitors, the batch-incremental UMAP-kNN did a better job of preserving density by capturing both of global and local structures, as shown in Fig. 1d. The fact that UMAP and kNN are both neighborhood-based methods arises as a key element in achieving a good accuracy. UMAP has not only the power of visualization but also the ability to reduce the dimensionality of data efficiently which makes it useful as pre-processing technique for machine learning.

Table 2 reports the comparison of UMAP-kNN against state-of-the-art classifiers. We highlight that our approach performs better on almost all datasets. It achieves similar accuracies to UMAP-SkNN on several datasets but in terms of resources, the latter is slower because of its drift detection mechanism.

Table 2. Comparison of UMAP-kNN, PCA-kNN, UMAP-SkNN, and UMAP-HAT.

Dataset	UMAP-kNN	PCA-kNN	UMAP-SkNN	UMAP-HAT
Accuracy (%)				
Tweets$_1$	75.71	69.89	75.37	66.47
Tweets$_2$	75.16	69.21	74.40	61.27
Tweets$_3$	71.01	70.81	70.47	66.98
Har	75.30	70.50	64.09	84.89
CNAE	76.67	67.41	75.18	40.18
Enron	92.24	93.41	91.89	91.77
IMDB	67.38	67.28	67.43	64.52
Nomao	91.92	91.13	91.63	83.75
Covt	61.29	66.73	53.08	55.43
Memory (MB)				
Tweets$_1$	1366.71	1354.24	1373.15	2738.32
Tweets$_2$	2530.30	2518.76	2532.95	4891.23
Tweets$_3$	3706.99	3706.55	3722.68	7144.77
Har	311.58	310.48	312.84	381.49
CNAE	254.17	246.94	260.29	262.52
Enron	269.00	267.31	271.56	288.74
IMDB	3012.85	3013.28	3018.04	7471.64
Nomao	289.81	285.50	290.60	508.50
Covt	700.69	689.97	704.46	3788.54
Time (Sec)				
Tweets$_1$	558.56	217.44	1396.32	2163.14
Tweets$_2$	616.50	350.63	908.59	3453.21
Tweets$_3$	667.43	400.62	1066.98	6273.19
Har	75.20	24.37	77.99	82.47
CNAE	8.89	4.81	13.17	19.78
Enron	12.80	9.52	17.26	32.84
IMDB	715.68	407.60	1038.77	4691.07
Nomao	248.79	20.46	327.36	228.00
Covt	2311.21	137.62	3756.41	2297.01

UMAP-kNN has a better performance than PCA-kNN, e.g. the Tweets datasets at the cost of being slower. We also observe the UMAP-HAT failed to overcome our approach (in terms of accuracy, memory, and time) due to the integration of a neighborhood-based technique (UMAP) to a tree structure (HAT).

Figure 5 reports detailed results for Tweet$_1$ dataset with five output dimensions. Figure 5a exhibits the accuracy of our approach which is consistently above

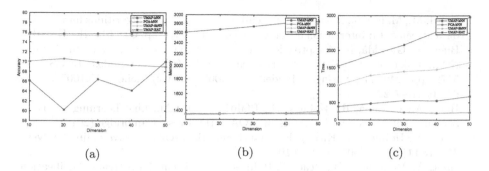

Fig. 5. Comparison of UMAP-kNN, PCA-kNN, UMAP-SkNN, and UMAP-HAT over different output dimensions on Tweet$_1$: (a) Accuracy. (b) Memory. (c) Time.

competitors whilst ensuring stability for different manifolds. Figures 5b and c show that kNN-based classifiers use much less resources than the tree-based UMAP-HAT. We see that UMAP-kNN requires less time than UMAP-HAT and UMAP-SkNN to execute the stream but PCA-kNN is fastest thanks to its simplicity. Still, the gain in accuracy with UMAP-kNN is more significant.

5   Concluding Remarks and Future Work

In this paper, we presented a novel batch-incremental approach for mining data streams using the kNN algorithm. UMAP-kNN combines the simplicity of kNN and the high performance of UMAP which is used as an internal pre-processing step to reduce the feature space of data streams. We showed that UMAP is capable of embedding efficiently data streams within a batch-incremental strategy in an extensive evaluation with well-known state-of-the-art algorithms using various datasets. We further demonstrated that the batch-incremental approach is just as effective as the offline approach in visualization and its accuracy outperforms reputed baselines while using reasonable resources.

We would like to pursue our promising approach further to enhance its runtime performance by applying a fast dimension reduction before using of UMAP. Another area for future work could be the use of a different mechanism, such as the application of UMAP for each incoming data inside a sliding window. We believe that this may be slow but will be suited for instance-incremental learning.

References

1. Aha, D.W.: Lazy Learning. Springer, Heidelberg (2013)
2. Anguita, D., Ghio, A., Oneto, L., Parra, X., Reyes-Ortiz, J.L.: Human activity recognition on smartphones using a multiclass hardware-friendly support vector machine. In: Bravo, J., Hervás, R., Rodríguez, M. (eds.) IWAAL 2012. LNCS, vol. 7657, pp. 216–223. Springer, Heidelberg (2012). https://doi.org/10.1007/978-3-642-35395-6_30

3. Bahri, M., Maniu, S., Bifet, A.: A sketch-based Naive Bayes algorithms for evolving data streams. In: International Conference on Big Data, pp. 604–613. IEEE (2018)
4. Bifet, A., Gavaldà, R.: Adaptive learning from evolving data streams. In: Adams, N.M., Robardet, C., Siebes, A., Boulicaut, J.-F. (eds.) IDA 2009. LNCS, vol. 5772, pp. 249–260. Springer, Heidelberg (2009). https://doi.org/10.1007/978-3-642-03915-7_22
5. Bifet, A., Gavaldà, R., Holmes, G., Pfahringer, B.: Machine Learning for Data Streams: with Practical Examples in MOA. MIT Press, Cambridge (2018)
6. Bifet, A., Holmes, G., Kirkby, R., Pfahringer, B.: MOA: massive online analysis. JMLR 11(May), 1601–1604 (2010)
7. Bifet, A., Pfahringer, B., Read, J., Holmes, G.: Efficient data stream classification via probabilistic adaptive windows. In: SIGAPP, pp. 801–806. ACM (2013)
8. Candillier, L., Lemaire, V.: Design and analysis of the nomao challenge active learning in the real-world. In: ALRA, Workshop ECML-PKDD. sn (2012)
9. Ciarelli, P.M., Oliveira, E.: Agglomeration and elimination of terms for dimensionality reduction. In: ISDA, pp. 547–552. IEEE (2009)
10. Cortes, C., Vapnik, V.: Support-vector networks. ML 20(3), 273–297 (1995)
11. Domingos, P., Hulten, G.: Mining high-speed data streams. In: SIGKDD, pp. 71–80. ACM (2000)
12. Gama, J., Sebastião, R., Rodrigues, P.P.: Issues in evaluation of stream learning algorithms. In: SIGKDD, pp. 329–338. ACM (2009)
13. Günter, S., Schraudolph, N.N., Vishwanathan, S.: Fast iterative kernel principal component analysis. JMLR 8(8), 1893–1918 (2007)
14. Hand, D.J., Mannila, H., Smyth, P.: Principles of Data Mining. MIT Press, Cambridge (2001)
15. Holmes, G., Kirkby, R.B., Bainbridge, D.: Batch-incremental learning for mining data streams (2004)
16. Hotelling, H.: Analysis of a complex of statistical variables into principal components. J. Educ. Psychol. 24(6), 417 (1933)
17. Klimt, B., Yang, Y.: The enron corpus: a new dataset for email classification research. In: Boulicaut, J.-F., Esposito, F., Giannotti, F., Pedreschi, D. (eds.) ECML 2004. LNCS (LNAI), vol. 3201, pp. 217–226. Springer, Heidelberg (2004). https://doi.org/10.1007/978-3-540-30115-8_22
18. Losing, V., Hammer, B., Wersing, H.: KNN classifier with self adjusting memory for heterogeneous concept drift. In: ICDM, pp. 291–300. IEEE (2016)
19. Maas, A.L., Daly, R.E., Pham, P.T., Huang, D., Ng, A.Y., Potts, C.: Learning word vectors for sentiment analysis. In: ACL-HLT, pp. 142–150. Association for Computational Linguistics (2011)
20. van der Maaten, L., Hinton, G.: Visualizing data using t-SNE. JMLR 9, 2579–2605 (2008)
21. McInnes, L., Healy, J., Melville, J.: UMAP: Uniform manifold approximation and projection for dimension reduction. arXiv preprint arXiv:1802.03426 (2018)
22. Montiel, J., Read, J., Bifet, A., Abdessalem, T.: Scikit-multiflow: a multi-output streaming framework. JMLR 19(1), 2914–2915 (2018)
23. Read, J., Bifet, A., Pfahringer, B., Holmes, G.: Batch-incremental versus instance-incremental learning in dynamic and evolving data. In: Hollmén, J., Klawonn, F., Tucker, A. (eds.) IDA 2012. LNCS, vol. 7619, pp. 313–323. Springer, Heidelberg (2012). https://doi.org/10.1007/978-3-642-34156-4_29

24. Ross, D.A., Lim, J., Lin, R.S., Yang, M.H.: Incremental learning for robust visual tracking. IJCV **77**(1–3), 125–141 (2008)
25. Sorzano, C.O.S., Vargas, J., Montano, A.P.: A survey of dimensionality reduction techniques. arXiv preprint arXiv:1403.2877 (2014)
26. Weng, J., Zhang, Y., Hwang, W.S.: Candid covariance-free incremental principal component analysis. TPAMI **25**(8), 1034–1040 (2003)

GraphMDL: Graph Pattern Selection Based on Minimum Description Length

Francesco Bariatti$^{(\boxtimes)}$, Peggy Cellier, and Sébastien Ferré

Univ Rennes, INSA, CNRS, IRISA,
Campus de Beaulieu, Rennes, France
{francesco.bariatti,peggy.cellier,sebastien.ferre}@irisa.fr

Abstract. Many graph pattern mining algorithms have been designed to identify recurring structures in graphs. The main drawback of these approaches is that they often extract too many patterns for human analysis. Recently, pattern mining methods using the *Minimum Description Length* (MDL) principle have been proposed to select a characteristic subset of patterns from transactional, sequential and relational data. In this paper, we propose an MDL-based approach for selecting a characteristic subset of patterns on labeled graphs. A key notion in this paper is the introduction of *ports* to encode connections between pattern occurrences without any loss of information. Experiments show that the number of patterns is drastically reduced. The selected patterns have complex shapes and are representative of the data.

Keywords: Pattern mining · Graph mining · Minimum Description Length

1 Introduction

Many fields have complex data that need labeled graphs, i.e. graphs where vertices and edges have labels, for an accurate representation. For instance, in chemistry and biology, molecules are represented as atoms and bonds; in linguistics, sentences are represented as words and dependency links; in the semantic web, knowledge graphs are represented as entities and relationships. Depending on the domain, graph datasets can be made of large graphs or large collections of graphs. Graphs are complex to analyze in order to extract knowledge, for instance to identify frequent structures in order to make them more intelligible.

In the field of pattern mining, there has been a number of proposals, namely *graph mining* approaches, to extract frequent subgraphs. Classical approaches to graph mining, e.g. gSpan [12] and Gaston [7], work on collections of graphs, and generate all patterns w.r.t. a frequency threshold. The major drawback of this kind of approach is the huge amount of generated patterns, which renders them difficult to analyze. Some approaches such as CloseGraph [13] reduce the number of patterns by only generating *closed patterns*. However, the set of closed patterns generally remains too large, with a lot of redundancy between

M. R. Berthold et al. (Eds.): IDA 2020, LNCS 12080, pp. 54–66, 2020.
https://doi.org/10.1007/978-3-030-44584-3_5

patterns. *Constraint-based* approaches, such as gPrune [14], reduce the number of extracted patterns by extracting only the patterns following a certain acceptance rule. These algorithms generally manage to reduce the number of patterns, however they also limit their type. Additionally, if the acceptance rule is user-provided, the user needs some background knowledge on the data.

More effective approaches to reduce the number of patterns are those based on the *Minimum Description Length* (MDL) principle [3]. The MDL principle comes from information theory, and states that the *model* that describes the data the best is the one that compresses the data the best. It has been shown on sets of items [10], sequences [9] and relations [4] that an MDL-based approach can select a small and descriptive subset of patterns. Few MDL-based approaches have been proposed for graphs. SUBDUE [1] iteratively compresses a graph by replacing each occurrence of a pattern by a single vertex. At each step, the chosen pattern is the one that compresses the most. The drawback of SUBDUE is that the replacement of pattern occurrences by vertices entails a loss of information. VoG [5] summarizes graphs as a composition of predefined families of patterns (e.g., paths, stars). Like SUBDUE, VoG aims to only extract "interesting" patterns, but instead of evaluating each pattern individually like SUBDUE, it evaluates the set of extracted patterns as a whole. This allows the algorithm to find a "good set of patterns" instead of a "set of good patterns". One limitation of VoG is that the type of patterns is restricted to predefined ones. Another limitation is that VoG works on unlabeled graphs, (e.g. network graphs), while we are interested in labeled graphs.

The contribution of this paper (Sect. 3) is a novel approach called GRAPH-MDL, leveraging the MDL principle to select graph patterns from labeled graphs. Contrary to SUBDUE, GRAPHMDL ensures that there is no loss of information thanks to the introduction of the notion of *ports* associated to graph patterns. Ports represent how adjacent occurrences of patterns are connected. We evaluate our approach experimentally (Sect. 4) on two datasets with different kinds of graphs: one on AIDS-related molecules (few labels, many cycles), and the other one on dependency trees (many labels, no cycles). Experiments validate our approach by showing that the data can be significantly compressed, and that the number of selected patterns is drastically reduced compared to the number of candidate patterns. More so, we observe that the patterns can have complex and varied shapes, and are representative of the data.

2   Background Knowledge

2.1   The MDL Principle

The *Minimum Description Length* (MDL) principle [3] is a technique from the domain of information theory that allows to select the model, from a family of models, that best describes some data. The MDL principle states that the best model M for describing some data D is the one that minimizes the *description length* $L(M, D) = L(M) + L(D|M)$, where $L(M)$ is the length of the model and $L(D|M)$ the length of the data encoded with the model. The MDL principle does

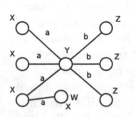

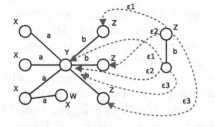

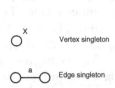

Fig. 1. A labeled undirected simple graph.

Fig. 2. Embeddings of a pattern in the graph of Fig. 1.

Fig. 3. Two singleton patterns.

not define how to compute every possible description length. However, common primitives exist for data and distributions [6]:

– An element $x \in \mathcal{X}$ with uniform distribution has a code of $\log(|\mathcal{X}|)$ bits.
– An element $x \in \mathcal{X}$, appearing $usage(x, D)$ times in some data D has a code of $L_{usage}^{\mathcal{X}}(x, D) = -\log\left(\frac{usage(x,D)}{\sum_{x_i \in \mathcal{X}} usage(x_i,D)}\right)$ bits. This encoding is optimal.
– An integer $n \in \mathbb{N}$ without a known upper bound can be encoded with a *universal integer encoding*, whose size in bits is noted $L_{\mathbb{N}}(n)$[1].

Description lengths of elements that are common to all models are usually ignored, since they do not affect their comparison.

Krimp [10] is a pattern mining algorithm using the MDL principle to select a "characteristic" set of itemset patterns from a transactional database. Because of its good performances, Krimp has been adapted to other types of data, such as sequences [9] and relational databases [4]. In our approach we redefine Krimp's key concepts on graphs, in order to apply a Krimp-like approach to graph mining.

2.2 Graphs and Graph Patterns

Definition 1. *A labeled graph $G = (V, E, l_V, l_E)$ over two label sets $\mathcal{L}_V$ and $\mathcal{L}_E$ is a data structure composed of a set of vertices V, a set of edges $E \subseteq V \times V$, and two labeling functions $l_V \in V \rightarrow 2^{\mathcal{L}_V}$ and $l_E \in E \rightarrow \mathcal{L}_E$ that associate a set of labels to vertices, and one label to edges.*

G is said undirected *if E is symmetric, and* simple *if E is irreflexive.*

Although our approach applies to all labeled graphs, in the following we only consider undirected simple graphs, so as to compare ourselves with existing tools and benchmarks. Figure 1 shows an example of graph, with 8 vertices and 7 edges, defined over vertex label set $\{W, X, Y, Z\}$ and edge label set $\{a, b\}$. In our definition vertices can have several or no labels, unlike usual definitions in graph mining, because it makes it applicable to more datasets.

---

[1] In our implementation we use *Elias gamma encoding* [2], shifted by 1 so that it can encode 0. Therefore $L_{\mathbb{N}}(n) = 2\lfloor \log(n+1) \rfloor + 1$.

P	G^P				v_π			c_π
	Pattern structure	Pattern usage	Pattern code	Pattern code length (bits)	Port count	Port ID	Port usage	Port code	Port code length (bits)
P1	X ①—a—② Y—b—③ Z	3	[P1]	1	2	v1 / v2	1 / 3	[v1] / [v2]	2 / 0.42
Pa	①—a—②	1	[Pa]	2.58	2	v1 / v2	1 / 1	[v1] / [v2]	1 / 1
Pw	① W	1	[Pw]	2.58	1	v1	1		0
Px	① X	1	[Px]	2.58	1	v1	1		0

Fig. 4. Example of a GRAPHMDL code table over the graph of Fig. 1. Pattern and port usages, and code lengths have been added as illustration and are not part of the table definition. Unused singleton patterns are omitted.

Definition 2. *Let G^P and G^D be graphs. An* embedding *(or* occurrence*) of G^P in G^D is an injective function $\varepsilon \in V^P \to V^D$ such that: (1) $l_V^P(v) \subseteq l_V^D(\varepsilon(v))$ for all $v \in V^P$; (2) $(\varepsilon(u), \varepsilon(v)) \in E^D$ for all $(u,v) \in E^P$; and (3) $l_E^P(e) = l_E^D(\varepsilon(e))$ for all $e \in E^P$.*

We define *graph patterns* as graphs G^P having some occurrences in the data graph G^D. Figure 2 shows the three embeddings $\varepsilon_1, \varepsilon_2, \varepsilon_3$ of a two-vertices graph pattern into the graph of Fig. 1. We define *singleton patterns* as the elementary patterns. A *vertex singleton pattern* is a graph with one vertex having one label. An *edge singleton pattern* is a graph with two unlabeled vertices, connected by a single labeled edge. Figure 3 shows examples of singleton patterns.

3   GRAPHMDL: MDL for Graphs

In this section we present our contribution: the GRAPHMDL approach. This approach takes as input a graph—the *original graph* G^o—and a set of patterns extracted from that graph—the *candidate patterns*—and outputs the most descriptive subset of candidate patterns according to the MDL principle. The candidates can be generated with any graph mining algorithm, e.g. gSpan [12].

The intuition behind GRAPHMDL is that since data and patterns are both graphs, the data can be seen as a composition of pattern embeddings. Informally, we want a user analyzing the output of GRAPHMDL to be able to say "the data is composed of one occurrence of pattern A, connected to one occurrence of pattern B, which is itself connected to one occurrence of pattern C". More so, we want the user to be able to tell *how* these structures are connected together: which vertices of each pattern are used to connect it to other patterns.

3.1   Model: A Code Table for Graph Patterns

Similarly to Krimp [10], we define our model as a *Code Table* (CT), i.e. a set $\mathcal{P}$ of patterns with associated coding information. A first difference with Krimp is that the patterns are graph patterns. A second difference is the need for additional coding information: a single code would not suffice since all the information related to connectivity between pattern occurrences would be lost.

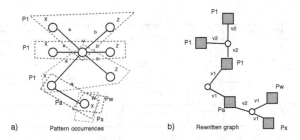

a)   Pattern occurrences   b)   Rewritten graph

Fig. 5. How the data graph of Fig. 1 is encoded with the code table of Fig. 4. *(a)* Retained occurrences of CT patterns. *(b)* The rewritten graph. Blue squares are pattern embeddings (their label indicates the pattern), white circles are port vertices. Edge labels represent which pattern port correspond to each port vertex. (Color figure online)

We therefore introduce the notion of *ports* in order to represent how pattern embeddings connect to each other to form the original graph. The set of ports of a pattern is a subset of the vertices of the pattern. Intuitively, a pattern vertex is a port if at least one pattern embedding maps this vertex to a vertex in the original graph that is also used by another embedding (be it of the same pattern or a different one). For example, in Fig. 5a the three occurrences of pattern $P1$ are inter-connected through their middle vertex: this vertex is a port. Since port information increases the description length, we expect our approach to select patterns with few ports.

Figure 4 shows an example of CT associated to the graph of Fig. 1. Every row of the CT is composed of three parts, and contains information about a pattern $P \in \mathcal{P}$ (e.g. the first row contains information about pattern $P1$). The first part of a row is the graph G^P, which represents the structure of the pattern (e.g. $P1$ is a pattern with three labeled vertices and two labeled edges). The second part of a row is the code c_P, associated to the pattern. The third part of a row is the description of the port set of the pattern, Π_P, (e.g. $P1$ has two ports, its first two vertices, with codes of 2 and 0.42 bits[2]). We note Π the set of all ports of all patterns. Like Krimp, the length of the code of a pattern or port depends on its usage in the encoding of the data, i.e. how many times it is used to describe the original graph G^o (e.g. $P1$ has a code of 1 bit because it is used 3 times and the sum of pattern usages in the CT is 6, see Sects. 3.2 and 3.3).

3.2 Encoding the Data with a Code Table

The intuition behind GRAPHMDL is that we can represent the original graph G^o (i.e. the data) as a set of pattern occurrences, connected via ports. Encoding the data with a CT consists in creating a structure that explicits which occurrences are used and how they interconnect to form the original graph. We call this structure the *rewritten graph* G^r.

---

[2] MDL approaches deal with *theoretical* code lengths, which may not be integers.

Definition 3. *A rewritten graph $G^r = (V^r, E^r, l_V^r, l_E^r)$ is a graph where the set of vertices is $V^r = V_{emb}^r \cup V_{port}^r$: V_{emb}^r is the set of pattern embedding vertices and V_{port}^r is the set of port vertices. $E^r \subseteq V_{emb}^r \times V_{port}^r$ is the set of edges from embeddings to ports, $l_V^r \in V_{emb}^r \to \mathcal{P}$ and $l_E^r \in E^r \to \Pi$ are the labelings.*

In order to compute the encoding of the data graph with a given CT, we start with an empty rewritten graph. One after another, we select patterns from the CT. For each pattern, we compute the occurrences of its graph G^P. Similarly to Krimp, we limit embeddings overlaps: we admit overlap on vertices (since it is the key notion behind ports), but we forbid edge overlaps.

Each retained embedding is represented in the rewritten graph by a *pattern embedding vertex*: a vertex $v_e \in V_{emb}^r$ with a label $P \in \mathcal{P}$ indicating which pattern it instantiates. Vertices that are shared by several embeddings are represented in the rewritten graph by a *port vertex* $v_p \in V_{port}^r$. We add an edge $(v_e, v_p) \in E^r$ between the pattern embedding vertex v_e of a pattern P and the port vertex v_p, when the embedding associated to v_e maps the pattern's port $v_\pi \in \Pi_P$ to v_p. We label this edge v_π.

We make sure that code tables always include all singleton patterns, so that they can always encode any vertex and edge of the original graph.

Figure 5 shows the graph of Fig. 1 encoded with the CT of Fig. 4. Embeddings of CT patterns become pattern embedding vertices in the rewritten graph (blue squares). Vertices that are at the boundary between multiple embeddings become port vertices in the rewritten graph (white circles). When an embedding has a port, its pattern embedding vertex in the rewritten graph is connected to the corresponding port vertex and the edge label indicates which pattern's port it is. For instance, the three retained occurrences of pattern $P1$ all share the same vertex labeled Y (middle of the original graph), thus in the rewritten graph the three corresponding pattern embedding vertices are connected to the same port vertex via port v_2.

3.3   Description Lengths

In this section we define how to compute the description length of the CT and the rewritten graph. Description lengths are used to compare CTs. Formulas are explained below and grouped in Fig. 6.

Code Table. The description length $L(M) = L(CT)$ of a CT is the sum of the description lengths of its rows (skipping rows with unused patterns), and every row is composed of three parts: the pattern graph structure, the pattern code, and the pattern port description.

To describe the structure $G = G^P$ of a pattern ($L(G)$) we start by encoding the number of vertices of the pattern. Then we encode the vertices one after the other. For each vertex v, we encode its labels then its adjacent edges. To encode the vertex labels ($L_V(v, G)$) we specify their number first, then the labels themselves. To encode the adjacent edges ($L_E(v, G)$) we specify their number (between 0 and $|V| - 1$ in a simple graph), then for each edge, its destination

$$L(c_P) = L_{usage}^{\mathcal{P}}(P, G^r) \quad \text{where } usage(P_i, G^r) = |\{v_e \in V_{emb}^r \mid l_V^r(v_e) = P_i\}|$$

$$L(c_\pi, P) = L_{usage}^{\Pi_P}(\pi, G^r) \quad \text{where } usage(\pi_i, G^r) = |\{e \in E_{emb}^r \mid l_E^r(e) = \pi_i\}|$$

$$L(M) = L(CT) = \underbrace{\sum_{\substack{P \in \mathcal{P} \\ usage(P) \neq 0}} L(G)}_{\text{structure}} + \underbrace{L(c_P)}_{\text{code}} + \underbrace{L(\Pi_P)}_{\text{ports}}$$

$$\left| \begin{array}{l} L(G) = \underbrace{L_{\mathbb{N}}(|V|)}_{\text{vertex count}} + \sum_{v \in V} [\, \underbrace{L_V(v, G)}_{\text{vertex labels}} + \underbrace{L_E(v, G)}_{\text{edges of vertex}} \,] \\[2ex] L_V(v, G) = \underbrace{L_{\mathbb{N}}(|l_V(v)|)}_{\text{label count}} + \sum_{l \in l_V(v)} \underbrace{L_{usage}^{\mathcal{L}_V}(l, G^o)}_{\text{label code}} \\[2ex] L_E(v, G) = \underbrace{\log(|V|)}_{\text{edge count}} + \sum_{(v,w) \in E \mid v < w} [\, \underbrace{\log(|V|)}_{\text{destination}} + \underbrace{L_{usage}^{\mathcal{L}_E}(l_E(v, w), G^o)}_{\text{label}}] \\[2ex] L(\Pi_P) = \underbrace{\log(|V| + 1)}_{\text{port count } |\Pi_P|} + \underbrace{\log\left(\binom{|V|}{|\Pi_P|} \right)}_{\text{port ids}} + \sum_{\pi \in \Pi_P} \underbrace{L(c_\pi, P)}_{\text{port code}} \end{array} \right.$$

$$L(D|M) = L(G^r) = \underbrace{L_{\mathbb{N}}(|V_{port}^r|)}_{\text{port vertex count}} + \sum_{v \in V_{emb}^r} L_{emb}(v, P, G^r) \quad \text{with} \quad P = l_V^r(v)$$

$$\left| L_{emb}(v, P, G^r) = \underbrace{L(c_P)}_{\text{pattern code}} + \underbrace{\log(|\Pi_P| + 1)}_{\text{edge count}} + \sum_{\substack{(v,w) \in E^r \\ \pi = l_E^r(v,w)}} \underbrace{\log(|V_{port}^r|)}_{\text{port vertex id}} + \underbrace{L(c_\pi, P)}_{\text{port code}} \right.$$

Fig. 6. Formulas used for computing description lengths. The structure $G^P = (V^P, E^P, l_V^P, l_E^P)$ is shortened to $G = (V, E, l_V, l_E)$ for ease of reading.

vertex and its label. To avoid encoding twice the same edge, we decide—in undirected graphs—to encode edges with the vertex with the smallest identifier. Vertex and edge labels are encoded based on their relative usage in the original graph G^o ($L_{usage}^{\mathcal{L}_V}(l, G^o)$ and $L_{usage}^{\mathcal{L}_E}(l_E(v, w), G^o)$). Since this encoding does not change between CTs, it is a meaningful way to compare them.

The second element of a CT row is the code c_P associated to the pattern ($L(c_P)$). This code is based on the usage of the pattern in the rewritten graph.

The last element of a CT row is the description of the pattern's ports ($L(\Pi_P)$). First, we encode the number of pattern's ports (between 0 and $|V|$). Then we specify which vertices are ports: if there are k ports, then there are $\binom{|V|}{k}$ possibilities. Finally, we encode the port codes ($L(c_\pi, P)$): their code is based on the usage of the port in the rewritten graph w.r.t. other ports of the pattern.

Rewritten Graph. The rewritten graph has two types of vertices: port vertices and pattern embedding vertices. Port vertices do not have any associated information, so we just need to encode their number. The description length $L(D|M) = L(G^r)$ of the rewritten graph is the length needed for encoding the number of vertex ports plus the sum of the description lengths $L_{emb}(v, P, G^r)$ of the pattern embedding vertices v. Every pattern embedding vertex has a label $l_V^r(v)$ specifying its pattern P, encoded with the code c_P of the pattern. We then encode the number of edges of the vertex i.e. the number of ports of this embedding in particular (between 0 and $|\Pi_P|$). Then for each edge we encode the port vertex to which it is connected and to which port it corresponds (using the port code c_π).

Table 1. Characteristics of the datasets used in the experiments

Dataset	Graph count	$|V|$	$|E|$	$|\mathcal{L}_V|$	$|\mathcal{L}_E|$
AIDS-CA	423	16714	17854	21	3
AIDS-CM	1082	34387	37033	26	3
UD-PUD-En	1000	21176	20176	17	46

3.4   The GRAPHMDL Algorithm

In previous subsections we presented the different MDL definitions that GRAPH-MDL uses to evaluate pattern sets (CT). A naive algorithm for finding the most descriptive pattern set (in the MDL sense) could be to create a CT for every possible subset of candidates and retain the one yielding the smallest description length, such an approach is often infeasible because of the large amount of possible subsets. That is why GRAPHMDL applies a greedy heuristic algorithm, adapting Krimp algorithm [10] to our MDL definitions.

Like Krimp, our algorithm starts with a CT composed of all singletons, which we call CT_0. One after the other, candidates are added to the CT if they allow to lower the description length. Two heuristics guide GRAPHMDL: the candidate order and the order of patterns in the CT. We use the same heuristics as Krimp, with the difference that we define the size of a pattern as its total number of labels (vertices and edges). We also implement Krimp's "post-acceptance pruning": after a pattern is accepted in the CT, GRAPHMDL verifies if the removal of some patterns from the CT allows to lower the description length $L(M, D)$.

4   Experimental Evaluation

In order to evaluate our proposal, we developed a prototype of GRAPHMDL. The prototype was developed in Java 1.8 and is available as a git repository[3].

4.1   Datasets

The first two datasets that we use, AIDS-CA and AIDS-CM, are part of the National Cancer Institute AIDS antiviral screen data[4]. They are collections of graphs often used to compare graph mining algorithms [11]. Graphs of this collection represent molecules: vertices are atoms and edges are bonds. We stripped all hydrogen atoms from the molecules, since their positions can be inferred.

We took our third dataset, UD-PUD-En, from the Universal Dependencies project[5]. This project curates a collection of trees describing dependency

---

[3] https://gitlab.inria.fr/fbariatt/graphmdl.
[4] https://wiki.nci.nih.gov/display/NCIDTPdata/AIDS+Antiviral+Screen+Data.
[5] https://universaldependencies.org/.

Table 2. Experimental results for different candidate sets

Dataset	gSpan support	Candidate count	Runtime	$|CT|$	$\frac{L(CT,D)}{L(CT_0,D)}$	Median label count	Median port count
AIDS-CA	20%	2194	19 m	115	24.42%	9	3
AIDS-CA	15%	7867	1 h 47 m	123	21.64%	10	4
AIDS-CA	10%	20596	3 h 36 m	148	19.03%	11	3
AIDS-CM	20%	433	22 m	111	28.91%	7	4
AIDS-CM	15%	779	32 m	131	27.44%	9	4
AIDS-CM	10%	2054	1 h 10 m	163	24.94%	9	4
AIDS-CM	5%	9943	5 h 02 m	225	20.43%	9	4
UD-PUD-En	10%	164	1 m	162	39.55%	5	2
UD-PUD-En	5%	458	3 m	249	34.45%	5	2
UD-PUD-En	1%	6021	19 m	523	28.14%	7	2
UD-PUD-En	0%	233434	9 h 57 m	773	26.25%	7	2

relationships between words of sentences of multiple corpora in multiple languages. We used the trees corresponding to the English version of the PUD corpus.

Table 1 presents the main characteristics of the three datasets that we use: the number of elementary graphs in the dataset, the total amount of vertices, the total amount of edges, the number of different vertex labels, and the number of different edge labels. Since GRAPHMDL works on a single graph instead of a collection, we aggregate collections into a single graph with multiple connected components when needed. We generate candidate patterns by using a gSpan implementation available on its author's website[6].

4.2 Quantitative Evaluation

Table 2 presents the results of the first experiment. For instance the first line tells that we ran GRAPHMDL on the AIDS-CA dataset, with as candidates the 2194 patterns generated by gSpan for a support threshold of 20%. It took 19 min for our approach to select a CT composed of 115 patterns, yielding a description length that is 24% of the description length obtained by the singleton-only CT_0. Selected patterns have a median of 9 labels and 3 ports.

We observe that the number of patterns of a CT is often significantly smaller than the number of candidates. This is particularly remarkable for experiments ran with small support thresholds, where GRAPHMDL reduces the number of patterns up to 300 times: patterns generated for these support thresholds probably contain a lot of redundancy, that GRAPHMDL avoids.

We also note that the description lengths of the CTs found by GRAPHMDL are between 20% and 40% of the lengths of the baseline code tables CT_0, which shows that our algorithm succeeds in finding regularities in the data. Description

---

[6] https://sites.cs.ucsb.edu/~xyan/software/gSpan.htm.

lengths are smaller when the number of candidates is higher: this may be because with more candidates, there are more chances of finding "good" candidates that allow to better reduce description lengths.

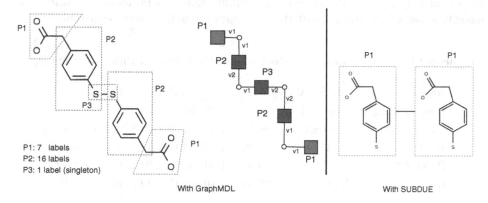

Fig. 7. How GRAPHMDL (left) and SUBDUE (right) encode one of AIDS-CM graphs.

We observe that GRAPHMDL can find patterns of non-trivial size, as shown by the median label count in Table 2. Also, most patterns have few ports, which shows that GRAPHMDL manages to find models in which the original graph is described as a set of components without many connections between them. We think that a human will interpret such a model with more ease, as opposed to a model composed of "entangled" components.

4.3   Qualitative Evaluations

Interpretation of Rewritten Graphs. Figure 7 shows how GRAPHMDL uses patterns selected on the AIDS-CM dataset to encode one of the graphs of the dataset (more results are available in our git repository). It illustrates the key idea behind our approach: find a set of patterns so that each one describes part of the data, and connect their occurrences via ports to describe the whole data.

We observe that GRAPHMDL selects bigger patterns (such as P2), describing big chunks of data, as well as smaller patterns (such as P3, edge singleton), that can form bridges between pattern occurrences. Big patterns increase the description length of the CT, but describe more of the data in a single occurrence, whereas small patterns do the opposite. Following the MDL principle, GRAPHMDL finds a good balance between the two types of patterns.

It is interesting to note that pattern P1 in Fig. 7 corresponds to the carboxylic acid functional group, common in organic chemistry. GRAPHMDL selected this pattern without any prior knowledge of chemistry, solely by using MDL.

Comparison with SUBDUE. On the right of Fig. 7 we can observe the encoding found by SUBDUE on the same graph. The main disadvantage of SUBDUE is information loss: we can see that the data is composed of two occurrences of pattern P1, but not how these two occurrences are connected. Thanks to the notion of ports, GRAPHMDL does not suffer from this problem: the user can exactly know which atoms lie at the boundary of each pattern occurrence.

Table 3. Classification accuracies. Results of methods marked with * are from [8].

Algorithm	AIDS-CA/CI	Mutag	PTC-MR	PTC-FR
Baseline-Largest	50.01 ± 0.03	66.50 ± 0.00	55.80 ± 0.00	65.50 ± 0.00
GRAPHMDL	71.61 ± 0.96	80.79 ± 1.51	57.38 ± 1.68	62.70 ± 1.86
WL*	N/A	87.26 ± 1.42	63.12 ± 1.44	67.64 ± 0.74
P-WL-C*	N/A	90.51 ± 1.34	64.02 ± 0.82	67.15 ± 1.09
RetGK*	N/A	90.30 ± 1.10	62.15 ± 1.60	67.80 ± 1.10

Assessing Patterns Through Classification. We showed in the previous experiments that GRAPHMDL manages to reduce the amount of patterns, and that the introduction of ports allows for a precise analysis of graphs. We now ask ourselves if the extracted patterns are *characteristic* of the data. To evaluate this aspect, we adopt the classification approach used by Krimp [10]. We apply GRAPHMDL independently on each class of a multi-class dataset, and then use the resulting CTs to classify each graph: we encode it with each of the CTs, and classify it in the class whose CT yields the smallest description length $L(D|M)$. Since GRAPHMDL is not designed with the goal of classification in mind, we would expect existing classifiers to outperform GRAPHMDL. In particular, note that patterns are selected on each class independently of other classes. Indeed, GRAPHMDL follows a descriptive approach whereas classifiers generally follow a discriminative approach. Table 3 presents the results of this new experiment. We compare GRAPHMDL with graph classification algorithms found in the literature [8], and a baseline that classifies all graphs as belonging to the largest class. The AIDS-CA/CI dataset is composed of the CA class of the AIDS dataset and a same-size same-labels random sample from the CI class (corresponding to negative examples). The other datasets[7] are from [8]. We performed a 10-fold validation repeated 10 times and report average accuracies and standard deviations.

GRAPHMDL clearly outperforms the baseline on two datasets, AIDS and Mutag, but is only comparable to the baseline for the PTC datasets. On Mutag, GRAPHMDL is less accurate than other classifiers but closer to them than to the baseline. On the PTC datasets, we hypothesize that the learned descriptions are not discriminative w.r.t. the chosen classes, although they are characteristic enough to reduce description length. Nonetheless results are still better than random guessing (accuracy would be 50%). An interesting point of GRAPHMDL

---

[7] For concision, we do not report on PTC-{MM,FM}, they yield similar results.

classification is that it is explainable: the user can look at how the patterns of the two classes encode a graph (similarly to Fig. 7) and understand *why* one class is chosen over another.

5   Conclusion

In this paper, we have proposed GRAPHMDL, an MDL-based pattern mining approach to select a representative set of graph patterns on labeled graphs. We proposed MDL definitions allowing to compute description lengths necessary to apply the MDL principle. The originality of our approach lies in the notion of *ports*, which guarantee that the original graph can be perfectly reconstructed, i.e., without any loss of information. Our experiments show that GRAPHMDL significantly reduces the amount of patterns w.r.t. complete approaches. Further, the selected patterns can have complex shapes with simple connections. The introduction of the notion of ports facilitates interpretation w.r.t. to SUBDUE. We plan to apply our approach to more complex graphs, e.g. knowledge graphs.

References

1. Cook, D.J., Holder, L.B.: Substructure discovery using minimum description length and background knowledge. J. Artif. Intell. Res. **1**, 231–255 (1993)
2. Elias, P.: Universal codeword sets and representations of the integers. IEEE Trans. Inf. Theory **21**(2), 194–203 (1975)
3. Grünwald, P.: Model selection based on minimum description length. J. Math. Psychol. **44**(1), 133–152 (2000)
4. Koopman, A., Siebes, A.: Characteristic relational patterns. In: Proceedings of the 15th ACM SIGKDD International Conference on Knowledge Discovery and Data Mining, KDD 2009, pp. 437–446. ACM (2009)
5. Koutra, D., Kang, U., Vreeken, J., Faloutsos, C.: Summarizing and understanding large graphs. Stat. Anal. Data Mining: ASA Data Sci. J. **8**(3), 183–202 (2015)
6. Lee, T.C.M.: An introduction to coding theory and the two-part minimum description length principle. Int. Stat. Rev. **69**(2), 169–183 (2001)
7. Nijssen, S., Kok, J.N.: The Gaston tool for frequent subgraph mining. Electron. Notes Theor. Comput. Sci. **127**(1), 77–87 (2005)
8. Rieck, B., Bock, C., Borgwardt, K.: A persistent Weisfeiler-Lehman procedure for graph classification. In: Proceedings of the 36th International Conference on Machine Learning, pp. 5448–5458. PMLR (2019)
9. Tatti, N., Vreeken, J.: The long and the short of it: summarising event sequences with serial episodes. In: Proceedings of the International Conference on Knowledge Discovery and Data Mining (KDD 2012), pp. 462–470. ACM (2012)
10. Vreeken, J., van Leeuwen, M., Siebes, A.: KRIMP: mining itemsets that compress. Data Min. Knowl. Discov. **23**(1), 169–214 (2011)
11. Wörlein, M., Meinl, T., Fischer, I., Philippsen, M.: A quantitative comparison of the subgraph miners MoFa, gSpan, FFSM, and Gaston. In: Jorge, A.M., Torgo, L., Brazdil, P., Camacho, R., Gama, J. (eds.) PKDD 2005. LNCS (LNAI), vol. 3721, pp. 392–403. Springer, Heidelberg (2005). https://doi.org/10.1007/11564126_39

12. Yan, X., Han, J.: gSpan: graph-based substructure pattern mining. In: Proceedings of the 2002 IEEE International Conference on Data Mining (ICDM 2002), pp. 721–724. IEEE Computer Society (2002)
13. Yan, X., Han, J.: CloseGraph: mining closed frequent graph patterns. In: ACM SIGKDD International Conference Knowledge Discovery and Data Mining (KDD), pp. 286–295. ACM (2003)
14. Zhu, F., Yan, X., Han, J., Yu, P.S.: gPrune: a constraint pushing framework for graph pattern mining. In: Zhou, Z.-H., Li, H., Yang, Q. (eds.) PAKDD 2007. LNCS (LNAI), vol. 4426, pp. 388–400. Springer, Heidelberg (2007). https://doi.org/10.1007/978-3-540-71701-0_38

Towards Content Sensitivity Analysis

Elena Battaglia, Livio Bioglio, and Ruggero G. Pensa$^{(\boxtimes)}$ (iD)

Department of Computer Science, University of Turin, Turin, Italy
{elena.battaglia,livio.bioglio,ruggero.pensa}@unito.it

Abstract. With the availability of user-generated content in the Web, malicious users dispose of huge repositories of private (and often sensitive) information regarding a large part of the world's population. The self-disclosure of personal information, in the form of text, pictures and videos, exposes the authors of such contents (and not only them) to many criminal acts such as identity thefts, stalking, burglary, frauds, and so on. In this paper, we propose a way to evaluate the harmfulness of any form of content by defining a new data mining task called *content sensitivity analysis*. According to our definition, a score can be assigned to any object (text, picture, video...) according to its degree of sensitivity. Even though the task is similar to sentiment analysis, we show that it has its own peculiarities and may lead to a new branch of research. Thanks to some preliminary experiments, we show that content sensitivity analysis can not be addressed as a simple binary classification task.

Keywords: Privacy · Text mining · Text categorization

1   Introduction

Internet privacy has gained much attention in the last decade due to the success of online social networks and other social media services that expose our lives to the wide public. In addition to personal and behavioral data collected more or less legitimately by companies and organizations, many websites and mobile/web applications store and publish tons of user-generated content in the form of text posts and comments, pictures and videos which, very often, capture and represent private moments of our life. The availability of user-generated content is a huge source of relatively easy-to-access private (and often very sensitive) information concerning habits, preferences, families and friends, hobbies, health and philosophy of life, which expose the authors of such contents (or any other individual referenced by them) to many (cyber)criminal risks, including identity theft, stalking, burglary, frauds, cyberbullying or "simply" discrimination in workplace or in life in general. Sometimes users are not aware of the dangers due to the uncontrolled diffusion of their sensitive information and would probably avoid publishing it if only someone told them how harmful it could be.

In this paper, we address exactly this problem by proposing a way to measure the degree of sensitivity of any type of user-generated content. To this purpose,

© The Author(s) 2020
M. R. Berthold et al. (Eds.): IDA 2020, LNCS 12080, pp. 67–79, 2020.
https://doi.org/10.1007/978-3-030-44584-3_6

we define a new data mining task that we call *content sensitivity analysis* (CSA), inspired by sentiment analysis [13]. The goal of CSA is to assign a score to any object (text, picture, video...) according to the amount of sensitive information it potentially discloses. The problem of private content analysis has already been investigated as a way to characterize anonymous vs. non anonymous content posting in specific social media [5,15,16] or question-and-answer platforms [14]. However, the link between anonymity and sensitive contents is not that obvious: users may post anonymously because, for instance, they are referring to illegal matters (e.g., software/steaming piracy, black market and so on); conversely, fully identifiable persons may post very sensitive contents simply because they are underestimating the visibility of their action [18,19]. Although CSA has some points in common with anonymous content analysis and the well-known sentiment analysis task, we show that it has its own peculiarities and may lead to a brand new branch of research, opening many intriguing challenges in several computer science and linguistics fields.

Through some preliminary but extensive experiments on a large annotated corpus of social media posts, we show that content sensitivity analysis can not be addressed straightforwardly. In particular, we design a simplified CSA task leveraging binary classification to distinguish between sensitive and non sensitive posts by testing several bag-of-words and word embedding models. According to our experiments, the classification performances achieved by the most accurate models are far from being satisfactory. This suggests that content sensitivity analysis should consider more complex linguistic and semantic aspects, as well as more sophisticated machine learning models.

The remainder of the paper is organized as follows: we report a short analysis of the related scientific literature in Sect. 2 and Sect. 3 provides the definition of content sensitivity analysis and presents some challenging aspects of this new task together with some hints for possible solutions; the preliminary experiments are reported and discussed in Sect. 4; finally, Sect. 5 concludes by also presenting some open problems and suggestions for future research.

2   Related Work

With the success of online social networks and content sharing platforms, understanding and measuring the exposure of user privacy in the Web has become crucial [11,12]. Thus, many different metrics and methods have been proposed with the goal of assessing the risk of privacy leakage in posting activities [1,23]. Most research efforts, however, focus on measuring the overall exposure of users according to their privacy settings [8,19] or position within the network [18].

Very few research works address the problem of measuring the amount of sensitivity of user-generated content, and yet different definitions of sensitivity are adopted. In [5], for instance, the authors define sensitivity of a social media post as the extent to which users think the post should be anonymous. Then, they try to understand the nature of content posted anonymously and analyze the differences between content posted on anonymous (e.g., Whisper) and non-anonymous (e.g., Twitter) social media sites. They also find significant linguistic

differences between anonymous and non-anonymous content. A similar approach
has been applied on posts collected from a famous question-and-answer website
[14]. The authors of this work identify categories of questions for which users are
more likely to exercise anonymity and analyze different machine learning model
to predict whether a particular answer will be written anonymously. They also
show that post sensitivity should be viewed as a nuanced measure rather than as
a binary concept. In [2], the authors propose a ranking-based method for assessing the privacy risk emerging from textual contents related to sensitive topics,
such as depression. They use latent topic models to capture the background
knowledge of an hypothetical rational adversary who aims at targeting the most
exposed users. Additionally, the results are exploited to inform and alert users
about their risk of being targeting.

Similarly to sentiment analysis [13], valuable linguistic resources are needed
to identify sensitive content in texts. To the best of our knowledge, the only
works addressing this issue are [6, 22], where the authors leverage prototype theory and traditional theoretical approaches to describe and evaluate a dictionary
of privacy designed for content analysis. Dictionary categories are evaluated
according to privacy-related categories from an existing content analysis tool,
using a variety of text corpora.

The problem of sensitive content detection has been investigated as a pattern
recognition problem in images as well. In [25], the authors leverage massive
social images and their privacy settings to learn the object-privacy correlation
and identify categories of privacy-sensitive object automatically. To increase the
accuracy and speed of the classifier, they propose a deep multi-task learning
architecture that learn more representative deep convolutional neural networks
and more discriminative tree classifier. Additionally, they use the outcomes of
such model to identify the most suitable privacy settings and/or blur sensitive
objects automatically. This framework is further improved in [24], where the
authors add a clustering-based approach to also incorporate trustworthiness of
users being granted to see the images in the prediction model.

Contrary to the above-mentioned works, in this paper we formally define
the general task of *content sensitivity analysis* independently from the type of
data to be analyzed. Additionally, we provide some suggestions for improving
the accuracy of the results and show experimentally that the task is challenging,
and deserves further investigation and greater research efforts.

3   Content Sensitivity Analysis

In this section, we introduce the new data mining task that we call *content
sensitivity analysis* (CSA), aimed at determining the amount of privacy-sensitive
content expressed in any user-generated content. We first distinguish two cases,
namely *basic CSA* and *continuous CSA*, according to the outcome of the analysis
(binary or continuous). Then, we identify a set of subtasks and discuss their
theoretical and technical details. Before introducing the technical details of CSA,
we briefly provide the intuition behind CSA by describing a motivating example.

3.1 Motivating Example

To explain the main objectives of CSA and the scientific challenges associated to them, we consider the example in Fig. 1. To decide whether (and to what extent) the sentence is sensitive, an inference algorithm should be able to answer the following questions:

1. **Subjects**: whose information is going to be disclosed?
2. **Information types**: does the post refer to any potentially sensitive information type?
3. **Terms**: does the post mention any sensitive term?
4. **Topics**: does the post mention any sensitive topic?
5. **Relations**: is sensitive information referred to any of the subjects?

Fig. 1. An example of a potentially privacy-sensitive post.

By observing the post in Fig. 1, it is clear that: the post discloses information about the author and his friend Alice Green (1); the post contains spatiotemporal references ("now" and "General Hospital"), which are generally considered intrinsically sensitive; the post mentions "chemo", a potentially sensitive term (3); the sentence is related to "cancer", a potentially sensitive topic (4); the sentence structure suggests that the two subjects of disclosure have cancer and they are both about to start their first course of chemotherapy (5).

It is clear that, reducing sensitivity to anonymity, as done in previous research work [5,14], is only one side of the coin. Instead, CSA has much more in common with the famous *sentiment analysis* (SA) task, where the objective is to measure the "polarity" or "sentiment" of a given text [7,13]. However, while SA has already a well-established theory and may count on a set of easy-to-access and easy-to-use tools, CSA has never been defined before. Therefore, apart from the known open problems in SA (such as sarcasm detection), CSA involves three new scientific challenges:

1. **Definition of sensitivity**. A clear definition of sensitivity is required. Sensitivity is often defined in the legal systems, such as in the EU General Data Protection Regulation (GDPR), as a characteristic of some personal data (e.g., criminal or medical records), but a cognitive and perceptive explanation of what can be defined as "sensitive" is still missing [22].
2. **Sensitivity-annotated corpora**. Large text corpora need to be annotated according to sensitivity and at multiple levels: at the sentence level ("I got cancer" is more sensitive than "I got some nice volleyball shorts"), at the topic level ("health" is more sensitive than "sports") and at the term level ("cancer" is more sensitive than "shorts").
3. **Context-aware sensitivity**. Due to its subjectivity, a clear evaluation of the context is needed. The fact that a medical doctor talks about cancer is not sensitive per se, but if she talks about some of her patients having cancer, she could disclose very sensitive information.

In the following, we will provide the formal definitions concerning CSA and provide some preliminary ideas on how to address the problem.

3.2   Definitions

Here, we provide the details regarding the formal framework of *content sensitivity analysis*. To this purpose, we consider generic user-generated contents, without specifying their nature (whether textual, visual or audiovisual). We will propose a definition of "sensitivity" further in this section. The simplest way to define CSA is as follows:

Definition 1 (basic content sensitivity analysis). *Given a user-generated object $o_i \in \mathcal{O}$, with $\mathcal{O}$ being the domain of all user-generated contents, the basic content sensitivity analysis task consists in designing a function $f_s : \mathcal{O} \to \{sens, na, ns\}$, such that $f_s(o_i) = sens$ iff o_i is privacy-sensitive, $f_s(o_i) = ns$ iff o_i is not sensitive, otherwise $f_s(o_i) = na$.*

The *na* value is required since the assignment of a correct sensitivity value could be problematic when dealing with controversial contents or borderline topics. In some cases, assessing the sensitivity of a content object is simply impossible without some additional knowledge, i.e., the conversation a post is part of, the identity of the author of a post, and so on. In addition, sensitivity is not the same for all sensitive objects: a post dealing with health is certainly more sensitive than a post dealing with vacations, although both can be considered as sensitive. This suggests that, instead of considering sensitivity as a binary feature of a text, a more appropriate definition of CSA should take into account different degrees of sensitivity, as follows:

Definition 2 (continuous content sensitivity analysis). *Let $o_i \in \mathcal{O}$ be a user-generated object, with $\mathcal{O}$ being the domain of all user-generated contents. The continuous content sensitivity analysis task consists in designing a function $f_s : \mathcal{O} \to [-1, 1]$, such that $f_s(o_i) = 1$ iff o_i is maximally privacy-sensitive,*

$f_s(o_i) = -1$ *iff o_i is minimally privacy-sensitive,* $f_s(o_i) = 0$ *iff o_i has unknown sensitivity. The value $\sigma_i = f_s(o_i)$ is the* **sensitivity score** *of object o_i.*

According to this definition, sensitive objects have $0 < \sigma \leq 1$, while non-sensitive posts have $-1 \leq \sigma < 0$. In general, when $\sigma \approx 0$ the sensitivity of an object cannot be assessed confidently. Of course, by setting appropriate thresholds, a continuous CSA can be easily turned into a basic CSA task.

At this point, a congruent definition of "sensitivity" is required to set up the task correctly. Although different characterizations of privacy-sensitivity exist, there is no consistent and uniform theory [22]; so, in this work, we consider a more generic, flexible and application-driven definition of privacy-sensitive content.

Definition 3 (privacy-sensitive content). *A generic user-generated content object is privacy-sensitive if it makes* **the majority of users** *feel uncomfortable in writing or reading it because it may reveal some aspects of their own or others' private life to unintended people.*

Notice that "uncomfortableness" should not be guided by some moral or ethical judgement about the disclosed fact, but uniquely by its harmfulness towards privacy. Such a definition allows the adoption of the "wisdom of the crowd" principle in contexts where providing an objective definition of what is sensitive (and what is not sensitive) is particularly hard. Moreover, it has also an intuitive justification. Different social media may have different meaning of sensitivity. For instance, in a professional social networking site, revealing details about one's own job is not only tolerated, but also encouraged, while one may want to hide detailed information about her professional life in a generic photo-video sharing platform. Similarly, in a closed message board (or group), one may decide to disclose more private information than in open ones. Sensitivity towards certain topics also varies from country to country. As a consequence, function f_s can be learnt according to an annotated corpus of content objects as follows.

Definition 4 (sensitivity function learning). *Let $O = \{(o_i, \sigma_i)\}_{i=1}^{N}$ be a set of N annotated objects $o_i \in \mathcal{O}$ with the related sensitivity score $\sigma_i \in [-1, 1]$. The goal of a sensitivity function learning algorithm is to search for a function $f_s : \mathcal{O} \to [-1, 1]$, such that $\sum_{i=1}^{N} (f_s(o_i) - \sigma_i)^2$ is minimum.*

The simplest way to address this problem is by setting a regression (or classification, in the case of basic CSA) task. However, we will show in Sect. 4 that such an approach is unable to capture the actual manifold of sensitivity accurately. Hence, in the following sections, we present a fine-grained definition of CSA together with a list of open subproblems related to CSA and provide some hints on how to address them.

3.3   Fine-Grained Content Sensitivity Analysis

In the previous section, we have considered contents as monolithic objects with a sensitivity score associated to them. However, in general, any user-generated content object (text, video, picture) may contain both privacy-sensitive and privacy-unsensitive elements. For instance, a long text post (or video) may deal with some unsensitive topic but the author may insert some references to her or his private life. Similarly, a user may post a picture of her own desk deemed to be anonymous but some elements may disclose very private information (e.g., the presence of train tickets, drug paraphernalia, someone else's photo and so on). Moreover, the same object (or some of its elements) may violate the privacy of multiple subjects, including the author and other people mentioned in the corpus, in a different way. For all these reasons, here we propose a fine-grained definition of content sensitivity analysis. The definition is as follows:

Definition 5 (fine-grained content sensitivity analysis). *Let $o_i \in \mathcal{O}$ be a user-generated content object. Let $E_i = \{e_j^i\}_{j=1}^{m_i} \subset \mathcal{E}$ be a set of $m_i \geq 1$ elements (or components) that constitutes the object o_i, with $\mathcal{E}$ being the domain of all possible elements. Let $P_i = \{p_k^i\}_{j=1}^{n_i} \subset \mathcal{P}$ be the set of $n_i \geq 1$ persons (or subjects) mentioned in o_i, with $\mathcal{P}$ being the domain of all subjects. The fine-grained content sensitivity analysis task consists in designing a function $f_s : \mathcal{E} \times \mathcal{P} \to [-1, 1]$, such that $f_s(e_j^i, p_k^i) = 1$ iff e_j^i is maximally privacy-sensitive for subject p_k^i, $f_s(e_j^i, p_k^i) = -1$ iff e_i^i is minimally privacy-sensitive for subject p_k^i, $f_s(e_j^i, p_k^i) = 0$ iff e_j^i has unknown sensitivity for subject p_k^i. The value $\sigma_{jk}^i = f_s(e_j^i, p_k^i)$ is the **sensitivity score** of element e_j^i towards subject p_k^i.*

Notice that $|E_i| \geq 1$ since each object contains at least one element (when $|E_i| = 1$, the only element e_1^i corresponds the object o_i itself). Similarly $|P_i| \geq 1$ because each object has at least the author as subject. In the example reported in Fig. 1, the post contains only one element (there is only one sentence) and concerns two subjects (the author and Alice Green). According to Definition 5 (and to what we said in Sect. 3.1), the sensitivity score of the post towards both the author and Alice Green will be high.

3.4   Challenges and Possible Solutions

Fine-grained content sensitivity analysis presents many scientific and technical challenges, and may benefit of the cross-fertilization of computational linguistics, machine learning and semantic analysis. Addressing the problem of connecting sensitivity to specific subjects in texts requires the solution of many NLP tasks such as named entity recognition, relation extraction [21], and coreference resolution [4]. Additionally, concept extraction and topic modeling are important to understand whether a given text deals with sensitive content. To this purpose, privacy dictionaries [22] could provide a valid support for tagging certain topics/terms as sensitive or non-sensitive. Sentiment analysis and emotion detection could also reveal private personality traits if related to contents associated to certain topics, persons or categories of persons. Furthermore, elements in a sentence cannot be simply considered as separated entities, but the connection between different parts of a text play an important role in determining the correct fine-grained sensitivity. It is clear that such a complex problem requires the availability of massive annotated text corpora and the design of robust machine learning algorithms to cope with the sparsity of the feature space. All these considerations apply to the case of visual and audiovisual content as well, but, in addition, the intrinsic difficulty of handling multimedia data makes the above mentioned challenge even harder and more computationally expensive.

In the next section, we will show how the basic content sensitivity analysis settings can be modeled as a binary classification problem on text data using different approaches with scarce or moderate success, thus showing the necessity of a more systematic and in-depth investigation of the problem.

4   Preliminary Experiments

In this section, we report the results of some preliminary experiments aimed at showing the feasibility of content sensitivity analysis together with its difficulties. The experiments are conducted under the basic CSA framework (see Definition 1 in Sect. 3) with the only difference that we do not consider the "na" class. We set up a binary classification task to distinguish whether a given input text is privacy-sensitive or not. Before presenting the results, in the following, we first introduce the data, then we provide the details of our experimental protocol.

4.1   Annotated Corpus

Since all previous attempts of identifying sensitive text have leveraged user anonymity as a discriminant for sensitive content [5,14], there is no reliable annotated corpus that we can use as benchmark. Hence, we construct our own dataset by leveraging a crowdsourcing experiment. We use one of the datasets described in [3], consisting of 9917 anonymized social media posts, mostly written in English, with a minimum length of 2 characters and a maximum length of 435 (the average length is 80). Thus, they well represent typical social media short posts. On the other hand, they are not annotated for the specific purpose of our experiment and, because of their shortness, they are also very difficult to analyze. Consequently, after discarding all useless posts (mostly uncomprehensible ones) we have set up a crowdsourcing experiment by using a Telegram bot that, for each post, asks whether it is sensitive or not. As third option, it was also possible to select "unable to decide". We collected the annotations of 829 posts from 14 distinct annotators. For each annotated post, we retain the most frequently chosen annotation. Overall, 449 posts where tagged as non sensitive, 230 as sensitive, 150 as undecidable. Thus, the final dataset consists of 679 posts of the first two categories (we discarded all 150 undecidable posts).

4.2   Datasets

We consider two distinct document representations for the dataset, a bag-of-words and four word vector models. To obtain the bag-of-word representation we perform the following steps. First, we remove all punctuation characters of terms contained in the input posts as well as short terms (less than two characters) and terms containing digits. Then, we build the bag-of-words model with all remaining 2584 terms weighted by their *tfidf* score. Differently from classic text mining approaches, we deliberately exclude lemmatization, stemming and stop word removal from text preprocessing, since those common steps would affect content sensitivity analysis negatively. Indeed, inflections (removed by lemmatization and stemming) and stop words (like "me", "myself") are important to decide whether a sentence reproduces some personal thoughts or private action/status. Hereinafter, the bag-of-words representation is referred to as *BW2584*.

The word vector representation, instead, is built using word vectors pre-trained with two billion tweets (corresponding to 42 billion tokens) using the *GloVe* (Global Vector) model [17]. We use this word embedding method as it consistently outperforms both continuous bag-of-words and skip-gram model architectures of word2vec [10]. In detail, we use three representation, here called *WV25*, *WV50* and *WV100* with, respectively, 25, 50 and 100 dimensions[1]. Additionally, we build an ensemble by considering the concatenation of the three vector spaces. The latter representation is named *WVEns*.

Finally, from all five datasets we removed all posts having an empty bag-of-words or word vector representation. Such preprocessing step further reduces the size of the dataset down to 611 posts (221 sensitive and 390 non sensitive), but allows for a fair performance comparison.

4.3 Experimental Settings

Each dataset obtained as described beforehand is given in input to a set of six classifiers. In details, we use *k*-NN, decision tree (DT), Multi-layer Perceptron (MLP), SVM, Random Forest (RF), and Gradient Boosted trees (GBT). We do not execute any systematic parameter selection procedure since our main goal is not to compare the performances of classifiers, but, rather, to show the overall level of accuracy that can be achieved in a basic content sensitivity analysis task. Hence, we use the following default parameter for each classifier:

- **kNN**: we set $k = 3$ in all experiments;
- **DT**: for all datasets, we use C4.5 with Gini Index as split criterion, allowing a minimum of two records per node and minimum description length as pruning strategy;
- **MLP**: we train a shallow neural network with one hidden layer; the number of neurons of the hidden layer is 30 for the bag-of-words representation and 20 for all word vector representations;
- **SVM**: for all datasets, we use the polynomial kernel with default parameters;
- **RF**: we train 100 models with Gini index as splitting criterion in all experiments;
- **GBT**: for all datasets, we use 100 models with 0.1 as learning rate and 4 as maximum tree depth.

---

[1] Pre-trained vectors are available at https://nlp.stanford.edu/projects/glove/.

All experiments are conducted by performing ten-fold cross-validation, using, for each iteration, nine folds as training set and the remaining fold as test set.

4.4   Results and Discussion

The summary of the results, in terms of average F1-score, are reported in Table 1. It is worth noting that the scores are, in general, very low (between 0.5826, obtained by the neural network on the bag-of-words model, and 0.6858, obtained by Random Forest on the word vector representation with 50 dimensions). Of course, these results are biased by the fact that data are moderately unbalanced (64% of posts fall in the non-sensible class). However they are not completely negative, meaning that there is space for improvement. We observe that the winning model-classifier pair (50-dimensional word vector processed with Random Forest) exhibits high recall on the non-sensitive class (0.928) and rather similar results in terms of precision for the two classes (0.671 and 0.688 for the sensitive and non-sensitive classes respectively). The real negative result is the low recall on the sensitive class (only 0.258), due to the high number of false negatives[2]. We recall that the number of annotated sensitive posts is only 221, i.e., the number of examples is not sufficiently large for training a prediction model accurately.

Table 1. Classification in terms of average F1-score for different post representations.

Dataset	Type	kNN	DT	MLP	SVM	RF	GBT
BW2584	bag-of-words	0.6579	0.6743	0.5826	0.6481	0.6776	0.6678
WV25	word vector	0.6203	0.6317	0.6497	0.6383	0.6628	0.6268
WV50	word vector	0.6121	0.6105	0.6530	0.6448	**0.6858**	0.6399
WV100	word vector	0.6367	0.6088	0.6497	0.6563	0.6694	0.6497
WVEns	word vector	0.6432	0.5859	0.6481	0.6547	0.6628	0.6416

These results highlight the following issues and perspectives. First, negative (or not-so-positive) results are certainly due to the lack of annotated data (especially for the sensitive class). Sparsity is certainly a problem in our settings. Hence, a larger annotated corpus is needed, although this objective is not trivial. In fact, private posts are often difficult to obtain, because social media platforms (luckily, somehow) do not allow users to get them using their API. As a consequence, all previous attempts to guess the sensitivity of text or construct privacy dictionaries strongly leverage user anonymity in public post sharing activities [5,14], or rely on focus groups and surveys [22]. Moreover, without a sufficiently large corpus, not even the application of otherwise successful deep learning techniques (e.g., RNNs for sentiment analysis [9]) would produce valid results. Second, simple classifiers, even when applied to rather complex and rich representations, can not capture the manifold of privacy sensitivity accurately.

---

[2] Due to space limitations, we do not report detailed precision/recall results.

So, more complex and heterogenous models should be considered. Probably, an accurate sensitivity content analysis tool should consider lexical, semantic as well as grammatical features. Topics are certainly important, but sentence construction and lexical choices are also fundamental. Therefore, reliable solutions would consist of a combination of computational linguistic techniques, machine learning algorithms and semantic analysis. Third, the success of picture and video sharing platforms (such as Instagram and TikTok), implies that any successful sensitivity content analysis tool should be able to cope with audiovisual contents and, in general, with multimodal/multimedia objects (an open problem in sentiment analysis as well [20]). Finally, provided that a taxonomy of privacy categories in everyday life exists (e.g., health, location, politics, religious belief, family, relationships, and so on) a more complex CSA setting might consider, for a given content object, the privacy sensitivity degree in each category.

5   Conclusions

In this paper, we have addressed the problem of determining whether a given content object is privacy-sensitive or not by defining the generic task of content sensitivity analysis (CSA). Then, we have declined it according to increasing complexity of the problem settings. Although the task promises to be challenging, we have shown that it is not unfeasible by presenting a simplified formulation of CSA based on text categorization. With some preliminary but extensive experiments, we have showed that, no matter the data representation, the accuracy of such classifiers can not be considered satisfactory. Thus, it is worth investigating more complex techniques borrowed from machine learning, computational linguistics and semantic analysis. Moreover, without a strong effort in building massive and reliable annotated corpora, the performances of any CSA tool would be barely sufficient, no matter the complexity of the learning model.

Acknowledgments. The authors would like to thank Daniele Scanu for implementing the Telegram bot used by the annotators. This work is supported by Fondazione CRT (grant number 2019-0450).

References

1. Alemany, J., del Val Noguera, E., Alberola, J.M., García-Fornes, A.: Metrics for privacy assessment when sharing information in online social networks. IEEE Access **7**, 143631–143645 (2019)
2. Biega, J.A., Gummadi, K.P., Mele, I., Milchevski, D., Tryfonopoulos, C., Weikum, G.: R-Susceptibility: an IR-centric approach to assessing privacy risks for users in online communities. In: Proceedings of ACM SIGIR 2016, pp. 365–374 (2016)
3. Celli, F., Pianesi, F., Stillwell, D., Kosinski, M.: Workshop on computational personality recognition: shared task. In: Proceedings of ICWSM 2013 (2013)
4. Clark, K., Manning, C.D.: Improving coreference resolution by learning entity-level distributed representations. In: Proceedings of ACL 2016 (2016)

5. Correa, D., Silva, L.A., Mondal, M., Benevenuto, F., Gummadi, K.P.: The many shades of anonymity: characterizing anonymous social media content. In: Proceedings of ICWSM **2015**, pp. 71–80 (2015)
6. Gill, A.J., Vasalou, A., Papoutsi, C., Joinson, A.N.: Privacy dictionary: a linguistic taxonomy of privacy for content analysis. In: Proceedings of ACM CHI 2011, pp. 3227–3236 (2011)
7. Liu, B., Zhang, L.: A survey of opinion mining and sentiment analysis. In: Aggarwal, C.C., Zhai, C. (eds.) Mining Text Data, pp. 415–463. Springer, Heidelberg (2012). https://doi.org/10.1007/978-1-4614-3223-4_13
8. Liu, K., Terzi, E.: A framework for computing the privacy scores of users in online social networks. TKDD **5**(1), 6:1–6:30 (2010)
9. Ma, Y., Peng, H., Khan, T., Cambria, E., Hussain, A.: Sentic LSTM: a hybrid network for targeted aspect-based sentiment analysis. Cogn. Comput. **10**(4), 639–650 (2018). https://doi.org/10.1007/s12559-018-9549-x
10. Mikolov, T., Sutskever, I., Chen, K., Corrado, G.S., Dean, J.: Distributed representations of words and phrases and their compositionality. In: Proceedings of NIPS 2013, pp. 3111–3119 (2013)
11. Oukemeni, S., Rifà-Pous, H., i Puig, J.M.M.: IPAM: information privacy assessment metric in microblogging online social networks. IEEE Access **7**, 114817–114836 (2019)
12. Oukemeni, S., Rifà-Pous, H., i Puig, J.M.M.: Privacy analysis on microblogging online social networks: a survey. ACM Comput. Surv. **52**(3), 60:1–60:36 (2019)
13. Pang, B., Lee, L.: Opinion mining and sentiment analysis. Found. Trends Inf. Retrieval **2**(1–2), 1–135 (2007)
14. Peddinti, S.T., Korolova, A., Bursztein, E., Sampemane, G.: Cloak and swagger: understanding data sensitivity through the lens of user anonymity. In: Proceedings of IEEE SP 2014, pp. 493–508 (2014)
15. Peddinti, S.T., Ross, K.W., Cappos, J.: Finding sensitive accounts on Twitter: an automated approach based on follower anonymity. In: Proceedings of ICWSM 2016, pp. 655–658 (2016)
16. Peddinti, S.T., Ross, K.W., Cappos, J.: User anonymity on Twitter. IEEE Secur. Privacy **15**(3), 84–87 (2017)
17. Pennington, J., Socher, R., Manning, C.D.: Glove: global vectors for word representation. In: Proceedings of EMNLP 2014, pp. 1532–1543 (2014)
18. Pensa, R.G., di Blasi, G., Bioglio, L.: Network-aware privacy risk estimation in online social networks. Soc. Netw. Analys. Mining **9**(1), 15:1–15:15 (2019)
19. Pensa, R.G., Blasi, G.D.: A privacy self-assessment framework for online social networks. Expert Syst. Appl. **86**, 18–31 (2017)
20. Poria, S., Majumder, N., Hazarika, D., Cambria, E., Gelbukh, A.F., Hussain, A.: Multimodal sentiment analysis: addressing key issues and setting up the baselines. IEEE Intell. Syst. **33**(6), 17–25 (2018)
21. Surdeanu, M., McClosky, D., Smith, M., Gusev, A., Manning, C.D.: Customizing an information extraction system to a new domain. In: Proceedings of RELMS@ACL 2011, pp. 2–10 (2011)
22. Vasalou, A., Gill, A.J., Mazanderani, F., Papoutsi, C., Joinson, A.N.: Privacy dictionary: a new resource for the automated content analysis of privacy. JASIST **62**(11), 2095–2105 (2011)
23. Wagner, I., Eckhoff, D.: Technical privacy metrics: a systematic survey. ACM Comput. Surv. **51**(3), 57:1–57:38 (2018)

24. Yu, J., Kuang, Z., Zhang, B., Zhang, W., Lin, D., Fan, J.: Leveraging content sensitiveness and user trustworthiness to recommend fine-grained privacy settings for social image sharing. IEEE Trans. Inf. Forensics Secur. **13**(5), 1317–1332 (2018)
25. Yu, J., Zhang, B., Kuang, Z., Lin, D., Fan, J.: iPrivacy: image privacy protection by identifying sensitive objects via deep multi-task learning. IEEE Trans. Inf. Forensics Secur. **12**(5), 1005–1016 (2017)

Gibbs Sampling Subjectively Interesting Tiles

Anes Bendimerad[1]([✉]), Jefrey Lijffijt[2], Marc Plantevit[3], Céline Robardet[1], and Tijl De Bie[2]

[1] Univ Lyon, INSA, CNRS UMR 5205, 69621 Villeurbanne, France
ahmed-anes.bendimerad@insa-lyon.fr
[2] IDLab, ELIS Department, Ghent University, Ghent, Belgium
[3] Univ Lyon, UCBL, CNRS UMR 5205, 69621 Lyon, France

Abstract. The local pattern mining literature has long struggled with the so-called pattern explosion problem: the size of the set of patterns found exceeds the size of the original data. This causes computational problems (enumerating a large set of patterns will inevitably take a substantial amount of time) as well as problems for interpretation and usability (trawling through a large set of patterns is often impractical).

Two complementary research lines aim to address this problem. The first aims to develop better measures of interestingness, in order to reduce the number of uninteresting patterns that are returned [6,10]. The second aims to avoid an exhaustive enumeration of all 'interesting' patterns (where interestingness is quantified in a more traditional way, e.g. frequency), by directly sampling from this set in a way that more 'interesting' patterns are sampled with higher probability [2].

Unfortunately, the first research line does not reduce computational cost, while the second may miss out on the most interesting patterns. In this paper, we combine the best of both worlds for mining interesting tiles [8] from binary databases. Specifically, we propose a new pattern sampling approach based on Gibbs sampling, where the probability of sampling a pattern is proportional to their subjective interestingness [6]—an interestingness measure reported to better represent true interestingness.

The experimental evaluation confirms the theory, but also reveals an important weakness of the proposed approach which we speculate is shared with any other pattern sampling approach. We thus conclude with a broader discussion of this issue, and a forward look.

Keywords: Pattern mining · Subjective interestingness · Pattern sampling · Gibbs sampling

1 Introduction

Pattern mining methods aim to select elements from a given language that bring to the user "implicit, previously unknown, and potentially useful information

M. R. Berthold et al. (Eds.): IDA 2020, LNCS 12080, pp. 80–92, 2020.
https://doi.org/10.1007/978-3-030-44584-3_7

from data" [7]. To meet the challenge of selecting the appropriate patterns for a user, several lines of work have been explored: (1) Many constraints on some measures that assess the quality of a pattern using exclusively the data have been designed [4,12,13]; (2) Preference measures have been considered to only retrieve patterns that are non dominated in the dataset; (3) Active learning systems have been proposed that interact with the user to explicit her interest on the patterns and guide the exploration toward those she is interested in; (4) Subjective interestingness measures [6,10] have been introduced that aim to take into account the implicit knowledge of a user by modeling her prior knowledge and retrieving the patterns that are unlikely according to the background model.

The shift from threshold-constraints on objective measures toward the use of subjective measures provides an elegant solution to the so-called pattern explosion problem by considerably reducing the output to only truly interesting patterns. Unfortunately, the discovery of subjectively interesting patterns with exact algorithms remains computationally challenging.

In this paper we explore another strategy that is pattern sampling. The aim is to reduce the computational cost while identifying the most important patterns, and allowing for distributed computations. There are two families of local pattern sampling techniques.

The first family uses Metropolis Hastings [9], a Markov Chain Monte Carlo (MCMC) method. It performs a random walk over a transition graph representing the probability of reaching a pattern given the current one. This can be done with the guarantee that the distribution of the considered quality measure is proportional on the sample set to the one of the whole pattern set [1]. However, each iteration of the random walk is accepted only with a probability equal to the acceptance rate α. This can be very small, which may result in a prohibitively slow convergence rate. Moreover, in each iteration the part of the transition graph representing the probability of reaching patterns given the current one, has to be materialized in both directions, further raising the computational cost. Other approaches [5,11] relax this constraint but lose the guarantee.

Methods in the second family are referred to as direct pattern sampling approaches [2,3]. A notable example is [2], where a two-step procedure is proposed that samples frequent itemsets without simulating stochastic processes. In a first step, it randomly selects a row according to a first distribution, and from this row, draws a subset of items according to another distribution. The combination of both steps follows the desired distribution. Generalizing this approach to other pattern domains and quality measures appeared to be difficult.

In this paper, we propose a new pattern sampling approach based on Gibbs sampling, where the probability of sampling a pattern is proportional to their Subjective Interestingness (SI) [6]. Gibbs sampling – described in Sect. 3 – is a special case of Metropolis Hastings where the acceptance rate α is always equal to 1. In Sect. 4, we show how the random walk can be simulated without materializing any part of the transition graph, except the currently sampled pattern. While we present this approach particularly for mining tiles in rectangular databases, applying it for other pattern languages can be relatively easily achieved. The experimental evaluation (Sect. 5) confirms the theory, but also reveals a weakness of the proposed approach which we speculate is shared by other direct pattern sampling approaches. We thus conclude with a broader discussion of this issue (Sect. 6), and a forward look (Sect. 7).

2 Problem Formulation

2.1 Notation

Input Dataset. A dataset $\mathbf{D}$ is a Boolean matrix with m rows and n columns. For $i \in [\![1, m]\!]$ and $j \in [\![1, n]\!]$, $\mathbf{D}(i, j) \in \{0, 1\}$ denotes the value of the cell corresponding to the i-th row and the j-th column. For a given set of rows $I \subseteq [\![1, m]\!]$, we define the support function $supp_C(I)$ that gives all the columns having a value of 1 in all the rows of I, i.e., $supp_C(I) = \{j \in [\![1, n]\!] \mid \forall i \in I : \mathbf{D}(i, j) = 1\}$. Similarly, for a set of columns $J \subseteq [\![1, n]\!]$, we define the function $supp_R(J) = \{i \in [\![1, m]\!] \mid \forall j \in J : \mathbf{D}(i, j) = 1\}$. Table 1 shows a toy example of a Boolean matrix, where for $I = \{4, 5, 6\}$ we have that $supp_C(I) = \{2, 3, 4\}$.

Table 1. Example of a binary dataset $\mathbf{D}$.

#	1	2	3	4	5
1	0	1	0	1	0
2	0	1	1	0	0
3	1	0	1	0	1
4	0	1	1	1	0
5	1	1	1	1	1
6	0	1	1	1	0
7	0	1	1	1	1

Pattern Language. This paper is concerned with a particular kind of pattern known as a tile [8], denoted $\tau = (I, J)$ and defined as an ordered pair of a set of rows $I \subseteq \{1, ..., m\}$ and a set of columns $J \subseteq \{1, ...n\}$. A tile τ is said to be contained (or present) in $\mathbf{D}$, denoted as $\tau \in \mathbf{D}$, iff $\mathbf{D}(i, j) = 1$ for all $i \in I$ and $j \in J$. The set of all tiles present in the dataset is denoted as T and is defined as: $T = \{(I, J) \mid I \subseteq \{1, ..., m\} \wedge J \subseteq \{1, ...n\} \wedge (I, J) \in \mathbf{D}\}$. In Table 1, the tile $\tau_1 = (\{4, 5, 6, 7\}, \{2, 3, 4\})$ is present in $\mathbf{D}$ ($\tau_1 \in T$), because each of its cells has a value of 1, but $\tau_2 = (\{1, 2\}, \{2, 3\})$ is not present ($\tau_2 \notin T$) since $\mathbf{D}(1, 3) = 0$.

2.2 The Interestingness of a Tile

In order to assess the quality of a tile τ, we use the framework of subjective interestingness SI proposed in [6]. We briefly recapitulate the definition of this measure for tiles, denoted $SI(\tau)$ for a tile τ, and refer the reader to [6] for more details. $SI(\tau)$ measures the quality of a tile τ as the ratio of its subjective information content $IC(\tau)$ and its description length $DL(\tau)$:

$$SI(\tau) = \frac{IC(\tau)}{DL(\tau)}.$$

Tiles with large $SI(\tau)$ thus compress subjective information in a short description. Before introducing IC and DL, we first describe the background model—an important component required to define the subjective information content IC.

Background Model. The SI is subjective in a sense that it accounts for prior knowledge of the current data miner. A tile τ is informative for a particular user if this tile is somehow surprising for her, otherwise, it does not bring new information. The most natural way for formalizing this is to use a background distribution representing the data miner's prior expectations, and to compute the probability $\Pr(\tau \in \mathbf{D})$ of this tile under this distribution. The smaller $\Pr(\tau \in \mathbf{D})$, the more information this pattern contains. Concretely, the background model consists of a value $\Pr(\mathbf{D}(i,j) = 1)$ associated to each cell $\mathbf{D}(i,j)$ of the dataset, and denoted p_{ij}. More precisely, p_{ij} is the probability that $\mathbf{D}(i,j) = 1$ under user prior beliefs. In [6], it is shown how to compute the background model and derive all the values p_{ij} corresponding to a given set of considered user priors. Based on this model, the probability of having a tile $\tau = (I, J)$ in $\mathbf{D}$ is:

$$\Pr(\tau \in \mathbf{D}) = \Pr\left(\bigwedge_{i \in I, j \in J} \mathbf{D}(i,j) = 1 \right) = \prod_{i \in I, j \in J} p_{ij}.$$

Information Content IC. This measure aims to quantify the amount of information conveyed to a data miner when she is told about the presence of a tile in the dataset. It is defined for a tile $\tau = (I, J)$ as follows:

$$\mathrm{IC}(\tau) = -\log(\Pr(\tau \in \mathbf{D})) = \sum_{i \in I, j \in J} -\log(p_{ij}).$$

Thus, the smaller $\Pr(\tau \in \mathbf{D})$, the higher $\mathrm{IC}(\tau)$, and the more informative τ. Note that for $\tau_1, \tau_2 \in \mathbf{D} : \mathrm{IC}(\tau_1 \cup \tau_2) = \mathrm{IC}(\tau_1) + \mathrm{IC}(\tau_2) - \mathrm{IC}(\tau_1 \cap \tau_2)$.

Description Length DL. This function should quantify how difficult it is for a user to assimilate the pattern. The description length of a tile $\tau = (I, J)$ should thus depend on how many rows and columns it refers to: the larger are $|I|$ and $|J|$, the larger is the description length. Thus, $\mathrm{DL}(\tau)$ can be defined as:

$$\mathrm{DL}(\tau) = a + b \cdot (|I| + |J|),$$

where a and b are two constants that can be handled to give more or less importance to the contributions of $|I|$ and $|J|$ in the description length.

2.3   Problem Statement

Given a Boolean dataset $\mathbf{D}$, the goal is to sample a tile τ from the set of all the tiles T present in $\mathbf{D}$, with a probability of sampling P_S proportional to $\mathrm{SI}(\tau)$, that is: $P_S(\tau) = \dfrac{\mathrm{SI}(\tau)}{\sum_{\tau' \in T} \mathrm{SI}(\tau')}$.

A naïve approach to sample a tile pattern according to this distribution is to generate the list $\{\tau_1, ..., \tau_N\}$ of all the tiles present in $\mathbf{D}$, sample $x \in [0, 1]$ uniformly at random, and return the tile τ_k with $\dfrac{\sum_{i=1}^{k-1} \mathrm{SI}(\tau_i)}{\sum_i \mathrm{SI}(\tau_i)} \leq x < \dfrac{\sum_{i=1}^{k} \mathrm{SI}(\tau_i)}{\sum_i \mathrm{SI}(\tau_i)}$.

However, the goal behind using sampling approaches is to avoid materializing the pattern space which is generally huge. We want to sample without exhaustively enumerating the set of tiles. In [2], an efficient procedure is proposed to directly sample patterns according to some measures such as the frequency and the area. However, this procedure is limited to only some specific measures. Furthermore, it is proposed for pattern languages defined on only the column dimension, for example, itemset patterns. In such language, the rows related to an itemset pattern $F \subseteq \{1, ..., n\}$ are uniquely identified and they correspond to all the rows containing the itemset, that are $supp_R(F)$. In our work, we are interested in tiles which are defined by both columns and rows indices. In this case, it is not clear how the direct procedure proposed in [2] can be applied.

For more complex pattern languages, a generic procedure based on Metropolis Hasting algorithm has been proposed in [9], and illustrated for subgraph patterns with some quality measures. While this approach is generic and can be extended relatively easily to different mining tasks, a major drawback of using Metropolis Hasting algorithm is that the random walk procedure contains the acceptance test that needs to be processed in each iteration, and the acceptance rate α can be very small, which makes the convergence rate practically extremely slow. Furthermore, Metropolis Hasting can be computationally expensive, as the part of the transition graph representing the probability of reaching patterns given the current one, has to be materialized.

Interestingly, a very useful MCMC technique is Gibbs sampling, which is a special case of Metropolis-Hasting algorithm. A significant benefit of this approach is that the acceptante rate α is always equal to 1, i.e., the proposal of each sampling iteration is always accepted. In this work, we use Gibbs sampling to draw patterns with a probability distribution that converges to P_S. In what follows, we will first generically present the Gibbs sampling approach, and then we show how we efficiently exploit it for our problem. Unlike Metropolis Hasting, the proposed procedure performs a random walk by materializing in each iteration only the currently sampled pattern.

3   Gibbs Sampling

Suppose we have a random variable $X = (X_1, X_2, ..., X_l)$ taking values in some domain Dom. We want to sample a value $x \in Dom$ following the joint distribution $P(X = x)$. Gibbs sampling is suitable when it is hard to sample directly from P but known how to sample just one dimension x_k ($k \in [\![1, l]\!]$) from the conditional probability $P(X_k = x_k \mid X_1 = x_1, ..., X_{k-1} = x_{k-1}, X_{k+1} = x_{k+1}, ..., X_l = x_l)$. The idea of Gibbs sampling is to generate samples by sweeping through each variable (or block of variables) to sample from its conditional distribution with the remaining variables fixed to their current values. Algorithm 1 depicts a generic Gibbs Sampler. At the beginning, x is set to its initial values (often values sampled from a prior distribution q). Then, the algorithm performs a random walk of p iterations. In each iteration, we sample $x_1 \sim P(X_1 = x_1^{(i_1)} \mid X_2 = x_2^{(i_1)}, ..., X_l = x_l^{(i_1)})$ (while fixing the other dimensions), then we follow the same procedure to sample x_2, ..., until x_l.

Algorithm 1: Gibbs sampler

---

1 Initialize $x^{(0)} \sim q(x)$
2 **for** $k \in [\![1, p]\!]$ **do**
3 $\quad$ draw $x_1^{(k)} \sim P\left(X_1 = x_1 \mid X_2 = x_2^{(k-1)}, X_3 = x_3^{(k-1)}, ..., X_l = x_l^{(k-1)}\right)$
4 $\quad$ draw $x_2^{(k)} \sim P\left(X_2 = x_2 \mid X_1 = x_1^{(k)}, X_3 = x_3^{(k-1)}, ..., X_l = x_l^{(k-1)}\right)$
5 $\quad$ ...
6 $\quad$ draw $x_l^{(k)} \sim P\left(X_l = x_l \mid X_1 = x_1^{(k)}, X_2 = x_2^{(k)}, ..., X_{l-1} = x_{l-1}^{(k)}\right)$
7 **return** $x^{(p)}$

The random walk needs to satisfy some constraints to guarantee that the Gibbs sampling procedure converges to the stationary distribution P. In the case of a finite number of states (a finite space *Dom* in which X takes values), sufficient conditions for the convergence are irreducibility and aperiodicity:

Irreducibility. A random walk is irreducible if, for any two states $x, y \in Dom$ s.t. $P(x) > 0$ and $P(y) > 0$, we can get from x to y with a probability > 0 in a finite number of steps. I.e. the entire state space is reachable.

Aperiodicity. A random walk is aperiodic if we can return to any state $x \in Dom$ at any time. I.e. revisiting x is not conditioned to some periodicity constraint.

One can also use blocked Gibbs sampling. This consists in growing many variables together and sample from their joint distribution conditioned to the remaining variables, rather than sampling each variable x_i individually. Blocked Gibbs sampling can reduce the problem of slow mixing that can be due to the high number of dimensions used to sample from.

4 Gibbs Sampling of Tiles with Respect to SI

In order to sample a tile $\tau = (I, J)$ with a probability proportional to $SI(\tau)$, we propose to use Gibbs sampling. The simplest solution is to consider a tile τ as $m + n$ binary random variables $(x_1, ..., x_m, ..., x_{m+n})$, each of them corresponds to a row or a column, and then apply the procedure described in Algorithm 1. In this case, an iteration of Gibbs sampling requires to sample from each column and row separately while fixing all the remaining rows and columns. The drawback of this approach is the high number of variables $(m + n)$ which may lead to a slow mixing time. In order to reduce the number of variables, we propose to split $\tau = (I, J)$ into only two separated blocks of random variables I and J, we then directly sample from each block while fixing the value of the other block. This means that an iteration of the random walk contains only two sampling operations instead of $m+n$ ones. We will explain in more details how this Blocked Gibbs sampling approach can be applied, and how to compute the distributions used to directly sample a block of rows or columns.

Algorithm 2: Gibbs-SI

---
1 Initialize $(I, J)^{(0)} \sim q(x)$
2 **for** $k \in [\![1, p]\!]$ **do**
3   draw $I^{(k)} \sim P\left(\mathbf{I} = I \mid \mathbf{J} = J^{(k-1)}\right)$, draw $J^{(k)} \sim P\left(\mathbf{J} = J \mid \mathbf{I} = I^{(k)}\right)$
4 **return** $(I, J)^{(p)}$

Algorithm 2 depicts the main steps of Blocked Gibbs sampling for tiles. We start by initializing $(I, J)^{(0)}$ with a distribution q proportional to the area ($|I| \times |J|$) following the approach proposed in [2]. This choice is mainly motivated by its linear time complexity of sampling. Then, we need to efficiently sample from $P(\mathbf{I} = I \mid \mathbf{J} = J)$ and $P(\mathbf{J} = J \mid \mathbf{I} = I)$. In the following, we will explain how to sample I with $P(\mathbf{I} = I | \mathbf{J} = J)$, and since the SI is symmetric w.r.t. rows and columns, the same strategy can be used symmetrically to sample a set of columns with $P(\mathbf{J} = J \mid \mathbf{I} = I)$.

Sampling a Set of Rows I Conditioned to Columns J. For a specific $J \subseteq \{1, ..., n\}$, the number of tiles (I, J) present in the dataset can be huge, and can go up to 2^m. This means that naïvely generating all these candidate tiles and then sampling from them is not a solution. Thus, to sample a set of rows I conditioned to a fixed set of columns J, we propose an iterative algorithm that builds the sampled I by drawing each $i \in I$ separately, while ensuring that the joint distribution of all the drawings is equal to $P(\mathbf{I} = I | \mathbf{J} = J)$. I is built using two variables: $R_1 \subseteq \{1, ..., m\}$ made of rows that belong to I, and $R_2 \subseteq \{1, ..., m\} \setminus R_1$ that contains candidate rows that can possibly be sampled and added to R_1. Initially, we have $R_1 = \emptyset$ and $R_2 = supp_R(J)$. At each step, we take $i \in R_2$, do a random draw to determine whether i is added to R_1 or not, and remove it from R_2. When $R_2 = \emptyset$, the sampled set of rows I is set equal to R_1. To apply this strategy, all we need is to compute $P\left(i \in \mathbf{I} \mid R_1 \subseteq \mathbf{I} \subseteq R_1 \cup R_2 \wedge \mathbf{J} = J\right)$, the probability of sampling i considering the current sets R_1, R_2 and J:

$$P\left(i \in \mathbf{I} \mid R_1 \subseteq \mathbf{I} \subseteq R_1 \cup R_2 \wedge \mathbf{J} = J\right) = \frac{P\left(R_1 \cup \{i\} \subseteq \mathbf{I} \subseteq R_1 \cup R_2 \wedge \mathbf{J} = J\right)}{P\left(R_1 \subseteq \mathbf{I} \subseteq R_1 \cup R_2 \wedge \mathbf{J} = J\right)}$$

$$= \frac{\sum_{F \subseteq R_2 \setminus \{i\}} SI(R_1 \cup \{i\} \cup F, J)}{\sum_{F \subseteq R_2} SI(R_1 \cup F, J)} = \frac{\sum_{F \subseteq R_2 \setminus \{i\}} \frac{IC_{(R_1 \cup \{i\}, J)} + IC_{(F, J)}}{a + b \cdot (|R_1| + |F| + 1 + |J|)}}{\sum_{F \subseteq R_2} \frac{IC_{(R_1, D_i)} + IC_{(F, D_i)}}{a + b \cdot (|R_1| + |F| + |J|)}}$$

$$= \frac{\sum_{k=0}^{|R_2|-1} \frac{1}{a + b \cdot (|R_1| + k + 1 + |J|)} \sum_{\substack{F \subseteq R_2 \setminus \{i\} \\ |F| = k}} (IC(R_1 \cup \{i\}, J) + IC(F, J))}{\sum_{k=0}^{|R_2|} \frac{1}{a + b \cdot (|R_1| + k + |J|)} \sum_{\substack{F \subseteq R_2 \\ |F| = k}} (IC(R_1, J) + IC(F, J))}$$

$$= \frac{\sum_{k=0}^{|R_2|-1} \frac{1}{a + b \cdot (|R_1| + k + 1 + |J|)} \left(\binom{|R_2|-1}{k} \cdot IC(R_1 \cup \{i\}, J) + \binom{|R_2|-2}{k-1} \cdot IC(R_2 \setminus \{i\}, J)\right)}{\sum_{k=0}^{|R_2|} \frac{1}{a + b \cdot (|R_1| + k + |J|)} \left(\binom{|R_2|}{k} \cdot IC(R_1, J) + \binom{|R_2|-1}{k-1} \cdot IC(R_2, J)\right)}$$

$$= \frac{IC(R_1 \cup \{i\}, J) \cdot f(|R_2| - 1, |R_1| + 1) + IC(R_2 \setminus \{i\}, J) \cdot f(|R_2| - 2, |R_1| + 1)}{IC(R_1, J) \cdot f(|R_2|, |R_1|) + IC(R_2, J) \cdot f(|R_2| - 1, |R_1|)},$$

with $f(x, y) = \sum_{k=0}^{x} \frac{\binom{x}{k}}{a + b \cdot (y + k + |J|)}$.

Complexity. Let's compute the complexity of sampling I with a probability $P(\mathbf{I} = I | \mathbf{J} = J)$. Before starting the sampling of rows from R_2, we first compute the value of $IC(\{i\}, J)$ for each $i \in R_2$ (in $\mathcal{O}(n \cdot m)$). This will allow to compute in $\mathcal{O}(1)$ the values of IC that appear in $P(i \in \mathbf{I} | R_1 \subseteq \mathbf{I} \subseteq R_1 \cup R_2 \wedge \mathbf{J} = J)$, based on the relation $IC(I_1 \cup I_2, J) = IC(I_1, J) + IC(I_2, J)$ for $I_1, I_2 \subseteq [\![1, m]\!]$. In addition to that, sampling each element $i \in R_2$ requires to compute the corresponding values of $f(x, y)$. These values are computed once for the first sampled row $i \in R_2$ with a cost of $\mathcal{O}(m)$, and then they can be updated directly when sampling the next rows, using the following relation:

$$f(x - 1, y) = f(x, y) - \frac{1}{a + b \cdot (x + y + |J|)} \cdot f(x - 1, y + 1).$$

This means that the overall cost of sampling the whole set of rows I with a probability $P(\mathbf{I} = I | \mathbf{J} = J)$ is $\mathcal{O}(n \cdot m)$. Following the same approach, sampling J conditionned to I is done in $\mathcal{O}(n \cdot m)$. As we have p sampling iterations, the worst case complexity of the whole Gibbs sampling procedure of a tile τ is $\mathcal{O}(p \cdot n \cdot m)$.

Convergence Guarantee. In order to guarantee the convergence to the stationary distribution proportional to the SI measure, the Gibbs sampling procedure needs to satisfy some constraints. In our case, the sampling space is finite, as the number of tiles is limited to at most 2^{m+n}. Then, the sampling procedure converges if it satisfies the aperiodicity and the irreducibility constraints. The Gibbs sampling for tiles is indeed aperiodic, as in each iteration it is possible to remain in exactly the same state. We only have to verify if the irreducibility property is satisfied. We can show that, in some cases, the random walk is reducible, we will show how to make Gibbs sampling irreducible in those cases.

Theorem 1. *Let us consider the bipartite graph $G = (U, V, E)$ derived from the dataset $\mathbf{D}$, s.t., $U = \{1, .., m\}$, $V = \{1, ..., n\}$, and $E = \{(i, j) | i \in [\![1, m]\!] \wedge j \in [\![1, n]\!] \wedge \mathbf{D}(i, j) = 1\}$. A tile $\tau = (I, J)$ present in $\mathbf{D}$ corresponds to a complete bipartite subgraph $G_\tau = (I, J, E_\tau)$ of G. If the bipartite graph G is connected, then the Gibbs sampling procedure on tiles of $\mathbf{D}$ is irreducible.*

Proof. We need to prove that for all pair of tiles $\tau_1 = (I_1, J_1), \tau_2 = (I_2, J_2)$ present in $\mathbf{D}$, the Gibbs sampling procedure can go from τ_1 to τ_2. Let G_{τ_1}, G_{τ_2} be the complete bipartite graphs corresponding to τ_1 and τ_2. As G is connected, there is a path from any vertex of G_{τ_1} to any vertex of G_{τ_2}. The probability that the sampling procedure walks through one of these paths is not 0, as each step of these paths constitutes a tile present in $\mathbf{D}$. After walking on one of these paths, the procedure will find itself on a tile $\tau' \subseteq \tau_2$. Reaching τ_2 from τ' is probable after one iteration by sampling the right rows and then the right columns.

Thus, if the bipartite graph G is connected, the Gibbs sampling procedure converges to a stationary distribution. To make the random walk converge when G is not connected, we can compute the connected components of G, and then apply Gibbs sampling separately in each corresponding subset of the dataset.

Table 2. Dataset characteristics.

Dataset	# rows	# columns	Avg. |row|
mushrooms	8124	120	24
chess	3196	76	38
kdd	843	6159	65.3

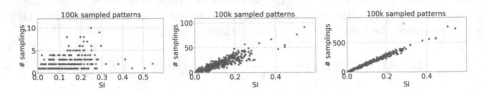

Fig. 1. Distribution of sampled patterns in synthetic data with 10 rows and 10 columns.

5 Experiments

We report our experimental study to evaluate the effectiveness of Gibbs-SI. Java source code is made available[1]. We consider three datasets whose characteristics are given in Table 2. *mushrooms* and *chess* from the UCI repository[2] are commonly used for evaluation purposes. *kdd* contains a set of SIGKDD paper abstracts between 2001 and 2008 downloaded from the ACM website. Each abstract is represented by a row and words correspond to columns, after stop word removal and stemming. For each dataset, the user priors that we represent in the SI background model are the row and column margins. In other terms, we consider that user knows (or, is already informed about) the following statistics: $\sum_j D(i,j)$ for all $i \in I$, and $\sum_i D(i,j)$ for all $j \in J$.

Empirical Sampling Distribution. First, we want to experimentally evaluate how the Gibbs sampling distribution matches with the desired distribution. We need to run Gibbs-SI in small datasets where the size of T is not huge. Then, we take a sufficiently large number of samples so that the sampling distribution can be created. To this aim, we have synthetically generated a dataset containing 10 rows, 10 columns, and 855 tiles. We run Gibbs-SI with three different numbers of iterations p: $1k$, $10k$, and $100k$, for each case, we keep all the visited tiles, and we study their distribution w.r.t. their SI values. Figure 1 reports the results. For $1k$ sampled patterns, the proportionality between the number of sampling and SI is not clearly established yet. For higher numbers of sampled patterns, a linear relation between the two axis is evident, especially for the case of $100k$ sampled patterns, which represents around 100 times the total number of all the tiles in the dataset. The two tiles with the highest SI are sampled the most, and the number of sampling clearly decreases with the SI value.

---

[1] http://tiny.cc/g5zmgz.
[2] https://archive.ics.uci.edu/ml/.

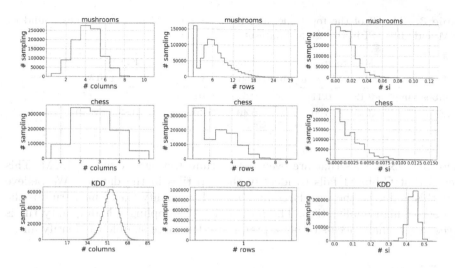

Fig. 2. Distributions of the sampled patterns w.r.t. # rows, # columns and SI.

Characteristics of Sampled Tiles. To investigate which kind of patterns are sampled by Gibbs-SI, we show in Fig. 2 the distribution of sampled tiles w.r.t their number of rows, columns, and their SI, for each of the three datasets given in Table 2. For *mushrooms* and *chess*, Gibbs-SI is able to return patterns with a diverse number of rows and columns. It samples much more patterns with low SI than patterns with high SI values. In fact, even if we are sampling proportionally to SI, the number of tiles in T with poor quality are significantly higher than the ones with high quality values. Thus, the probability of sampling one of low quality patterns is higher than sampling one of the few high quality patterns. For *kdd*, although the number of columns in sampled tiles varies, all the sampled tiles unfortunately cover only one row. In fact, the particularity of this dataset is the existence of some very large transactions (max = 180).

Quality of the Sampled Tiles. In this part of the experiment, we want to study whether the quality of the top sampled tiles is sufficient. As mining exhaustively the best tiles w.r.t. SI is not feasible, we need to find some strategy that identifies high quality tiles. We propose to use LCM [14] to retrieve the closed tiles corresponding to the top 10k frequent closed itemsets. A closed tile $\tau = (I, J)$ is a tile that is present in $\mathbf{D}$ and whose I and J cannot be extended anymore. Although closed tiles are not necessarily the ones with the highest SI, we make the hypothesis that at least some of them have high SI values as they maximize the value of IC function. For each of the three real world datasets, we compare between the SI of the top closed tiles identified with LCM and the ones identified with Gibbs-SI. In Table 3, we show the SI of the top-1 tile, and the average SI of the top-10 tiles, for each of LCM and Gibbs-SI.

Unfortunately, the scores of tiles retrieved with LCM are substantially larger than the ones of Gibbs-SI, especially for *mushrooms* and *chess*. Importantly,

Table 3. The SI of the top-1 tile, and the average SI of the top-10 tiles, found by LCM and Gibbs-SI in the studied datasets.

	Mushrooms		Chess		KDD	
	Top 1 SI	Avg(top 10 SI)	Top 1 SI	Avg(top 10 SI)	Top 1 SI	Avg(top 10 SI)
Gibbs sampling	0.12	0.11	0.015	0.014	0.54	0.54
LCM	3.89	3.20	0.40	0.40	0.83	0.70

there may exist tiles that are even better than the ones found by LCM. This means that Gibbs-SI fails to identify the top tiles in the dataset. We believe that this is due to the very large number of low quality tiles which trumps the number of high quality tiles. The probability of sampling a high-quality tile is exceedingly small, necessitating a practically too large sample to identify any.

6 Discussion

Our results show that efficiently sampling from the set of tiles with a sampling probability proportional to the tiles' subjective interestingness is possible. Yet, they also show that if the purpose is to identify some of the most interesting patterns, direct pattern sampling may not be a good strategy. The reason is that the number of tiles with low subjective interestingness is vastly larger that those with high subjective interestingness. This imbalance is not sufficiently offset by the relative differences in their interestingness and thus in their sampling probability. As a result, the number of tiles that need to be sampled in order to sample one of the few top interesting ones is of the same order as the total number of tiles.

To mitigate this, one could attempt to sample from alternative distributions that attribute an even higher probability to the most interesting patterns, e.g. with probabilities proportional to the *square* or other high powers of the subjective interestingness. We speculate, however, that the computational cost of sampling from such more highly peaked distributions will also be larger, undoing the benefit of needing to sample fewer of them. This intuition is supported by the fact that direct sampling schemes according to itemset support are computationally cheaper than according to the square of their support [2].

That said, the use of sampled patterns as features for downstream machine learning tasks, even if these samples do not include the most interesting ones, may still be effective as an alternative to exhaustive pattern mining.

7 Conclusions

Pattern sampling has been proposed as a computationally efficient alternative to exhaustive pattern mining. Yet, existing techniques have been limited in terms of which interestingness measures they could handle efficiently.

In this paper, we introduced an approach based on Gibbs sampling, which is capable of sampling from the set of tiles proportional to their subjective interestingness. Although we present this approach for a specific type of pattern language and quality measure, we can relatively easily follow the same scheme to apply Gibbs sampling for other pattern mining settings. The empirical evaluation demonstrates effectiveness, yet, it also reveals a potential weakness inherent to pattern sampling: when the number of interesting patterns is vastly outnumbered by the number of non-interesting ones, a large number of samples may be required, even if the samples are drawn with a probability proportional to the interestingness. Investigating our conjecture that this problem affects all approaches for sampling interesting patterns (for sensible measures of interestingness) seems a fruitful avenue for further research.

Acknowledgements. This work was supported by the ERC under the EU's Seventh Framework Programme (FP7/2007-2013)/ERC Grant Agreement no. 615517, the Flemish Government under the "Onderzoeksprogramma Artificiële Intelligentie (AI) Vlaanderen" programme, the FWO (project no. G091017N, G0F9816N, 3G042220), and the EU's Horizon 2020 research and innovation programme and the FWO under the Marie Sklodowska-Curie Grant Agreement no. 665501, and by the ACADEMICS grant of the IDEXLYON, project of the Université de Lyon, PIA operated by ANR-16-IDEX-0005.

References

1. Boley, M., Gärtner, T., Grosskreutz, H.: Formal concept sampling for counting and threshold-free local pattern mining. In: Proceedings of SDM, pp. 177–188 (2010)
2. Boley, M., Lucchese, C., Paurat, D., Gärtner, T.: Direct local pattern sampling by efficient two-step random procedures. In: Proceedings of KDD, pp. 582–590 (2011)
3. Boley, M., Moens, S., Gärtner, T.: Linear space direct pattern sampling using coupling from the past. In: Proceedings of KDD, pp. 69–77 (2012)
4. Boulicaut, J., Jeudy, B.: Constraint-based data mining. In: Maimon, O., Rokach, L. (eds.) Data Mining and Knowledge Discovery Handbook, pp. 339–354. Springer, Heidelberg (2010). https://doi.org/10.1007/978-0-387-09823-4_17
5. Chaoji, V., Hasan, M.A., Salem, S., Besson, J., Zaki, M.J.: ORIGAMI: a novel and effective approach for mining representative orthogonal graph patterns. SADM 1(2), 67–84 (2008)
6. De Bie, T.: Maximum entropy models and subjective interestingness: an application to tiles in binary databases. DMKD 23(3), 407–446 (2011)
7. Frawley, W.J., Piatetsky-Shapiro, G., Matheus, C.J.: Knowledge discovery in databases: an overview. AI Mag. 13(3), 57–70 (1992)
8. Geerts, F., Goethals, B., Mielikäinen, T.: Tiling databases. In: Suzuki, E., Arikawa, S. (eds.) DS 2004. LNCS (LNAI), vol. 3245, pp. 278–289. Springer, Heidelberg (2004). https://doi.org/10.1007/978-3-540-30214-8_22
9. Hasan, M.A., Zaki, M.J.: Output space sampling for graph patterns. PVLDB 2(1), 730–741 (2009)
10. Kontonasios, K.N., Spyropoulou, E., De Bie, T.: Knowledge discovery interestingness measures based on unexpectedness. Wiley IR: DMKD 2(5), 386–399 (2012)

11. Moens, S., Goethals, B.: Randomly sampling maximal itemsets. In: Proceedings of KDD-IDEA, pp. 79–86 (2013)
12. Pei, J., Han, J., Wang, W.: Constraint-based sequential pattern mining: the pattern-growth methods. JIIS **28**(2), 133–160 (2007)
13. Raedt, L.D., Zimmermann, A.: Constraint-based pattern set mining. In: Proceedings of SDM, pp. 237–248 (2007)
14. Uno, T., Asai, T., Uchida, Y., Arimura, H.: An efficient algorithm for enumerating closed patterns in transaction databases. In: Suzuki, E., Arikawa, S. (eds.) DS 2004. LNCS (LNAI), vol. 3245, pp. 16–31. Springer, Heidelberg (2004). https://doi.org/10.1007/978-3-540-30214-8_2

Even Faster Exact k-Means Clustering

Christian Borgelt[1,2(✉)]

[1] Department of Mathematics/Computer Sciences, Paris-Lodron-University
of Salzburg, Hellbrunner Straße 34, 5020 Salzburg, Austria
christian.borgelt@sbg.ac.at
[2] Department of Computer and Information Science, University of Konstanz,
Universitätsstraße 10, 78457 Konstanz, Germany
christian@borgelt.net

Abstract. A naïve implementation of k-means clustering requires computing for each of the n data points the distance to each of the k cluster centers, which can result in fairly slow execution. However, by storing distance information obtained by earlier computations as well as information about distances between cluster centers, the triangle inequality can be exploited in different ways to reduce the number of needed distance computations, e.g. [3–5,7,11]. In this paper I present an improvement of the Exponion method [11] that generally accelerates the computations. Furthermore, by evaluating several methods on a fairly wide range of artificial data sets, I derive a kind of map, for which data set parameters which method (often) yields the lowest execution times.

Keywords: Exact k-means · Triangle inequality · Exponion

1 Introduction

The k-means algorithm [9] is, without doubt, the best known and (among) the most popular clustering algorithm(s), mainly because of its simplicity. However, a naïve implementation of the k-means algorithm requires $O(nk)$ distance computations in each update step, where n is the number of data points and k is the number of clusters. This can be a severe obstacle if clustering is to be carried out on truly large data sets with hundreds of thousands or even millions of data points and hundreds to thousands of clusters, especially in high dimensions.

Hence, in our "big data" age, considerable effort was spent on trying to accelerate the computations, mainly by reducing the number of needed distance computations. This led to several very clever approaches, including [3–5,7,11]. These methods exploit that for assigning data points to cluster centers knowing actual distances is not essential (in contrast to e.g. fuzzy c-means clustering [2]). All one really needs to know is which center is closest. This, however, can sometimes be determined without actually computing (all) distances.

A core idea is to maintain, for each data point, bounds on its distance to different centers, especially to the closest center. These bounds are updated by

M. R. Berthold et al. (Eds.): IDA 2020, LNCS 12080, pp. 93–105, 2020.
https://doi.org/10.1007/978-3-030-44584-3_8

exploiting the triangle inequality, and can enable us to ascertain that the center that was closest before the most recent update step is still closest. Furthermore, by maintaining additional information, tightening these bounds can sometimes be done by looking at only a subset of the cluster centers.

In this paper I present an improvement of one of the most sophisticated of such schemes: the Exponion method [11]. In addition, by comparing my new approach to other methods on several (artificial) data sets with a wide range of number of dimensions and number of clusters, I derive a kind of map, for which data set parameters which method (often) yields the lowest execution times.

2 k-Means Clustering

The k-means algorithm is a very simple, yet effective clustering scheme that finds a user-specified number k of clusters in a given data set. This data set is commonly required to consist of points in a metric space. The algorithm starts by choosing an initial set of k cluster centers, which may naïvely be obtained by sampling uniformly at random from the given data points. In the subsequent cluster center optimization phase, two steps are executed alternatingly: (1) each data point is assigned to the cluster center that is closest to it (that is, closer than any other cluster center) and (2) the cluster centers are recomputed as the vector means of the data points assigned to them (to enable these mean computations, the data points are supposed to live in a metric space).

Using $\nu_m(x)$ to denote the cluster center m-th closest to a point x in the data space, this update scheme can be written (for n data points $x_1, \ldots, x_n$) as

$$\forall i; 1 \leq i \leq k: \qquad c_i^{t+1} = \frac{\sum_{j=1}^{n} \mathbb{1}(\nu_1^t(x_j) = c_i^t) \cdot x_j}{\sum_{j=1}^{n} \mathbb{1}(\nu_1^t(x_j) = c_i^t)},$$

where the indices t and $t+1$ indicate the update step and the function $\mathbb{1}(\phi)$ yields 1 if ϕ is true and 0 otherwise. Here $\nu_1^t(x_j)$ represents the assignment step and the fraction computes the mean of the data points assigned to center c_i.

It can be shown that this update scheme must converge, that is, must reach a state in which another execution of the update step does not change the cluster centers anymore [14]. However, there is no guarantee that the obtained result is optimal in the sense that it yields the smallest sum of squared distances between the data points and the cluster centers they are assigned to. Rather, it is very likely that the optimization gets stuck in a local optimum. It has even been shown that k-means clustering is NP-hard for 2-dimensional data [10].

Furthermore, the quality of the obtained result can depend heavily on the choice of the initial centers. A poor choice can lead to inferior results due to a local optimum. However, improvements of naïvely sampling uniformly at random from the data points are easily found, for example the Maximin method [8] and the k-means++ procedure [1], which has become the de facto standard.

3   Bounds-Based Exact k-Means Clustering

Some approaches to accelerate the k-means algorithm rely on approximations, which may lead to different results, e.g. [6,12,13]. Here, however, I focus on methods to accelerate *exact* k-means clustering, that is, methods that, starting from the same initialization, produce the same result as a naïve implementation.

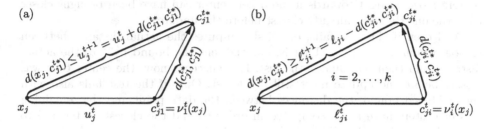

Fig. 1. Using the triangle inequality to update the distance bounds for a data point x_j.

The core idea of these methods is to compute for each update step the distance each center moved, that is, the distance between the new and the old location of the center. Applying the triangle inequality one can then derive how close or how far away an updated center can be from a data point in the worst possible case. For this we distinguish between the center closest (before the update) to a data point x_j on the one hand and all other centers on the other.

k Distance Bounds. The first approach along these lines was developed in [5] and maintains one distance bound for each of the k cluster centers.

For the center closest to a data point x_j an upper bound u_j^t on its distance is updated as shown in Fig. 1(a): If we know before the update that the distance between x_j and its closest center $c_{j1}^t = \nu_1^t(x_j)$ is (at most) u_j^t, and the update moved the center c_{j1}^t to the new location c_{j1}^{t*}, then the distance $d(x_j, c_{j1}^t)$ between the data point and the new location of this center[1] cannot be greater than $u_j^{t+1} = u_j^t + d(c_{j1}^t, c_{j1}^{t*})$. This bound is actually reached if before the update the bound was tight and the center c_{j1}^t moves away from the data point x_j on the straight line through x_j and c_{j1}^t (that is, if the triangle is "flat").

For all other centers, that is, centers that are *not* closest to the point x_j, lower bounds ℓ_{ji}, $i = 2, \ldots, k$, are updated as shown in Fig. 1(b): If we know before the update that the distance between x_j and a center $c_{ji}^t = \nu_i^t(x_j)$, is (at least) ℓ_{ji}^t, and the update moved the center c_{ji}^t to the new location c_{ji}^{t*}, then the distance $d(x_j, c_{ji}^t)$ between the data point and the new location of this center cannot be less than $\ell_{ji}^{t+1} = \ell_{ji}^t - d(c_{ji}^t, c_{ji}^{t*})$. This bound is actually reached if before the update the bound was tight and the center c_{ji}^t moves towards the data point x_j on the straight line through x_j and c_{ji}^t ("flat" triangle).

---

[1] Note that it may be $c_{j1}^{t*} \neq c_{j1}^{t+1}$ (although equality is not ruled out either), because the update may have changed which cluster center is closest to the data point x_j.

These bounds are easily exploited to avoid distance computations for a data point x_j: If we find that $u_j^{t+1} < \ell_j^{t+1} = \min_{i=2}^{k} \ell_{ji}^{t+1}$, that is, if the upper bound on the distance to the center that was closest before the update (in step t) is less than the smallest lower bound on the distances to any other center, the center that was closest before the update must still be closest after the update (that is, in step $t+1$). Intuitively: even if the worst possible case happens, namely if the formerly closest center moves straight away from the data point and the other centers move straight towards it, no other center can have been brought closer than the one that was already closest before the update.

And even if this test fails, one first computes the actual distance between the data point x_j and c_{j1}^{t*}. That is, one tightens the bound u_j^{t+1} to the actual distance and then reevaluates the test. If it succeeds now, the center that was closest before the update must still be closest. Only if the test fails also with the tightened bound, the distances between the data point and the remaining cluster centers have to be computed in order to find the closest center and to reinitialize the bounds (all of which are tight after such a computation).

This scheme leads to considerable acceleration, because the cost of computing the distances between the new and the old locations of the cluster centers as well as the cost of updating the bounds is usually outweighed by the distance computations that are saved in those cases in which the test succeeds.

2 **Distance Bounds.** A disadvantage of the scheme just described is that k bound updates are needed for each data point. In order to reduce this cost, in [7] only two bounds are kept per data point: u_j^t and ℓ_j^t, that is, all non-closest centers are captured by a single lower bound. This bound is updated according to $\ell_j^{t+1} = \ell_j^t - \max_{i=2}^{k} d(c_{ji}^t, c_{ji}^{t*})$. Even though this leads to worse lower bounds for the non-closest centers (since they are all treated as if they moved by the maximum of the distances any one of them moved), the fact that only two bounds have to be updated leads to faster execution, at least in many cases.

YinYang Algorithm. Instead of having either one distance bound for each center (k bounds) or capturing all non-closest centers by a single bound (2 bounds), one may consider a hybrid approach that maintains lower bounds for subsets of the non-closest centers. This improves the quality of bounds over the 2 bounds approach, because bounds are updated only by the maximum distance a center in the corresponding group moved (instead of the global maximum). On the other hand, (considerably) fewer than k bounds have to be updated.

This is the idea of the YinYang algorithm [4], which forms the groups of centers by clustering the initial centers with k-means clustering. The number of groups is chosen as $k/10$ in [4], but other factors may be tried. The groups found initially are maintained, that is, there is no re-clustering after an update.

However, apart from fewer bounds (compared to k bounds) and better bounds (compared to 2 bounds), grouping the centers has yet another advantage: If the bounds test fails, even with a tightened bound u_j^t, the groups and their bounds may be used to limit the centers for which a distance recomputation is needed. Because if the test succeeds for some group, one can infer that the closest center

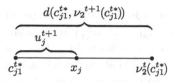

Fig. 2. If $2u_j^{t+1} < d(c_{j1}^{t*}, \nu_2^{t+1}(c_{j1}^{t*}))$, then the center c_{j1}^{t*} must still be closest to the data point x_j, due to the triangle inequality.

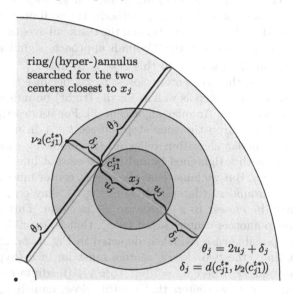

Fig. 3. Annular algorithm [3]: If even after the upper bound u_j for the distance from data point x_j to its (updated) formerly closest center c_{j1}^{t*} has been made tight, the lower bound ℓ_j for distances to other centers is still lower, it is necessary to recompute the two closest centers. Exploiting information about the distance between c_{j1}^{t*} and another center $\nu_2(c_{j1}^{t*})$ closest to it, these two centers are searched in a (hyper-)annulus around the origin (dot in the bottom left corner) with c_{j1}^{t*} in the middle and thickness $2\theta_j$, where $\theta_j = 2u_j + \delta_j$ and $\delta_j = d(c_{i1}^{t*}, \nu_2(c_{j1}^{t*}))$. (Color figure online)

cannot be in that group. Only centers in groups, for which the group-specific test fails, need to be considered for recomputation.

Cluster to Cluster Distances. The described bounds test can be improved by not only computing the distance each center moved, but also the distances between (updated) centers, to find for each center another center that is closest to it [5]. With my notation I can denote such a center as $\nu_2^{t+1}(c_{j1}^{t*})$, that is, the center that is second closest[2] to the point c_{j1}^{t*}. Knowing the distances $d(c_{j1}^{t*}, \nu_2^{t+1}(c_{j1}^{t*}))$, one can test whether $2u_l^{t+1} < d(c_{j1}^{t*}, \nu_2^{t+1}(c_{j1}^{t*}))$. If this is the case, the center that was closest to the data point x_j before the update must still be closest after, as

---

[2] Note that $\nu_1^{t+1}(c_{j1}^{t*}) = c_{j1}^{t*}$, because a center is certainly the center closest to itself.

is illustrated in Fig. 2 for the worst possible case (namely x_j, c_{ji}^{t*} and $\nu_2^{t+1}(c_{j1}^{t*})$ lie on a straight line with c_{ji}^{t*} and $\nu_2^{t+1}(c_{j1}^{t*})$ on opposite sides of x_j).

Note that this second test can be used with k as well as with 2 bounds. However, it should also be noted that, although it can lead to an acceleration, if used in isolation it may also make an algorithm slower, because of the $O(k^2)$ distance computations needed to find the k distances $d(c_i^{t+1}, \nu_2^{t+1}(c_i^{t+1}))$.

Annular Algorithm. With the YinYang algorithm an idea appeared on the scene that is at the focus of all following methods: try to limit the centers that need to be considered in the recomputations if the tests fail even with a tightened bound u_j^{t+1}. Especially, if one uses the 2 bounds approach, significant gains may be obtained: all we need to achieve in this case is to find $c_{i1}^{t+1} = \nu_1^{t+1}(x_j)$ and $c_{i2}^{t+1} = \nu_2^{t+1}(x_j)$, that is, the two centers closest to x_j, because these are all that is needed for the assignment step as well as for the (tight) bounds u_j^{t+1} and ℓ_j^{t+1}.

One such approach is the Annular algorithm [3]. For its description, as generally in the following, I drop the time step indices $t + 1$ in order to simplify the notation. The Annular algorithm relies on the following idea: if the tests described above fail with a tightened bound u_j, we cannot infer that c_{ji}^{t*} is still the center closest to x_j. But we know that the closest center must lie in (hyper-) ball with radius u_j around x_j (darkest circle in Fig. 3). Any center outside this (hyper-)ball cannot be closest to x_j, because c_{ji}^{t*} is closer. Furthermore, if we know the distance to another center closest to c_{ji}^{t*}, that is, $\nu_2(c_{j1}^{t*})$, we know that even in the worst possible case (which is depicted in Fig. 3: x_j, c_{ji}^{t*} and $\nu_2(c_{j1}^{t*})$ lie on a straight line), the two closest centers must lie in a (hyper-)ball with radius $u_j + \delta_j$ around x_j, where $\delta_j = d(c_{i1}^{t*}, \nu_2(c_{j1}^{t*}))$ (medium circle in Fig. 3), because we already know two centers that are this close, namely c_{ji}^{t*} and $\nu_2(c_{j1}^{t*})$. Therefore, if we know the distances of the centers from the origin, we can easily restrict the recomputations to those centers that lie in a (hyper-)annulus (hence the name of this algorithm) around the origin with c_{j1}^{t*} in the middle and thickness $2\theta_j$, where $\theta_j = 2u_j + \delta_j$ with $\delta_j = d(c_{i1}^{t*}, \nu_2(c_{j1}^{t*}))$ (see Fig. 3, light gray ring section, origin in the bottom left corner; note that the green line is perpendicular to the red/blue lines only by accident/for drawing convenience).

Exponion Algorithm. The Exponion algorithm [11] improves over the Annular algorithm by switching from annuli around the origin to (hyper-)balls around the (updated) formerly closest center c_{j1}^{t*}. Again we know that the center closest to x_j must lie in a (hyper-)ball with radius u_j around x_j (darkest circle in Fig. 4) and that the two closest centers must lie in a (hyper-)ball with radius $u_j + \delta_j$ around x_j, where $\delta_j = d(c_{i1}^{t*}, \nu_2(c_{j1}^{t*}))$ (medium circle in Fig. 4). Therefore, if we know the pairwise distances between the (updated) centers, we can easily restrict the recomputations to those centers that lie in the (hyper-)ball with radius $r_j = 2u_j + \delta_j$ around c_{j1}^{t*} (lightest circle in Fig. 4).

The Exponion algorithm also relies on a scheme with which it is avoided having to sort, for each cluster center, the lists of the other centers by their distance. For this concentric annuli, one set centered at a each center, are created, with each annulus further out containing twice as many centers as the preceding

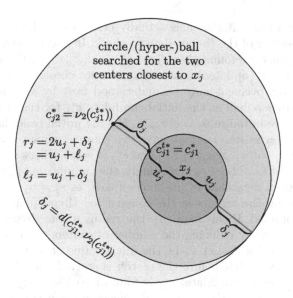

Fig. 4. Exponion algorithm [11]: If even after the upper bound u_j for the distance from a data point x_j to its (updated) formerly closest center c_{j1}^{t*} has been made tight, the lower bound ℓ_j for distance to other centers is still lower, it is necessary to recompute the two closest centers. Exploiting information about the distance between c_{j1}^{t*} and another center $\nu_2(c_{j1}^{t*})$ closest to it, these two centers are searched in a (hyper-)sphere around center c_{j1}^{t*} with radius $r_j = 2u_j + \delta_j$ where $\delta_j = d(c_{j1}^{t*}, \nu_2(c_{j1}^{t*}))$. (Color figure online)

one. Clearly this creates an onion-like structure, with an exponentially increasing number of centers in each layer (hence the name of the algorithm).

However, avoiding the sorting comes at a price, namely that more centers may have to be checked (although at most twice as many [11]) for finding the two closest centers and thus additional distance computations ensue. In my implementation I avoided this complication and simply relied on sorting the distances, since the gains achievable by concentric annuli over sorting are somewhat unclear (in [11] no comparisons of sorting versus concentric annuli are provided).

Shallot Algorithm. The Shallot algorithm is the main contribution of this paper. It starts with the same considerations as the Exponion algorithm, but adds two improvements. In the first place, not only the closest center c_{j1} and the two bounds u_j and ℓ_j are maintained for each data point (as for Exponion), but also the second closest center c_{j2}. This comes at practically no cost (apart from having to store an additional integer per data point), because the second closest center has to be determined anyway in order to set the bound ℓ_j.

If a recomputation is necessary, because the tests fail even for a tightened u_j, it is *not* automatically assumed that c_{j1}^{t*} is the best center z for a (hyper-)ball to search. As it is plausible that the formerly second closest center c_{j2}^{t*} may now be closer to x_j than c_{j1}^{t*}, the center c_{j2}^{t*} is processed first among the centers c_{ji}^{t*},

$i = 2, \ldots, k$. If it turns out that it is actually closer to x_j than c_{j1}^{t*}, then c_{j2}^{t*} is chosen as the center z of the (hyper-)ball to check. In this case the (hyper-)ball will be smaller (since we found that $d(x_j, c_{j2}^{t*}) < d(x_j, c_{j1}^{t*})$). For the following, let p denote the other (updated) center that was not chosen as the center z.

The second improvement may be understood best by viewing the chosen center z of the (hyper-)ball as the initial candidate c_{j1}^* for the closest center in step $t+1$. Hence we initialize $u_j = d(x_j, z)$. For the initial candidate c_{j2}^* for the second closest center in step $t+1$ we have two choices, namely p and $\nu_2(z)$. We choose $c_{j2}^* = p$ if $u_j + d(x_j, p) < 2u_l + \delta_j$ and $c_{j2}^* = \nu_2(z)$ otherwise, and initialize $\ell_j = u_j + d(x_j, p)$ or $\ell_j = 2u_j + \delta_j$ accordingly, thus minimizing the radius, which then can be written, regardless of the choice taken, as $r_j = u_j + \ell_j$.

While traversing the centers in the constructed (hyper-)ball, better candidates may be obtained. If this happens, the radius of the (hyper-)ball may be reduced, thus potentially reducing the number of centers to be processed. This idea is illustrated in Fig. 5. Let u_j° be the initial value of u_j when the (hyper-)ball center was chosen, but before the search is started, that is $u_j^\circ = d(x_j, z)$. If a new closest center (candidate) c_{j1}^* is found (see Fig. 5(a)), we can update $u_j = d(x_j, c_{j1}^*)$ and $\ell_j = d(x_j, c_{j2}^*) = u_j^\circ$. Hence we can shrink the radius to $r_j = 2u_j^\circ = u_j^\circ + \ell_j$. If then an even closer center is found (see Fig. 5(b)), the radius may be shrunk further as u_j and ℓ_j are updated again. As should be clear from these examples, the radius is always $r_j = u_j^\circ + \ell_j$.

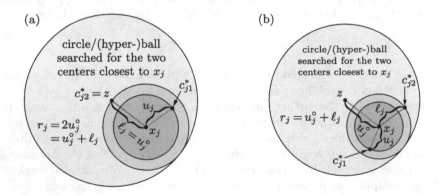

Fig. 5. Shallot algorithm: If a center closer to the data point than the two currently closest centers is found, the radius of the (hyper-)ball to be searched can be shrunk.

A *shallot* is a type of onion, smaller than, for example, a bulb onion. I chose this name to indicate that the (hyper-)ball that is searched for the two closest centers tends to be smaller than for the Exponion algorithm. The reference to an onion may appear misguided, because I rely on sorting the list of other centers by their distance for each cluster center, rather than using concentric annuli. However, an onion reference may also be justified by the fact that my algorithm may shrink the (hyper-)ball radius during the traversal of centers in the (hyper-)ball, as this also creates a layered structure of (hyper-)balls.

4   Experiments

In order to evaluate the performance of the different exact k-means algorithms I generated a large number of artificial data sets. Standard benchmark data sets proved to be too small to measure performance differences reliably and would also not have permitted drawing "performance maps" (see below). I fixed the number of data points in these data sets at $n = 100\,000$. Anything smaller renders the time measurements too unreliable, anything larger requires an unpleasantly long time to run all benchmarks. Thus I varied only the dimensionality m of the data space, namely as $m \in \{2, 3, 4, 5, 6, 8, 10, 15, 20, 25, 30, 35, 40, 45, 50\}$, and the number k of clusters, from 20 to 300 in steps of 20. For each parameter combination I generated 10 data sets, with clusters that are (roughly, due to random deviations) equally populated with data points and that may vary in size by a factor of at most ten per dimension. All clusters were modeled as isotropic normal (or Gaussian) distributions. Each data set was then processed 10 times with different initializations. All optimization algorithms started from the same initializations, thus making the comparison as fair as possible.

The clustering program is written in C (however, there is also a Python version, see the link to the source code below). All implementations of the different algorithms are entirely my own and use the same code to read the data and to write the clustering results. This adds to the fairness of the comparison, as in this way any differences in execution time can only result from differences of the actual algorithms. The test systems was an Intel Core 2 Quad Q9650@3GHz with 8 GB of RAM running Ubuntu Linux 18.04 64bit.

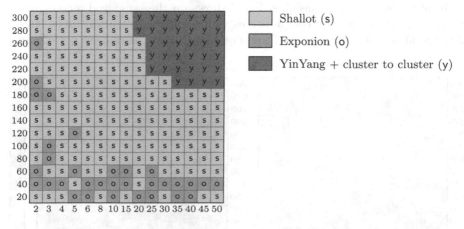

Fig. 6. Map of the algorithms that produced the best execution times over number of dimensions (horizontal) and number of clusters (vertical), showing fairly clear regions of algorithm superiority. Enjoyably, the Shallot algorithm that was developed in this paper yields the best results for the largest number of parameter combinations.

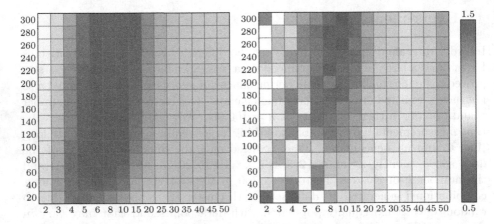

Fig. 7. Relative comparison between the Shallot algorithm and the Exponion algorithm. The left diagram refers to the number of distance computations, the right diagram to execution time. Blue means that Shallot is better, red that Exponion is better. (Color figure online)

The results of these experiments are visualized in Figs. 6, 7 and 8. Figure 6 shows on a grid spanned by the number of dimensions (horizontal axis) and the number of clusters inducted into the data set (vertical axis) which algorithm performed best (in terms of execution time) for each combination. Clearly, the Shallot algorithm wins most parameter combinations. Only for larger numbers of dimensions and larger numbers of clusters the YinYang algorithm is superior.

In order to get deeper insights, Fig. 7 shows on the same grid a comparison of the number of distance computations (left) and the execution times (right) of the Shallot algorithm and the Exponion algorithm. The relative performance

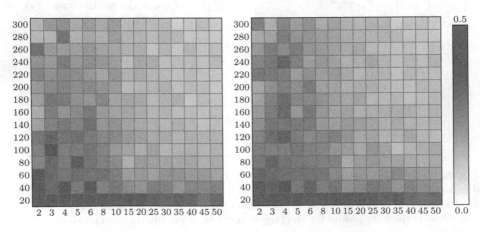

Fig. 8. Variation of the execution times over number of dimensions (horizontal) and number of clusters (vertical). The left diagram refers to the Shallot algorithm, the right diagram to the Exponion algorithm. The larger variation for fewer clusters and fewer dimensions may explain the speckled look of Figs. 6 and 7.

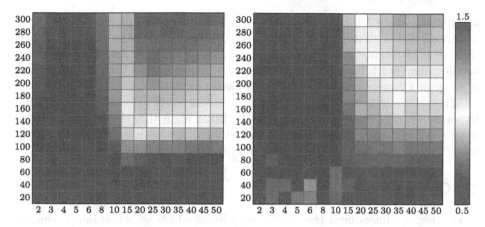

Fig. 9. Relative comparison between the Shallot algorithm and the YinYang algorithm using the cluster to cluster distance test (pure YinYang is very similar, though). The left diagram refers to the number of distance computations, the right diagram to execution time. Blue means that Shallot is better, red that YinYang is better. (Color figure online)

is color-coded: saturated blue means that the Shallot algorithm needed only half the distance computations or half the execution time of the Exponion algorithm, saturated red means that it needed 1.5 times the distance computations or execution time compared to the Exponion algorithm.

W.r.t. distance computations there is no question who is the winner: the Shallot algorithm wins all parameter combinations, some with a considerable margin. W.r.t. execution times, there is also a clear region towards more dimensions and more clusters, but for fewer clusters and fewer dimensions the diagram looks a bit speckled. This is a somewhat strange result, as a smaller number of distance computations should lead to lower execution times, because the effort spent on organizing the search, which is also carried out in exactly the same situations, is hardly different between the Shallot and the Exponion algorithm.

The reason for this speckled look could be that the benchmarks were carried out with heavy parallelization (in order to minimize the total time), which may have distorted the measurements. As a test of this hypothesis, Fig. 8 shows the standard deviation of the execution times relative to their mean. White means no variation, fully saturated blue indicates a standard deviation half as large as the mean value. The left diagram refers to the Shallot, the right diagram to the Exponion algorithm. Clearly, for a smaller number of dimensions and especially for a smaller number of clusters the execution times vary more (this may be, at least in part, due to the generally lower execution times for these parameter combinations). It is plausible to assume that this variability is the explanation for the speckled look of the diagrams in Fig. 6 and in Fig. 7 on the right.

Finally, Fig. 9 shows, again on the same grid, a comparison of the number of distance computations (left) and the execution times (right) of the Shallot algorithm and the YinYang algorithm (using the test based on cluster to cluster distances, although a pure YinYang algorithm performs very similarly). The relative performance is color-coded in the same way as in Fig. 7. Clearly, the smaller number of distance computations explains why the YinYang algorithm is superior for more clusters and more dimensions.

The reason is likely that grouping the centers leads to better bounds. This hypothesis is confirmed by the fact that the Elkan algorithm (k distance bounds) always needs the fewest distance computations (not shown as a grid) and loses on execution time only due to having to update so many distance bounds.

5   Conclusion

In this paper I introduced the Shallot algorithm, which adds two improvements to the Exponion algorithm [11], both of which can potentially shrink the (hyper-) ball that has to be searched for the two closest centers if recomputation becomes necessary. This leads to a measurable, sometimes even fairly large speedup compared to the Exponion algorithm due to fewer distance computations. However, for high-dimensional data and large numbers of clusters the YinYang algorithm [4] (with or without the cluster to cluster distance test) is superior to both algorithms. Yet, since clustering in high dimensions is problematic anyway due to the curse of dimensionality, it may be claimed reasonably confidently that the Shallot algorithm is the best choice for standard clustering tasks.

Software. My implementation of the described methods (C and Python), with which I conducted the experiments, can be obtained under the MIT License at
http://www.borgelt.net/cluster.html.

Complete Results. A table with the complete experimental results I obtained can be retrieved as a simple text table at
http://www.borgelt.net/docs/clsbench.txt.

More maps comparing the performance of the algorithms can be found at
http://www.borgelt.net/docs/clsbench.pdf.

References

1. Arthur, D., Vassilvitskii, S.: k-Means++: the advantages of careful seeding. In: Proceedings of 18th Annual SIAM Symposium on Discrete Algorithms, SODA 2007, New Orleans, LA, pp. 1027–1035. Society for Industrial and Applied Mathematics, Philadelphia (2007)
2. Bezdek, J.C.: Pattern Recognition with Fuzzy Objective Function Algorithms. Plenum Press, New York (1981)
3. Drake, J.: Faster k-means clustering, Master's thesis, Baylor University, Waco, TX, USA (2013)
4. Ding, Y., Zhao, Y., Shen, Y., Musuvathi, M., Mytkowicz, T.: YinYang k-means: a drop-in replacement of the classic k-means with consistent speedup. In: Proceedings of 32nd International Conference on Machine Learning, ICML 2015, Lille, France, JMLR Workshop and Conference Proceedings, vol. 37, pp. 579–587 (2015)

5. Elkan, C.: Using the triangle inequality to accelerate k-means. In: Proceedings 20th International Conference on Machine Learning, ICML 2003, Washington, DC, pp. 147–153. AAAI Press, Menlo Park (2003)
6. Frahling, G., Sohler, C.: A fast k-means implementation using coresets. In: Proceedings of 22nd Annual Symposium on Computational Geometry, SCG 2006, Sedona, AZ, pp. 135–143. ACM Press, New York (2006)
7. Hamerly, G.: Making k-means even faster. In: Proceedings of SIAM International Conference on Data Mining, SDM 2010, Columbus, OH, pp. 130–140. Society for Industrial and Applied Mathematics, Philadelphia (2010)
8. Hathaway, R.J., Bezdek, J.C., Huband, J.M.: Maximin initialization for cluster analysis. In: Martínez-Trinidad, J.F., Carrasco Ochoa, J.A., Kittler, J. (eds.) CIARP 2006. LNCS, vol. 4225, pp. 14–26. Springer, Heidelberg (2006). https://doi.org/10.1007/11892755_2
9. Lloyd, S.P.: Least square quantization in PCM. IEEE Trans. Inf. Theory **28**, 129–137 (1982)
10. Mahajan, M., Nimbhorkar, P., Varadarajan, K.: The planar k-means problem is NP-hard. Theor. Comput. Sci. **442**, 13–21 (2009)
11. Newling, J., Fleuret, F.: Fast k-means with accurate bounds. In: Proceedings of 33rd International Conference on Machine Learning, ICML 2016, New York, NY, JMLR Workshop and Conference Proceedings, vol. 48, pp. 936–944 (2016)
12. Philbin, J., Chum, O., Isard, M., Sivic, J., Zisserman, A.: Object retrieval with large vocabularies and fast spatial matching. In: Proceedings of IEEE International Conference on Computer Vision and Pattern Recognition, CVPR 2007, Minneapolis, MN. IEEE Press, Piscataway (2007)
13. Sculley, D.: Web-scale k-means clustering. In: Proceedings of 19th International Conference on World Wide Web, WWW 2010, Raleigh, NC, pp. 1177–1178. ACM Press, New York (2010)
14. Selim, S.Z., Ismail, M.A.: k-means-type algorithms: a generalized convergence theorem and characterization of local optimality. IEEE Trans. Pattern Anal. Mach. Intell. **1**(6), 81–87 (1984)

Ising-Based Consensus Clustering on Specialized Hardware

Eldan Cohen[1](✉), Avradip Mandal[2], Hayato Ushijima-Mwesigwa[2], and Arnab Roy[2]

[1] University of Toronto, Toronto, Canada
ecohen@mie.utoronto.ca
[2] Fujitsu Laboratories of America, Inc., Sunnyvale, USA
{amandal,hayato,aroy}@us.fujitsu.com

Abstract. The emergence of specialized optimization hardware such as CMOS annealers and adiabatic quantum computers carries the promise of solving hard combinatorial optimization problems more efficiently in hardware. Recent work has focused on formulating different combinatorial optimization problems as Ising models, the core mathematical abstraction used by a large number of these hardware platforms, and evaluating the performance of these models when solved on specialized hardware. An interesting area of application is data mining, where combinatorial optimization problems underlie many core tasks. In this work, we focus on consensus clustering (clustering aggregation), an important combinatorial problem that has received much attention over the last two decades. We present two Ising models for consensus clustering and evaluate them using the Fujitsu Digital Annealer, a quantum-inspired CMOS annealer. Our empirical evaluation shows that our approach outperforms existing techniques and is a promising direction for future research.

1 Introduction

The increasingly challenging task of scaling the traditional Central Processing Unit (CPU) has lead to the exploration of new computational platforms such as quantum computers, CMOS annealers, neuromorphic computers, and so on (see [3] for a detailed exposition). Although their physical implementations differ significantly, adiabatic quantum computers, CMOS annealers, memristive circuits, and optical parametric oscillators all share Ising models as their core mathematical abstraction [3]. This has lead to a growing interest in the formulation of computational problems as Ising models and in the empirical evaluation of these models on such novel computational platforms. This body of literature includes clustering and community detection [14,19,23], graph partitioning [26], and many NP-Complete problems such as covering, packing, and coloring [17].

Consensus clustering is the problem of combining multiple 'base clusterings' of the same set of data points into a single consolidated clustering [9]. Consensus clustering is used to generate robust, stable, and more accurate clustering

---

E. Cohen—Work done while at Fujitsu Laboratories of America.

© The Author(s) 2020
M. R. Berthold et al. (Eds.): IDA 2020, LNCS 12080, pp. 106–118, 2020.
https://doi.org/10.1007/978-3-030-44584-3_9

results compared to a single clustering approach [9]. The problem of consensus clustering has received significant attention over the last two decades [9], and was previously considered under different names (clustering aggregation, cluster ensembles, clustering combination) [10]. It has applications in different fields including data mining, pattern recognition, and bioinformatics [10] and a number of algorithmic approaches have been used to solve this problem. The consensus clustering is, in essence, a combinatorial optimization problem [28] and different instances of the problem have been proven to be NP-hard (e.g., [6,25]).

In this work, we investigate the use of special purpose hardware to solve the problem of consensus clustering. To this end, we formulate the problem of consensus clustering using Ising models and evaluate our approach on a specialized CMOS annealer. We make the following contributions:

1. We present and study two Ising models for consensus clustering that can be solved on a variety of special purpose hardware platforms.
2. We demonstrate how our models are embedded on the Fujitsu Digital Annealer (DA), a quantum-inspired specialized CMOS hardware.
3. We present an empirical evaluation based on seven benchmark datasets and show our approach outperforms existing techniques for consensus clustering.

2 Background

2.1 Problem Definition

Let $X = \{x_1, ..., x_n\}$ be a set of n data points. A *clustering* of X is a process that partitions X into subsets, referred to as *clusters*, that together cover X. A clustering is represented by the mapping $\pi : X \to \{1, \ldots, k_\pi\}$ where k_π is the number of clusters produced by clustering π. Given X and a set $\Pi = \{\pi_1, \ldots, \pi_m\}$ of m clusterings of the points in X, the *Consensus Clustering Problem* is to find a new clustering, π^*, of the data X that best summarizes the set of clusterings Π. The new clustering π^* is referred to as the *consensus* clustering.

Due to the ambiguity in the definition of an optimal consensus clustering, several approaches have been proposed to measure the solution quality of consensus clustering algorithms [9]. In this work, we focus on the approach of determining a consensus clustering that agrees the most with the original clusterings. As an objective measure to determine this agreement, we use the mean Adjusted Rand Index (ARI) metric (Eq. 14). However, we also consider clustering quality measured by mean Silhouette Coefficient [22] and clustering accuracy based on true labels. In Sect. 4 these evaluation criteria are discussed in more details.

2.2 Existing Criteria and Methods

Various criteria or objectives have been proposed for the Consensus Clustering Problem. In this work we mainly focus on two well-studied criteria, one based on the pairwise similarity of the data points, and the other based on the different assignments of the base clusterings. Other well-known criteria and objectives for the Consensus Clustering Problem can be found in the excellent surveys of [9,27], with most defining NP-Hard optimization problems.

Pairwise Similarity Approaches: In this approach, a similarity matrix S is constructed such that each entry in S represents the fraction of clusterings in which two data points belong to the same cluster [20]. In particular,

$$S_{uv} = \frac{1}{m} \sum_{i=1}^{m} \mathbb{1}(\pi_i(u) = \pi_i(v)), \tag{1}$$

with $\mathbb{1}$ being the indicator function. The value S_{uv} lies between 0 and 1, and is equal to 1 if all the base clusterings assign points u and v to the same cluster. Once the pairwise similarity matrix is constructed, one can use any similarity-based clustering algorithm on S to find a consensus clustering with a fixed number of clusters, K. For example, [16] proposed to find a consensus clustering π^* with exactly K clusters that minimizes the within-cluster dissimilarity:

$$\min \sum_{\substack{u,v \in X: \\ \pi^*(u) = \pi^*(v)}} (1 - S_{uv}). \tag{2}$$

Partition Difference Approaches: An alternative formulation is based on the different assignments between clustering. Consider two data points $u, v \in X$, and two clusterings $\pi_i, \pi_j \in \Pi$. The following binary indicator tests if π_i and π_j disagree on the clustering of u and v:

$$d_{u,v}(\pi_i, \pi_j) = \begin{cases} 1, & \text{if } \pi_i(u) = \pi_i(v) \text{ and } \pi_j(u) \neq \pi_j(v) \\ 1, & \text{if } \pi_i(u) \neq \pi_i(v) \text{ and } \pi_j(u) = \pi_j(v) \\ 0, & \text{otherwise.} \end{cases} \tag{3}$$

The distance between two clusterings is then defined based on the number of pairwise disagreements:

$$d(\pi_i, \pi_j) = \frac{1}{2} \sum_{u,v \in X} d_{u,v}(\pi_i, \pi_j) \tag{4}$$

with the $\frac{1}{2}$ factor to take care of double counting and can be ignored. This measure is defined as the number of pairs of points that are in the same cluster in one clustering and in different clusters in the other, essentially considering the (unadjusted) Rand index [9]. Given this measure, a common objective is to find a consensus clustering π^* with respect to the following optimization problem:

$$\min \sum_{i=1}^{m} d(\pi_i, \pi^*). \tag{5}$$

Methods and Algorithms: The two different criteria given above define fundamentally different optimization problems, thus different algorithms have been proposed. One key difference between the two approaches inherently lies in determining the number of clusters k_{π^*} in π^*. The pairwise similarity approaches (e.g.,

Eq. (2)) require an input parameter K that fixes the number of clusters in π^*, whereas the partition difference approaches such as Eq. (5) do not have this requirement and determining k_{π^*} is part of the objective of the problem. Therefore, for example, Eq. (2) will have a minimum value in the case when $k_{\pi^*} = n$, however this does not hold for Eq. (5).

The Cluster-based Similarity Partitioning Algorithm (CSPA) is proposed in [24] for solving the pairwise similarity based approach. The CSPA constructs a similarity-based graph with each edge having a weight proportional to the similarity given by S. Determining the consensus clustering with exactly K clusters is treated as a K-way graph partitioning problem, which is solved by methods such as METIS [12]. In [20], the authors experiment with different clustering algorithms including hierarchical agglomerative clustering (HAC) and iterative techniques that start from an initial partition and iteratively reassign points to clusters based on their pairwise similarities. For the partition difference approach, Li et al. [15] proposed to solve Eq. (5) using nonnegative matrix factorization (NMF). Gionis et al. [10] proposed several algorithms that make use of the connection between Eq. (5) and the problem of correlation clustering. CSPA, HAC, NMF: these three approaches are considered as baseline in our empirical evaluation section (Sect. 4).

2.3 Ising Models

Ising models are graphical models that include a set of nodes representing spin variables and a set of edges corresponding to the interactions between the spins. The energy level of an Ising model which we aim to minimize is given by:

$$E(\sigma) = \sum_{(i,j)\in\mathcal{E}} J_{i,j}\sigma_i\sigma_j + \sum_{i\in\mathcal{N}} h_i\sigma_i, \tag{6}$$

where the variables $\sigma_i \in \{-1,1\}$ are the spin variables and the couplers, $J_{i,j}$, represent the interaction between the spins.

A Quadratic Unconstrained Binary Optimization (QUBO) model includes binary variables $q_i \in \{0,1\}$ and couplers, $c_{i,j}$. The objective to minimize is:

$$E(\mathbf{q}) = \sum_{i=1}^{n} c_i q_i + \sum_{i<j} c_{i,j}q_i q_j. \tag{7}$$

QUBO models can be transformed to Ising models by setting $\sigma_i = 2q_i - 1$ [2].

3 Ising Approach for Consensus Clustering on Specialized Hardware

In this section, we present our approach for solving consensus clustering on specialized hardware using Ising models. We present two Ising models that correspond to the two approaches in Sect. 2.2. We then demonstrate how they can be solved on the Fujitsu Digital Annealer (DA), a specialized CMOS hardware.

3.1   Pairwise Similarity-Based Ising Model

For each data point $u \in X$, let $q_{uc} \in \{0, 1\}$ be the binary variable such that $q_{uc} = 1$ if π^* assigns u to cluster c, and 0 otherwise. Then the constraints

$$\sum_{c=1}^{K} q_{uc} = 1, \quad \text{for each } u \in X \tag{8}$$

ensure π^* assigns each point to exactly one cluster. Subject to the constraints (8), the sum of quadratic terms $\sum_{c=1}^{K} q_{uc} q_{vc}$ is 1 if π^* assigns both $u, v \in X$ to the same cluster, and is 0 if assigned to different clusters. Therefore the value

$$\sum_{\substack{u,v \in X: \\ \pi^*(u) = \pi^*(v)}} (1 - S_{uv}) = \sum_{u,v \in X} (1 - S_{uv}) \sum_{c=1}^{K} q_{uc} q_{vc} \tag{9}$$

represents the sum of within-cluster dissimilarities in π^*: $(1 - S_{uv})$ is the fraction of clusterings in Π that assign u and v to different clusters while π^* assigns them to the same cluster. We therefore reformulate Eq. (2) as QUBO:

$$\min \sum_{u,v \in X} (1 - S_{uv}) \sum_{c=1}^{K} q_{uc} q_{vc} + \sum_{u \in X} A\left(\sum_{c=1}^{K} q_{uc} - 1\right)^2. \tag{10}$$

where the term $\sum_{u \in X} A(\sum_{c=1}^{K} q_{uc} - 1)^2$ is added to the objective function to ensure that the constraints (8) are satisfied. A is positive constant that penalizes the objective for violations of constraints (8). One can show that if $A \geq n$, the optimal solution of the QUBO in Eq. (10) does not violate the constraints (8). The proof is very similar to proof of Theorem 1 and a similar result in [14].

3.2   Partition Difference Ising Model

The partition difference approach essentially considers the (unadjusted) Rand Index [9] and therefore can be expected to perform better. The *Correlation Clustering Problem* is another important problem in data mining. Gionis et al. [10] showed that Eq. (5) is a restricted case of the Correlation Clustering Problem, and that Eq. (5) can be expressed as the following equivalent form of the Correlation Clustering Problem

$$\min_{\pi^*} \sum_{\substack{u,v \in X: \\ \pi^*(u) = \pi^*(v)}} (1 - S_{uv}) + \sum_{\substack{u,v \in X: \\ \pi^*(u) \neq \pi^*(v)}} S_{uv}. \tag{11}$$

We take advantage of this equivalence to model Eq. (5) as a QUBO. In a similar fashion to the QUBO formulated in the preceding subsection, the terms

$$\sum_{\substack{u,v \in X: \\ \pi^*(u) \neq \pi^*(v)}} S_{uv} = \sum_{u,v \in X} S_{uv} \sum_{1 \leq c \neq l \leq K} q_{uc} q_{vl} \tag{12}$$

measure the similarity between points in *different* clusters, where K represents an *upper bound* for the number of clusters in π^*. This then leads to the minimizing the following QUBO:

$$\sum_{u,v \in X} (1 - S_{uv}) \sum_{c=1}^{K} q_{uc} q_{vc} + \sum_{u,v \in X} S_{uv} \sum_{1 \leq c \neq l \leq K} q_{uc} q_{vl} + \sum_{u \in X} B (\sum_{c=1}^{K} q_{uc} - 1)^2.$$
(13)

Intuitively, Eq. (13) measures the disagreement between the consensus clustering and the clusterings in Π. This disagreement is due to points that are clustered together in the consensus clustering but not in the clusterings in Π, however it is also due to points that are assigned to different clusters in the consensus partition but in the same cluster in some of the partitions in Π.

Formally, we can show that Eq. (13) is equivalent to the correlation clustering formulation in Eq. (11) when setting $B \geq n$. Consistent with other methods that optimize Eq. (5) (e.g., [15]), our approach takes as an input K, an *upper bound* on the number of clusters in π^*, however the obtained solution can use smaller number of clusters. In our proof, we assume K is large enough to represent the optimal solution, i.e., greater than the number of clusters in optimal solutions to the correlation clustering problem in Eq. (11).

Theorem 1. *Let $\bar{q}$ be the optimal solution to the QUBO given by Eq. (13). If $B \geq n$, for a large enough $K \leq n$, an optimal solution to the Correlation Clustering Problem in Eq. (11), $\bar{\pi}$, can be efficiently evaluated from $\bar{q}$.*

Proof. First we show the optimal solution to the QUBO in Eq. (13) satisfies the one-hot encoding ($\sum_k q_{uk} = 1$). This would imply given $\bar{q}$ we can create a valid clustering $\bar{\pi}$. Note, the optimal solution will never have $\sum_c q_{uc} > 1$ as it can only increase the cost. The only case in which an optimal solution will have $\sum_c q_{uc} < 1$ is when the cost of assigning a point to a cluster is higher than the cost of not assigning it to a cluster (i.e., the penalty B). Assigning a point u to a cluster will incur a cost of $(1 - S_{uv})$ for each point v in the same cluster and S_{uv} for each point v that is not in the cluster. As there is additional $n - 1$ points in total, and both $(1 - S_{uv})$ and S_{uv} are less or equal to one (Eq. (1)), setting $B \geq n$ guarantees the optimal solution satisfies the one-hot encoding.

Now we assume that $\bar{\pi}$ is not optimal, i.e., there exists an optimal solution $\hat{\pi}$ to Eq. (11) that has a strictly lower cost than $\bar{\pi}$. Let $\hat{q}$ be the corresponding QUBO solution to $\hat{\pi}$, such that $\bar{\pi}(u) = k$ if and only if $\bar{q}_{uk} = 1$. This is possible because K is large enough to accomodate all clusters in $\hat{\pi}$. As both $\bar{q}$ and $\hat{q}$ satisfy that one-hot encoding (penalty terms are zero), their cost is identical to the cost of $\bar{\pi}$ and $\hat{\pi}$. Since the cost of $\hat{\pi}$ is strictly lower than $\bar{\pi}$, and the cost of $\bar{q}$ is lower or equal to $\hat{q}$, we have a contradiction.   □

3.3 Solving Consensus Clustering on the Fujitsu Digital Annealer

The Fujitsu Digital Annealer (DA) is a recent CMOS hardware for solving combinatorial optimization problems formulated as QUBO [1,8]. We use the second generation of the DA that is capable of representing problems with up to 8192 variables with up to 64 bits of precision. The DA has previously been used to solve problems in areas such as communication [18] and signal processing [21].

The DA algorithm [1] is based on simulated annealing (SA) [13], while taking advantage of the massive parallelization provided by the CMOS hardware [1]. It has several key differences compared to SA, most notably a *parallel-trial* scheme in which each MC step considers all possible one-bit flips in parallel and *dynamic offset* mechanism that increase the energy of a state to escape local minima [1].

Encoding Consensus Clustering on the DA. When embedding our Ising models on the DA, we need to consider the hardware specification and adapt the representation of our model accordingly. Due to hardware precision limit, we need to embed the couplers and biases on an integer scale with limited granularity. In our experiments, we normalize the pairwise costs S_{uv} in the discrete range $[0, 100]$, $D_{ij} = [S_{uv} \cdot 100]$, and accordingly $(1 - S_{uv})$ is replaced by $(100 - D_{uv})$. Note that the theoretical bound $B = n$ is adjusted accordingly to be $B = 100 \cdot n$.

The theoretical bound guarantees that all constraints are satisfied if problems are solved to optimality. In practice, the DA does not necessarily solve problems to optimality and due to the nature of annealing-based algorithms, using very high weights for constraints is likely to create deep local minima and result in solutions that may satisfy the constraints but are often of low-quality. This is especially relevant to our pairwise similarity model where the bound tends to become loose as the number of clusters grows. In our experiments, we use constant, reasonably high, weights that were empirically found to perform well across datasets. For the pairwise similarity-based model (Eq. (10)) we use $A = 2^{14}$, and for the partition difference model (Eq. (13)) we use $B = 2^{15}$. While we expect to get better performance by tuning the weights per-dataset, our goal is to demonstrate the performance of our approach in a general setting. Automatic tuning of the weight values for the DA is a direction for future work.

Unlike many of the existing consensus clustering algorithms that run until convergence, our method runs for a given time limit (defined by the number of runs and iterations) and returns the best solution encountered. In our experiments, we arbitrarily choose *three seconds* as a (reasonably short) time limit to solve our Ising models. As with the weights, we employ a single temperature schedule across all datasets, and *do not* tune it per dataset.

4   Empirical Evaluation

We perform an extensive empirical evaluation of our approach using a set of seven benchmark datasets. We first describe how we generate the set of clusterings, Π. Next, we describe the baselines, the evaluation metrics, and the datasets.

Generating Partitions. We follow [7] and generate a set of clusterings by randomizing the parameters of the K-Means algorithm, namely the number of clusters K and the initial cluster centers. In this work, we only use labelled datasets for which we know the number of clusters, $\widetilde{K}$, based on the true labels. To generate the base clusterings we run the K-Means algorithm with random cluster centers and we randomly choose K from the range $[2, 3\widetilde{K}]$. For each dataset, we generate 100 clusterings to serve as the clustering set Π.

Baseline Algorithms. We compare our pairwise similarity-based Ising model, referred to as DA-Sm, and our correlation clustering Ising model, referred to as DA-Cr, to three popular algorithms for consensus clustering:

1. The cluster-based similarity partitioning algorithm (CSPA) [24] solved as a K-way graph partitioning problem using METIS [12].
2. The nonnegative matrix factorization (NMF) formulation in [15].
3. Hierarchical agglomerative clustering (HAC) starts with all points in single-ton clusters and repeatedly merges the two clusters with the largest average similarity based on S, until reaching the desired number of clusters [20].

Evaluation. We evaluate the different methods using three measures. Our main concern in this work is the level of agreement between the consensus clustering and the set of input clusterings. To this end, one requires a metric measuring the similarity of two clusterings that can be used to measure how close the consensus clustering π^* to each base clustering $\pi_i \in \Pi$ is. One of popularly used metrics to measure the similarity between two clusterings is the Rand Index (RI) and Adjusted Rand Index (ARI) [11]. The Rand Index of two clustering lies between 0 and 1, obtaining the value 1 when both clusterings perfectly agree. Likewise, the maximum score of ARI, which is corrected-for-chance version of RI, is achieved when both clusterings perfectly agree. $ARI(\pi_i, \pi^*)$ can be viewed as measure of *agreement* between the consensus clustering π^* and some base clusterings $\pi_i \in \Pi$. We use the mean ARI as the main evaluation criteria:

$$\frac{1}{m} \sum_{i=1}^{m} ARI(\pi_i, \pi^*) \tag{14}$$

We also evaluate π^* based on clustering quality and accuracy. For clustering quality, we use the mean Silhouette Coefficient [22] of all data points (computed using the Euclidean distance between the data points). For clustering accuracy, we compute the ARI between the consensus partition π^* and the true labels.

Benchmark Datasets. We run experiments on seven datasets with differ-ent characteristics: *Iris, Optdigits, Pendigits, Seeds, Wine* from the UCI reposi-tory [5] as well as *Protein* [29] and *MNIST*.[1] *Optdigits-389* is a randomly sampled subset of Optdigits containing only the digits $\{3, 8, 9\}$. Similarly, *MNIST-3689* and *Pendigits-149* are subsets of the MNIST and Pendigits datasets.

---

[1] http://yann.lecun.com/exdb/mnist/.

Table 1 provides statistics on each of the data set, with the coefficient of variation (CV) [4] describing the degree of class imbalance: zero indicates perfectly balanced classes, while higher values indicate higher degree of class imbalance.

Table 1. Datasets

Dataset	# Instances	# Features	# Clusters	CV
Iris	150	4	3	0.000
MNIST-3689	389	784	4	0.015
Optdigits-389	537	64	3	0.021
Pendigits-149	532	16	3	0.059
Protein	116	20	6	0.301
Seeds	210	7	3	0.000
Wine	178	13	3	0.158

4.1 Results

We compare the baseline algorithms to the two Ising models in Sect. 3 solved using the Fujitsu Digital Annealer described in Sect. 3.3.

Clustering is typically an unsupervised task and the number of clusters is unknown. The number of clusters in the true labels, $\widetilde{K}$, is not available in real scenarios. Furthermore, $\widetilde{K}$ is not necessarily the best value for clustering tasks (e.g., in many cases it is better to have smaller clusters that are more pure). We therefore test the algorithms in two configurations: when the number of clusters is set to $\widetilde{K}$, as in the true labels, and when the number of clusters is set to $2\widetilde{K}$.

Table 2. Consensus performance measured by mean ARI across partitions

Dataset	$\widetilde{K}$ clusters					$2\widetilde{K}$ clusters				
	CSPA	NMF	HAC	DA-Sm	DA-Cr	CSPA	NMF	HAC	DA-Sm	DA-Cr
Iris	0.555	**0.618**	**0.618**	**0.619**	**0.621**	0.536	0.614	0.627	0.608	**0.642**
MNIST	0.459	0.449	0.469	**0.474**	**0.474**	0.456	0.511	0.517	0.490	**0.521**
Optdig.	0.528	**0.550**	0.541	**0.550**	**0.551**	0.492	0.596	0.608	0.576	**0.612**
Pendig.	0.546	0.546	0.507	**0.555**	**0.555**	0.531	0.629	**0.642**	0.605	**0.644**
Protein	0.344	0.393	0.379	0.390	**0.405**	0.324	0.419	**0.423**	0.378	0.415
Seeds	0.558	**0.577**	0.534	**0.575**	**0.577**	0.484	0.602	0.602	0.580	**0.612**
Wine	0.481	**0.536**	0.535	**0.537**	**0.538**	0.502	**0.641**	**0.641**	**0.641**	**0.643**
# Best	0	4	1	6	**7**	0	1	3	1	**6**

Consensus Criteria. Table 2 shows the mean ARI between π^* and the clusterings in Π. To avoid bias due to very minor differences, we consider all the methods that achieved Mean ARI that is within a threshold of 0.0025 from the best method to be equivalent and highlight them in bold. We also summarize the number of times each method was considered best across the different datasets.

The results show that DA-Cr is the best performing method for both $\tilde{K}$ and $2\tilde{K}$ clusters. The results of DA-Sm are not consistent: DA-Sm and NMF are performing well for $\tilde{K}$ clusters and HAC is performing better for $2\tilde{K}$ clusters.

Clustering Quality. Table 3 report the mean Silhouette Coefficient of all data points. Again, DA-Cr is the best performing method across datasets, followed by HAC. NMF seems to be equivalent to HAC for $2\tilde{K}$.

Table 3. Clustering quality measured by Silhouette

Dataset	$\tilde{K}$ clusters					$2\tilde{K}$ clusters				
	CSPA	NMF	HAC	DA-Sm	DA-Cr	CSPA	NMF	HAC	DA-Sm	DA-Cr
Iris	0.519	**0.555**	**0.555**	0.551	**0.553**	0.289	0.366	**0.371**	0.343	**0.373**
MNIST	0.075	0.072	**0.078**	**0.079**	**0.078**	0.069	**0.082**	0.074	0.074	**0.082**
Optdig.	0.127	0.120	0.120	**0.130**	**0.130**	0.088	**0.119**	**0.119**	0.112	**0.121**
Pendig.	0.307	0.307	**0.315**	0.310	0.310	0.305	0.332	**0.375**	0.368	0.364
Protein	0.074	**0.106**	0.095	0.094	**0.104**	0.068	0.111	0.115	**0.119**	**0.118**
Seeds	0.461	0.468	0.410	0.469	**0.472**	0.275	**0.343**	0.304	**0.344**	0.302
Wine	0.453	0.542	**0.571**	0.547	0.545	0.452	**0.543**	**0.541**	0.539	**0.542**
# Best	0	2	4	2	**5**	0	4	4	2	**5**

Clustering Accuracy. Table 4 shows the clustering accuracy measured by the ARI between π^* and the true labels. For $\tilde{K}$, we find DA-Sm to be best-performing solution (followed by DA-Cr). For $2\tilde{K}$, DA-Cr outperforms the other methods. Interestingly, there is no clear winner between CSPA, NMF, and HAC.

Experiments with Higher K. In partition difference approaches, increasing K does not necessarily lead to a π^* that has more clusters. Instead, K serves as an upper bound and new clusters will be used in case they reduce the objective.

To demonstrate how different algorithms handle different K values, Table 5 shows the consensus criteria and the actual number of clusters in π^* for different values of K (note that $\tilde{K} = 3$ in Iris). The results show that the performance of the pairwise similarity methods (CSPA, HAC, DA-Sm) degrades as we increase K. This is associated with the fact the actual number of clusters in π^* is equal to K which is significantly higher compared to the clusterings in Π. Methods based on partition difference (NMF and DA-Cr) do not exhibit significant degradation and the actual number of clusters does not grow beyond 5 for DA-Cr and 6 for NMF. Note that the average number of clusters in Π is 5.26.

Table 4. Clustering accuracy measured by ARI compared to true labels

Dataset	$\widetilde{K}$ clusters					$2\widetilde{K}$ clusters				
	CSPA	NMF	HAC	DA-Sm	DA-Cr	CSPA	NMF	HAC	DA-Sm	DA-Cr
Iris	**0.868**	0.746	0.746	0.716	0.730	0.438	0.463	0.447	0.433	**0.521**
MNIST	0.684	0.518	0.704	**0.730**	0.720	0.412	0.484	**0.545**	0.440	0.484
Optdig.	0.712	0.642	0.675	0.734	**0.738**	0.380	0.513	**0.630**	0.481	0.623
Pendig.	0.674	**0.679**	0.499	0.668	0.668	0.398	0.614	0.625	0.490	**0.639**
Protein	0.365	0.298	0.363	0.349	**0.376**	0.237	0.332	0.301	0.308	**0.345**
Seeds	0.705	0.710	0.704	**0.764**	0.717	0.424	0.583	0.573	0.500	**0.619**
Wine	0.324	0.395	0.371	**0.402**	0.398	0.231	0.245	0.240	**0.248**	0.238
# Best	1	1	0	**3**	2	0	0	2	1	4

Table 5. Results for Iris dataset with different number of clusters

K	Consensus Criteria					# of clusters in consensus clustering				
	CSPA	NMF	HAC	DA-Sm	DA-Cr	CSPA	NMF	HAC	DA-Sm	DA-Cr
3	0.555	**0.618**	**0.618**	**0.619**	**0.621**	3	3	3	3	3
6	0.536	0.614	0.627	0.608	**0.642**	6	6	6	6	5
9	0.447	0.614	0.591	0.497	**0.642**	9	6	9	9	5
12	0.370	0.614	0.507	0.414	**0.642**	12	6	12	12	5

5   Conclusion

Motivated by the recent emergence of specialized hardware platforms, we present a new approach to the consensus clustering problem that is based on Ising models and solved on the Fujitsu Digital Annealer, a specialized CMOS hardware. We perform an extensive empirical evaluation and show that our approach outperforms existing methods on a set of seven datasets. These results shows that using specialized hardware in core data mining tasks can be a promising research direction. As future work, we plan to investigate additional problems in data mining that can benefit from the use of specialized optimization hardware as well as experimenting with different types of specialized hardware platforms.

References

1. Aramon, M., Rosenberg, G., Valiante, E., Miyazawa, T., Tamura, H., Katzgraber, H.G.: Physics-inspired optimization for quadratic unconstrained problems using a digital annealer. Front. Phys. **7**, 48 (2019)
2. Bian, Z., Chudak, F., Macready, W.G., Rose, G.: The Ising model: teaching an old problem new tricks. D-Wave Syst. **2** (2010)

3. Coffrin, C., Nagarajan, H., Bent, R.: Evaluating Ising processing units with integer programming. In: Rousseau, L.-M., Stergiou, K. (eds.) CPAIOR 2019. LNCS, vol. 11494, pp. 163–181. Springer, Cham (2019). https://doi.org/10.1007/978-3-030-19212-9_11

4. DeGroot, M.H., Schervish, M.J.: Probability and Statistics. Pearson, London (2012)

5. Dua, D., Graff, C.: UCI machine learning repository (2017). http://archive.ics.uci.edu/ml

6. Filkov, V., Skiena, S.: Integrating microarray data by consensus clustering. Int. J. Artif. Intell. Tools **13**(04), 863–880 (2004)

7. Fred, A.L., Jain, A.K.: Combining multiple clusterings using evidence accumulation. IEEE TPAMI **27**(6), 835–850 (2005)

8. Fujitsu: Digital annealer. https://www.fujitsu.com/jp/digitalannealer/

9. Ghosh, J., Acharya, A.: Cluster ensembles. Wiley Interdisc. Rev.: Data Mining Knowl. Discov. **1**(4), 305–315 (2011)

10. Gionis, A., Mannila, H., Tsaparas, P.: Clustering aggregation. TKDD **1**(1), 4 (2007)

11. Hubert, L., Arabie, P.: Comparing partitions. J. Classif. **2**(1), 193–218 (1985)

12. Karypis, G., Kumar, V.: Multilevelk-way partitioning scheme for irregular graphs. J. Parallel Distrib. Comput. **48**(1), 96–129 (1998)

13. Kirkpatrick, S., Gelatt, C.D., Vecchi, M.P.: Optimization by simulated annealing. Science **220**(4598), 671–680 (1983)

14. Kumar, V., Bass, G., Tomlin, C., Dulny, J.: Quantum annealing for combinatorial clustering. Quantum Inf. Process. **17**(2), 1–14 (2018). https://doi.org/10.1007/s11128-017-1809-2

15. Li, T., Ding, C., Jordan, M.I.: Solving consensus and semi-supervised clustering problems using nonnegative matrix factorization. In: ICDM, pp. 577–582 (2007)

16. Li, T., Ogihara, M., Ma, S.: On combining multiple clusterings: an overview and a new perspective. Appl. Intell. **33**(2), 207–219 (2010)

17. Lucas, A.: Ising formulations of many np problems. Front. Phys. **2**, 5 (2014)

18. Naghsh, Z., Javad-Kalbasi, M., Valaee, S.: Digitally annealed solution for the maximum clique problem with critical application in cellular v2x. In: ICC, pp. 1–7 (2019)

19. Negre, C.F.A., Ushijima-Mwesigwa, H., Mniszewski, S.M.: Detecting multiple communities using quantum annealing on the d-wave system. PLoS ONE **15**, 1–14 (2020)

20. Nguyen, N., Caruana, R.: Consensus clusterings. In: ICDM, pp. 607–612 (2007)

21. Rahman, M.T., Han, S., Tadayon, N., Valaee, S.: Ising model formulation of outlier rejection, with application in WiFi based positioning. In: ICASSP, pp. 4405–4409 (2019)

22. Rousseeuw, P.J.: Silhouettes: a graphical aid to the interpretation and validation of cluster analysis. J. Comput. Appl. Math. **20**, 53–65 (1987)

23. Shaydulin, R., Ushijima-Mwesigwa, H., Safro, I., Mniszewski, S., Alexeev, Y.: Network community detection on small quantum computers. Adv. Quantum Technol. **2**, 1900029 (2019)

24. Strehl, A., Ghosh, J.: Cluster ensembles-a knowledge reuse framework for combining multiple partitions. JMLR **3**(Dec), 583–617 (2002)

25. Topchy, A., Jain, A.K., Punch, W.: Clustering ensembles: models of consensus and weak partitions. IEEE TPAMI **27**(12), 1866–1881 (2005)

26. Ushijima-Mwesigwa, H., Negre, C.F., Mniszewski, S.M.: Graph partitioning using quantum annealing on the d-wave system. In: PMES, pp. 22–29 (2017)

27. Vega-Pons, S., Ruiz-Shulcloper, J.: A survey of clustering ensemble algorithms. IJPRAI **25**(03), 337–372 (2011)
28. Wu, J., Liu, H., Xiong, H., Cao, J., Chen, J.: K-means-based consensus clustering: a unified view. IEEE TKDE **27**(1), 155–169 (2014)
29. Xing, E.P., Jordan, M.I., Russell, S.J., Ng, A.Y.: Distance metric learning with application to clustering with side-information. In: NIPS, pp. 521–528 (2003)

Transfer Learning by Learning Projections from Target to Source

Antoine Cornuéjols[1]([envelope]) [ORCID], Pierre-Alexandre Murena[1,2], and Raphaël Olivier[3]

[1] UMR MIA-Paris, AgroParisTech, INRA, Université Paris-Saclay,
75005 Paris, France
antoine.cornuejols@agroparistech.fr
[2] Telecom ParisTech - Université Paris-Saclay, 75013 Paris, France
[3] University Carnegie-Mellon, Pittsburgh, USA
http://www.agroparistech.fr/mia/equipes:membres:page:antoine

Abstract. Using transfer learning to help in solving a new classification task where labeled data is scarce is becoming popular. Numerous experiments with deep neural networks, where the representation learned on a source task is transferred to learn a target neural network, have shown the benefits of the approach. This paper, similarly, deals with hypothesis transfer learning. However, it presents a new approach where, instead of transferring a representation, the source hypothesis is kept and this is a translation from the target domain to the source domain that is learned. In a way, a change of representation is learned. We show how this method performs very well on a classification of time series task where the space of time series is changed between source and target.

Keywords: Transfer learning · Boosting

1 Introduction

While transfer learning has a long history, dating back at least to the study of analogy reasoning, it has enjoyed a spectacular rise of interest in recent years, thanks largely to its use and effectiveness in learning new tasks with deep neural networks using an architecture learned on a source task. This approach is called Hypothesis Transfer Learning [6]. The justification for this strategy is that, in the absence of enough data in the target domain to learn anew a good hypothesis, it might be effective to transfer the intermediate representations learned on the source task. This is indeed the case, for instance, in face analysis when the source task is to guess the age of the person, and the target task is to recognize the gender. Technically, with neural networks, this amounts to keeping the first layers of the source neural network in the target network and learning only the last layers, the ones that combine intermediate representations of the examples in order to make a prediction.

Let $\mathcal{X}$, $\mathcal{Y}$ and $\mathcal{Z}$ be the input, output and feature spaces respectively. Let F be a class of representation functions, where $f \in F: \mathcal{X} \rightarrow \mathcal{Z}$. Let G be a class

M. R. Berthold et al. (Eds.): IDA 2020, LNCS 12080, pp. 119–131, 2020.
https://doi.org/10.1007/978-3-030-44584-3_10

of decision functions that use descriptions of the examples in the feature space: $g \in G : \mathcal{Z} \to \mathcal{Y}$. Then, in the context of deep neural networks, the hypothesis class is $\mathcal{H} := \{h : \exists f \in F, g \in G \text{ st. } h = g \circ f\}$ and while f is kept (at least approximately) from the source problem to the target one, only g remains to be learned to solve the target problem.

In this paper, we adopt a dual perspective: we propose to keep the decision function g fixed, and learn *translation functions* from the target input space to the source input space, $\pi : \mathcal{X}_T \to \mathcal{X}_S$, such that the target hypothesis space becomes $\mathcal{H}_T := \{h_T : \exists \pi \in \Pi, f \in F, g \in G \text{ st. } h_T = g \circ f \circ \pi\}$, which, given that $h_S = g \circ f$ might be considered as the source hypothesis, may be re-expressed as: $\mathcal{H}_T := \{h_T : \exists \pi \in \Pi, f \in F, g \in G \text{ st. } h_T = h_S \circ \pi\}$.

Indeed, for some problems, it might be much more easy to learn a translation (also called *projections* in this paper) from the target input space $\mathcal{X}_T$ to the source input space $\mathcal{X}_S$ than to learn a new target decision function. Furthermore, this allows one to tackle problems with different input spaces $\mathcal{X}_S$ and $\mathcal{X}_T$.

In the following, Sect. 2 presents `TransBoost` a new algorithm for transfer learning. The theoretical analysis of Sect. 3 provides a PAC-learning bound on the generalization error on the target domain. Controlled experiments are described in Sect. 4 together with an analysis of the results. The new approach is put in perspective in Sect. 5 before we conclude in Sect. 6.

2   A New Algorithm for Transfer Learning

Suppose that we have a system that is able to recognize poppy fields in satellite images. We might imagine that knowing how to translate a biopsy image into a satellite image, we could, using the recognition function defined on satellite image, decide if there is cancerous cells in the biopsy.

Ideally then, one could translate a target query: "what is the label of $\mathbf{x}^T \in \mathcal{X}_T$" into a source query "what is the label of $\pi(\mathbf{x}^T) \in \mathcal{X}_S$" where h_S is the source hypothesis which, applied to $\pi(\mathbf{x}^T) \in \mathcal{X}_S$, provides the answer we are looking for. Notice here that we suppose that $\mathcal{Y}_S = \mathcal{Y}_T$, but not $\mathcal{X}_S = \mathcal{X}_T$.

The goal is then to learn a good translation $\pi : \mathcal{X}_T \to \mathcal{X}_S$. However, defining a proper space of candidate projections Π might be problematic, not to mention the risk of overfitting if the space of functions $h_S \circ \Pi$ has too high a capacity. It might be more easy and manageable to discover "weak projections" from $\mathcal{X}_T$ to $\mathcal{X}_S$ using a boosting learning scheme.

Definition 1. *A **weak projection** w.r.t. source decision function h_S is a function $\pi : \mathcal{X}_T \to \mathcal{X}_S$ such that the decision function $h_S(\pi(\mathbf{x}^T))$ has better than random classification performance on the target training set S_T.*

In this setting, the training set $S_T = \{(\mathbf{x}_i^T, y_i^T)\}_{1 \leq i \leq m}$ is used to learn *weak projections* (Fig. 1).

Once the concept of weak projection is assumed, it is natural to use a boosting algorithm in order to learn a set of such weak projections and to combine them to get a final good classification on elements of $\mathcal{T}$. This is what does the

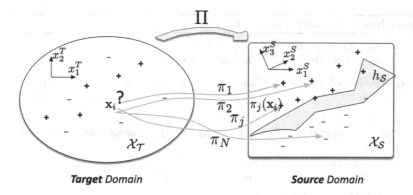

Fig. 1. The principle of prediction using `TransBoost`. A given target example $\mathbf{x}_i^T$ is projected in the source domain using a set of identified weak projections π_j and the prediction for $\mathbf{x}_i^T$ is computed as: $H_T(\mathbf{x}_i^T) = \mathbf{sign}\left\{\sum_{j=1}^{N} \alpha_j h_{\mathcal{S}}\left(\pi_j(\mathbf{x}_i^T)\right)\right\}$.

`TransBoost` algorithm (see Algorithm 1). It does rely on the property of the boosting algorithm to find and combine weak rules to get a strong(er) rule.

3 Theoretical Analysis

Here, we study the question: can we get *guarantees* about the performance of the learned decision function H_T in the target space using `TransBoost`?

We tackle this question in two steps. First, we suppose that we learn a single projection function $\pi \in \Pi : \mathcal{X}_T \to \mathcal{X}_{\mathcal{S}}$ so that $h_T = h_{\mathcal{S}} \circ \pi$, and we find bounds on the generalization error on the target domain given the generalization error on the source domain. Second, we turn to the `TransBoost` algorithm in order to justify the use of a boosting approach.

3.1 Generalization Error Bounds When Using a Single Projection

For this analysis, we suppose the existence of a source input distribution $\mathbf{P}_{\mathcal{X}_{\mathcal{S}}}$ in addition to the target input distribution $\mathbf{P}_{\mathcal{X}_T}$. We consider the binary classification setting $\mathcal{Y} = \{-1, +1\}$, and we note $\bar{h}_{\mathcal{S}}$ and $\bar{h}_T$ respectively the source and the target labelling functions. We note $R_{\mathcal{S}}(h)$ (resp. $R_T(h)$) the risk of a hypothesis h on the source (resp. target) domain: $R_{\mathcal{S}}(h) = \mathbb{E}_{\mathbf{x}_{\mathcal{S}} \sim \mathbf{P}_{\mathcal{X}_{\mathcal{S}}}}[h_{\mathcal{S}}(\mathbf{x}^{\mathcal{S}}) \neq \bar{h}_{\mathcal{S}}(\mathbf{x}^{\mathcal{S}})]$ (resp. $R_T(h) = \mathbb{E}_{\mathbf{x}_T \sim \mathbf{P}_{\mathcal{X}_T}}[h_T(\mathbf{x}^T) \neq \bar{h}_T(\mathbf{x}^T)]$). Let $\widehat{R}_{\mathcal{S}}(h)$ and $\widehat{R}_T(h)$ be the corresponding empirical risks, with $m_{\mathcal{S}}$ training points for $\mathcal{S}$ and m_T training points for T. Let $d_{\mathcal{H}}$ be the VC dimension of the hypothesis space $\mathcal{H}$.

In the following, what is learned is a projection $\pi \in \Pi : \mathcal{X}_T \to \mathcal{X}_{\mathcal{S}}$ in order to get a target hypothesis of the form $h_T = \widehat{h}_{\mathcal{S}} \circ \pi$, where $\widehat{h}_{\mathcal{S}} = \mathrm{ArgMin}_{h \in \mathcal{H}_{\mathcal{S}}} \widehat{R}_{\mathcal{S}}(h)$ is the source hypothesis. Our aim is to upper-bound $R_T(\widehat{h}_T)$, the risk of the learned hypothesis on the target domain in terms of:

Algorithm 1. Transfer learning by boosting

Input: $h_S : \mathcal{X}_S \to \mathcal{Y}_S$ the source hypothesis
$\quad\quad S_T = \{(\mathbf{x}_i^T, y_i^T)\}_{1 \leq i \leq m}$: the target training set

Initialization of the distribution on the training set: $D_1(i) = 1/m$ for
$i = 1, \ldots, m$;

for $n = 1, \ldots, N$ **do**
$\quad$ Find a projection $\pi_i : \mathcal{X}_T \to \mathcal{X}_S$ st. $h_S(\pi_i(\cdot))$ performs better than random
$\quad$ on $D_n(S_T)$;
$\quad$ Let ε_n be the error rate of $h_S(\pi_i(\cdot))$ on $D_n(S_T)$:
$\quad$ $\varepsilon_n \doteq \mathbf{P}_{i \sim D_n}[h_S(\pi_n(\mathbf{x}_i)) \neq y_i]$ (with $\varepsilon_n < 0.5$) ;
$\quad$ Computes $\alpha_i = \frac{1}{2}\log_2\left(\frac{1 - \varepsilon_i}{\varepsilon_i}\right)$;
$\quad$ Update, for $i = 1 \ldots, m$:

$$D_{n+1}(i) = \frac{D_n(i)}{Z_n} \times \begin{cases} e^{-\alpha_n} & \text{if } h_S(\pi_n(\mathbf{x}_i^T)) = y_i^T \\ e^{\alpha_n} & \text{if } h_S(\pi_n(\mathbf{x}_i^T)) \neq y_i^T \end{cases}$$

$$= \frac{D_n(i)\, \exp\left(-\alpha_n\, y_i^{(T)}\, h_S(\pi_n(\mathbf{x}_i^{(T)}))\right)}{Z_n}$$

$\quad$ where Z_n is a normalization factor chosen so that D_{n+1} be a distribution on
$\quad$ S_T ;
end

Output: the final target hypothesis $H_T : \mathcal{X}_T \to \mathcal{Y}_T$:

$$H_T(\mathbf{x}^T) = \text{sign}\left\{\sum_{n=1}^{N} \alpha_n\, h_S(\pi_n(\mathbf{x}^T))\right\} \tag{1}$$

- the empirical risk $\widehat{R}_S(\widehat{h}_S)$ of the source hypothesis,
- the generalization error of a hypothesis $\widehat{h}_S$ in $\mathcal{H}_S$ learned from m_S examples, which depends on $d_{\mathcal{H}_S}$,
- the generalization error of a hypothesis $\widehat{h}_T = h_S \circ \pi$ in $\mathcal{H}_T$ learned from m_T examples, which depends on $d_{\mathcal{H}_T} = d_{h_S \circ \pi}$,
- a term that expresses the "proximity" between the source and the target problems.

For the latter term, we adapt the theoretical study of McNamara and Balcan [9] on the transfer of representation in deep neural networks. We suppose that P_S, P_T, h_S, $h_T = \widehat{h}_S \circ \pi$ ($\pi \in \Pi$), $\widehat{h}_S$ and Π have the property:

$$\forall\, \widehat{h}_S \in \mathcal{H}_S : \quad \underset{\pi \in \Pi}{\text{Min}}\, R_T(\widehat{h}_S \circ \pi) \leq \omega(R_S(h_S)) \tag{2}$$

where $\omega : \mathbb{R} \to \mathbb{R}$ is a non-decreasing function.

Equation (2) means that the best target hypothesis expressed using the learned source hypothesis has a true risk bounded by a non-decreasing function of the true risk on the source domain of the learned source hypothesis.

We are now in position to get the desired theorem.

Theorem 1. *Let* $\omega : \mathbb{R} \to \mathbb{R}$ *be a non-decreasing function. Suppose that* P_S, P_T, h_S, $h_T = \hat{h}_S \circ \pi(\pi \in \Pi)$, $\hat{h}_S$ *and* Π *have the property given by Eq. (2). Let* $\hat{\pi} := ArgMin_{\pi \in \Pi} \hat{R}_T(\hat{h}_S \circ \pi)$, *be the best apparent projection.*

Then, with probability at least $1 - \delta$ ($\delta \in (0,1)$) *over pairs of training sets for tasks* S *and* T:

$$R_T(\hat{h}_T) \leq \omega(\hat{R}_S(\hat{h}_S)) + 2\sqrt{\frac{2\,d_{\mathcal{H}_S}\log(2em_S/d_{\mathcal{H}_S}) + 2\log(8/\delta)}{m_S}}$$
$$+ 4\sqrt{\frac{2\,d_{h_{S\circ\Pi}}\log(2em_T/d_{h_{S\circ\Pi}}) + 2\log(8/\delta)}{m_T}} \tag{3}$$

Proof. Let $\pi^* = ArgMin_{\pi \in \Pi} R_T(h_S \circ \pi)$. With probability at least $1 - \delta$:

$$R_T(h_S \circ \hat{\pi}) \leq \hat{R}_T(h_S \circ \hat{\pi}) + 2\sqrt{\frac{2\,d_{h_{S\circ\Pi}}\log(2em_T/d_{h_{S\circ\Pi}}) + 2\log(8/\delta)}{m_T}}$$
$$\leq \hat{R}_T(h_S \circ \pi^*) + 2\sqrt{\frac{2\,d_{h_{S\circ\Pi}}\log(2em_T/d_{h_{S\circ\Pi}}) + 2\log(8/\delta)}{m_T}}$$
$$\leq R_T(h_S \circ \pi^*) + 4\sqrt{\frac{2\,d_{h_{S\circ\Pi}}\log(2em_T/d_{h_{S\circ\Pi}}) + 2\log(8/\delta)}{m_T}}$$
$$\leq \omega(R_S(\hat{h}_S)) + 4\sqrt{\frac{2\,d_{h_{S\circ\Pi}}\log(2em_T/d_{h_{S\circ\Pi}}) + 2\log(8/\delta)}{m_T}}$$
$$\leq \omega(\hat{R}_S(\hat{h}_S)) + 2\sqrt{\frac{2\,d_{\mathcal{H}_S}\log(2em_S/d_{\mathcal{H}_S}) + 2\log(8/\delta)}{m_S}}$$
$$+ 4\sqrt{\frac{2\,d_{h_{S\circ\Pi}}\log(2em_T/d_{h_{S\circ\Pi}}) + 2\log(8/\delta)}{m_T}}$$

This follows from the fact that [10] (p. 48) using m training points and a hypothesis class of VC dimension d, with probability at least $1 - \delta$, for all hypotheses h simultaneously, the true risk $R(h)$ and empirical risk $\hat{R}(h)$ satisfy $|(R(h) - \hat{R}(h)| \leq 2\sqrt{\frac{2\,d\log(2em/d)+2\log(4/\delta)}{m}}$. For $h_S \circ \Pi$, this yields the first and third inequalities with probabilities at least $1 - \delta/2$. For $\mathcal{H}_S$, this yields the fifth inequality with probability at least $1 - \delta/2$. Applying the union bound archives the desired results. The second inequality follows from the definition of $\hat{\pi}$, and the fourth inequality is where we inject our assumption about the transferability (or proximity) between the source and the target problem. $\qquad\square$

We can thus control the generalization error on the transfer domain by controlling $d_{h_{S\circ\Pi}}$, m_S and ω which measures the link between the domain and the target domain. The number of target training data m_T is typically supposed to be small in transfer learning and thus cannot be employed to control the error.

3.2 Boosting Projections from Target to Source

The above analysis bounds the generalization error of the learned target hypothesis $h_S \circ \widehat{\pi}$ in terms, among others, of the VC dimension of the space $h_S \circ \Pi$. The problem of controlling the capacity of such a space of functions in order to prevent under or over-fitting is the same as in the traditional supervised learning setting. The difficulty lies in choosing the right space Π of projection functions from $\mathcal{X}_T$ to $\mathcal{X}_S$.

The space of hypothesis functions considered is:

$$L(h_S \circ \Pi_B) := \left\{ \mathbf{x} \mapsto \text{sign}\left[\sum_{n=1}^{N} \alpha_n \left(h_S \circ \pi_n(\mathbf{x}^T) \right) \right] : \forall n, \alpha_n \in \mathbb{R}, \text{ and } \pi_n \in \Pi_B \right\}$$

where Π_B is a space of weak projections satisfying definition (1).

Now, from [11] (p. 109), the VC dimension of the space $h_S \circ \Pi_B$ satisfies:

$$d_{L(h_S \circ \Pi_B)} \leq N(d_{h_S \circ \Pi_B} + 1)(3\log(N(d_{h_S \circ \Pi_B} + 1)) + 2)$$

If $d_{h_S \circ \Pi_B} \ll d_{h_S \circ \Pi}$, then $d_{L(h_S \circ \Pi_B)}$ can also be much less than $d_{h_S \circ \Pi}$, and theorem (1) provides tighter bounds.

Using the `TransBoost` method, we can thus gain both on the theoretical bounds on the generalization error and on the ease of finding an appropriate space of projections $\mathcal{X}_T \to \mathcal{X}_S$.

4 Design of the Experiments

4.1 The Main Dimensions of Experiments in Transfer Learning

There are two dimensions that can be expected to govern the efficiency of transfer learning:

1. The *level of signal* in the target data.
2. The *relatedness* between the source and the target domains.

Regarding the *first dimension*, one can expect that if there is no signal in the target data (i.e. the examples are labelled randomly), then no regularity can be extracted, directly or using transfer. In fact, only overfitting of the training data can potentially occur. If, on the contrary, the target learning task is easy, then there cannot be much advantage in using transfer learning. A question therefore arises as to whether there might be an optimal level of signal in the target data so as to maximally benefit from transfer learning.

The *second dimension* is tricky. Here, we intuitively expect that the closer the source and target domains (and problems), the more profitable transfer learning should be. However, how should we measure the "relatedness" of the source and target problems? In the domain adaptation setting, closeness can be measured through a measure of the divergence between the source distribution and the target one, since they are defined on the same input space. In transfer learning, the input spaces can be different, so that it is much more difficult to define a divergence between distributions. This is why we resorted to the function ω in our theoretical analysis. In our experiments, we control relatedness through the information shared between source and target (see below).

4.2   Experimental Setup

In our study, we devised an experimental setup that would allow us to control the two dimensions above.

In the **target domain**, the learning task is to classify time series of length t_T into two classes: $h_T : \mathbb{R}^{t_T} \to \{-1, +1\}$. By controlling the level of noise and the difference between the distributions governing the two classes, we can control the signal level, that is the difficulty of extracting information from the target training data. We control the amount of information by varying the size m_T of the target training set.

Likewise, the source input space is the space of sequences of real measurements of length t_S. Therefore, we have $h_S : \mathbb{R}^{t_S} \to \{-1, +1\}$.

Varying $|t_S - t_T|$ is a way of controlling the information potentially shared in the two domains. With $t_S = t_T$, the two input domains are the same.

Note that learning to classify times series is not a trivial task. It has many applications, some of them involving to classify time series of length different from the length for which exists a classifier.

4.3   Description of the Experiments

Time series were generated according to the following equation:

$$x_t = \underbrace{t \times \text{slope} \times \text{class}}_{\text{information gain}} + \underbrace{x_{max} \sin(\omega_i \times t + \varphi_j)}_{\text{sub shape within class}} + \underbrace{\eta(t)}_{\text{noise factor}} \tag{4}$$

The fact that the noise factor is generated according to a Gaussian distribution induces a distribution over the data (class $\in \{-1, +1\}$).

The **level of signal in the training data** is governed by:

1. the *slope factor*: the higher the value of the *slope* factor, the easier the discrimination between the two classes at each additional time step
2. the *number of different shapes* in each class of sequences, each shape controlled by ω_i and ϕ_j, and the importance of this factor in the equation being weighted by x_{max}
3. the *noise factor* $\eta(t)$
4. the *length* of the time series, that is the number of measurements
5. the *size* of the training set

In our experiments, the noise factor is generated according to a Gaussian distribution of mean = 0 and standard deviation in $\{0.001, 0.002, 0.02, 0.2, 1\}$.

Figure 2 illustrates what can be obtained with slope = 0.01 with 3 subclasses in the +1 class, and 2 subclasses in the −1 class.

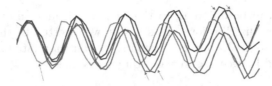

Fig. 2. A synthetic data set $\mathcal{S}$ with 5 times series where η is Gaussian ($\mu = 0, \sigma = 0.2$).

In the experiments reported here, we kept the size of the training set constant. In each experiment, 900 times series of length 200 were generated according to the equation described above: 450 times series in each class -1 or $+1$. We varied the difficulty of learning by varying the slope from almost non existent: 0.001 to significant: 0.01. Similarly, we varied the length t_T of the target training set in $\{20, 50, 70, 100\}$ thus providing increasing levels of signal.

A *target training data set* of 300 time series was drawn equally balanced between the two classes. Note that this relatively small number corresponds to transfer learning scenarios where the training data is limited in the target domain. The remaining 600 time series were used as a *test set*. The source hypothesis was learned using the complete time series generated as explained above.

In these experiments, the *set of projections* Π was chosen as a set of "hinge functions", defined by three parameters, the slope of the first linear part, the time t where the hinge takes place, and the slope of the second linear part. The set is explored randomly by the algorithm and a projection is retained if its error rate on the current weighted data is lower than 0.45. We explored other, richer, spaces of projections without gaining superior performances. This simple set seems to be sufficient for this learning task.

In order to better assess the value of `TransBoost`, its performance was compared (1) to a classifier (Gaussian SVM as implemented in Scikit Learn) acting directly on the target training data, (2) to a boosting algorithm operating in the target domain with base classifiers being Gaussian SVMs, and (3) to a baseline transfer learning method that consists in finding a regression from the target input space to the source input space using a SVR regression. In this last method the regression acts as a translation from $\mathcal{X}_T$ to $\mathcal{X}_S$ and the class of an example $\mathbf{x}^T$ is given by $h_S(\text{regression}(\mathbf{x}^T))$.

Table 1 provides representative examples of the results obtained. Each cell of the table shows the average performance (and the standard deviations) computed from 100 experiments repeated under the same conditions. The experimental conditions are organized according to the level of signal in the training data. In the experiments corresponding to this table, the source hypotheses were learned according to the first protocol defined above.

Several lessons can be drawn. First of all, in most situations, `TransBoost` brings *very significant gains* over learning without transfer or using transfer learning with regression. Figures 3 and 4 that sum up a larger set of experimental

Table 1. Comparison of the error rate (lower is better) between: learning directly in the target domain (columns h_T (train) and h_T (test)), using `TransBoost` (columns H_T (train) and H_T (test)), learning in the source domain (column h_S (test)) and, finally, mapping the time series with a SVR regression and using h_S (naïve transfer, column $H'_T(test)$). Test errors are highlighted in the orange columns. Bold numbers indicate where `TransBoost` significantly dominates both learning without transfer and learning with naïve transfer.

slope, noise, t_T	h_T (train)	h_T (test)	H_T (train)	H_T (test)	h_S (test)	H'_T (test)
0.001, 0.001, 20	0.46 ± 0.02	0.50 ± 0.08	0.08 ± 0.03	**0.08 ± 0.02**	0.05	0.49 ± 0.01
0.005, 0.001, 20	0.46 ± 0.02	0.49 ± 0.01	0.01 ± 0.01	**0.01 ± 0.01**	0.01	0.45 ± 0.01
0.005, 0.002, 20	0.46 ± 0.02	0.49 ± 0.03	0.03 ± 0.02	**0.04 ± 0.02**	0.02	0.43 ± 0.01
0.005, 0.02, 20	0.44 ± 0.02	0.48 ± 0.03	0.09 ± 0.01	**0.10 ± 0.01**	0.01	0.47 ± 0.01
0.001, 0.2, 20	0.46 ± 0.02	0.50 ± 0.01	0.46 ± 0.02	0.51 ± 0.02	0.11	0.49 ± 0.01
0.01, 0.2, 20	0.42 ± 0.03	0.47 ± 0.03	0.34 ± 0.02	0.35 ± 0.02	0.02	0.35 ± 0.01
0.001, 0.001, 50	0.46 ± 0.02	0.50 ± 0.01	0.08 ± 0.03	**0.08 ± 0.02**	0.06	0.41 ± 0.01
0.005, 0.001, 50	0.25 ± 0.07	0.28 ± 0.09	0.01 ± 0.01	**0.01 ± 0.01**	0.01	0.28 ± 0.01
0.005, 0.002, 50	0.27 ± 0.07	0.30 ± 0.08	0.02 ± 0.01	**0.02 ± 0.01**	0.02	0.28 ± 0.01
0.005, 0.02, 50	0.26 ± 0.07	0.30 ± 0.08	0.04 ± 0.01	**0.04 ± 0.01**	0.01	0.31 ± 0.01
0.001, 0.2, 50	0.44 ± 0.02	0.50 ± 0.01	0.38 ± 0.03	0.44 ± 0.02	0.15	0.43 ± 0.01
0.01, 0.2, 50	0.10 ± 0.03	0.12 ± 0.04	0.10 ± 0.02	0.11 ± 0.02	0.03	0.15 ± 0.02
0.001, 0.001, 100	0.43 ± 0.03	0.47 ± 0.03	0.07 ± 0.02	**0.07 ± 0.02**	0.02	0.23 ± 0.01
0.005, 0.001, 100	0.06 ± 0.03	0.07 ± 0.03	0.01 ± 0.01	**0.01 ± 0.01**	0.01	0.07 ± 0.02
0.005, 0.002, 100	0.08 ± 0.03	0.10 ± 0.04	0.02 ± 0.01	**0.02 ± 0.01**	0.02	0.07 ± 0.01
0.005, 0.02, 100	0.08 ± 0.03	0.09 ± 0.03	0.02 ± 0.01	**0.03 ± 0.01**	0.01	0.07 ± 0.01
0.001, 0.2, 100	0.04 ± 0.03	0.46 ± 0.02	0.28 ± 0.02	0.31 ± 0.01	0.16	0.31 ± 0.01
0.01, 0.2, 100	0.03 ± 0.01	0.05 ± 0.02	0.04 ± 0.01	0.05 ± 0.01	0.02	0.05 ± 0.01

conditions make this even more striking. In both tables, the x-axis reports the error rate obtained using `TransBoost`, while the y-axis reports the error rate of the competing algorithm: either the hypothesis h_T learnt on the target training data alone (Fig. 3), or the hypothesis H'_T learned on the target data projected on the source input space using a SVR regression (Fig. 4). The remarkable efficiency of `TransBoost` in a large spectrum of situations is readily apparent.

Secondly, as expected, `Transboost` is less dominant when either the data is so noisy that no method can learn from the data (high level of noise or low slope): this is apparent on the right part of the graphs 3 and 4 (near the diagonal), or when the task is so easy (large slope and/or low noise) that nothing can be gained from transfer learning (left part of the two graphs).

We did not report here the results obtained with boosting directly in the target input space $\mathcal{X}_T$ since the learning performance was almost the same as the performance as the one of the SVM classifier. This shows that this is not boosting in itself that brings a gain.

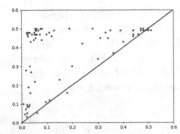

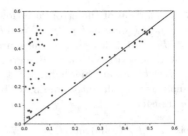

Fig. 3. Comparison of error rates. y-axis: test error of the SVM classifier (without transfer). x-axis: test error of the TransBoost classifier with 10 boosting steps. The results of 75 experiments (each one repeated 100 times) are summed up in this graph.

Fig. 4. Comparison of error rates. y-axis: test error of the "naïve" transfer method. x-axis: test error of the Trans-Boost classifier with 10 boosting steps. The results of 75 experiments (each one repeated 100 times) are summed up in this graph.

4.4   Additional Experiments

We show here, in Figs. 5, 6 and 7 qualitative results obtained on the classical half-moon problem. It is apparent that `Transboost` brings satisfying results.

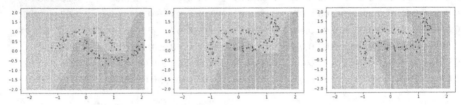

(a) kNN source model trained on the data source : it fits to the data source

(b) kNN source model trained on the data source : it does not fit to the data target

(c) kNN source model trained on the data source transBoosted to the data target

Fig. 5. Experiments on the half-moon problem.

5   Comparison to Previous Works

In the theoretical analysis of Ben-David *et al.* [1,2], one central idea is that a *common representation space* should be found in which the projections of the source data $\{(\mathbf{x}_i^S)\}_{1 \leq i \leq m}$ and of the target data $\{(\mathbf{x}_i^T)\}_{1 \leq i \leq m}$ should be as undistinguishable as possible using discriminative functions from the hypothesis space $\mathcal{H}$. The intuition is that if the domains become indistinguishable, a classifier constructed for the source domain should work also for the target domain. It has been at the core of many proposed methods so far [3,5,7,12].

In [8] a scenario in which multiple sources are available for a single target domain is studied. For each source $i \in \{1, \ldots, k\}$, the input distribution D_i is

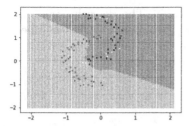
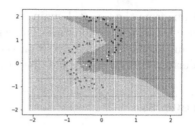

Fig. 6. A KNN model trained on the few target data points (in yellow). (Color figure online)

Fig. 7. A KNN model transboosted on the few target data points.

known as well as a hypothesis h_i with loss bounded by ε on D_i. It is further assumed that the target input distribution is a mixture of the k source distributions D_i. The adaptation problem is thus seen as finding a combination of the hypotheses h_i. It is shown that guarantees on the loss of the combined target hypothesis can be given for some forms of combinations. However, the authors do not show how to learn the parameters of these combinations. In [4], the authors present a system called `TrAdaboost`, which uses a boosting scheme to eliminate data points that seem irrelevant for the new task defined over the same space $\mathcal{X}$. Despite the use of boosting, the scope is quite different from ours.

Finally, the authors in [6] study a scheme seemingly very close to ours. They define *Hypothesis Transfer Learning algorithms* as algorithms taking as input a training set in the target domain and a source hypothesis in the source domain, and producing a target hypothesis:

$$A^{\mathrm{htl}} : (\mathcal{X}_T \times \mathcal{Y}_T)^m \times \mathcal{H}_S \rightarrow \mathcal{H}_T \subseteq \mathcal{Y}^{\mathcal{X}}$$

One goal of the paper is to identify the effect of the source hypothesis on the generalization properties of A^{htl}. However, the scope of the analysis is limited in several ways. First, it focusses on linear regression with the Regularized Least Square algorithm. Second, the formal framework necessitates that in fact $\mathcal{X}_T = \mathcal{X}_S$ and $\mathcal{Y}_T = \mathcal{Y}_S$. It is thus more an analysis of domain adaptation than of transfer learning. Third, the transfer learning algorithm in effect tries to find a weight vector $\mathbf{w}^T$ as close as possible to the source weight vector $\mathbf{w}^S$ while fitting the target data set. There is therefore a parameter λ to set. More importantly, the consequence is that the analysis singles out the performance of the source hypothesis on the target domain as the most significant factor controlling the expected error on the target problem. Again, therefore, the target hypothesis cannot be much different from the source one, which seems to defeat the whole purpose of transfer learning.

6   Conclusion

This paper has presented a new transfer learning algorithm, `TransBoost`, that uses the boosting mechanism in an original way by selecting and combining weak projections from the target domain to the source domain. The algorithm inherits some nice features from boosting. There is only one parameter to set: the number of boosting steps, and guarantees on the training error an on the test error are easily derived from the ones obtained in the theory of boosting.

References

1. Ben-David, S., Blitzer, J., Crammer, K., Kulesza, A., Pereira, F., Vaughan, J.W.: A theory of learning from different domains. Mach. Learn. **79**(1), 151–175 (2010). https://doi.org/10.1007/s10994-009-5152-4
2. Ben-David, S., Blitzer, J., Crammer, K., Pereira, F., et al.: Analysis of representations for domain adaptation. In: Advances in Neural Information Processing Systems, vol. 19, p. 137 (2007)
3. Bickel, S., Brückner, M., Scheffer, T.: Discriminative learning for differing training and test distributions. In: Proceedings of the 24th International Conference on Machine Learning, pp. 81–88. ACM (2007)
4. Dai, W., Yang, Q., Xue, G.R., Yu, Y.: Boosting for transfer learning. In: Proceedings of the 24th International Conference on Machine Learning, pp. 193–200. ACM (2007)
5. Jiang, J., Zhai, C.: Instance weighting for domain adaptation in NLP. In: ACL, vol. 7, pp. 264–271 (2007)
6. Kuzborskij, I., Orabona, F.: Stability and hypothesis transfer learning. In: ICML (3), pp. 942–950 (2013)
7. Mansour, Y., Mohri, M., Rostamizadeh, A.: Domain adaptation: learning bounds and algorithms. arXiv preprint arXiv:0902.3430 (2009)
8. Mansour, Y., Mohri, M., Rostamizadeh, A.: Domain adaptation with multiple sources. In: Advances in Neural Information Processing Systems, pp. 1041–1048 (2009)
9. McNamara, D., Balcan, M.F.: Risk bounds for transferring representations with and without fine-tuning. In: International Conference on Machine Learning, pp. 2373–2381 (2017)
10. Mohri, M., Rostamizadeh, A., Talwalkar, A.: Foundations of Machine Learning. MIT Press, Cambridge (2012)
11. Shalev-Shwartz, S., Ben-David, S.: Understanding Machine Learning: From Theory to Algorithms. Cambridge University Press, Cambridge (2014)
12. Sugiyama, M., Nakajima, S., Kashima, H., Buenau, P.V., Kawanabe, M.: Direct importance estimation with model selection and its application to covariate shift adaptation. In: Advances in Neural Information Processing Systems, pp. 1433–1440 (2008)

Computing Vertex-Vertex Dissimilarities Using Random Trees: Application to Clustering in Graphs

Kevin Dalleau[✉], Miguel Couceiro, and Malika Smail-Tabbone

Universite de Lorraine, CNRS, Inria, LORIA, 54000 Nancy, France
{kevin.dalleau,miguel.couceiro,malika.smail}@loria.fr

Abstract. A current challenge in graph clustering is to tackle the issue of complex networks, *i.e*, graphs with attributed vertices and/or edges. In this paper, we present GraphTrees, a novel method that relies on random decision trees to compute pairwise dissimilarities between vertices in a graph. We show that using different types of trees, it is possible to extend this framework to graphs where the vertices have attributes. While many existing methods that tackle the problem of clustering vertices in an attributed graph are limited to categorical attributes, GraphTrees can handle heterogeneous types of vertex attributes. Moreover, unlike other approaches, the attributes do not need to be preprocessed. We also show that our approach is competitive with well-known methods in the case of non-attributed graphs in terms of quality of clustering, and provides promising results in the case of vertex-attributed graphs. By extending the use of an already well established approach – the random trees – to graphs, our proposed approach opens new research directions, by leveraging decades of research on this topic.

Keywords: Graph clustering · Attributed graph · Random tree · Dissimilarity · Heterogeneous data

1 Introduction

Identifying community structure in graphs is a challenging task in many applications: computer networks, social networks, etc. Graphs have an expressive power that enables an efficient representation of relations between objects as well as their properties. Attributed graphs where vertices or edges are endowed with a set of attributes are now widely available, many of them being created and curated by the semantic web community. While these so-called knowledge graphs[1] contain a lot of information, their exploration can be challenging in practice. In particular, common approaches to find communities in such graphs rely on rather complex transformations of the input graph.

---

[1] Although many definitions can be found in the literature [9].

Funded by the RHU FIGHT-HF (ANR-15-RHUS-0004) and the Region Grand Est (France).

M. R. Berthold et al. (Eds.): IDA 2020, LNCS 12080, pp. 132–144, 2020.
https://doi.org/10.1007/978-3-030-44584-3_11

In this paper, we propose a decision tree based method that we call Graph-Trees (GT) to compute dissimilarities between vertices in a straightforward manner. The paper is organized as follows. In Sect. 2, we briefly survey related work. We present our method in Sect. 3, and we discuss its performance in Sect. 4 through an empirical study on real and synthetic datasets. In the last section of the paper, we present a brief discussion of our results and state some perspectives for future research.

Main Contributions of the Paper:

1. We propose a first step to bridge the gap between random decision trees and graph clustering and extend it to vertex attributed graphs (Subsect. 4.1).
2. We show that the vertex-vertex dissimilarity is meaningful and can be used for clustering in graphs (Subsect. 4.2).
3. Our method GT applies directly on the input graph without any preprocessing, unlike the many community detection in vertex-attributed graphs that rely on the transformation of the input graph.

2   Related Work

Community detection aims to find highly connected groups of vertices in a graph. Numerous methods have been proposed to tackle this problem [1,8,24]. In the case of vertex-attributed[2] graph, clustering aims at finding homogeneous groups of vertices sharing (i) common neighbourhoods and structural properties, and (ii) common attributes. A *vertex-attributed graph* is thought of as a finite structure $G = (V, E, A)$, where

- $V = \{v_1, v_2, \ldots, v_n\}$ is the set of *vertices* of G,
- $E \subseteq V \times V$ is the set of edges between the vertices of V, and
- $A = \{x_1, x_2, \ldots, x_n\}$ is the set of feature tuples, where each x_i represents the attribute value of the vertex v_i.

In the case of vertex-attributed graphs, the problem of clustering refers to finding communities (*i.e.*, clusters), where vertices in the same cluster are densely connected, whereas vertices that do not belong to the same cluster are sparsely connected. Moreover, as attributes are also taken into account, the vertices in the same cluster should be similar w.r.t. attributes.

In this section, we briefly recall existing approaches to tackle this problem.

Weight-Based Approaches. The weight-based approach consists in transforming the attributed graphs in weighted graphs. Standard clustering algorithms that focus on structural properties can then be applied.

The problem of mapping attribute information into edge weight have been considered by several authors. Neville *et al.* define a matching coefficient [20] as

---

[2] To avoid terminology-related issues, we will exclusively use the terms vertex for graphs and node for random trees throughout the paper.

a similarity measure S between two vertices v_i and v_j based on the number of attribute values the two vertices have in common. The value S_{v_i,v_j} is used as the edges weight between v_i and v_j. Although this approach leads to good results using Min-Cut [15], MajorClust [26] and spectral clustering [25], only nominal attributes can be handled. An extended matching coefficient was proposed in [27] to overcome this limitation, based on a combination of normalized dissimilarities between continuous attributes and increments of the resulting weight per pair of common categorical attributes.

Optimization of Quality Functions. A second type of methods aim at finding an optimal clustering of the vertices by optimizing a quality function over the partitions (clusters).

A commonly used quality function is *modularity* [21], that measures the density differences between vertices within the same cluster and vertices in different clusters. However, modularity is only based on the structural properties of the graph. In [6], the authors use entropy as the quality metric to optimize between attributes, combined with a modularity-based optimization. Another method, recently proposed by Combe *et al.* [5], groups similar vertices by maximizing both modularity and *inertia*.

However, these methods suffer from the same drawbacks as any other modularity optimization based methods in simple graphs. Indeed, it was shown by [17] that these methods are biased, and do not always lead to the best clustering. For instance, such methods fail to detect small clusters in graphs with clusters of different sizes.

Aggregated Distance Measures. Another type of methods used to find communities in vertex-attributed graphs is to define an aggregated vertex-vertex distance between the topological distance and the symbolic distance. All these methods express a distance d_{v_i,v_j} between two vertices v_i and v_j as $d_{v_i,v_j} = \alpha d_T(v_i, v_j) + (1 - \alpha)d_S(v_i, v_j)$ where d_T is a structural distance and d_S is a distance in the attribute space. These structural and attribute distances represent the two different aspects of the data. These distances can be chosen from the vast number of available ones in the literature. For instance, in [4] a combination of geodesic distance and cosine similarities are used by the authors. The parameter α is useful to control the importance of each aspect of the overall similarity in each use case. These methods are appealing because once the distances between vertices are obtained, many clustering algorithms that cannot be applied to structures such as graphs can be used to find communities.

Miscellaneous. There is yet another family of methods that enable the use of common clustering methods on attributed graphs. SA-cluster [3,32] is a method performing the clustering task by adding new vertices. The virtual vertices represent possible values of the attributes. This approach, although appealing by its simplicity, has some drawbacks. First, continuous attributes cannot be taken into account. Second, the complexity can increase rapidly as the number of added vertices depends on the number of attributes and values for each attribute. However, the authors proposed an improvement of their method named *Inc-Cluster* in [33], where they reduce its complexity.

Some authors have worked on model-based approaches for clustering in vertex-attributed settings. In [29], the authors proposed a method based on a bayesian probabilistic model that is used to perform the clustering of vertex-attributed graphs, by transforming the clustering problem into a probabilistic inference problem. Also, graph embeddings can be used for this task of vertex-attributed graph clustering. Examples of these techniques include node2vec [13] or deepwalk [23], and aim to efficiently learn a low dimensional vector representation of each vertex. Some authors focused on extending vertex embeddings to vertex-attributed networks [11, 14, 30].

In this paper, we take a different approach and present a tree-based method enabling the computation of vertex-vertex dissimilarities. This method is presented in the next section.

3 Method

Previous works [7, 28] have shown that random partitions of data can be used to compute a similarity between the instances. In particular, in Unsupervised Extremely Randomized Trees (UET), the idea is that all instances ending up in the same leaves are more similar to each other than to other instances. The pairwise similarities $s(i, j)$ are obtained by increasing $s(i, j)$ for each leaf where both i and j appear. A normalisation is finally performed when all trees have been constructed, so that values lie in the interval $[0, 1]$. Leaves, and, more generally, nodes of the trees can be viewed as partitions of the original space. Enumerating the number of co-occurrences in the leaves is then the same as enumerating the number of co-occurrence of instances in the smallest regions of a specific partition.

So far, this type of approach has not been applied to graphs. The intuition behind our proposed method, GT, is to leverage a similar partition in the vertices of a graph. Instead of using the similarity computation that we described previously, we chose to use the mass-based approach introduced by Ting *et al.* [28] instead. The key property of their measure is that the dissimilarity between two instances in a dense region is higher than the same interpoint dissimilarity between two instances in a sparse region of the same space. One of the interesting aspects of this approach is that a dissimilarity is obtained without any post-processing.

Let $H \in \mathcal{H}(D)$ be a hierarchical partitioning of the original space of a dataset D into non-overlapping and non-empty regions, and let $R(x, y|H)$ be the smallest local region covering x and y with respect to H. The mass-based dissimilarity m_e estimated by a finite number t of models – here, random trees – is given by the following equation:

$$m_e(x, y|D) = \frac{1}{t} \sum_{i=1}^{t} \tilde{P}(R(x, y|H_i)) \tag{1}$$

where $\tilde{P}(R) = \frac{1}{|D|} \sum_{z \in D} \mathbb{1}(z \in R)$. Figure 1 presents an example of a hierarchical partition H of a dataset D containing 8 instances. These instances are vertices in our case. For the sake of the example, let us compute $m_e(1, 4)$ and $m_e(1, 8)$. We have $m_e(1, 4) = \frac{1}{8}(2) = 0.25$, as the smallest region where instances 1 and 4 co-appear contains 2 instances. However, $m_e(1, 8) = \frac{1}{8}(8) = 1$, since instances 1 and 8 only appear in one region of size 8, the original space. The same approach can be applied to graphs.

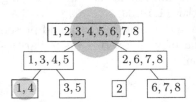

Fig. 1. Example of partitioning of 8 instances in non-overlapping non-empty regions using a random tree structure. The blue and red circles denote the smallest nodes (*i.e.*, regions) containing vertices 1 and 4 and vertices 1 and 8, respectively. (Color figure online)

Our proposed method is based on two steps: (i) obtain several partitions of the vertices using random trees, (ii) use the trees to obtain a relevant dissimilarity measure between the vertices. The Algorithm 1 describes how to build one tree, describing one possible partition of the vertices. Each tree corresponds to a model of (1). Finally, the dissimilarity can be obtained using Eq. 1.

The computation of pairwise vertex-vertex dissimilarities using Graph Trees and the mass-based dissimilarity we just described has a time complexity of $O(t \cdot \Psi log(\Psi) + n^2 t log(\Psi))$ [28], where t is the number of trees, Ψ the maximum height of the trees, and n is the number of vertices. When $\Psi << n$, this time complexity becomes $O(n^2)$.

To extend this approach to vertex-attributed graphs, we propose to build a forest containing trees obtained by GT over the vertices and trees obtained by UET on the vertex attributes. We can then compute the dissimilarity between vertices by averaging the dissimilarities obtained by both types of trees.

In the next section, we evaluate GT on both real-world and synthetic datasets.

4   Evaluation

This section is divided into 2 subsections. First, we assess GT's performance on graphs without vertex attributes (Subsect. 4.1). Then we present

---

Algorithm 1. Algorithm describing how to build a random tree partitioning the vertices of a graph.

Data: A graph $G(V, E)$, an uninitialized stack S
root_node $= V$;   // *The root node contains all the vertices of G*
$v_s =$ a vertex sampled without replacement from V;
$V_{left} = \mathcal{N}(v_s) \cup \{v_s\}$;   //$\mathcal{N}(v)$ *returns the set of neighbours of v*
$V_{right} = V \setminus V_{left}$;
Push V_{left} and V_{right} to S ;
leaves $= []$; //*leaves is an empty list*
while S *is not empty* **do**
$\quad V_{node} =$ pop the last element of S;
$\quad$ **if** $|V_{node}| < n_{min}$ **then**
$\quad\quad$ Append V_{node} to *leaves*;   //*node size in lower than n_{min}, it is a leaf node*
$\quad$ **end**
$\quad$ **else**
$\quad\quad v_s =$ a vertex sampled without replacement from V_{node};
$\quad\quad V_{left} = (V_{node} \cap \mathcal{N}(v_s)) \cup \{v_s\}$;
$\quad\quad V_{right} = V_{node} \setminus V_{left}$;
$\quad\quad$ Push V_{left} to S;
$\quad\quad$ Push V_{right} to S;
$\quad$ **end**
end
return *leaves*;

---

the performance of our proposed method in the case of vertex-attributed graphs (Subsect. 4.2). An implementation of GT, as well as these benchmarks are available on https://github.com/jdalleau/gt.

4.1   Graph Trees on Simple Graphs

We first evaluate our approach on simple graphs with no attributes, in order to assess if our proposed method is able to discriminate clusters in such graphs. This evaluation is performed on both synthetic and real-world graphs, presented Table 1.

Table 1. Datasets used for the evaluation of clustering on simple graphs using graph-trees

Dataset	# vertices	# edges	Average degree	# clusters
Football	115	1226	10.66	10
Email-Eu-Core	1005	25571	33.24	42
Polbooks	105	441	8.40	3
SBM	450	65994	293.307	3

The graphs we call *SBM* are synthetic graphs generated using stochastic block models composed of k blocks of a user-defined size, that are connected by edges depending on a specific probability which is a parameter. The Football graph represents a network of American football games during a given season [12]. The Email-Eu-Core graph [18,31] represents relations between members of a research institution, where edges represents communication between those members. We also use a random graph in our first experiment. This graph is an Erdos-Renyi graph [10] generated with the parameters $n = 300$ and $p = 0.2$. Finally, the PolBooks data [16] is a graph where nodes represent books about US politics sold by an online merchant and edges books that were frequently purchased by the same buyers.

Our first empirical setting aims to compare the differences between the mean intracluster and the mean intercluster dissimilarities. These metrics enable a comparison that is agnostic to a subsequent clustering method.

The mean difference is computed as follows. First, the arithmetic mean of the pairwise similarities between all vertices with the same label is computed, corresponding to the mean intracluster dissimilarity μ_{intra}. The same process is performed for vertices with a different label, giving the mean intercluster similarity μ_{inter}. We finally compute the difference $\Delta = |\mu_{intra} - \mu_{inter}|$. In our experiments, this difference Δ is computed 20 times. $\bar{\Delta}$ denotes the mean of differences between runs, and σ its standard deviation. The results are presented Table 2. We observe that in the case of the random graph, $\bar{\Delta}$ is close to 0, unlike the graphs where a cluster structure exists. A projection of the vertices based on their pairwise dissimilarity obtained using GT is presented Fig. 2.

Table 2. Mean difference between intercluster and intracluster similarities in different settings.

Dataset	$\bar{\Delta}$	σ
Random graph	0.0003	0.0002
SBM	0.29	0.005
Football	0.25	0.002

We then compare the Normalized Mutual Information (NMI) obtained using GT with the NMI obtained using two well-known clustering methods on simple graphs, namely MCL [8] and Louvain [1]. *NMI* is a clustering quality metric when a ground truth is available. Its values lie in the range $[0,1]$, with a value of 1 being a perfect matching between the computed clusters and the reference one. The empirical protocol is the following:

1. Compute the dissimilarity matrices using GT, with a total number of trees $n_{trees} = 200$.
2. Obtain a 2D projection of the points using t-SNE [19] ($k = 2$).
3. Apply k-means on the points of the projection and compute the NMI.

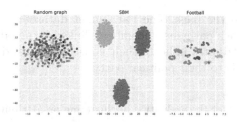

Fig. 2. Projection of the vertices obtained using GT on (left) a random graph, (middle) an SBM generated graph (middle) and (right) the football graph. Each cluster membership is denoted by a different color. Note how in the case of the random graph, no clear cluster can be observed. (Color figure online)

We repeated this procedure 20 times and computed means and standard deviations of the NMI.

The results are presented Table 3. We compared the mean NMI using the t-test, and checked that the differences between the obtained values are statistically significant.

We observe that our approach is competitive with the two well-known methods we chose in the case of non-attributed graphs on the benchmark datasets. In one specific case, we even observe that Graph trees significantly outperforms state of the art results, on the graphs generated by the SBM model. Since the dissimilarity computation is based on the method proposed by [28] to find clusters in regions of varying densities, this may indicate that our approach performs particularly well in the case of clusters of different size.

Table 3. Comparison of NMI on benchmark graph datasets. Best results are in boldface.

Dataset	Graph-trees	Louvain	MCL
Football	**0.923 (0.007)**	**0.924 (0.000)**	0.879 (0.015)
Email-Eu-Core	**0.649 (0.008)**	0.428 (0.000)	0.589 (0.012)
Polbooks	0.524 (0.012)	0.521 (0.000)	**0.544 (0.02)**
SBM	**0.998 (0.005)**	0.684 (0.000)	0.846 (0.000)

4.2 Graph Trees on Attributed Graphs

Now that we have tested GT on simple graphs, we can assess its performance on vertex-attributed graphs. The datasets that we used in this subsection are presented Table 4.

WebKB represents relations between web pages of four universities, where each vertex label corresponds to the university and the attributes represent the words that appear in the page. The Parliament dataset is a graph where the vertices represent french parliament members, linked by an edge if they cosigned a bill. The vertex attributes indicate their constituency, and each vertex has a label that corresponds to their political party.

Table 4. Datasets used for the evaluation of clustering on attributed graphs using GT

Dataset	# vertices	# edges	# attributes	# clusters
WebKB	877	1480	1703	4
Parliament	451	11646	108	7
HVR	307	6526	6	2

The empirical setup is the following. We first compute the vertex-vertex dissimilarities using GT, and the vertex-vertex dissimilarities using UET. In this first step, a forest of trees on the structures and a forest of trees on the attributes of each vertex are constructed. We then compute the average of the pairwise dissimilarities. Finally, we then apply t-SNE and use the k-means algorithm on the points in the embedded space. We set k to the number of clusters, since we have the ground truths. We repeat these steps 20 times and report the means and standard deviations. During our experiments, we found out that preprocessing the dissimilarities prior to the clustering phase may lead to better results, in particular with *Scikit learn*'s [22] *QuantileTransformer*. This transformation tends to spread out the most frequent values and to reduce the impact of outliers. In our evaluations, we performed this quantile transformation prior to every clustering, with $n_{quantile} = 10$.

The NMI obtained after the clustering step are presented in Table 5.

Table 5. NMI using GT on the structure only, UET on the attributes only and GT+UET. Best results are indicated in boldface.

Dataset	GT	UET	GT+UET
WebKB	0.64 (0.07)	0.73 (0.08)	**0.98 (0.01)**
HVR	0.58 (0.06)	0.58 (0.00)	**0.89 (0.06)**
Parliament	**0.65 (0.02)**	0.03 (0.00)	**0.66 (0.02)**

We observe that for two datasets, namely *WebKB* and *HVR*, considering both structural and attribute information leads to a significant improvement in NMI. For the other dataset considered in this evaluation, while the attribute information does not improve the NMI, we observe that is does not decrease it either. Here, we give the same weight to structural and attribute information.

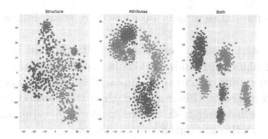

Fig. 3. Projection of the WebKB data based on the dissimilarities computed (left) using GT on structural data, (middle) using UET on the attributes data and (right) using the aggregated dissimilarity. Each cluster membership is denoted by a different color. (Color figure online)

In Fig. 3 we present the projection of the WebKB dataset, where we observe that the structure and attribute information both bring a different view of the data, each with a strong cluster structure.

HVR and Parliament datasets are extracted from [2]. Using their proposed approach, they obtain an NMI of 0.89 and 0.78, respectively. Although the NMI we obtained using our approach are not consistently better in this first assessment, the methods still seems to give similar results without any fine tuning.

5   Discussion and Future Work

In this paper, we presented a method based on the construction of random trees to compute dissimilarities between graph vertices, called GT. For vertex clustering purposes, our proposed approach is plug-and-play, since any clustering algorithm that can work on a dissimilarity matrix can then be used. Moreover, it could find application beyond graphs, for instance in relational structures in general.

Although the goal of our empirical study was not to show a clear superiority in terms of clustering but rather to assess the vertex-vertex dissimilarities obtained by GT, we showed that our proposed approach is competitive with well-known clustering methods, Louvain and MCL. We also showed that by computing forests of graph trees and other trees that specialize in other types of input data, *e.g*, feature vectors, it is then possible to compute pairwise dissimilarities between vertices in attributed graphs.

Some aspects are still to be considered. First, the importance of the vertex attributes is dataset dependent and, in some cases, considering the attributes can add noise. Moreover, the aggregation method between the graph trees and the attribute trees can play an essential role. Indeed, in all our experiments, we gave the same importance to the attribute and structural dissimilarities. This choice implies that both the graph trees and the attribute trees have the same weight, which may not always be the case. Finally, we chose here a specific algorithm to compute the dissimilarity in the attribute space, namely, UET. The poor results we obtained for some datasets may be caused by some limitations of UET in these cases.

It should be noted that our empirical results depend on the choice of a specific clustering algorithm. Indeed, GT is not a clustering method *per se*, but a method to compute pairwise dissimilarities between vertices. Like other dissimilarity-based methods, this is a strength of the method we propose in this paper. Indeed, the clustering task can be performed using many algorithms, leveraging their respective strengths and weaknesses.

As a future work, we will explore an approach where the choice of whether to consider the attribute space in the case of vertex-attributed graphs is guided by the distribution of the variables or the visualization of the embedding. We also plan to apply our methods on bigger graphs than the ones we used in this paper.

References

1. Blondel, V.D., Guillaume, J.L., Lambiotte, R., Lefebvre, E.: Fast unfolding of communities in large networks. J. Stat. Mech: Theory Exp. **2008**(10), P10008 (2008)
2. Bojchevski, A., Günnemann, S.: Bayesian robust attributed graph clustering: joint learning of partial anomalies and group structure (2018)
3. Cheng, H., Zhou, Y., Yu, J.X.: Clustering large attributed graphs: a balance between structural and attribute similarities. ACM Trans. Knowl. Discov. Data (TKDD) **5**(2), 12 (2011)
4. Combe, D., Largeron, C., Egyed-Zsigmond, E., Géry, M.: Combining relations and text in scientific network clustering. In: 2012 IEEE/ACM International Conference on Advances in Social Networks Analysis and Mining, pp. 1248–1253. IEEE (2012)
5. Combe, D., Largeron, C., Géry, M., Egyed-Zsigmond, E.: I-Louvain: an attributed graph clustering method. In: Fromont, E., De Bie, T., van Leeuwen, M. (eds.) IDA 2015. LNCS, vol. 9385, pp. 181–192. Springer, Cham (2015). https://doi.org/10.1007/978-3-319-24465-5_16
6. Cruz, J.D., Bothorel, C., Poulet, F.: Entropy based community detection in augmented social networks. In: 2011 International Conference on Computational Aspects of Social Networks (CASoN), pp. 163–168. IEEE (2011)
7. Dalleau, K., Couceiro, M., Smail-Tabbone, M.: Unsupervised extremely randomized trees. In: Phung, D., Tseng, V.S., Webb, G.I., Ho, B., Ganji, M., Rashidi, L. (eds.) PAKDD 2018. LNCS (LNAI), vol. 10939, pp. 478–489. Springer, Cham (2018). https://doi.org/10.1007/978-3-319-93040-4_38
8. Dongen, S.: A cluster algorithm for graphs (2000)
9. Ehrlinger, L., Wöß, W.: Towards a definition of knowledge graphs. In: SEMANTiCS (Posters, Demos, SuCCESS) (2016)
10. Erdös, P., Rényi, A.: On the evolution of random graphs. Publ. Math. Inst. Hung. Acad. Sci. **5**(1), 17–60 (1960)
11. Fan, M., Cao, K., He, Y., Grishman, R.: Jointly embedding relations and mentions for knowledge population. In: Proceedings of the International Conference Recent Advances in Natural Language Processing, pp. 186–191 (2015)
12. Girvan, M., Newman, M.E.J.: Community structure in social and biological networks. Proc. Natl. Acad. Sci. **99**(12), 7821–7826 (2002). https://doi.org/10.1073/pnas.122653799

13. Grover, A., Leskovec, J.: node2vec: scalable feature learning for networks. In: Proceedings of the 22nd ACM SIGKDD International Conference on Knowledge Discovery and Data Mining, pp. 855–864. ACM (2016)
14. Huang, X., Li, J., Hu, X.: Accelerated attributed network embedding. In: Proceedings of the 2017 SIAM International Conference on Data Mining, pp. 633–641. SIAM (2017)
15. Karger, D.R.: Global min-cuts in RNC, and other ramifications of a simple min-cut algorithm. In: SODA 1993, pp. 21–30 (1993)
16. Krebs, V.: Political books network (2004, Unpublished). Retrieved from Mark Newman's website. www-personal.umich.edu/mejn/netdata
17. Lancichinetti, A., Fortunato, S.: Limits of modularity maximization in community detection. Phys. Rev. E **84**(6), 066122 (2011)
18. Leskovec, J., Kleinberg, J., Faloutsos, C.: Graph evolution: densification and shrinking diameters. ACM Trans. Knowl. Discov. Data (TKDD) **1**(1), 2 (2007)
19. van der Maaten, L., Hinton, G.: Visualizing data using t-SNE. J. Mach. Learn. Res. **9**(Nov), 2579–2605 (2008)
20. Neville, J., Adler, M., Jensen, D.: Clustering relational data using attribute and link information. In: Proceedings of the Text Mining and Link Analysis Workshop, 18th International Joint Conference on Artificial Intelligence, pp. 9–15. Morgan Kaufmann Publishers, San Francisco (2003)
21. Newman, M.E.: Modularity and community structure in networks. Proc. Nat. Acad. Sci. **103**(23), 8577–8582 (2006)
22. Pedregosa, F., et al.: Scikit-learn: machine learning in Python. J. Mach. Learn. Res. **12**(Oct), 2825–2830 (2011)
23. Perozzi, B., Al-Rfou, R., Skiena, S.: Deepwalk: online learning of social representations. In: Proceedings of the 20th ACM SIGKDD International Conference on Knowledge Discovery and Data Mining, pp. 701–710. ACM (2014)
24. Schaeffer, S.E.: Graph clustering. Comput. Sci. Rev. **1**(1), 27–64 (2007)
25. Shi, J., Malik, J.: Normalized cuts and image segmentation. IEEE Trans. Pattern Anal. Mach. Intell. **22**(8), 888–905 (2000)
26. Stein, B., Niggemann, O.: On the nature of structure and its identification. In: Widmayer, P., Neyer, G., Eidenbenz, S. (eds.) WG 1999. LNCS, vol. 1665, pp. 122–134. Springer, Heidelberg (1999). https://doi.org/10.1007/3-540-46784-X_13
27. Steinhaeuser, K., Chawla, N.V.: Community detection in a large real-world social network. In: Liu, H., Salerno, J.J., Young, M.J. (eds.) Social Computing, Behavioral Modeling, and Prediction, pp. 168–175. Springer, Boston (2008). https://doi.org/10.1007/978-0-387-77672-9_19
28. Ting, K.M., Zhu, Y., Carman, M., Zhu, Y., Zhou, Z.H.: Overcoming key weaknesses of distance-based neighbourhood methods using a data dependent dissimilarity measure. In: Proceedings of the 22nd ACM SIGKDD International Conference on Knowledge Discovery and Data Mining, pp. 1205–1214. ACM (2016)
29. Xu, Z., Ke, Y., Wang, Y., Cheng, H., Cheng, J.: A model-based approach to attributed graph clustering. In: Proceedings of the 2012 ACM SIGMOD International Conference on Management of Data, pp. 505–516. ACM (2012)
30. Yang, Z., Tang, J., Cohen, W.: Multi-modal Bayesian embeddings for learning social knowledge graphs. arXiv preprint arXiv:1508.00715 (2015)
31. Yin, H., Benson, A.R., Leskovec, J., Gleich, D.F.: Local higher-order graph clustering. In: Proceedings of the 23rd ACM SIGKDD International Conference on Knowledge Discovery and Data Mining, pp. 555–564. ACM (2017)

32. Zhou, Y., Cheng, H., Yu, J.X.: Graph clustering based on structural/attribute similarities. Proc. VLDB Endow. **2**(1), 718–729 (2009). https://doi.org/10.14778/1687627.1687709
33. Zhou, Y., Cheng, H., Yu, J.X.: Clustering large attributed graphs: an efficient incremental approach. In: 2010 IEEE International Conference on Data Mining, pp. 689–698. IEEE (2010)

Evaluation of CNN Performance in Semantically Relevant Latent Spaces

Jeroen van Doorenmalen[✉] and Vlado Menkovski[✉]

Eindhoven University of Technology, Eindhoven, The Netherlands
j.v.doorenmalen@student.tue.nl, v.menkovski@tue.nl

Abstract. We examine deep neural network (DNN) performance and behavior using contrasting explanations generated from a semantically relevant latent space. We develop a semantically relevant latent space by training a variational autoencoder (VAE) augmented by a metric learning loss on the latent space. The properties of the VAE provide for a smooth latent space supported by a simple density and the metric learning term organizes the space in a semantically relevant way with respect to the target classes. In this space we can both linearly separate the classes and generate meaningful interpolation of contrasting data points across decision boundaries. This allows us to examine the DNN model beyond its performance on a test set for potential biases and its sensitivity to perturbations of individual factors disentangled in the latent space.

Keywords: Deep learning · VAE · Metric learning · Interpretability · Explanation

1 Introduction

Advances in machine learning and deep learning have had a profound impact on many tasks involving high dimensional data such as object recognition and behavior monitoring. The domain of Computer Vision especially has been witnessing a great growth in bridging the gap between the capabilities of humans and machines. This field tries to enable machines to view the world as humans do, perceive it similar and even use the knowledge for a multitude of tasks such as Image & Video Recognition, Image Analysis and Classification, Media Recreation, recommender systems, etc. And, has since been implemented in high-level domains like COMPAS [8], healthcare [3] and politics [17]. However, as blackbox models inner workings are still hardly understood, can lead to dangerous situations [3], such as racial bias [8], gender inequality [1].

The need for confidence, certainty, trust and explanations when using supervised black-box models is substantial in domains with high responsibility. This paper provides an approach towards better understanding of a model's predictions by investigating its behavior on semantically relevant (contrastive) explanations. The build a semantically relevant latent space we need a smooth space

© The Author(s) 2020
M. R. Berthold et al. (Eds.): IDA 2020, LNCS 12080, pp. 145–157, 2020.
https://doi.org/10.1007/978-3-030-44584-3_12

that corresponds well with the generating factors of the data (i.e. regions well-supported by the associated density should correspond to realistic data points) and with a distance metric that conveys semantic information about the target task. The vanilla VAE without any extra constraints is insufficient as is does not necessarily deliver a distance metric that corresponds to the semantics of the target class assignment (in our task). Our target is to develop semantically relevant decision boundaries in the latent space, which we can use to examine our target classification model. Therefore, we propose to use a weakly-supervised VAE that uses a combination of metric learning and VAE disentanglement to create a semantically relevant, smooth and well separated space. And, we show that we can use this VAE and semantically relevant latent space can be used for various interpretability/explainability tasks, such as validate predictions made by the CNN, generate (contrastive) explanations when predictions are odd and being able to detect bias. The approach we propose for these tasks is more specifically explained using Fig. 1.

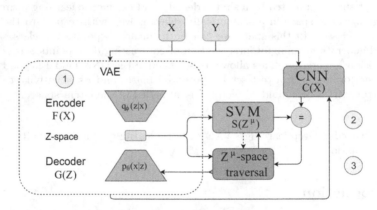

Fig. 1. The diagnostics approach to validate and understand the behavior of the CNN. (1) extra constraints, loss functions are applied during training of the VAE in order to create semantically relevant latent spaces. The generative model captures the essential semantics within the data and is used by (2) A linear Support Vector Machine. The linear SVM is trained on top of the latent space to classify input on semantics rather than the direct mapping from input data X and labels Y. If the SVM and CNN do not agree on a prediction then (3) we traverse the latent space in order to generate and capture semantically relevant synthetic images, tested against the CNN, in order to check what elements have to change in order to change its prediction from a to b, where a and b are different classes.

In this paper, the key contributions are: (1) an approach that can be used in order to validate and check predictions made by a CNN by utilizing a weakly-supervised generative model that is trained to create semantically relevant latent spaces. (2) The semantically relevant latent spaces are then used in order to train a linear support vector machine to capture decision rules that define a class assignment. The SVM is then used to check predictions based on semantics rather than the direct mapping of the CNN. (3) if there is a misalignment in the predictions (i.e. the CNN and SVM do not agree) then we posit the top k best candidates (classes) and for these candidates traverse the latent spaces in order to generate semantically relevant (contrastive) explanations by utilizing the decision boundaries of the SVM.

To conclude, This paper posits a method that allows for the validation of CNN performance by comparing it against the linear classifier that is based on semantics and provides a framework that generates explanations when the classifiers do not agree. The explanations are provided qualitatively to an expert within the field. This explanation encompasses the original image, reconstructed images and the path towards its most probable answers. Additionally, it shows the minimal difference that makes the classifiers change its prediction to one of the most probable answers. The expert can then check these results to make a quick assessment to which class the image actually belongs to. Additionally, the framework provides the ability to further investigate the model mathematically using the linear classifier as a proxy model.

2 Related Work

Interest in interpretability and explainability studies has significantly grown since the inception of "Right to Explanation" [20] and ethicality studies into the behavior of machine learning models [1,3,8,17]. As a result, developers of AI are promoted and required, amongst others, to create algorithms that are transparent, non-discriminatory, robust and safe. Interpretability is most commonly used as an umbrella term and stands for providing insight into the behavior and thought processes behind machine-learning algorithms and many other terms are used for this phenomenon, such as, Interpretable AI, Explainable machine learning, causality, safe AI, computational social science, etc. [5]. We posit our research as an interpretability study, but it does not necessarily mean that other interpretability studies are directly closely related to this work.

There have been many approaches that all work towards the goal of understanding black-box models: Linear Proxy Models: Lime [18] are approaches that locally approximate complex models using linear fits, Decision trees and Rule extraction methods, such as deepred [21] are also considered highly explainable, but quickly become intractable as complexity increases and salience mapping [19] that provide visual information as to which part of an image is most likely used in its prediction, however, it has been demonstrated to be unreliable if not strongly conditioned [10]. Additionally, another approach to interpretability is explaining the role of each part within a black-box models such as the role of a layer or individual neurons [2] or representation vectors within the activation space [9].

Most of the approaches stated above assume that there has to be a trade-off between model performance and explainability. Additionally, as the current interpretable methods for black-box models are still insufficient and approximated can cause more harm than good when communicated as a method that solves all problems. A lot of the interpretability methods do not take into account the actual needs that stakeholders require [13]. Or, fail to take into account the vast research into explanations or interpretability of the field of psychology [14] and social sciences [15]. The "Explanation in Artificial Intelligence" study by Miller [15] describes the current state of interpretable and explainable algorithms, how most of the techniques currently fail to capture the essence of an explanation and how to improve: an interpretability or explainability method should at least include, but is not limited to, a non-disputable textual- and/or mathematical- and/or visual explanation that is selective, social and depending on the proof, contrastive.

For this reason, our approach focuses on providing selective (contrastive) explanations that combines visual aspects as well as the ability to further investigate the model mathematically using a proxy model that does not impact the CNN directly. Usually, generative models such as the Variational Autoencoders (VAE) [11] and Generative Adversarial Networks (GAN)s are unsupervised and used in order to sample and generate images from a latent space, provided by training the generative network. However, we posit to use a weakly-supervised generative network in order to impose (discriminative) structure in addition to variational inference to the latent space of said model using metric learning [6].

This approach and method is therefore most related to the interpretability area of sub-sampling proxy generative models to answer questions about a discriminative black box model. The two closest studies that attempt similar research is a preprint of CDeepEx [4] by Amir Feghahati et al. and xGEMs [7] by Joshi et al. Both cDeepEx and xGEMS propose the use of a proxy generative model in order to explain the behavior of a black-box model, primarily using generative adversarial networks (GANs). The xGems paper presents a framework to characterize and explaining binary classification models by generating manifold guided examples using a generative model. The behavior of the black box model is summarized by quantitatively perturbing data samples along the manifold. And, xGEMS detects and quantifies bias during model training to understand how bias affects black box models. The xGEMS approach is similar to our approach as in using a generative model in order to explain a black box model. Similarly, the cDeepEx paper posits their work as generating contrastive explanations using a proxy generative model. The generated explanations focus on answering the question "why a and not b?" with GANs, where a is the class of an input example I and b is a chosen class to which to capture the differences.

However, both of these papers do not state that in a multi-class (discriminative) classification problem if the generative models' latent space is not smooth, well separated and semantically relevant then unexpected behavior can happen. For instance, when traversing the latent space it is possible to can pass from a to any number of classes before reaching class b because the space is not well separated and smooth. This will create ineffective explanations, as depending on how they generate explanations will give information on 'why class a and not b using properties of c'. An exact geodesic path along the manifold would require great effort, especially in high dimensions. Also, our approach is different in the fact that we utilize a weakly-supervised generative model as well as an extra linear classifier on top of the latent space to provide us with extra information on the data and the latent space. Some approaches we take, however, are very similar, such as using a generative model as a proxy to explain a black-box model as well as sub-sampling the latent space to probe the behavior of a black-box model and generate explanations using the predictions.

3 Methodology

This paper posits its methodology as a way to explain and validate decisions made by a CNN. The predictions made by the CNN are validated and explained utilizing the properties of a weakly-supervised proxy generative model, more specifically, a triplet-vae. There are three main factors that contribute to the validation and explanation of the CNN. First, a triplet-vae is trained in order to provide a semantically relevant and well separated latent space. Second, this latent space is then used to train an interpretable linear support vector machine and is used to validate decisions by the CNN by comparison. Third, when a CNN decision is misaligned with the decision boundaries in the latent space, we generate explanations through stating the K most probable answers as well as provide a qualitative explanation to validate the top K most probable answers. Each of these factors respectively refer to the number stated in Fig. 1 as well as link to each section: (1) triplet-vae Sect. 3.1, (2) CNN Decision Validation, Sect. 3.2, (3) Generating (contrastive) Explanations, Sect. 3.3.

3.1 Semantically Relevant Latent Space

Typically, a triplet network consists of three instances of a neural network that share parameters. These three instances are separately fed differences types of input: an anchor, positive sample and negative sample. These are then used to learn useful representations by distance comparisons. We propose to incorporate this notion of a triplet network to semantically structure and separate the latent space of the VAE using the available input and labels. A triplet VAE consists of three instances of the encoder with shared parameters that are each fed pre-computed triplets: an anchor, positive sample and negative sample; x_a, x_p and x_n. The anchor x_a and positive sample x_p are of the same class but not the same image, whereas negative sample x_n is from a different class. In each iteration of training, the input triplet is fed to the encoder network to get their mean latent embedding: $\mathcal{F}(x_a)^\mu = z_a^\mu$, $\mathcal{F}(x_p)^\mu = z_p^\mu$, $\mathcal{F}(x_n)^\mu = z_n^\mu$. These are then used to compute a similarity loss function as to induce loss when a negative sample z_n^μ is closer to z_a^μ than z_p^μ distance-wise. i.e. $\delta_{ap}(z_a^\mu, z_p^\mu) = ||z_a^\mu - z_p^\mu||$ and $\delta_{an}(z_a^\mu, z_n^\mu) = ||z_a^\mu - z_n^\mu||$ and, provides us with three possible situations: $\delta_{ap} > \delta_{an}$, $\delta_{ap} < \delta_{an}$ and $\delta_{ap} = \delta_{an}$ [6].

We wish to find an embedding where samples of a certain class lie close to each other in the latent space of the VAE. For this reason, we wish to add loss the algorithm when we arrive in the situation where $\delta_{ap} > \delta_{an}$. In other words,

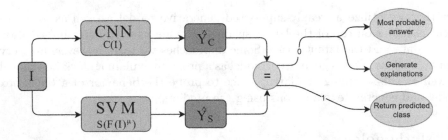

Fig. 2. Given an input image I we check the prediction of the CNN as well as the SVM. If both classifiers predict the same class, we return the predicted class. In contrast, if the classifiers do not predict the same class, we propose to return the top k most probable answers as well as an explanation why those classes are the most probable.

we wish to push x_n further away, such that we ultimately arrive in the situation where $\delta_{ap} < \delta_{an}$ or $\delta_{ap} = \delta_{an}$ with some margin ϕ. As such we arrive at the triplet loss function that we'll use in addition to the KL divergence and reconstruction loss within the VAE: $L(z_a^\mu, z_p^\mu, z_n^\mu) = \alpha * \text{argmax}\{||z_a^\mu - z_p^\mu|| - ||z_a^\mu - z_n^\mu|| + \phi, 0\}$. Where ϕ will provide leeway when $\delta_{ap} = \delta_{an}$ and push the negative sample away even when the distances are equal.

We have an already present CNN which we would like to validate, and is trained by input data $X : x_i...x_n$ and labels $Y : y_i...y_n$ where each y_i states the true class of x_i. We then use the same X and Y to train the triplet-VAE. (1) First, we compute triplets of the form $x_a, x_p x_n$ from the input data X and labels Y which are then used to train the triplet VAE. A typical VAE consists of an $\mathcal{F}(x) = Encoder(x) \sim q(z|x)$ which compresses the data into a latent space Z, a $\mathcal{G}(z) = Decoder(z) \sim p(x|z)$ which reconstructs the data given the latent space Z and a prior $p(z)$, in our case a gaussian $\mathcal{N}(0,1)$, imposed on the model. In order for the VAE to train a latent space similar to its prior and be able to reconstruct images it is trained by minimizing the Evidence Lower Bound (ELBO). $ELBO = -\mathbb{E}_{z\sim\mathcal{Q}(z|X)}[\log P(x|z)] + \mathcal{KL}[\mathcal{Q}(z|X)||P(z)]$ This can be explained as the reconstruction loss or expected negative loglikelihood: $-\mathbb{E}_{z\sim\mathcal{Q}(z|X)}[\log P(x|z)]$ and the KL divergence loss $\mathcal{KL}[\mathcal{Q}(z|X)||P(z)]$, to which we add the triplet loss:

$$\mathcal{L}(z_a^\mu, z_p^\mu, z_n^\mu) = \alpha * \text{argmax}\{||z_a^\mu - z_p^\mu|| - ||z_a^\mu - z_n^\mu|| + \phi, 0\}$$

This compound loss semi-forces the latent space of the VAE to be well separated due to the triplet loss, disentangled due to the KL divergence loss combined with β scalar, and provides a means of (reasonably) reconstructing images by the reconstruction loss. And, thus results in the following loss function for training the VAE:

$$loss = -\mathbb{E}_{z\sim\mathcal{Q}(z|X)}[\log P(x|z)] + \beta * \mathcal{KL}[\mathcal{Q}(z|X)||P(z)] + \mathcal{L}(z_a^\mu, z_p^\mu, z_n^\mu).$$

3.2   Decision Validation

Afterwards, given a semantically relevant latent space we can use it for step two and three as indicated in Fig. 1. (2) Second step - CNN Decision Validation, we train an additional classifier on top of the triplet-VAE latent space, specifically z^μ. We train the linear Support Vector Machine using Z^μs as input data and Y as labels where $[Z^\mu, Z^\sigma] = \mathcal{F}(\mathcal{X})$. The goal of the linear support vector machine is two-fold. It provides a means of validating each prediction made by the CNN by using the encoder and the linear classifier. i.e. given an input example I, we have $\mathcal{C}(I) = \hat{y}_{\mathcal{C}(I)}$ and $\mathcal{S}(\mathcal{F}(I)^\mu) = \hat{y}_{\mathcal{S}(I)}$, and compare them against each other $\hat{y}_{\mathcal{C}(I)} = \hat{y}_{\mathcal{S}(I)}$. And, as the linear classifier is a simpler model than the highly complex CNN it will function as the ground-truth base for the predictions that are made. As such, we arrive at two possible cases:

$$\text{Comparison(I)} = \begin{cases} \text{Positive} & \text{if } (\hat{y}_{\mathcal{C}(I)} = \hat{y}_{\mathcal{S}(I)}) \\ \text{Negative} & \text{if } (\hat{y}_{\mathcal{C}(I)} \neq \hat{y}_{\mathcal{S}(I)}) \end{cases} \tag{1}$$

First, If both classifiers agree then we arrive at an optimal state, meaning that the prediction is based on semantics and the direct mapping found by the CNN. In this way, we can say with high confidence that the prediction is correct. In the second case, if the classifiers do not agree, three cases can occur: the SVM is correct and the CNN is incorrect, the SVM is incorrect and the CNN is correct, or both the SVM and the CNN is incorrect. In each of these cases we can suggest a most probable answer as well as a selective (contrastive) explanation indicated as step 3 of the framework as explained in Fig. 2.

3.3   Generating (contrastive) Explanations

An explanation consists of (1) the most probable answers and (2) a qualitative investigation of latent traversal towards the most probable answers The most probable answer is presented by the averaged sum rule [12] over the predicted probabilities per class for both the CNN and SVM and selecting the top K answers, where K can be appropriately selected. Additionally, originally an SVM does not return a probabilistic answer, however, applying Platts [16] method we apply an additional sigmoid function to map the SVM outputs into probabilities. These top k answers are then used in order to present and generate selected contrastive explanations.

The top K predictions or classes will be used in order to traverse and sub-sample the latent space from the initial representation or Z_I^μ location towards another class. We can find a path by finding the closest point within the latent space such that the decision boundary is crossed and the SVM predicts the target class. Alternatively we could use the closest data point in the latent space that adheres to the training set $\operatorname{argmin} \mathcal{F}(x_i)^\mu - Z_I^\mu$ for every $x_i \in X$. Traversing and sub-sampling the latent space will change the semantics minimally to change the class prediction. We capture the minimal change needed in order to change both the SVM and CNN prediction to the target class. This information is then

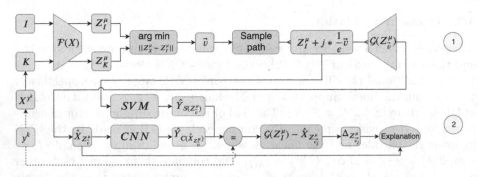

Fig. 3. Generating (contrastive) explanations consist of several steps: First, given an input image I in question and the K top most probable answer. K denotes training data X for class k labeled with $y = k$. We feed both I and K through the encoder $\mathcal{F}(X)$ to receive their respective semantic location in the latent space. We then find the closest training point that belongs to the target class k and find the vector v; the direction of that point. Afterwards, uniformly sample ϵ data points along this vector v, where j iterates over $0 \cdots j \cdots \epsilon$ and is denoted as Z_v^μ. Z_v^μ is then used to check these against the SVM and use them to generate images $X_{Z_v^\mu}$ using the decoder $\mathcal{G}(Z_v^\mu)$. The generated images are then fed to the CNN to make a prediction and as the images will semantically change along the vector the prediction will change as well. Afterwards, we can compare the predictions from both the CNN and SVM. Subsequently, we use the first moment where both predictions are equal to target class k, denoted as moment l for generating an explanation - minimal semantic difference necessary to be equal to the target class, ΔU_l.

presented to the domain expert for verification and answers the following question: The most probable answer is a because the input image I is semantically closest to the following features, where the features are presented qualitatively. The explanations are generated as follows: see Fig. 3.

The decision boundaries around the clusters within the latent space are fitted by the SVM and can be used to answer questions of the form 'why a and not b?'. If $\hat{y}_{\mathcal{C}(I)}$ and $\hat{y}_{\mathcal{S}(I)}$ do not predict the same class, then, we assume that $\hat{y}_{\mathcal{S}(I)}$ is correct. We then use the find a path, indicated by v from $\hat{y}_{\mathcal{S}(I)}$ to $\hat{y}_{\mathcal{C}(I)}$, Z_I^μ to the target class. This can be done by calculating a vector orthogonal to the hyper-plane fitted by the SVM towards the target class. Alternatively, we can find the closest $z^\mu \in Z^\mu$ that satisfies $\hat{y}_{S(z^\mu)} = \hat{y}_{\mathcal{C}(z^\mu)}$ that are not the same as the initial prediction $\hat{y}_{\mathcal{C}(I)}$. This means that v is the vector from I to the closest data point of the target class, with respect to Euclidean distance.

We then uniformly sample points along vector v and check them against the SVM as well as the CNN. The sampled points can directly be fed to the SVM to get a prediction $\hat{y}(\mathcal{f}(v_i)$ for every $v_i \in V$. Similarly, we can get predictions of the CNN by transforming the images using the decoder $\mathcal{D}$. The images are then fed to the CNN to get a prediction $\hat{y}(\mathcal{C}(\mathcal{D}(v_i))$ for every $v_i \in V$. The predictions of both classifiers will change as the images start looking more and more like the target class as generative factors change along the vector. If we capture the changes that make the change happen, we can show the minimal difference required in order to change the prediction of the CNN. In this way we can generate contrastive examples: For the top 'close' class that is not $\hat{y}_I$ we answer the question: 'why $\hat{y}_I$ and not the other semantically close class?'. Hence, we find the answer to the question "why a and not b?", as the answer is the shortest approximate changes between the two classes that make the CNN change its prediction. As a result, we have found a way to validate the inner workings of the CNN. If there are doubts about a prediction it can be investigated and checked.

4 Results

In this paper we show experimental results on MNIST by generating (contrastive) explanations to provide extra information to predictions made by the CNN and evaluate its performance. The creation of these explanations requires a semantically relevant and well separated latent space. Therefore, we first show the difference between the latent space of the vanilla VAE and the triplet-VAE and its effects on training a linear classifier on top of the latent space. The Figs. 4 and 5 show a tSNE visualization of the separation of classes within the latent space. Not surprisingly the triplet-VAE separated the data in a far more semantically relevant way and this is also reflected with respect to the accuracy of training a linear model on the data.

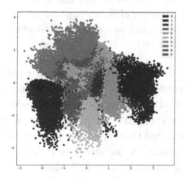 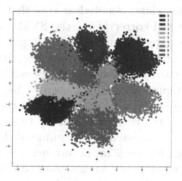

Fig. 4. Visualization of a two-dimensional latent space of a vanilla VAE on MNIST

Fig. 5. Visualization of a two-dimensional latent space of a $\mathcal{T}$-VAE on MNIST

Second, the percentages show as to know how much both classifiers agree by showing the percentage per possible case, as shown in Table 1. Not surprisingly case four happens more often than case three and can mean two things, our latent space is too simple to capture the full complexity of the class assignment and the CNN is not constraint by extra loss functions. However, in three of the four cases where $Y_S \neq Y_C$ we can explain

Table 1. This table shows the percentages of agreement with respect to all possible cases.

Case	Percentage
(1) $\hat{Y}_S = \hat{Y}_C = Y$	0.9586
(2) $\hat{Y}_S = \hat{Y}_C \neq Y$	0.003
(3) $(\hat{Y}_S = Y) \neq \hat{Y}_C$	0.0086
(4) $\hat{Y}_S \neq (\hat{Y}_C = Y)$	0.0314
(5) $\hat{Y}_S \neq \hat{Y}_C \neq Y$	0.0044

the most probable predictions and provide a generated (contrastive) explanation. The only case we cannot check or know about is case two, where both Y_S and Y_C predict the same class but is wrong. The only way to capture this behavior is by explaining every single decision by generating explanations for everything. Nevertheless, as an example for generating explanations we use an example: 6783 (case 5) as shown in Fig. 6.

Generating explanations consists of three parts: First, we propose the top K probable answers: for this example the true label is 1, the most probable answers are 6, 8 and then 1 with averaged probabilities 0,512332, 0.3382, 0.1150. Second, Then for those most probable target classes, 6, 8, 1 we traverse the latent space from the initial location Z_I^μ to the closest point of that class, denoted as $v \in$ that is predicted correctly i.e. the SVM and CNN agree. Figure 7 shows the generated images from the uniformly sam-

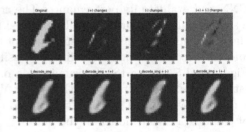

Fig. 6. Once the SVM and the CNN both predict the target class we capture the minimal changes that are necessary to change their predictions

pled data points along vectors $v_k \in V$ where $k \in K$ stand for 6, 8, 1 in this case. The figures show which changes happen when traversing the latent space and at which points both the SVM and the CNN agree with respect to their decision.

For the traversal from Z_I^μ to class 6 it can be seen that rather quickly both classifiers agree and only minimal changes are required to change the predictions. Third, for such an occurrence we can further zoom in on what is happening and what really makes that the most probable answer. Figure 6 shows these minimal changes required to change its prediction as well as the transformed image on which the classifiers agree. The first row shows the original image, positive changes, negative changes and the changes combined. The second row shows the reconstructed image and the reconstructed images with the positive changes, negative changes and positive and negative changes respectively. In this way, for each probable answer it shows its closest representative and the changes required to be part of that class.

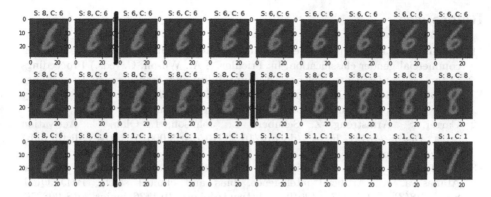

Fig. 7. Per top k probable answers we traverse and sample the latent space to generate images that can be used to test the behavior of the CNN. The red line indicates the moment where both the SVM and the CNN predict the target class (Color figure online)

5 Conclusion

This paper examines deep neural network's behaviour and performance by utilizing a weakly-supervised generative model as a proxy. The weakly-supervised generative model aims to uncover the generative factors underlying the data and separate abstract classes by applying metric learning. The proxy's goal is three-fold: the semantically meaningful space will be the base for a linear support vector machine; The model's generative capabilities will be used to generate images that can be probed against the black box in question; the latent space is traversed and sampled from an anchor I to another class k in order to find the minimal important difference that changes both classifier's predictions. The goal of the framework is to be sure of the predictions made by the black box by better understanding the behaviour of the CNN by simulating questions of the form 'Why a and not b?' where a and b are different classes.

We examine deep neural network (DNN) performance and behaviour using contrasting explanations generated from a semantically relevant latent space. The results show that each of the above goals can be achieved and the framework performs as expected. We develop a semantically relevant latent space by training an variational autoencoder (VAE) augmented by a metric learning loss on the latent space. The properties of the VAE provide for a smooth latent space supported by a simple density and the metric learning term organizes the space in a semantically relevant way with respect to the target classes. In this space we can both linearly separate the classes and generate relevant interpolation of contrasting data points across decision boundaries and find the minimal important difference that changes the classifier's predictions. This allows us to examine the DNN model beyond its performance on a test set for potential biases and its sensitivity to perturbations of individual factors in the latent space.

References

1. Buolamwini, J., Gebru, T.: Gender shades: intersectional accuracy disparities in commercial gender classification. In: Friedler, S.A., Wilson, C. (eds.) Proceedings of the 1st Conference on Fairness, Accountability and Transparency. Proceedings of Machine Learning Research, vol. 81, pp. 77–91. PMLR, New York, February 2018. http://proceedings.mlr.press/v81/buolamwini18a.html
2. Carter, S., Armstrong, Z., Schubert, L., Johnson, I., Olah, C.: Activation atlas. Distill (2019). https://doi.org/10.23915/distill.00015, https://distill.pub/2019/activation-atlas
3. Challen, R., Denny, J., Pitt, M., Gompels, L., Edwards, T., Tsaneva-Atanasova, K.: Artificial intelligence, bias and clinical safety. BMJ Qual. Saf. **28**(3), 231–237 (2019). https://doi.org/10.1136/bmjqs-2018-008370. https://qualitysafety.bmj.com/content/28/3/231
4. Feghahati, A., Shelton, C.R., Pazzani, M.J., Tang, K.: CDeepEx: contrastive deep explanations (2019). https://openreview.net/forum?id=HyNmRiCqtm
5. Gilpin, L.H., Bau, D., Yuan, B.Z., Bajwa, A., Specter, M., Kagal, L.: Explaining explanations: an approach to evaluating interpretability of machine learning. CoRR abs/1806.00069 (2018). http://arxiv.org/abs/1806.00069
6. Ishfaq, H., Hoogi, A., Rubin, D.: TVAE: triplet-based variational autoencoder using metric learning (2018)
7. Joshi, S., Koyejo, O., Kim, B., Ghosh, J.: xGEMs: generating examplars to explain black-box models. CoRR abs/1806.08867 (2018). http://arxiv.org/abs/1806.08867
8. Angwin, J., Larson, J., Mattu, S., Kirchner, L.: Machine bias, May 2016. https://www.propublica.org/article/machine-bias-risk-assessments-in-criminal-sentencing
9. Kim, B., et al.: Interpretability beyond feature attribution: quantitative testing with concept activation vectors (TCAV). In: Dy, J., Krause, A. (eds.) Proceedings of the 35th International Conference on Machine Learning, vol. 80, pp. 2668–2677, July 2018
10. Kindermans, P.J., et al.: The (un)reliability of saliency methods. CoRR abs/1711.00867 (2017). http://dblp.uni-trier.de/db/journals/corr/corr1711.html#abs-1711-00867
11. Kingma, D.P., Welling, M.: An introduction to variational autoencoders. CoRR abs/1906.02691 (2019). http://arxiv.org/abs/1906.02691
12. Kittler, J., Hatef, M., Duin, R.P.W., Matas, J.: On combining classifiers. IEEE Trans. Pattern Anal. Mach. Intell. **20**(3), 226–239 (1998). https://doi.org/10.1109/34.667881
13. Lipton, Z.C.: The doctor just won't accept that! In: NIPS Proceedings 2017, no. 24, pp. 1–3, November 2017. https://arxiv.org/pdf/1711.08037.pdf
14. Lombrozo, T.: Explanation and abductive inference. In: Oxford Handbook of Thinking and Reasoning, pp. 260–276 (2012). https://doi.org/10.1093/oxfordhb/9780199734689.013.0014
15. Miller, T.: Explanation in artificial intelligence: insights from the social sciences. CoRR abs/1706.07269 (2017). http://arxiv.org/abs/1706.07269
16. Platt, J.C.: Probabilistic outputs for support vector machines and comparisons to regularized likelihood methods. In: Advances in Large Margin Classifiers, pp. 61–74. MIT Press (1999)
17. Prakash, A.: Ai-politicians: a revolution in politics, August 2018. https://medium.com/politics-ai/ai-politicians-a-revolution-in-politics-11a7e4ce90b0

18. Ribeiro, M.T., Singh, S., Guestrin, C.: Why should I trust you?: explaining the predictions of any classifier. CoRR abs/1602.04938 (2016). http://arxiv.org/abs/1602.04938
19. Selvaraju, R.R., Das, A., Vedantam, R., Cogswell, M., Parikh, D., Batra, D.: Grad-CAM: why did you say that? Visual explanations from deep networks via gradient-based localization. CoRR abs/1610.02391 (2016). http://arxiv.org/abs/1610.02391
20. European Union: Official journal of the European union: Regulations (2016). https://eur-lex.europa.eu/legal-content/EN/TXT/PDF/?uri=CELEX:32016R0679&from=EN
21. Zilke, J.R., Loza Mencía, E., Janssen, F.: DeepRED – rule extraction from deep neural networks. In: Calders, T., Ceci, M., Malerba, D. (eds.) DS 2016. LNCS (LNAI), vol. 9956, pp. 457–473. Springer, Cham (2016). https://doi.org/10.1007/978-3-319-46307-0_29

Vouw: Geometric Pattern Mining
Using the MDL Principle

Micky Faas$^{(\boxtimes)}$ and Matthijs van Leeuwen

LIACS, Leiden University, Leiden, The Netherlands
micky@edukitty.org, m.van.leeuwen@liacs.leidenuniv.nl

Abstract. We introduce geometric pattern mining, the problem of finding recurring local structure in discrete, geometric matrices. It differs
from existing pattern mining problems by identifying complex spatial
relations between elements, resulting in arbitrarily shaped patterns.
After we formalise this new type of pattern mining, we propose an
approach to selecting a set of patterns using the Minimum Description
Length principle. We demonstrate the potential of our approach by introducing Vouw, a heuristic algorithm for mining exact geometric patterns.
We show that Vouw delivers high-quality results with a synthetic benchmark.

1 Introduction

Frequent pattern mining [1] is the well-known subfield of data mining that aims
to find and extract recurring substructures from data, as a form of knowledge
discovery. The generic concept of pattern mining has been instantiated for many
different types of patterns, e.g., for item sets (in Boolean transaction data) and
subgraphs (in graphs/networks). Little research, however, has been done on pattern mining for raster-based data, i.e., geometric matrices in which the row and
column orders are fixed. The exception is geometric tiling [4,11], but that problem only considers tiles, i.e., rectangular-shaped patterns, in Boolean data.

In this paper we generalise this setting in two important ways. First, we
consider geometric patterns *of any shape* that are geometrically connected, i.e.,
it must be possible to reach any element from any other element in a pattern by
only traversing elements in that pattern. Second, we consider *discrete geometric
data* with any number of possible values (which includes the Boolean case). We
call the resulting problem *geometric pattern mining*.

Figure 1 illustrates an example of geometric pattern mining. Figure 1a shows
a 32×24 grayscale 'geometric matrix', with each element in $[0, 255]$, apparently
filled with noise. If we take a closer look at all horizontal pairs of elements,
however, we find that the pair $(146, 11)$ is, amongst others, more prevalent than
expected from 'random noise' (Fig. 1b). If we would continue to try all combinations of elements that 'stand out' from the background noise, we would eventually
find four copies of the letter 'I' set in 16 point Garamond Italic (Fig. 1c).

© The Author(s) 2020
M. R. Berthold et al. (Eds.): IDA 2020, LNCS 12080, pp. 158–170, 2020.
https://doi.org/10.1007/978-3-030-44584-3_13

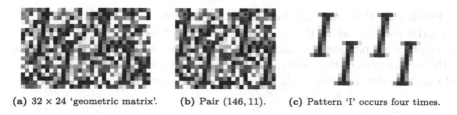

(a) 32 × 24 'geometric matrix'.   (b) Pair (146, 11).   (c) Pattern 'I' occurs four times.

Fig. 1. Geometric pattern mining example. Each element is in $[0, 255]$.

The 35 elements that make up a single 'I' in the example form what we call a *geometric pattern*. Since its four occurrences jointly cover a substantial part of the matrix, we could use this pattern to describe the matrix more succinctly than by 768 independent values. That is, we could describe it as the pattern 'I' at locations $(5, 4), (11, 11), (20, 3), (25, 10)$ plus 628 independent values, hereby separating structure from accidental (noise) data. Since the latter description is shorter, we have compressed the data. At the same time we have learned something about the data, namely that it contains four I's. This suggests that we can use compression as a criterion to find patterns that describe the data.

Approach and Contributions. Our first contribution is that we introduce and formally define *geometric pattern mining*, i.e., the problem of finding recurring local structure in geometric, discrete matrices. Although we restrict the scope of this paper to two-dimensional data, the generic concept applies to higher dimensions. Potential applications include the analysis of satellite imagery, texture recognition, and (pattern-based) clustering.

We distinguish three types of geometric patterns: (1) *exact* patterns, which must appear exactly identical in the data to match; (2) *fault-tolerant* patterns, which may have noisy occurrences and are therefore better suited to noisy data; and (3) *transformation-equivalent* patterns, which are identical after some transformation (such as mirror, inverse, rotate, etc.). Each consecutive type makes the problem more expressive and hence more complex. In this initial paper we therefore restrict the scope to the first, exact type.

As many geometric patterns can be found in a typical matrix, it is crucial to find a compact set of patterns that together describe the structure in the data well. We regard this as a model selection problem, where a model is defined by a set of patterns. Following our observation above, that geometric patterns can be used to compress the data, our second contribution is the formalisation of the model selection problem by using the *Minimum Description Length (MDL) principle* [5,8]. Central to MDL is the notion that 'learning' can be thought of as 'finding regularity' and that regularity itself is a property of data that is exploited by *compressing* said data. This matches very well with the goals of pattern mining, as a result of which the MDL principle has proven very successful for MDL-based pattern mining [7,12].

Finally, our third contribution is Vouw, a heuristic algorithm for MDL-based geometric pattern mining that (1) finds compact yet descriptive sets of patterns, (2) requires no parameters, and (3) is tolerant to noise in the data (but not in the occurrences of the patterns). We empirically evaluate Vouw on synthetic data and demonstrate that it is able to accurately recover planted patterns.

2 Related Work

As the first pattern mining approach using the MDL principle, Krimp [12] was one of the main sources of inspiration for this paper. Many papers on pattern-based modelling using MDL have appeared since, both improving search, e.g., Slim [10], and extensions to other problems, e.g., Classy [7] for rule-based classification.

The problem closest to ours is probably that of geometric tiling, as introduced by Gionis et al. [4] and later also combined with the MDL principle by Tatti and Vreeken [11]. Geometric tiling, however, is limited to Boolean data and rectangularly shaped patterns (tiles); we strongly relax both these limitations (but as of yet do not support patterns based on densities or noisy occurrences).

Campana et al. [2] also use matrix-like data (textures) in a compression-based similarity measure. Their method, however, has less value for *explanatory* analysis as it relies on generic compression algorithms that are essentially a black box.

Geometric pattern mining is different from graph mining, although the concept of a matrix can be redefined as a grid-like graph where each node has a fixed degree. This is the approach taken by Deville et al. [3], solving a problem similar to ours but using an approach akin to bag-of-words instead of the MDL principle.

3 Geometric Pattern Mining Using MDL

We define geometric pattern mining on bounded, discrete and two-dimensional raster-based data. We represent this data as an $M \times N$ matrix A whose rows and columns are finite and in a fixed ordering (i.e., reordering rows and columns semantically alters the matrix). Each element $a_{i,j} \in S$, where row $i \in [0; N)$, column $j \in [0; M)$, and S is a finite set of symbols, i.e., the alphabet of A.

According to the MDL principle, the shortest (optimal) description of A reveals all structure of A in the most succinct way possible. This optimal description is only optimal if we can unambiguously reconstruct A from it and nothing more—the compression is both minimal and lossless. Figure 2 illustrates how an example matrix could be succinctly described using patterns: matrix A is decomposed into patterns X and Y. A set of such patterns constitutes the **model** for a matrix A, denoted H_A (or H for short when A is clear from the context). In order to reconstruct A from this model, we also need a mapping from the H_A back to A. This mapping represents what (two-part) MDL calls **the data given the model** H_A. In this context we can think of this as a set of all instructions required to rebuild A from H_A, which we call the **instantiation** of H_A and is denoted by I in the example. These concepts allow us to express matrix A as a decomposition into sets of local and global spatial information, which we will next describe in more detail.

$$A = \begin{bmatrix} 1 & \cdot & \cdot & \cdot & 1 & 1 \\ \cdot & 1 & 1 & 1 & 1 & \cdot \\ 1 & 1 & \cdot & \cdot & \cdot & 1 \end{bmatrix}, \ I = \begin{bmatrix} X & \cdot & \cdot & \cdot & Y & \cdot \\ \cdot & \cdot & Y & \cdot & X & \cdot \\ Y & \cdot & \cdot & \cdot & \cdot & \cdot \end{bmatrix}, \ H = \left\{ X = \begin{bmatrix} 1 & \cdot \\ \cdot & 1 \end{bmatrix}, Y = \begin{bmatrix} 1 & 1 \end{bmatrix} \right\}$$

Fig. 2. Example decomposition of A into instantiation I and patterns X, Y.

3.1   Patterns and Instances

▷ *We define a **pattern** as an $M_X \times N_X$ submatrix X of the original matrix A. Elements of this submatrix may be $\cdot$, the empty element, which gives us the ability to cut-out any irregular-shaped part of A. We additionally require the elements of X to be adjacent (horizontal, vertical or diagonal) to at least one non-empty element and that no rows and columns are empty.*

From this definition, the dimensions $M_X \times N_X$ give the smallest rectangle around X (the *bounding box*). We also define the cardinality $|X|$ of X as the number of non-empty elements. We call a pattern X with $|X| = 1$ a **singleton pattern**, i.e., a pattern containing exactly one element of A.

Each pattern contains a special **pivot** element: $pivot(X)$ is the first non-empty element of X. A pivot can be thought of as a fixed point in X which we can use to position its elements in relation to A. This translation, or **offset**, is a tuple $q = (i, j)$ that is on the same domain as an index in A. We realise this translation by placing all elements of X in an empty $M \times X$ size matrix such that the pivot element is at (i, j). We formalise this in the **instantiation operator** $\otimes$:

▷ *We define the **instance** $X \otimes (i, j)$ as the $M \times N$ matrix containing all elements of X such that $\text{pivot}(X)$ is at index (i, j) and the distances between all elements are preserved. The resulting matrix contains no additional non-empty elements.*

Since this does not yield valid results for arbitrary offsets (i, j), we enforce two constraints: (1) an instance must be **well-defined**: placing $\text{pivot}(X)$ at index (i, j) must result in an instance that contains all elements of X, and (2) elements of instances cannot *overlap*, i.e., each element of A can be described only once.

▷ *Two pattern instances $X \otimes q$ and $Y \otimes r$, with $q \neq r$ are **non-overlapping** if $|(X \otimes q) + (Y \otimes r)| = |X| + |Y|$.*

From here on we will use the same letter in lower case to denote an arbitrary instance of a pattern, e.g., $x = X \otimes q$ when the exact value of q is unimportant. Since instances are simply patterns projected onto an $M \times N$ matrix, we can reverse $\otimes$ by removing all completely empty rows and columns:

▷ *Let $X \otimes q$ be an instance of X, then by definition we say that $\oslash(X \otimes q) = X$.*

We briefly introduced the instantiation I as a set of 'instructions' of where instances of each pattern should be positioned in order to obtain A. As Fig. 2 suggests, this mapping has the shape of an $M \times N$ matrix.

▷ Given a set of patterns H, the **instantiation (matrix)** I is an $M \times N$ matrix such that $I_{i,j} \in H \cup \{\cdot\}$ for all (i,j), where $\cdot$ denotes the empty element. For all non-empty $I_{i,j}$ it holds that $I_{i,j} \otimes (i,j)$ is a non-overlapping instance of $I_{i,j}$ in A.

3.2 The Problem and Its Solution Space

Larger patterns can be naturally constructed by joining (or merging) smaller patterns in a bottom-up fashion. To limit the considered patterns to those relevant to A, instances can be used as an intermediate step. As Fig. 3 demonstrates, we can use a simple element-wise matrix addition to sum two instances and use $\oslash$ to obtain a joined pattern. Here we start by instantiating X and Y with offsets $(1,0)$ and $(1,1)$, respectively. We add the resulting x and y to obtain $\oslash z$, the union of X and Y with relative offset $(1,1) - (1,0) = (0,1)$.

$$x = X \otimes (1,0) = \begin{bmatrix} \cdot & \cdot \\ 1 & \cdot \\ \cdot & 1 \end{bmatrix}, \ y = Y \otimes (1,1) = \begin{bmatrix} \cdot & \cdot \\ \cdot & 1 \\ \cdot & \cdot \end{bmatrix}, x + y = \begin{bmatrix} \cdot & \cdot \\ 1 & 1 \\ \cdot & 1 \end{bmatrix}, \ Z = \oslash(x+y) = \begin{bmatrix} 1 & 1 \\ \cdot & 1 \end{bmatrix}$$

Fig. 3. Example of joining patterns X and Y to construct a new pattern Z.

The Sets $\mathcal{H}_A$ and $\mathcal{I}_A$. We define the **model class** $\mathcal{H}$ as the set of all possible models for all possible inputs. Without any prior knowledge, this would be the search space. To simplify the search, however, we only consider the more bounded subset $\mathcal{H}_A$ of all possible models for A, and $\mathcal{I}_A$, the set of all possible instantiations for these models. To this end we first define H_A^0 to be the model with only singleton patterns, i.e., $H_A^0 = S$, and denote its corresponding instantiation matrix by I_A^0. Given that each element of I_A^0 must correspond to exactly one element of A in H_A^0, we see that each $I_{i,j} = a_{i,j}$ and so we have $I_A^0 = A$.

Using H_A^0 and I_A^0 as base cases we can now inductively define $\mathcal{I}_A$:

Base case $I_A^0 \in \mathcal{I}_A$

By induction If I is in $\mathcal{I}_A$ then take any pair $I_{i,j}, I_{k,l} \in I$ such that $(i,j) \leq (k,l)$ in lexicographical order. Then the set I' is also in $\mathcal{I}_A$, providing I' equals I except:

$$I'_{i,j} := \oslash\big(I_{i,j} \otimes (i,j) + I_{k,l} \otimes (k,l)\big)$$
$$I'_{k,l} := \cdot$$

This shows we can add any two instances together, in any order, as they are by definition always non-overlapping and thus valid in A, and hereby obtain another element of $\mathcal{I}_A$. Eventually this results in just one big instance that is equal to A. Note that when we take two elements $I_{i,j}, I_{k,l} \in I$ we force $(i,j) \leq (k,l)$, not only to eliminate different routes to the same instance matrix, but also so that the pivot of the new pattern coincides with $I_{i,j}$. We can then leave $I_{k,l}$ empty.

The construction of $\mathcal{I}_A$ also implicitly defines $\mathcal{H}_A$. While this may seem odd—defining models for instantiations instead of the other way around—note that there is no unambiguous way to find one instantiation for a given model. Instead we find the following definition by applying the inductive construction:

$$\mathcal{H}_A = \{\{\oslash(x) \mid x \in I\} \mid I \in \mathcal{I}_A\}. \tag{1}$$

So for any instantiation $I \in \mathcal{I}_A$ there is a corresponding set in $\mathcal{H}_A$ of all patterns that occur in I. This results in an interesting symbiosis between model and instantiation: increasing the complexity of one decreases that of the other. This construction gives a tightly connected lattice as shown in Fig. 4.

3.3   Encoding Models and Instances

From all models in $\mathcal{H}_A$ we want to select the model that describes A best. Two-part MDL [5] tells us to choose that model that minimises the sum of $L_1(H_A) + L_2(A|H_A)$, where L_1 and L_2 are two functions that give the length of the model and the length of 'the data given the model', respectively. In this context, the data given the model is given by I_A, which represents the accidental information needed to reconstruct the data A from H_A.

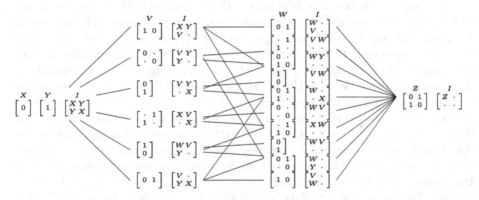

Fig. 4. Model space lattice for a 2×2 Boolean matrix. The V, W, and Z columns show which pattern is added in each step, while I depicts the current instantiation.

In order to compute their lengths, we need to decide how to encode H_A and I. As this encoding is of great influence on the outcome, we should adhere to the conditions that follow from MDL theory: (1) the model and data must be encoded losslessly; and (2) the encoding should be as concise as possible, i.e., it should be optimal. Note that for the purpose of model selection we only need the length functions; we do not need to actually encode the patterns or data.

Code Length Functions. Although the patterns in H and instantiation matrix I are all matrices, they have different characteristics and thus require different encodings. For example, the size of I is constant and can be ignored, while the

Table 1. Code length definitions. Each row specifies the code length given by the first column as the sum of the remaining terms.

	Matrix	Bounds	# Elements	Positions	Symbols
$L_p(X)$	Pattern	$\log(MN)$	$L_N\binom{M_X N_X}{|X|}$		$|X|\log(|S|)$
$L_1(H)$	Model	N/A	$L_N(|H|)$	N/A	$\sum_{X \in H} L_p(X)$
$L_2(I)$	Instantiation	$Constant$	$\log(MN)$	$Implicit$	$L_{pp}(I)$

sizes of the patterns vary and should be encoded. Hence we construct different length functions[1] for the different components of H and I, as listed in Table 1.

When encoding I, we observe that it contains each pattern $X \in H$ multiple times, given by the **usage** of X. Using the **prequential plug-in code** [5] to encode I enables us to omit encoding these usages separately, which would create unwanted bias. The prequential plug-in code gives us the following length function for I. We use $\epsilon = 0.5$ and elaborate on its derivation in the Appendix[2].

$$
L_{pp}(I \mid P_{plugin}) = - \sum_{X_i \in h}^{|H|} \left[\log \frac{\Gamma(\text{usage}(X_i) + \epsilon)}{\Gamma(\epsilon)} \right] + \log \frac{\Gamma(|I| + \epsilon|H|)}{\Gamma(\epsilon|H|)} \quad (2)
$$

Each length function has four terms. First we encode the total size of the matrix. Since we assume MN to be known/constant, we can use this constant to define the uniform distribution $\frac{1}{MN}$, so that $\log MN$ encodes an arbitrary index of A. Next we encode the number of elements that are non-empty. For patterns this value is encoded together with the third term, namely the positions of the non-empty elements. We use the previously encoded $M_X N_X$ in the binominal function to enumerate the ways we can place the $|X|$ elements onto a grid of $M_X N_X$. This gives us both *how many* non-empties there are as well as *where* they are. Finally the fourth term is the length of the actual symbols that encode the elements of the matrix. In case we encode single elements of A, we assume that each unique value in A occurs with equal probability; without other prior knowledge, using the uniform distribution has minimax regret and is therefore optimal. For the instance matrix, which encodes symbols to patterns, the prequential code is used as demonstrated before. Note that L_N is the universal prior for the integers [9], which can be used for arbitrary integers and penalises larger integers.

4   The Vouw Algorithm

Pattern mining often yields vast search spaces and geometric pattern mining is no exception. We therefore use a heuristic approach, as is common in MDL-based approaches [7,10,12]. We devise a greedy algorithm that exploits the inductive

---

[1] We calculate code lengths in bits and therefore all logarithms have base 2.
[2] The appendix is available on https://arxiv.org/abs/1911.09587.

definition of the search space as shown by the lattice in Fig. 4. We start with a completely underfit model (leftmost in the lattice), where there is one instance for each matrix element. Next, in each iteration we combine two patterns, resulting in one or more pairs of instances to be merged (i.e., we move one step right in the lattice). In each step we merge the pair of patterns that improves compression most, and we repeat this until no improvement is possible.

4.1 Finding Candidates

The first step is to find the 'best' **candidate** pair of patterns for merging (Algorithm 1). A candidate is denoted as a tuple (X, Y, δ), where X and Y are patterns and δ is the relative offset of X and Y as they occur in the data. Since we only need to consider pairs of patterns and offsets that actually occur in the instance matrix, we can directly enumerate candidates from the instantiation matrix and never even need to consider the original data.

Algorithm 1 FindCandidates	**Algorithm 2** Vouw
Input: I	**Input:** H, I
Output: C	1: $C \leftarrow$ FindCandidates(I)
1: **for all** $x \in I$ **do**	2: $(X, Y, \delta) \in C : \forall_{c \in C} \Delta L((X, Y, \delta)) \leq \Delta L(c)$
2: **for all** $y \in \text{POST}(x)$ **do**	3: $\Delta L_{best} = \Delta L((X, Y, \delta))$
3: $X \leftarrow \oslash(x), Y \leftarrow \oslash(y)$	4: **if** $\Delta L_{best} > 0$ **then**
4: $\delta \leftarrow \text{dist}(X, Y)$	5: $Z \leftarrow \oslash(X \otimes (0,0) + (Y \otimes \delta))$
5: **if** $X = Y$ **then**	6: $H \leftarrow H \cup \{Z\}$
6: **if** $V(x)[e] = 1$ **continue**	7: **for all** $x_i \in I \mid \oslash(x_i) = X$ **do**
7: $V(y)[e] \leftarrow 1$	8: **for all** $y \in \text{POST}(x_i) \mid \oslash(y) = Y$ **do**
8: **end if**	9: $x_i \leftarrow Z, y \leftarrow \cdot$
9: $C \leftarrow C \cup (X, Y, \delta)$	10: **end for**
10: $\sup(X, Y, \delta) \mathrel{+}= 1$	11: **end for**
11: **end for**	12: **end if**
12: **end for**	13: **repeat until** $\Delta L_{best} < 0$

The **support** of a candidate, written $\sup(X, Y, \delta)$, tells how often it is found in the instance matrix. Computing support is not completely trivial, as one candidate occurs multiple times in 'mirrored' configurations, such as (X, Y, δ) and $(Y, X, -\delta)$, which are equivalent but can still be found separately. Furthermore, due to the definition of a pattern, many potential candidates cannot be considered by the simple fact that their elements are not adjacent.

Peripheries. For each instance x we define its *periphery*: the set of instances which are positioned such that their union with x produces a valid pattern. This set is split into *anterior* $\text{ANT}(X)$ and *posterior* $\text{POST}(X)$ peripheries, containing instances that come before and after x in lexicographical order, respectively. This enables us to scan the instance matrix once, in lexicographical order. For each instance x, we only consider the instances $\text{POST}(x)$ as candidates, thereby eliminating any (mirrored) duplicates.

Self-overlap. Self-overlap happens for candidates of the form (X, X, δ). In this case, too many or too few copies may be counted. Take for example a straight line of five instances of X. There are four unique pairs of two X's, but only two can be merged at a time, in three different ways. Therefore, when considering candidates of the form (X, X, δ), we also compute an *overlap coefficient*. This coefficient e is given by $e = (2N_X + 1)\delta_i + \delta_j + N_X$, which essentially transforms δ into a one-dimensional coordinate space of all possible ways that X could be arranged *after* and *adjacent* to itself. For each instance x_1 a vector of bits $V(x)$ is used to remember if we have already encountered a combination x_1, x_2 with coefficient e, such that we do not count a combination x_2, x_3 with an equal e. This eliminates the problem of incorrect counting due to self-overlap.

4.2   Gain Computation

After candidate search we have a set of candidates C and their respective supports. The next step is to select the candidate that gives the best *gain*: the improvement in compression by merging the candidate pair of patterns. For each candidate $c = (X, Y, \delta)$ the gain $\Delta L(A', c)$ is comprised of two parts: (1) the negative gain of adding the union pattern Z to the model H, resulting in H', and (2) the gain of replacing all instances x, y with relative offset δ by Z in I, resulting in I'. We use length functions L_1, L_2 to derive an equation for gain:

$$\Delta L(A', c) = \Big(L_1(H') + L_2(I') \Big) - \Big(L_1(H) + L_2(I) \Big)$$
$$= L_N(|H|) - L_N(|H| + 1) - L_p(Z) + \Big(L_2(I') - L_2(I) \Big) \tag{3}$$

As we can see, the terms with L_1 are simplified to $-L_p(Z)$ and the model's length because L_1 is simply a summation of individual pattern lengths. The equation of L_2 requires the recomputation of the entire instance matrix' length, which is expensive considering we need to perform it for *every candidate, every iteration*. However, we can rework the function L_{pp} in Eq. (2) by observing that we can isolate the logarithms and generalise them into

$$\log_G(a, b) = \log \frac{\Gamma(a + b\epsilon)}{\Gamma(b\epsilon)} = \log \Gamma(a + b\epsilon) - \log \Gamma(b\epsilon), \tag{4}$$

which can be used to rework the second part of Eq. (3) in such way that the gain equation can be computed in constant time complexity.

$$L_2(I') - L_2(I) = \log_G(U(X), 1) + \log_G(U(Y), 1)$$
$$- \log_G(U(X) - U(Z), 1) - \log_G(U(Y) - U(Z), 1) \tag{5}$$
$$- \log_G(U(Z), 1) + \log_G(|I|, |H|) - \log_G(|I'|, |H'|)$$

Notice that in some cases the usages of X and Y are equal to that of Z, which means additional gain is created by removing X and Y from the model.

4.3 Mining a Set of Patterns

In the second part of the algorithm, listed in Algorithm 2, we select the candidate (X, Y, δ) with the largest gain and merge X and Y to form Z, as explained in Sect. 3.2. We linearly traverse I to replace all instances x and y with relative offset δ by instances of Z. (X, Y, δ) was constructed by looking in the posterior periphery of all x to find Y and δ, which means that Y always comes after X in lexicographical order. The pivot of a pattern is the first element in lexicographical order, therefore $\text{pivot}(Z) = \text{pivot}(X)$. This means that we can replace all matching x with an instance of Z and all matching y with $\cdot$.

4.4 Improvements

Local Search. To improve the efficiency of finding large patterns without sacrificing the underlying idea of the original heuristics, we add an optional local search. Observe that without local search, Vouw generates a large pattern X

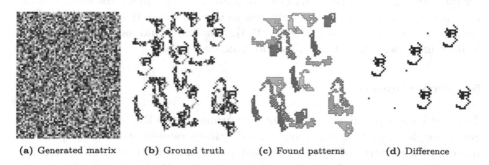

(a) Generated matrix   (b) Ground truth   (c) Found patterns   (d) Difference

Fig. 5. Synthetic patterns are added to a matrix filled with noise. The difference between the ground truth and the matrix reconstructed by the algorithm is used to compute precision and recall.

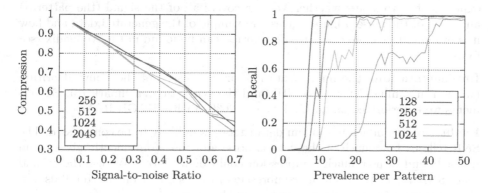

Fig. 6. The influence of SNR in the ground truth (left) and prevalence on recall (right)

by adding small elements to an incrementally growing pattern, resulting in a behaviour that requires up to $|X| - 1$ steps. To speed this up, we can try to 'predict' which elements will be added to X and add them immediately. After selecting candidate (X, Y, δ) and merging X and Y into Z, for all m resulting instances $z_i \in z_0, \ldots, z_{m-1}$ we try to find pattern W and offset δ such that

$$\forall_{i \in 0 \ldots m} \exists_w \in \text{ANT}(z_i) \cup \text{POST}(z_i) \cdot \oslash(w) = W \wedge dist(z_i, w) = \delta. \quad (6)$$

This yields zero or more candidates (Z, W, δ), which are then treated as any set of candidates: candidates with the highest gain are iteratively merged until no candidates with positive gain exist. This essentially means that we run the baseline algorithm only on the peripheries of all z_i, with the condition that the support of the candidates is equal to that of Z.

Reusing Candidates. We can improve performance by reusing the candidate set and slightly changing the search heuristic of the algorithm. The **Best-*** heuristic selects multiple candidates on each iteration, as opposed to the baseline **Best-1** heuristic that only selects a single candidate with the highest gain. Best-* selects candidates in descending order of gain until no candidates with positive gain are left. Furthermore we only consider candidates that are all *disjoint*, because when we merge candidate (X, Y, δ), remaining candidates with X and/or Y have unknown support and therefore unknown gain.

5   Experiments

To asses Vouw's practical performance we primarily use Ril, a synthetic dataset generator developed for this purpose. Ril utilises random walks to populate a matrix with patterns of a given size and prevalence, up to a specified density, while filling the remainder of the matrix with noise. Both the pattern elements and the noise are picked from the same uniform random distribution on the interval $[0, 255]$. The *signal-to-noise ratio* (SNR) of the data is defined as the number of pattern elements over the matrix size MN. The objective of the experiment is to assess whether Vouw recovers all of the signal (the patterns) and none of the noise. Figure 5 gives an example of the generated data and how it is evaluated. A more extensive description can be found in the Appendix (see footnote 2).

Implementation. The implementation[3] used consists of the Vouw algorithm (written in vanilla C/C++), a GUI, and the synthetic benchmark Ril. Experiments were performed on an Intel Xeon-E2630v3 with 512 GB RAM.

Evaluation. Completely random data (noise) is unlikely to be compressed. The SNR tells us how much of the data is noise and thus conveniently gives us an upper bound of how much compression could be achieved. We use the ground truth SNR versus the resulting compression ratio as a benchmark to tell us how close we are to finding all the structure in the ground truth.

---

[3] https://github.com/mickymuis/libvouw

In addition, we also compare the ground truth matrix to the obtained model and instantiation. As singleton patterns do not yield any compression over the baseline model, we reconstruct the matrix omitting any singleton patterns. Ignoring the actual values, this gives us a Boolean matrix with 'positives' (pattern occurrence = signal) and 'negatives' (no pattern = noise). By comparing each element in this matrix with the corresponding element in the ground truth matrix, *precision* and *recall* can be calculated and evaluated.

Figure 6 (left) shows the influence of ground truth SNR on compression ratio for different matrix sizes. Compression ratio and SNR are clearly strongly correlated. Figure 6 (right) shows that patterns with a low prevalence (i.e., number of planted occurrences) have a lower probability of being 'detected' by the algorithm as they are more likely to be accidental/noise. Increasing the matrix size also increases this threshold. In Table 2 we look at the influence of the two improvements upon the baseline algorithm as described in Sect. 4.4. In terms of quality, local search can improve the results quite substantially while Best-* notably *lowers* precision. Both improve speed by an order of magnitude.

Table 2. Performance measurements for the baseline algorithm and its optimisations.

Size	SNR	Precision/Recall				Average time			
		None	Local	Best-*	Both	None	Local	Best-*	Both
256	.05	.98/.98	.99/.99	.93/.98	.95/.99	29 s	1 s	2 s	1 s
	.3	.99/.8	.99/.88	.96/.82	.99/.89	2 m 32 s	9 s	5 s	5 s
512	.05	.98/.97	.99/.99	.87/.97	.93/.98	5 m 26 s	8 s	20 s	6 s
	.3	.97/.93	.99/.99	.94/.91	.97/.90	26 m 52 s	2 m 32 s	24 s	65 s
1024	.05	.97/.98	.99/.99	.84/.98	.92/.96	21 m 34 s	44 s	37 s	34 s
	.3	.98/.98	.99/.99	.93/.96	.98/.97	116 m 4 s	7 m 31 s	1 m 49 s	3 m 31 s

6   Conclusions

We introduced geometric pattern mining, the problem of finding recurring structures in discrete, geometric matrices, or raster-based data. Further, we presented Vouw, a heuristic algorithm for finding sets of geometric patterns that are good descriptions according to the MDL principle. It is capable of accurately recovering patterns from synthetic data, and the resulting compression ratios are on par with the expectations based on the density of the data. For the future, we think that extensions to fault-tolerant patterns and clustering have large potential.

References

1. Aggarwal, C.C., Han, J.: Frequent Pattern Mining. Springer, Cham (2014). https://doi.org/10.1007/978-3-319-07821-2
2. Bilson, J.L.C., Keogh, E.J.: A compression-based distance measure for texture. Statistical Analysis and Data Mining **3**(6), 381–398 (2010)

3. Deville, R., Fromont, E., Jeudy, B., Solnon, C.: GriMa: a grid mining algorithm for bag-of-grid-based classification. In: Robles-Kelly, A., Loog, M., Biggio, B., Escolano, F., Wilson, R. (eds.) S+SSPR 2016. LNCS, vol. 10029, pp. 132–142. Springer, Cham (2016). https://doi.org/10.1007/978-3-319-49055-7_12
4. Gionis, A., Mannila, H., Seppänen, J.K.: Geometric and combinatorial tiles in 0–1 data. In: Boulicaut, J.-F., Esposito, F., Giannotti, F., Pedreschi, D. (eds.) PKDD 2004. LNCS (LNAI), vol. 3202, pp. 173–184. Springer, Heidelberg (2004). https://doi.org/10.1007/978-3-540-30116-5_18
5. Grünwald, P.D.: The Minimum Description Length Principle. MIT press, Cambridge (2007)
6. Li, M., Vitányi, P.: An Introduction to Kolmogorov Complexity and Its Applications. TCS, vol. 3. Springer, New York (2008). https://doi.org/10.1007/978-0-387-49820-1
7. Proença, H.M., van Leeuwen, M.: Interpretable multiclass classification by MDL-based rule lists. Inf. Sci. 12, 1372–1393 (2020)
8. Rissanen, J.: Modeling by shortest data description. Automatica 14(5), 465–471 (1978)
9. Rissanen, J.: A universal prior for integers and estimation by minimum description length. Ann. Stat. 11, 416–431 (1983)
10. Smets, K., Vreeken, J.: Slim: directly mining descriptive patterns. In: Proceedings of the 2012 SIAM International Conference on Data Mining, SIAM, pp. 236–247 (2012)
11. Tatti, N., Vreeken, J.: Discovering descriptive tile trees - by mining optimal geometric subtiles. In: Proceedings of ECML PKDD 2012, pp. 9–24 (2012)
12. Vreeken, J., van Leeuwen, M., Siebes, A.: KRIMP: mining itemsets that compress. Data Min. Knowl. Disc. 23(1), 169–214 (2011)

A Consensus Approach to Improve NMF Document Clustering

Mickael Febrissy[✉] and Mohamed Nadif

LIPADE, Université de Paris, 75006 Paris, France
mickael.febrissy@u-paris.fr

Abstract. Nonnegative Matrix Factorization (NMF) which was originally designed for dimensionality reduction has received throughout the years a tremendous amount of attention for clustering purposes in several fields such as image processing or text mining. However, despite its mathematical elegance and simplicity, NMF has exposed a main issue which is its strong sensitivity to starting points, resulting in NMF struggling to converge toward an optimal solution. On another hand, we came to explore and discovered that even after providing a meaningful initialization, selecting the solution with the best local minimum was not always leading to the one having the best clustering quality, but somehow a better clustering could be obtained with a solution slightly off in terms of criterion. Therefore in this paper, we undertake to study the clustering characteristics and quality of a set of NMF best solutions and provide a method delivering a better partition using a consensus made of the best NMF solutions.

Keywords: NMF · Clustering · Clustering ensemble · Consensus

1 Introduction

When dealing with text data, document clustering techniques allow to divide a set of documents into groups so that documents assigned to the same group are more similar to each other than to documents assigned to other groups [12,18,21,22]. In information retrieval, the use of clustering relies on the assumption that if a document is relevant to a query, then other documents in the same cluster can also be relevant. This hypothesis can be used at different stages in the information retrieval process, the two most notable being: cluster-based retrieval to speed up search, and search result clustering to help users navigate and understand what is in the search results. The document clustering which still remains a hot topic can be tackled under different approaches. In our contribution we rely on the non-negative matrix factorization for its simplicity and popularity. We will not propose a new variant of NMF but rather a consensus approach that will boost its performance.

Unlike supervised learning, the evaluation of clustering algorithms - unsupervised learning - remains a difficult problem. When relying on generative models,

© The Author(s) 2020
M. R. Berthold et al. (Eds.): IDA 2020, LNCS 12080, pp. 171–183, 2020.
https://doi.org/10.1007/978-3-030-44584-3_14

it is easier to evaluate the performance of a given clustering algorithm based on the simulated partition. On real data already labeled, many papers evaluate the performance of clustering algorithms by relying on indices such as Accuracy (ACC), Normalized Mutual Information (NMI) [25] and Adjusted Rand Index (ARI) [14]. However, the algorithms commonly used which are of type k-means, EM [8], Classification EM [6], NMF [15] etc. are iterative and require several initializations; the resulting partition is the one optimizing the objective function. Sometimes in these works, we observe comparative studies between methods on the basis of maximum ACC/NMI/ARI measures obtained after several initializations and not optimizing the criterion used in the algorithm. Such a comparison is thereby not accurate, because in fact these measures cannot be calculated in practice and cannot be used in this way to evaluate the quality of a clustering algorithm.

A fair comparison can only be made on the basis of objective functions considered in a clustering purpose; for example, within-cluster inertia, likelihood, classification likelihood for mixture models, factorization, etc. Nonetheless, in our experiences, we realized that while the clustering results become better in terms of ACC/NMI/ARI when the objective function value increases, the best value is not necessarily associated with the best results. However, by ranking the objective values, the best partition tends to be among those leading to the first best scores. We illustrate this behavior in Fig. 4. This remark leads us to consider an *ensemble method* that is widely used in supervised learning [11,24] but a little less in unsupervised learning [25]. If this approach, referred to as *consensus clustering*, is often used in the context of comparing partitions obtained with different algorithms, it is less studied considering the same algorithm.

The paper is organized as follows. In Sect. 2, we review the nonnegative matrix factorization with the Frobenius norm and the Kullback–Leibler divergence. Section 3 is devoted to describe the ensemble method and the popular used algorithms. In Sect. 4, we perform comparisons on document-term matrices and propose a strategy to improve document clustering with NMF.

2   Nonnegative Matrix Factorization

Nonnegative Matrix Factorization (NMF) [15], aiming to deliver a lower rank decomposition of a nonnegative data matrix X has highlighted clustering properties for which strong connections with K-means or Spectral clustering can be drawn [16]. However, while several variants arise in order to accommodate its clustering property [10,29–31], its premier model formulation does not involve a clustering objective and was originally presented as a dimension reduction algorithm with exclusive nonnegative factors. More specifically in text mining where NMF produces a meaningful interpretation for document-term matrices in comparison with methods like Singular Value Decomposition (SVD) components or Latent Semantic Analysis (LSA) [7] arising factors with possible negative values. NMF seeks to approximate a matrix $X \in \mathbb{R}_+^{n \times d}$ by the product of two lower rank matrices $Z \in \mathbb{R}_+^{n \times g}$ and $W \in \mathbb{R}_+^{d \times g}$ with $g(n + d) < ng$. This problem can be formulated as a constrained optimization problem

$$\mathrm{F}(\boldsymbol{Z}, \boldsymbol{W}) = \min_{\boldsymbol{Z} \geq 0, \boldsymbol{W} \geq 0} D(\boldsymbol{X}, \boldsymbol{Z}\boldsymbol{W}^\top) \tag{1}$$

where D is a fitting error allowing to measure the quality of the approximation of $\boldsymbol{X}$ by $\boldsymbol{Z}\boldsymbol{W}^\top$, the most popular ones being the Frobenius norm and Kullback-Leibler (KL) divergence. For a clustering setup, $\boldsymbol{Z}$ will be referred to as the soft classification matrix while $\boldsymbol{W}$ will be the centers matrix. Despite its multiple applications benefits, NMF has a recurrent downside which takes place at its initialization. NMF provides a different solution for every different initialisation making it substantially sensitive to starting points as its convergence directly relies on the characteristics of the given entries. Several publications have shown interest in finding the best way to start a NMF algorithm by providing a structured initialization, in some cases obtained from results of clustering algorithms such as k-means or Spherical K-means [27,28] (especially for applying NMF on document-term matrices), Nonnegative Singular Value decomposition (NNDSVD) [4] or SVD based strategies [17]. The optimization procedures for D respectively equal to the Frobenius norm and the KL divergence, based on multiplicative update rules are given in Algorithms 1 and 2.

Algorithm 1. (NMF-F).	Algorithm 2. (NMF-KL).
Input: $\boldsymbol{X}, g, \boldsymbol{Z}^{(0)}; \boldsymbol{W}^{(0)}$.	**Input:** $\boldsymbol{X}, g, \boldsymbol{Z}^{(0)}; \boldsymbol{W}^{(0)}$.
Output: $\boldsymbol{Z}$ and $\boldsymbol{W}$.	**Output:** $\boldsymbol{Z}$ and $\boldsymbol{W}$.
repeat	**repeat**
1. $\boldsymbol{Z} \leftarrow \boldsymbol{Z} \odot \frac{\boldsymbol{X}\boldsymbol{W}}{\boldsymbol{Z}\boldsymbol{W}^\top\boldsymbol{W}}$;	1. $\boldsymbol{Z} \leftarrow \boldsymbol{Z} \odot \left(\frac{\boldsymbol{X}}{\boldsymbol{Z}\boldsymbol{W}^\top}\boldsymbol{W}\right)/\sum_j \boldsymbol{W}_{jk}$;
2. $\boldsymbol{W} \leftarrow \boldsymbol{W} \odot \frac{\boldsymbol{X}^\top\boldsymbol{Z}}{\boldsymbol{W}\boldsymbol{Z}^\top\boldsymbol{Z}}$;	2. $\boldsymbol{W} \leftarrow \boldsymbol{W} \odot \left(\frac{\boldsymbol{X}^\top}{\boldsymbol{W}\boldsymbol{Z}^\top}\boldsymbol{Z}\right)/\sum_i \boldsymbol{Z}_{ik}$;
until convergence	**until** convergence
5. Normalize $\boldsymbol{Z}$ so as it has unit-length column vectors.	5. Normalize $\boldsymbol{Z}$ so as it has unit-length column vectors.

3 Cluster Ensembles (CE)

In machine learning, the idea of utilizing multiple sources of data partitions firstly occurred with multi-learner systems where the output of several classifier algorithms where used together in order to improve the accuracy and robustness of a classification or regression, for which strong performances were acknowledged [24,25]. At this stage, very few approaches have worked toward applying a similar concept to unsupervised learning algorithms. In this sense, we denote the work of [5] who tried to combine several clustering partitions according to the combination of the cluster centers. In the early 2000, [25] were the first to consider an idea of combining several data partitions however, without accessing any original sources of information (features) or led computed centers. This approach is referred to as *cluster ensembles*. At the time, their idea was motivated by the possibilities of taking advantage of existing information such as a prior clustering partitions or an expert categorization (all regrouped under the terms

Knowledge Reuse), which may still be relevant or substantial for a user to consider in a new analysis on the same objects, whether or not the data associated with these objects may also be different than the ones used to define the prior partitions. Another motivation was *Distributed computing*, referring to analyzing different sources of data (which might be complicated to merge together for instance for privacy reasons) stored in different locations. In our concept, we will use *cluster ensembles* to improve the quality of the final partition (as opposed to selecting a unique one) and therefore extract all the possibilities offered by the miscellaneous best solutions created by NMF.

In [25], the authors introduced three consensus methods that can produce a partition. All of them consider the consensus problem on a hypergraph representation H of the set of partitions H^r. More specifically, each partition H^r equals a binary classification matrix (with objects in rows and clusters in columns) where the concatenation of all the set defines the hypergraph H.

- The first one is called Cluster-based Similarity Partitioning Algorithm (**CSPA**) and consists in performing a clustering on the hypergraph according to a similarity measure.
- The second is referred to as HyperGraph Partitioning Algorithm (**HGPA**) and aims at optimizing a minimum cut objective.
- The third one is called Meta-CLustering Algorithm (**MCLA**) and looks forward to identifying and constructing groups of clusters.

Furthermore, in [25] the authors proposed an objective function to characterize the *cluster ensembles* problem and therefore allowing a selection of the best consensus algorithm among the three to deliver its ensemble partition. Let $\Lambda = \{\lambda^{(q)}|q \in \{1, \ldots, r\}\}$ be a given set of r partitions $\lambda^{(q)}$ represented as labels vectors. The ensemble criterion denoted as $\lambda^{(k-opt)}$ is called the optimal combine clustering and aims at maximizing the Average Normalized Mutual Information (ANMI). It is defined as follows:

$$\lambda^{(k-opt)} = \underset{\widetilde{\lambda}}{argmax} \sum_{q=1}^{r} \mathrm{NMI}(\widetilde{\lambda}, \lambda^{(q)}) \qquad (2)$$

The ANMI is simply the average of the normalized mutual information of a labels vector $\widetilde{\lambda}$ with all labels vectors $\lambda^{(q)}$ in Λ:

$$\mathrm{ANMI}(\Lambda, \widetilde{\lambda}) = \frac{1}{r} \sum_{q=1}^{r} \mathrm{NMI}(\widetilde{\lambda}, \lambda^{(q)}) \qquad (3)$$

To cast with cases where the vector labels $\lambda^{(q)}$ have missing values, the authors have proposed a generalized expression of (2) not substantially different that viewers can refer to in the original paper [25].

4   Experiments

We conduct several experiences leading to emphasise the behavior of NMF regarding a clustering task compared to a dedicated clustering algorithm such as Spherical K-means referred to as S-Kmeans [9] which was introduced for clustering large sets of sparse text data (or directional data) and remains appealing for its low computational cost beside its good performances. It was also retained along side the random starting points (generated according to an uniform distribution $\mathcal{U}(0,1) \times mean(\boldsymbol{X})$) as initialization for NMF. We use two error measures frequently employed for NMF: the Frobenius norm (which will be referred to as NMF-F) and the Kullback-Leibler divergence (NMF-KL). Eventually, we compute the consensus partition by using the Cluster Ensemble Python package[1] which utilizes the consensus methods defined earlier [25].

4.1   Datasets

We apply NMF on 5 bench-marking document-term matrices for which the detailed characteristics are available in Table 1 where nz indicates the percentage of values other than 0 and the *balance* coefficient is defined as the ratio of the number of documents in the smallest class to the number of documents in the largest class. These datasets highlight several varieties of challenging situations such as the amount of clusters, the dimensions, the clusters balance, the degree of mixture of the different groups and the sparsity. We normalized each data matrix with TF-IDF and their respective documents-vectors to unit L_2-norm to remove the bias introduced by their length.

Table 1. Datasets description: # denotes the cardinality

Datasets	Characteristics				
	#Documents	#Words	#Clusters	$nz(\%)$	Balance
CSTR	475	1000	4	3.40	0.399
CLASSIC4	7095	5896	4	0.59	0.323
RCV1	6387	16921	4	0.25	0.080
NG5	4905	10167	5	0.92	0.943
NG20	18846	14390	20	0.59	0.628

4.2   NMF Raw Performances and Initialization

The results obtained by NMF-F and NMF-KL according to S-Kmeans and the random starting points are available in Table 2. The clustering quality of the

---

[1] https://pypi.org/project/Cluster_Ensembles/.

S-Kmeans partitions given as entry to both algorithms are also displayed. We make use of two relevant measures to quantify and assess the clustering quality of each algorithm. The first one is the NMI [25] which quantifies how much information the clustering partition shares with the true partition, the second is the ARI [14], sensitive to the clusters proportions and measures the degree of agreement between the clustering and the true partition. To replicate a relevant user experience achieving an unsupervised task, we refer to the criterion of each algorithm in order to select the 10 first best solutions (out of 30 runs) and report their average NMI and ARI with the true partition.

One can clearly see that **NMF-F** and **NMF-KL** do not react similarly to the different initializations. While **NMF-F** substantially benefits from the **S-kmeans** initialization on every datasets compared to the random initialization, **NMF-KL** does not seem to accommodate **S-kmeans** entries. In fact, **S-Kmeans** as starting values seems to worsen **NMF-KL** solutions, especially on CLASSIC4 and NG5. For this reason, we will avoid this initialization strategy for **NMF-KL** in the future although it improves on RCV1. Also, **NMF-KL** with a random initialization provides much better results than the other algorithms on almost all datasets.

Table 2. Mean and standard deviation of NMI and ARI computed over the 10 best solutions.

Datasets	Metrics	Skmeans	NMF-F (Random)	NMF-F (Skmeans)	NMF-KL (Random)	NMF-KL (Skmeans)
CSTR	NMI	0.76 ± 0.007	0.65 ± 0.002	0.73 ± 0.04	0.73 ± 0.03	0.76 ± 0.006
	ARI	0.80 ± 0.007	0.55 ± 0.002	0.75 ± 0.10	0.77 ± 0.04	0.80 ± 0.006
CLASSIC4	NMI	0.60 ± 0.001	0.53 ± 0.003	0.59 ± 0.002	0.71 ± 0.02	0.61 ± 0.03
	ARI	0.47 ± 0.0009	0.45 ± 0.003	0.47 ± 0.002	0.65 ± 0.06	0.47 ± 0.004
RCV1	NMI	0.38 ± 0.0003	0.35 ± 0.0005	0.38 ± 0.0002	0.47 ± 0.02	0.53 ± 0.002
	ARI	0.18 ± 0.0004	0.13 ± 0.0008	0.18 ± 0.0003	0.42 ± 0.02	0.46 ± 0.02
NG5	NMI	0.72 ± 0.02	$0.56 \pm 1.0e{-}05$	0.72 ± 0.02	0.80 ± 0.03	0.79 ± 0.003
	ARI	0.60 ± 0.01	$0.33 \pm 2.5e{-}05$	0.60 ± 0.01	0.82 ± 0.04	0.76 ± 0.005
NG20	NMI	0.49 ± 0.02	0.41 ± 0.01	0.49 ± 0.02	0.48 ± 0.02	0.51 ± 0.01
	ARI	0.30 ± 0.02	0.23 ± 0.01	0.30 ± 0.02	0.34 ± 0.02	0.32 ± 0.02

We reported in Figs. 1, 2, 3 and 4 the clustering quality of the algorithm's solutions ranked from the best one in terms of criterion to the poorest one. The respective criterion of each algorithm is normalized to belong to $[0, 1]$.

When one does have the real partition, a common practice to evaluate the clustering result, one relies on the best solution obtained by optimizing the objective function. Figures 1 and 3 highlight a critical behavior of **NMF-F** which tends to produce solutions with the lowest minima that do not fulfil the best clustering partitions, sometimes with a substantial gap (see CSTR, RCV1, NG5 in Fig. 1). Moreover, a surprising lesser but still similar behavior is delivered by **S-Kmeans** which compared to **NMF**, optimizes a clustering objective by definition. The results are displayed in Fig. 2. In reality, this behavior can be observed with several types of what we refer to clustering algorithms hosting an optimization procedure. Initializing **NMF-F** randomly as shown in Fig. 3 seems to lighten this

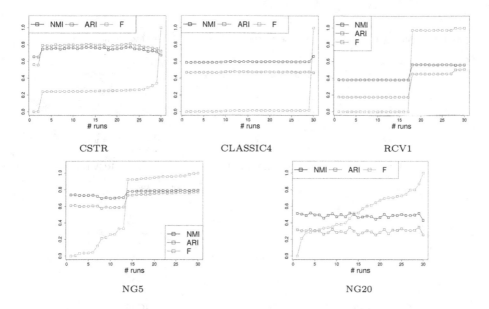

Fig. 1. NMF-F: NMI/ARI behaviour according to the objective function F (initializations by S-Kmeans)

effect (on CSTR, Classic4 and RCV1). On another hand, NMF-KL which to this day remains recognized as a relevant method for document clustering [13] seems to consistently deliver solutions with the lowest criteria aligned with the goodness of their clustering, sustaining the use of NMF for clustering purposes. Furthermore, compared to all, NMF-KL is the only method emphasizing a wide variety of solutions and therefore seems to explore way more possibilities than NMF-F or S-Kmeans. Its better behavior might almost comfort the idea of selecting the best partition in terms of criterion as the one to keep. However, it still fails on RCV1 which is the toughest dataset to partition mainly because of its scant density. Eventually, it remains concerning to select the best partition just based on the fact that, even with NMF-KL, the solution among the best ones providing the best clustering, is not necessarily the first one (see on CSTR, CLASSIC4 and NG5).

In addition, while the best solutions possibly share a similar amount of information with the true partition, they could be fairly distinct from each other, making their use appealing to deduce an even more exhaustive solution. Figure 5 shows results of pairwise NMI and ARI between the top 10 partitions (criterion-wise) of each algorithm. NMF-KL's best solutions appear to be fairly different among each other.

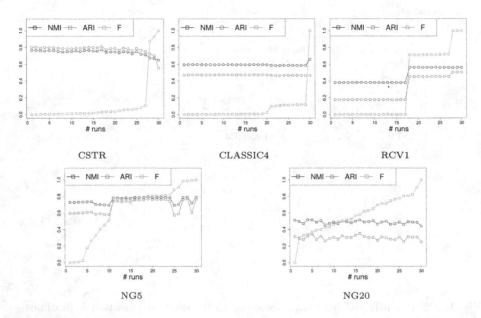

CSTR   CLASSIC4   RCV1

NG5   NG20

Fig. 2. S-Kmeans: NMI/ARI behaviour according to the objective function F (Random initializations)

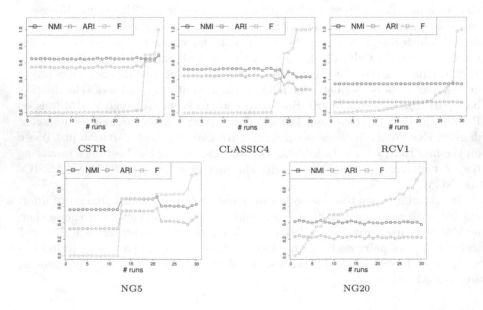

CSTR   CLASSIC4   RCV1

NG5   NG20

Fig. 3. NMF-F: NMI/ARI behaviour according to the objective function F (Random initializations)

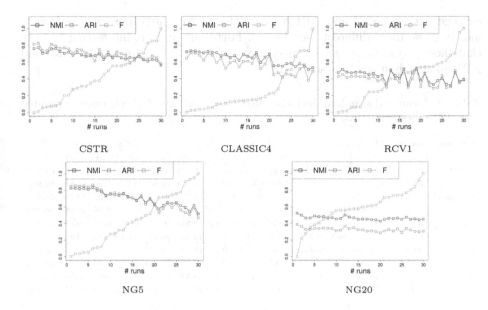

CSTR   CLASSIC4   RCV1

NG5   NG20

Fig. 4. NMF-KL: NMI/ARI behaviour according to the objective function F (Random initializations)

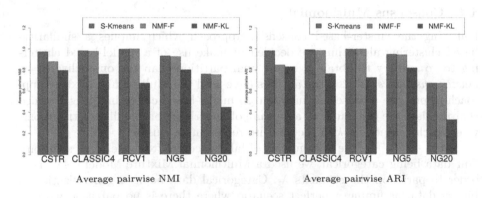

Average pairwise NMI   Average pairwise ARI

Fig. 5. Average pairwise NMI & ARI between top 10 solutions

4.3   Consensus Clustering

Following the previous statement, we went ahead and computed a cluster ensemble (CE) for NMF-F and NMF-KL according to their best initialization strategy as well as for S-Kmeans due to its pertinence for initializing NMF-F and the method being widely known as relevant for document clustering. The results are reported in Table 3. It appears that the consensus obtained with the top 10 results of each method generally outperforms the best solution. This result is even stronger for NMF-KL where the ensemble clustering increases the NMI and ARI by respectively 11 and 13 points on NG20. Note that NG20 is the dataset where the average pairwise NMI and ARI between the 10 top partitions are the lowest, meaning the most different (see Fig. 5). Furthermore, it is interesting to note that these performances are obtained from solutions giving an average NMI and ARI smaller than the best solution itself.

Table 3. Mean and standard deviation, first best result and CE consensus computed over the 10 best solutions.

Datasets	Metrics	NMF-F (Skmeans)			Skmeans			NMF-KL (Random)		
		Mean ± SD	(best)	CE	Mean ± SD	(best)	CE	Mean ± SD	(best)	CE
CSTR	NMI	0.73 ± 0.04	(0.65)	(0.76)	0.76 ± 0.007	(0.77)	(0.77)	0.73 ± 0.03	(0.76)	**(0.80)**
	ARI	0.75 ± 0.10	(0.56)	(0.80)	0.80 ± 0.007	(0.80)	(0.80)	0.77 ± 0.04	(0.81)	**(0.83)**
CLASSIC4	NMI	0.59 ± 0.002	(0.59)	(0.59)	0.60 ± 0.001	(0.59)	(0.60)	0.71 ± 0.02	(0.72)	**(0.74)**
	ARI	0.47 ± 0.002	(0.47)	(0.47)	0.47 ± 0.0009	(0.47)	(0.47)	0.65 ± 0.06	(0.65)	**(0.72)**
RCV1	NMI	0.38 ± 0.0002	(0.38)	(0.35)	0.38 ± 0.0003	(0.38)	(0.35)	0.47 ± 0.02	(0.47)	**(0.52)**
	ARI	0.18 ± 0.0003	(0.18)	(0.26)	0.18 ± 0.0004	(0.18)	(0.26)	0.42 ± 0.02	(0.43)	**(0.46)**
NG5	NMI	0.72 ± 0.02	(0.74)	(0.76)	0.72 ± 0.02	(0.73)	(0.75)	0.80 ± 0.03	(0.83)	**(0.86)**
	ARI	0.60 ± 0.01	(0.61)	(0.60)	0.60 ± 0.01	(0.60)	(0.64)	0.82 ± 0.04	(0.85)	**(0.88)**
NG20	NMI	0.49 ± 0.02	(0.51)	(0.50)	0.49 ± 0.02	(0.51)	(0.50)	0.48 ± 0.02	(0.50)	**(0.61)**
	ARI	0.30 ± 0.02	(0.32)	(0.34)	0.30 ± 0.02	(0.32)	(0.34)	0.34 ± 0.02	(0.36)	**(0.49)**

4.4 Consensus Multinomial

Following the cluster-based consensus approach which implies a similarity-based clustering algorithm, we decided to make use of a model-based clustering to go and try to obtain a better final partition than the one delivered by *cluster ensembles*. In [26], the authors have used the Multinomial mixture approach to propose a consensus function. In model-based clustering, it is assumed that the data are generated by a mixture of underlying probability distributions, where each component k of the mixture represents a cluster.

Let $\Lambda \in \mathbb{N}_0^{n \times r}$ be the data matrix of labels vectors from the top r solutions. Our data being categorical, we used a Multinomial Mixture Model (MMM) in order to partition the elements λ_i. Categorical data being a generalization of binary data; assuming a perfect scenario where there is no partition with an empty cluster, a disjunctive matrix $M \in \{0,1\}^{n \times rg}$ is usually used instead of Λ with value $m_{iq}^{(h)}$ where $h \in \{1, \ldots, g\}$ is a cluster label. Therefore, the data values $m_{iq}^{(h)}$ are assumed to be generated from a Multinomial distribution of parameter $\mathcal{M}(m_{iq}^{(h)}; \alpha_{kq}^{(h)})$ where $\alpha_{kq}^{(h)}$ is the probability that an element m_i in the group k takes the category h for the partition/variable λ_q. The density probability function of the model can be stated as:

$$f(M; \theta) = \prod_{i=1}^{n} \sum_{k=1}^{g} \pi_k \prod_{q,h}^{r,g} (\alpha_{kq}^{(h)})^{m_{iq}^{(h)}} \tag{4}$$

where $\theta = (\pi, \alpha)$ are the parameters of the model with $\pi = (\pi_1, \ldots, \pi_k)$ being the proportions and α the vector of the components parameters.

Table 4. MMM consensus results over the 10 best solutions

Datasets	Metrics	NMF-KL (Random)			
		Mean±SD	(best)	CE	MMM
CSTR	NMI	0.73 ± 0.03	(0.76)	**(0.80)**	(0.77)
	ARI	0.77 ± 0.04	(0.81)	**(0.83)**	(0.82)
CLASSIC4	NMI	0.71 ± 0.02	(0.72)	(0.74)	**(0.77)**
	ARI	0.65 ± 0.06	(0.65)	(0.72)	**(0.75)**
RCV1	NMI	0.47 ± 0.02	(0.47)	**(0.52)**	**(0.52)**
	ARI	0.42 ± 0.02	(0.43)	**(0.46)**	**(0.46)**
NG5	NMI	0.80 ± 0.03	(0.83)	**(0.86)**	**(0.86)**
	ARI	0.82 ± 0.04	(0.85)	(0.88)	**(0.89)**
NG20	NMI	0.48 ± 0.02	(0.50)	(0.61)	**(0.63)**
	ARI	0.34 ± 0.02	(0.36)	(0.49)	**(0.50)**

The Rmixmod package[2] is used to achieve our analysis. We employ the default settings to compute the clustering, allowing the selection between 10 parsimonious models according to the Bayesian information Criterion (BIC) [23]. With CSTR, the model mainly selected is the one keeping the proportions π_k free with the model also independent from the variables (labels vectors), meaning $\mathcal{M}(m_{iq}^{(h)}; \alpha_k)$. CSTR is the dataset with the highest pairwise NMI and ARI therefore with the most similar best solutions. On CLASSIC4 and RCV1 where the pairwise NMI & ARI are a little bit lower, it is the model with free proportions and parameters α depending on distinct components and labels vectors $(\mathcal{M}(m_{iq}^{(h)}; \alpha_{kq}^{(h)}))$ which is mainly chosen. On NG5 where the best solutions are fairly similar (high pairwise NMI & ARI), it is the model depending on the components and the labels vectors which has been retained. However, the proportions here were kept equal. For NG20 where the best solutions were fairly distinct, the model selected is the one depending on the components and the variables. As previously, the proportions π_k are kept equal. Following the characteristics in Table 1, it is notable to see that the datasets where the proportions are kept equal are actually those with the more balanced real clusters proportions. The results of the obtained consensus are displayed in Table 4 which only retains prior results of NMF-KL top 10 solutions and CE consensus, as they were the best overall. Apart from CSTR, we can see that MMM does a better job at computing a better partition from the top 10 solutions than CE.

5 Conclusion

In this paper, by using *cluster ensembles*, we have proposed a simple method to obtain a better clustering for the scope of NMF algorithms on text data. From its

---

[2] https://cran.r-project.org/web/packages/Rmixmod/Rmixmod.pdf.

gathering nature, this process should also alleviate the uncertainty based around the overall quality of the final partition compared to other selection practices such as keeping an unique solution according to the best criterion. Furthermore, we have shown that it was possible to improve the consensus quality through the use of finite mixture models, allowing more powerful underlying settings than cluster-based consensus involving plain similarities or distances. A future work will be to investigate the use of *cluster ensembles* for other recent clustering algorithms [1–3,19,20].

References

1. Ailem, M., Salah, A., Nadif, M.: Non-negative matrix factorization meets word embedding. In: SIGIR, pp. 1081–1084 (2017)
2. Allab, K., Labiod, L., Nadif, M.: A semi-NMF-PCA unified framework for data clustering. IEEE Trans. Knowl. Data Eng. **29**(1), 2–16 (2016)
3. Allab, K., Labiod, L., Nadif, M.: Simultaneous spectral data embedding and clustering. IEEE Trans. Neural Netw. Learn. Syst. **29**(12), 6396–6401 (2018)
4. Boutsidis, C., Gallopoulos, E.: SVD based initialization: a head start for nonnegative matrix factorization. Pattern Recogn. **41**(4), 1350–1362 (2008)
5. Bradley, P.S., Fayyad, U.M.: Refining initial points for k-means clustering. In: ICML, vol. 98, pp. 91–99. Citeseer (1998)
6. Celeux, G., Govaert, G.: A classification EM algorithm for clustering and two stochastic versions. Comput. Stat. Data Anal. **14**(3), 315–332 (1992)
7. Deerwester, S., Dumais, S.T., Furnas, G.W., Landauer, T.K., Harshman, R.: Indexing by latent semantic analysis. J. Am. Soc. Inf. Sci. **41**(6), 391–407 (1990)
8. Dempster, A.P., Laird, N.M., Rubin, D.B.: Maximum likelihood from incomplete data via the *EM* algorithm. J. Roy. Stat. Soc.: Ser. B (Methodol.) **39**(1), 1–22 (1977)
9. Dhillon, I.S., Modha, D.S.: Concept decompositions for large sparse text data using clustering. Mach. Learn. **42**(1–2), 143–175 (2001)
10. Ding, C., Li, T., Peng, W., Park, H.: Orthogonal nonnegative matrix t-factorizations for clustering. In: SIGKDD, pp. 126–135. ACM (2006)
11. Ghosh, J.: Multiclassifier systems: back to the future. In: Roli, F., Kittler, J. (eds.) MCS 2002. LNCS, vol. 2364, pp. 1–15. Springer, Heidelberg (2002). https://doi.org/10.1007/3-540-45428-4_1
12. Govaert, G., Nadif, M.: Mutual information, phi-squared and model-based co-clustering for contingency tables. Adv. Data Anal. Classif. **12**(3), 455–488 (2016). https://doi.org/10.1007/s11634-016-0274-6
13. Hosseini-Asl, E., Zurada, J.M.: Nonnegative matrix factorization for document clustering: a survey. In: Rutkowski, L., Korytkowski, M., Scherer, R., Tadeusiewicz, R., Zadeh, L.A., Zurada, J.M. (eds.) ICAISC 2014. LNCS (LNAI), vol. 8468, pp. 726–737. Springer, Cham (2014). https://doi.org/10.1007/978-3-319-07176-3_63
14. Hubert, L., Arabie, P.: Comparing partitions. J. Classif. **2**(1), 193–218 (1985)
15. Lee, D.D., Seung, H.S.: Algorithms for non-negative matrix factorization. In: Advances in Neural Information Processing Systems, pp. 556–562 (2001)
16. Li, T., Ding, C.: The relationships among various nonnegative matrix factorization methods for clustering. In: ICDM, pp. 362–371 (2006)
17. Qiao, H.: New SVD based initialization strategy for non-negative matrix factorization. Pattern Recogn. Lett. **63**, 71–77 (2015)

18. Role, F., Morbieu, S., Nadif, M.: Coclust: a Python package for co-clustering. J. Stat. Softw. **88**, 1–29 (2019)
19. Salah, A., Ailem, M., Nadif, M.: A way to boost SEMI-NMF for document clustering. In: CIKM, pp. 2275–2278 (2017)
20. Salah, A., Ailem, M., Nadif, M.: Word co-occurrence regularized non-negative matrix tri-factorization for text data co-clustering. In: AAAI, pp. 3992–3999 (2018)
21. Salah, A., Nadif, M.: Model-based von Mises-Fisher co-clustering with a conscience. In: Proceedings of the 2017 SIAM International Conference on Data Mining, pp. 246–254. SIAM (2017)
22. Salah, A., Nadif, M.: Directional co-clustering. Adv. Data Anal. Classif. **13**(3), 591–620 (2018). https://doi.org/10.1007/s11634-018-0323-4
23. Schwarz, G., et al.: Estimating the dimension of a model. Ann. Stat. **6**(2), 461–464 (1978)
24. Sharkey, A.J.: Multi-net systems. In: Sharkey, A.J.C. (ed.) Combining Artificial Neural Nets, pp. 1–30. Springer, London (1999). https://doi.org/10.1007/978-1-4471-0793-4_1
25. Strehl, A., Ghosh, J.: Cluster ensembles-a knowledge reuse framework for combining multiple partitions. J. Mach. Learn. Res. **3**(Dec), 583–617 (2002)
26. Topchy, A., Jain, A.K., Punch, W.: A mixture model for clustering ensembles. In: SDM, pp. 379–390. SIAM (2004)
27. Wild, S., Curry, J., Dougherty, A.: Improving non-negative matrix factorizations through structured initialization. Pattern Recogn. **37**(11), 2217–2232 (2004)
28. Wild, S., Wild, W.S., Curry, J., Dougherty, A., Betterton, M.: Seeding non-negative matrix factorizations with the spherical k-means clustering. Ph.D. thesis, University of Colorado (2003)
29. Yang, Z., Oja, E.: Linear and nonlinear projective nonnegative matrix factorization. IEEE Trans. Neural Netw. **21**(5), 734–749 (2010)
30. Yoo, J., Choi, S.: Orthogonal nonnegative matrix factorization: multiplicative updates on stiefel manifolds. In: Fyfe, C., Kim, D., Lee, S.-Y., Yin, H. (eds.) IDEAL 2008. LNCS, vol. 5326, pp. 140–147. Springer, Heidelberg (2008). https://doi.org/10.1007/978-3-540-88906-9_18
31. Yuan, Z., Oja, E.: Projective nonnegative matrix factorization for image compression and feature extraction. In: Kalviainen, H., Parkkinen, J., Kaarna, A. (eds.) SCIA 2005. LNCS, vol. 3540, pp. 333–342. Springer, Heidelberg (2005). https://doi.org/10.1007/11499145_35

Discriminative Bias for Learning Probabilistic Sentential Decision Diagrams

Laura Isabel Galindez Olascoaga[1]($\boxtimes$), Wannes Meert[2], Nimish Shah[1],
Guy Van den Broeck[3], and Marian Verhelst[1]

[1] Electrical Engineering Department, KU Leuven, Leuven, Belgium
{laura.galindez,nimish.shah,marian.verhelst}@esat.kuleuven.be
[2] Computer Science Department, KU Leuven, Leuven, Belgium
wannes.meert@cs.kuleuven.be
[3] Computer Science Department, University of California, Los Angeles, USA
guyvdb@cs.ucla.edu

Abstract. Methods that learn the structure of Probabilistic Sentential Decision Diagrams (PSDD) from data have achieved state-of-the-art performance in tractable learning tasks. These methods learn PSDDs incrementally by optimizing the likelihood of the induced probability distribution given available data and are thus robust against missing values, a relevant trait to address the challenges of embedded applications, such as failing sensors and resource constraints. However PSDDs are outperformed by discriminatively trained models in classification tasks. In this work, we introduce D-LEARNPSDD, a learner that improves the classification performance of the LEARNPSDD algorithm by introducing a discriminative bias that encodes the conditional relation between the class and feature variables.

Keywords: Probabilistic models · Tractable inference · PSDD

1 Introduction

Probabilistic machine learning models have shown to be a well suited approach to address the challenges inherent to embedded applications, such as the need to handle uncertainty and missing data [11]. Moreover, current efforts in the field of Tractable Probabilistic Modeling have been making great strides towards successfully balancing the trade-offs between model performance and inference efficiency: probabilistic circuits, such as Probabilistic Sentential Decision Diagrams (PSDDs), Sum-Product Networks (SPNs), Arithmetic Circuits (ACs) and Cutset Networks, posses myriad desirable properties [4] that make them amenable to application scenarios where strict resource budget constraints must be met [12]. But these models' robustness against missing data—from learning them generatively—is often at odds with their discriminative capabilities.

© The Author(s) 2020
M. R. Berthold et al. (Eds.): IDA 2020, LNCS 12080, pp. 184–196, 2020.
https://doi.org/10.1007/978-3-030-44584-3_15

We address such a conflict by proposing a discriminative-generative probabilistic circuit learning strategy, which aims to improve the models' discriminative capabilities, while maintaining their robustness against missing features.

We focus in particular on the PSDD [17], a state-of-the-art tractable representation that encodes a joint probability distribution over a set of random variables. Previous work [12] has shown how to learn hardware-efficient PSDDs that remain robust to missing data and noise. This approach relies largely on the LEARNPSDD algorithm [20], a generative algorithm that incrementally learns the structure of a PSDD from data. Moreover, it has been shown how to exploit such robustness to trade off resource usage with accuracy. And while the achieved accuracy is competitive when compared to Bayesian Network classifiers, discriminatively learned models perform consistently better than purely generative models [21] since the latter remain agnostic to the discriminative task they ought to perform. This begs the question of whether the discriminative performance of the PSDD could be improved while remaining robust and tractable.

In this work, we propose a hybrid discriminative-generative PSDD learning strategy, D-LEARNPSDD, that enforces the discriminative relationship between class and feature variables by capitalizing on the model's ability to encode domain knowledge as a logic formula. We show that this approach consistently outperforms the purely generative PSDD and is competitive compared to other classifiers, while remaining robust to missing values at test time.

2   Background

Notation. Variables are denoted by upper case letters X and their instantiations by lower case letters x. Sets of variables are denoted in bold upper case $\mathbf{X}$ and their joint instantiations in bold lower case $\mathbf{x}$. For the classification task, the feature set is denoted by $\mathbf{F}$ while the class variable is denoted by C.

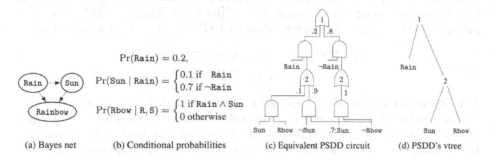

(a) Bayes net   (b) Conditional probabilities   (c) Equivalent PSDD circuit   (d) PSDD's vtree

Fig. 1. A Bayesian network and its equivalent PSDD (taken from [20]).

PSDD. Probabilistic Sentential Decision Diagrams (PSDDs) are circuit representations of joint probability distributions over binary random variables [17]. They were introduced as probabilistic extensions to Sentential Decision Diagrams (SDDs) [7], which represent Boolean functions as logical circuits. The inner nodes of a PSDD alternate between AND gates with two inputs and OR gates with arbitrary number of inputs; the root must be an OR node; and each leaf node encodes a distribution over a variable X (see Fig. 1c). The combination of an OR gate with its AND gate inputs is referred to as *decision* node, where the left input of the AND gate is called *prime* (p), and the right is called *sub* (s). Each of the n edges of a decision node are annotated with a normalized probability distribution $\theta_1, ..., \theta_n$.

PSDDs possess two important syntactic restrictions: (1) Each AND node must be *decomposable*, meaning that its input variables must be disjoint. This property is enforced by a *vtree*, a binary tree whose leaves are the random variables and which determines how will variables be arranged in primes and subs in the PSDD (see Fig. 1d): each internal vtree node is associated with the PSDD nodes at the same level, variables appearing in the left subtree $\mathbf{X}$ are the primes and the ones appearing in the right subtree $\mathbf{Y}$ are the subs. (2) Each decision node must be *deterministic*, thus only one of its inputs can be true.

Each PSDD node q represents a probability distribution. Terminal nodes encode a univariate distributions. Decision nodes, when normalized for a vtree node with $\mathbf{X}$ in its left subtree and $\mathbf{Y}$ in its right subtree, encode the following distribution over $\mathbf{XY}$ (see also Fig. 1a and c):

$$Pr_q(\mathbf{XY}) = \sum_i \theta_i Pr_{p_i}(\mathbf{X}) Pr_{s_i}(\mathbf{Y}) \tag{1}$$

Thus, each decision node decomposes the distribution into independent distributions over $\mathbf{X}$ and $\mathbf{Y}$. In general, prime and sub variables are independent at PSDD node q given the prime *base* $[q]$ [17]. This base is the support of the node's distribution, over which it defines a non-zero probability and it is written as a logical sentence using the recursion $[q] = \bigvee_i [p_i] \wedge [s_i]$. Kisa et al. [17] show that prime and sub variables are independent in PSDD node q given a prime base:

$$Pr_q(\mathbf{XY}|[p_i]) = Pr_{p_i}(\mathbf{X}|[p_i]) Pr_{s_i}(\mathbf{Y}|[p_i]) \tag{2}$$
$$= Pr_{p_i}(\mathbf{X}) Pr_{s_i}(\mathbf{Y})$$

This equation encodes *context specific independence* [2], where variables (or sets of variables) are independent given a logical sentence. The structural constraints of the PSDD are meant to exploit such independencies, leading to a representation that can answer a number of complex queries in polynomial time [1], which is not guaranteed when performing inference on Bayesian Networks, as they don't encode and therefore can't exploit such local structures.

LearnPSDD. The LEARNPSDD algorithm [20] generatively learns a PSDD by maximizing log-likelihood given available data. The algorithm starts by learning a *vtree* that minimizes the mutual information among all possible sets of variables. This vtree is then used to guide the PSDD structure learning stage, which relies on the iterative application of the Split and Clone operations [20]. These operations keep the PSDD syntactically sound while improving likelihood of the distribution represented by the PSDD. A problem with LEARNPSDD when using the resulting model for classification is that when the class variable is only weakly dependent on the features, the learner may choose to ignore that dependency, potentially rendering the model unfit for classification tasks.

3 A Discriminative Bias for PSDD Learning

Generative learners such as LEARNPSDD optimize the likelihood of the distribution given available data rather than the conditional likelihood of the class variable C given a full set of feature variables $\mathbf{F}$. As a result, their accuracy is often worse than that of simple models such as Naive Bayes (NB), and its close relative Tree Augmented Naive Bayes (TANB) [12], which perform surprisingly well on classification tasks even though they encode a simple—or naive—structure [10]. One of the main reasons for their performance, despite being generative, is that (TA)NB models have a discriminative bias that directly encodes the conditional dependence of all the features on the class variable.

We introduce D-LEARNPSDD, an extension to LEARNPSDD based on the insight that the learned model should satisfy the "class conditional constraint" present in Bayesian Network classifiers. That is, all feature variables must be conditioned on the class variable. This enforces a structure that is beneficial for classification while still allowing to generatively learn a PSDD that encodes the distribution over all variables using a state-of-the-art learning strategy [20].

3.1 Discriminative Bias

The classification task can be stated as a probabilistic query:

$$\Pr(C|\mathbf{F}) \sim \Pr(\mathbf{F}|C) \cdot \Pr(C). \tag{3}$$

Our goal is to learn a PSDD whose root decision node directly represents the conditional probability distribution $\Pr(\mathbf{F}|C)$. This can be achieved by forcing the primes of the first line in Eq. 2 to be $[p_0] = [\neg c]$ and $[p_1] = [c]$, where $[c]$ states that the propositional variable c representing the class variable is true (i.e. $C = 1$), and similarly $[\neg c]$ represents $C = 0$. For now we assume the class is binary and will show later how to generalize to a multi-valued class variable. For the feature variables we can assume they are binary without loss of generality since a multi-valued variable can be converted to a set of binary variables via a one-hot encoding (see, for example [20]). To achieve our goal we first need the following proposition:

Proposition 1. *Given (i) a vtree with a single variable C as the prime and variables $\mathbf{F}$ as the sub of the root node, and (ii) an initial PSDD where the root decision node decomposes the distribution as $[root] = ([p_0] \wedge [s_0]) \vee ([p_1] \wedge [s_1])$; applying the Split and Clone operators will never change the root decision decomposition $[root] = ([p_0] \wedge [s_0]) \vee ([p_1] \wedge [s_1])$.*

Proof. The D-LEARNPSDD algorithm iteratively applies two operations: Clone and Split (following the algorithm in [20]). First, the Clone operator requires a parent node, which is not available for the root node. Since the initial PSDD follows the logical formula described above, whose only restriction is on the root node, there is no parent available to clone and the root's base thus remains intact when applying the Clone operator. Second, the Split operator splits one of the subs to extend the sentence that is used to mutually exclusively and exhaustively define all children. Since the given vtree has only one variable, C, as the prime of the root node, there are no other variables available to add to the sub. The Split operator cant thus not be applied anymore and the root's base stays intact (see Figs. 1c and d).

We can now show that the resulting PSDD contains nodes that directly represent the distribution $\Pr(\mathbf{F}|C)$.

Proposition 2. *A PSDD of the form* $[root] = ([\neg c] \wedge [s_0]) \vee ([c] \wedge [s_1])$ *with* c *the propositional variable stating that the class variable is true, and* s_0 *and* s_1 *any formula with propositional feature variables* $f_0, \ldots, f_n$, *directly expresses the distribution* $\Pr(\mathbf{F}|C)$.

Proof. Applying this to Eq. 1 results in:

$$\begin{aligned} \Pr_q(C\mathbf{F}) &= \Pr_{\neg c}(C)\Pr_{s_0}(\mathbf{F}) + \Pr_c(C)\Pr_{s_1}(\mathbf{F}) \\ &= \Pr_{\neg c}(C|[\neg c]) \cdot \Pr_{s_0}(\mathbf{F}|[\neg c]) + \Pr_c(C|[c]) \cdot \Pr_{s_1}(\mathbf{F}|[c]) \\ &= \Pr_{\neg c}(C=0) \cdot \Pr_{s_0}(\mathbf{F}|C=0) + \Pr_c(C=1) \cdot \Pr_{s_1}(\mathbf{F}|C=1) \end{aligned}$$

The learned PSDD thus contains a node s_0 with distribution $\Pr_{s_0}$ that directly represents $\Pr(\mathbf{F}|C=0)$ and a node s_1 with distribution $\Pr_{s_1}$ that represents $\Pr(\mathbf{F}|C=1)$. The PSDD thus encodes $\Pr(\mathbf{F}|C)$ directly because the two possible value assignments of C are $C=0$ and $C=1$.

The following examples illustrate why both the specific vtree and initial PSDD are required.

Example 1. Figure 2b shows a PSDD that encodes a fully factorized probability distribution normalized for the vtree in Fig. 2a. The PSDD shown in this example initializes the incremental learning procedure of LEARNPSDD [20]. Note that the vtree does not connect the class variable C to all feature variables (e.g. F_1). Therefore, when initializing the algorithm on this vtree-PSDD combination, there are no guarantees that the conditional relations between certain features and the class will be learned.

Example 2. Figure 2e shows a PSDD that explicitly conditions the feature variables on the class variables by normalizing for the vtree in Fig. 2c and by following the logical formula from Proposition 2. This biased PSDD is then used to initialize the D-LEARNPSDD learner. Note that the vtree in Fig. 2c forces the prime of the root node to be the class variable C.

Example 3. Figure 2d shows, however, that only setting the vtree in Fig. 2c is not sufficient for the learner to condition the features on the class. When initializing on a PSDD that encodes a fully factorized formula, and then applying the Split and Clone operators, the relationship between the class variable and the features are not guaranteed to be learned. In this worst case scenario, the learned model could have an even worse performance than the case from Example 1. By applying Eq. 1 on the top split, we can give intuition why this is the case:

$$\Pr_q(\mathbf{CF}) = \Pr_{p_0}(C|[c \vee \neg c]) \cdot \Pr_{s_0}(\mathbf{F}|[c \vee \neg c])$$
$$= (\Pr_{p_1}(C|[c]) + \Pr_{p_2}(C|[\neg c])) \cdot \Pr_{s_0}(\mathbf{F}|[c \vee \neg c])$$
$$= (\Pr_{p_1}(C = 1) + \Pr_{p_2}(C = 0)) \cdot \Pr_{s_0}(\mathbf{F})$$

The PSDD thus encodes a distribution that assumes that the class variable is independent from all feature variables. While this model might still have a high likelihood, its classification accuracy will be low.

We have so far introduced the D-LEARNPSDD for a binary classification task. However, it can be easily generalized to an n-valued classification scenario: (1) The class variable C will be represented by multiple propositional variables $c_0, c_1, \ldots, c_n$ that represent the set $C = 0, C = 1, \ldots, C = n$, of which exactly one will be true at all times. (2) The vtree in Proposition 1 now starts as a right-linear tree over $c_0, \ldots, c_n$. The $\mathbf{F}$ variables are the sub of the node that has c_n as prime. (3) The initial PSDD in Proposition 2 now has a root the form $[root] = \bigvee_{i=0\ldots n}([c_i \bigwedge_{j:0\ldots n \wedge i \neq j} \neg c_j] \wedge [s_i])$, which remains the same after applying Split and Clone. The root decision node now represents the distribution $\Pr_q(\mathbf{CF}) = \sum_{i:0\ldots n} \Pr_{c_i \bigwedge_{j \neq i} \neg c_j}(C = i) \cdot \Pr_{s_i}(\mathbf{F}|C = i)$ and therefore has nodes at the top of the tree that directly represent the discriminative bias.

3.2   Generative Bias

Learning the distribution over the feature variables is a generative learning process and we can achieve this by applying the Split and Clone operators in the same way as the original LEARNPSDD algorithm. In the previous section we had not yet defined how should $\Pr(\mathbf{F}|C)$ from Proposition 2 be represented in the initial PSDD, we only explained how our constraint enforces it. So the question is how do we exactly define the nodes corresponding to s_0 and s_1 with distributions $\Pr(\mathbf{F}|C = 0)$ and $\Pr(\mathbf{F}|C = 1)$? We follow the intuition behind (TA)NB and start with a PSDD that encodes a distribution where all feature variables are independent given the class variable (see Fig. 2e). Next, the LEARNPSDD algorithm will incrementally learn the relations between the feature variables by applying the Split and Clone operations following the approach in [20].

3.3   Obtaining the Vtree

In LEARNPSDD, the decision nodes decompose the distribution into independent distributions. Thus, the vtree is learned from data by maximizing the approximate pairwise mutual information, as this metric quantifies the level of independence between two sets of variables. For D-LEARNPSDD we are interested in the level of conditional independence between sets of feature variables given the class variable. We thus obtain the vtree by optimizing for Conditional Mutual Information instead and replace mutual information in the approach in [20] with:

$$CMI(\mathbf{X}, \mathbf{Y}|\mathbf{Z}) = \sum_x \sum_y \sum_z \Pr(\mathbf{xy}) \log \frac{\Pr(\mathbf{z}) \Pr(\mathbf{xyz})}{\Pr(\mathbf{xz}) \Pr(\mathbf{yz})}.$$

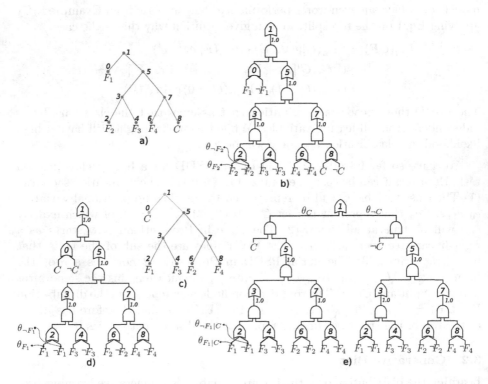

Fig. 2. Examples of vtrees and initial PSDDs.

4   Experiments

We compare the performance of D-LEARNPSDD, LEARNPSDD, two generative Bayesian classifiers (NB and TANB) and a discriminative classifier (logistic regression). In particular, we discuss the following research queries: (1) Sect. 4.2 examines whether the introduced discriminative bias improves classification performance on PSDDs. (2) Sect. 4.3 analyzes the impact of the vtree and the imposed structural constraints on model tractability and performance. (3) Finally, Sect. 4.4 compares the robustness to missing values for all classification approaches.

Table 1. Datasets

Dataset	$|F|$	$|C|$	$|N|$
Australian	40	2	690
Breast	28	2	683
Chess	39	2	3196
Cleve	25	2	303
Corral	6	2	160
Credit	42	2	653
Diabetes	11	2	768
German	54	2	1000
Glass	17	6	214
Heart	9	2	270
Iris	12	3	150
Mofn	10	2	1324
Pima	11	2	768
Vehicle	57	2	846
Waveform	109	3	5000

4.1 Setup

We ran our experiments on the suite of 15 standard machine learning benchmarks listed in Table 1. All of the datasets come from the UCI machine learning repository [8], with exception of "Mofn" and "Corral" [18]. As pre-processing steps, we applied the discretization method described in [9], and we binarized all variables using a one-hot encoding. Moreover, we removed instances with missing values and features whose value was always equal to 0. Table 1 summarizes the number of binary features $|\mathbf{F}|$, the number of classes $|C|$ and the available number of training samples $|N|$ per dataset.

4.2 Evaluation of DG-LearnPSDD

Table 2 compares D-LEARNPSDD, LEARNPSDD, Naive Bayes (NB), Tree Augmented Naive Bayes (TANB) and logistic regression (LogReg)[1] in terms of accuracy via five fold cross validation[2]. For LEARNPSDD, we incrementally learned a model on each fold until convergence on validation-data log-likelihood, following the methodology in [20].

For D-LEARNPSDD, we incrementally learned a model on each fold until likelihood converged but then selected the incremental model with the highest training set accuracy. For NB and TANB, we learned a model per fold and compiled them to Arithmetic Circuits[3], a more general form of PSDDs [6], which allows us to compare the size of these Bayes net classifiers and the PSDDs. Finally, we compare all probabilistic models with a discriminative classifier, a multinomial logistic regression model with a ridge estimator.

Table 2 shows that the proposed D-LEARNPSDD clearly benefits from the introduced discriminative bias, outperforming LEARNPSDD in all but two datasets, as the latter method is not guaranteed to learn significant relations between feature and class variables. Moreover, it outperforms Bayesian classifiers in most benchmarks, as the learned PSDDs are more expressive and allow to encode complex relationships among sets of variables or local dependencies such as context specific independence, while remaining tractable. Finally, note that the D-LEARNPSDD is competitive in terms of accuracy with respect to logistic regression (LogReg) a purely discriminative classification approach.

4.3 Impact of the Vtree on Discriminative Performance

The structure and size of the learned PSDD is largely determined by the vtree it is normalized for. Naturally, the vtree also has an important role in determining the quality (in terms of log-likelihood) of the probability distribution encoded by the learned PSDD [20]. In this section, we study the impact that the choice of vtree and learning strategy has on the trade-offs between model tractability, quality and discriminative performance.

---

[1] NB, TANB and LogReg are learned using Weka with default settings.

[2] In each fold, we hold 10% of the data for validation.

[3] Using the ACE tool Available at http://reasoning.cs.ucla.edu/ace/.

Table 2. Five cross fold accuracy and size in number of parameters

Dataset	D-LearnPSDD		LearnPSDD		NB		TANB		LogReg
	Accuracy	Size	Accuracy	Size	Accuracy	Size	Accuracy	Size	Accuracy
Australian	**86.2 ± 3.6**	367	84.9 ± 2.7	386	85.1 ± 3.1	161	85.8 ± 3.4	312	84.1 ± 3.4
Breast	97.1 ± 0.9	291	94.9 ± 0.5	491	**97.7 ± 1.2**	114	97.7 ± 1.2	219	96.5 ± 1.6
Chess	**97.3 ± 1.4**	2178	94.9 ± 1.6	2186	87.7 ± 1.4	158	91.7 ± 2.2	309	96.9 ± 0.7
Cleve	82.2 ± 2.5	292	81.9 ± 3.2	184	**84.9 ± 3.3**	102	79.9 ± 2.2	196	81.5 ± 2.9
Corral 6	**99.4 ± 1.4**	39	98.1 ± 2.8	58	89.4 ± 5.2	26	98.8 ± 1.7	45	86.3 ± 6.7
Credit	85.6 ± 3.1	693	86.1 ± 3.6	611	**86.8 ± 4.4**	170	86.1 ± 3.9	326	84.7 ± 4.9
Diabetes	**78.7 ± 2.9**	124	77.2 ± 3.3	144	77.4 ± 2.56	46	75.8 ± 3.5	86	78.4 ± 2.6
German	72.3 ± 3.2	1185	69.9 ± 2.3	645	73.5 ± 2.7	218	**74.5 ± 1.9**	429	74.4 ± 2.3
Glass	**79.1 ± 1.9**	214	72.4 ± 6.2	321	70.0 ± 4.9	203	69.5 ± 5.2	318	73.0 ± 5.7
Heart	**84.1 ± 4.3**	51	78.5 ± 5.3	75	84.0 ± 3.8	38	83.0 ± 5.1	70	84.0 ± 4.7
Iris	90.0 ± 0.1	76	94.0 ± 3.7	158	**94.7 ± 1.8**	75	94.7 ± 1.8	131	94.7 ± 2.9
Mofn	98.9 ± 0.9	260	97.1 ± 2.4	260	85.0 ± 5.7	42	92.8 ± 2.6	78	**100.0 ± 0**
Pima	**80.2 ± 0.3**	108	74.7 ± 3.2	110	77.6 ± 3.0	46	76.3 ± 2.9	86	77.7 ± 2.9
Vehicle	**95.0 ± 1.7**	1186	93.9 ± 1.69	1560	86.3 ± 2.00	228	93.0 ± 0.8	442	94.5 ± 2.4
Waveform	85.0 ± 1.0	3441	78.7 ± 5.6	2585	80.7 ± 1.9	657	83.1 ± 1.1	1296	**85.5 ± 0.7**

Figure 3a shows test-set log-likelihood and Fig. 3b classification accuracy as a function of model size (in number of parameters) for the "Chess" dataset. We display average log-likelihood and accuracy over logarithmically distributed ranges of model size. This figure contrasts the results of three learning approaches: D-LEARNPSDD when the vtree learning stage optimizes mutual information (MI, shown in light blue); when it optimizes conditional mutual information (CMI, shown in dark blue); and the traditional LEARNPSDD (in orange).

Figure 3a shows that likelihood improves at a faster rate during the first iterations of LEARNPSDD, but eventually settles to the same values as D-LEARNPSDD because both optimize for log-likelihood. However, the discriminative bias guarantees that classification accuracy on the initial model will be at least as high as that of a Naive Bayes classifier (see Fig. 3b). Moreover, this results in consistently superior accuracy (for the CMI case) compared to the purely generative LEARNPSDD approach as shown also in Table 2. The dip in accuracy during the second and third intervals are a consequence of the generative learning, which optimizes for log-likelihood and can therefore initially yield feature-value correlations that decrease the model's performance as a classifier.

Finally, Fig. 3b demonstrates that optimizing the vtree for conditional mutual information results in an overall better performance vs. accuracy trade-off when compared to optimizing for mutual information. Such a conditional mutual information objective function is consistent with the conditional independence constraint we impose on the structure of the PSDD and allows the model to consider the special status of the class variable in the discriminative task.

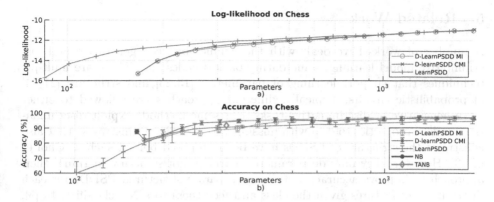

Fig. 3. Log-likelihood and accuracy vs. model size trade-off of the incremental PSDD learning approaches. MI and CMI denote mutual information and conditional mutual information vtree learning, respectively. (Color figure online)

4.4 Robustness to Missing Features

The generative models in this paper encode a joint probability distribution over all variables and therefore tend to be more robust against missing features than discriminative models, which only learn relations relevant to their discriminative task. In this experiment, we assessed this robustness aspect by simulating the random failure of 10% of the original feature set per benchmark and per fold in five-fold cross-validation. Figure 4 shows the average accuracy over 10 such feature failure trials in each of the 5 folds (flat markers) in relation to their full feature set accuracy reported in Table 2 (shaped markers). As expected, the performance of the discriminative classifier (LogReg) suffers the most during feature failure, while D-LEARNPSDD and LEARNPSDD are notably more robust than any other approach, with accuracy losses of no more than 8%. Note from the flat markers that the performance of D-LEARNPSDD under feature failure is the best in all datasets but one.

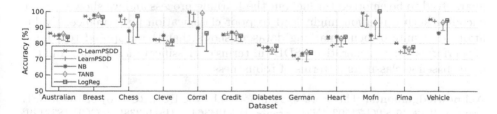

Fig. 4. Classification robustness per method.

5   Related Work

A number of works have dealt with the conflict between generative and discriminative model learning, some dating back decades [14]. There are multiple techniques that support learning of parameters [13,23] and structure [21,24] of probabilistic circuits. Typically, different approaches are followed to either learn generative or discriminative tasks, but some methods exploit discriminative models' properties to deal with missing variables [22]. Other works that also constraint the structure of PSDDs have been proposed before, such as Choi et al. [3]. However, they only do parameter learning, not structure learning: their approach to improve accuracy is to learn separate structured PSDDs for each distribution of features given the class and feed them to a NB classifier. In [5], Correira and de Campos propose a constrained SPN architecture that shows both computational efficiency and classification performance improvements. However, it focuses on decision robustness rather than robustness against missing values, essential to the application range discussed in this paper. There are also a number of methods that focus specifically on the interaction between discriminative and generative learning. In [15], Khosravi et al. provide a method to compute expected predictions of a discriminative model with respect to a probability distribution defined by an arbitrary generative model in a tractable manner. This combination allows to handle missing values using discriminative couterparts of generative classifiers [16]. More distant to this work is the line of hybrid discriminative and generative models [19], their focus is on semisupervised learning and deals with missing labels.

6   Conclusion

This paper introduces a PSDD learning technique that improves classification performance by introducing a discriminative bias. Meanwhile, robustness against missing data is kept by exploiting generative learning. The method capitalizes on PSDDs' domain knowledge encoding capabilities to enforce the conditional relation between the class and the features. We prove that this constraint is guaranteed to be enforced throughout the learning process and we show how not encoding such a relation might lead to poor classification performance. Evaluation on a suite of benchmarking datasets shows that the proposed technique outperforms purely generative PSDDs in terms of classification accuracy and the other baseline classifiers in terms of robustness.

Acknowledgements. This work was supported by the EU-ERC Project Re-SENSE grant ERC-2016-STG-71503; NSF grants IIS-1943641, IIS-1633857, CCF-1837129, DARPA XAI grant N66001-17-2-4032, gifts from Intel and Facebook Research, and the "Onderzoeksprogramma Artificiële Intelligentie Vlaanderen" programme from the Flemish Government.

References

1. Bekker, J., Davis, J., Choi, A., Darwiche, A., Van den Broeck, G.: Tractable learning for complex probability queries. In: Advances in Neural Information Processing Systems (2015)
2. Boutilier, C., Friedman, N., Goldszmidt, M., Koller, D.: Context-specific independence in Bayesian networks. In: Proceedings of the International Conference on Uncertainty in Artificial Intelligence (1996)
3. Choi, A., Tavabi, N., Darwiche, A.: Structured features in naive bayes classification. In: Thirtieth AAAI Conference on Artificial Intelligence (2016)
4. Choi, Y., Vergari, A., Van den Broeck, G.: Lecture Notes: Probabilistic Circuits: Representation and Inference (2020). http://starai.cs.ucla.edu/papers/LecNoAAAI20.pdf
5. Correia, A.H.C., de Campos, C.P.: Towards scalable and robust sum-product networks. In: Ben Amor, N., Quost, B., Theobald, M. (eds.) SUM 2019. LNCS (LNAI), vol. 11940, pp. 409–422. Springer, Cham (2019). https://doi.org/10.1007/978-3-030-35514-2_31
6. Darwiche, A.: Modeling and Reasoning with Bayesian Networks. Cambridge University Press, Cambridge (2009)
7. Darwiche, A.: SDD: a new canonical representation of propositional knowledge bases. In: International Joint Conference on Artificial Intelligence (2011)
8. Dua, D., Graff, C.: UCI machine learning repository (2017). http://archive.ics.uci.edu/ml
9. Fayyad, U., Irani, K.: Multi-interval discretization of continuous-valued attributes for classification learning. In: Proceedings of the International Joint Conference on Artificial Intelligence (IJCAI) (1993)
10. Friedman, N., Geiger, D., Goldszmidt, M.: Bayesian network classifiers. J. Mach. Learn. **29**(2), 131–163 (1997)
11. Galindez, L., Badami, K., Vlasselaer, J., Meert, W., Verhelst, M.: Dynamic sensor-frontend tuning for resource efficient embedded classification. IEEE J. Emerg. Sel. Top. Circuits Syst. **8**(4), 858–872 (2018)
12. Galindez Olascoaga, L., Meert, W., Shah, N., Verhelst, M., Van den Broeck, G.: Towards hardware-aware tractable learning of probabilistic models. In: Advances in Neural Information Processing Systems, pp. 13726–13736 (2019)
13. Gens, R., Domingos, P.: Discriminative learning of sum-product networks. In: Advances in Neural Information Processing Systems (2012)
14. Jaakkola, T., Haussler, D.: Exploiting generative models in discriminative classifiers. In: Advances in Neural Information Processing Systems (1999)
15. Khosravi, P., Choi, Y., Liang, Y., Vergari, A., Van den Broeck, G.: On tractable computation of expected predictions. In: Advances in Neural Information Processing Systems, pp. 11167–11178 (2019)
16. Khosravi, P., Liang, Y., Choi, Y., Van den Broeck, G.: What to expect of classifiers? Reasoning about logistic regression with missing features. In: Proceedings of the 28th International Joint Conference on Artificial Intelligence (IJCAI), (2019)
17. Kisa, D., den Broeck, G.V., Choi, A., Darwiche, A.: Probabilistic sentential decision diagrams. In: International Conference on the Principles of Knowledge Representation and Reasoning (2014)
18. Kohavi, R., John, G.H.: Wrappers for feature subset selection. Artif. Intell. **97**(1–2), 273–324 (1997)

19. Lasserre, J.A., Bishop, C.M., Minka, T.P.: Principled hybrids of generative and discriminative models. In: IEEE Computer Society Conference on Computer Vision and Pattern Recognition (CVPR) (2006)
20. Liang, Y., Bekker, J., Van den Broeck, G.: Learning the structure of probabilistic sentential decision diagrams. In: Proceedings of the Conference on Uncertainty in Artificial Intelligence (UAI) (2017)
21. Liang, Y., Van den Broeck, G.: Learning logistic circuits. In: Proceedings of the Conference on Artificial Intelligence (AAAI) (2019)
22. Peharz, R., et al.: Random sum-product networks: a simple and effective approach to probabilistic deep learning. In: Conference on Uncertainty in Artificial Intelligence (UAI) (2019)
23. Poon, H., Domingos, P.: Sum-product networks: a new deep architecture. In: IEEE International Conference on Computer Vision Workshops (2011)
24. Rooshenas, A., Lowd, D.: Discriminative structure learning of arithmetic circuits. In: Artificial Intelligence and Statistics, pp. 1506–1514 (2016)

Widening for MDL-Based Retail Signature Discovery

Clément Gautrais[1]([✉]) [iD], Peggy Cellier[2], Matthijs van Leeuwen[3],
and Alexandre Termier[2]

[1] Department of Computer Science, KU Leuven, Leuven, Belgium
clement.gautrais@cs.kuleuven.be
[2] Univ Rennes, Inria, INSA, CNRS, IRISA, Rennes, France
[3] LIACS, Leiden University, Leiden, The Netherlands

Abstract. *Signature patterns* have been introduced to model repetitive behavior, e.g., of customers repeatedly buying the same set of products in consecutive time periods. A disadvantage of existing approaches to signature discovery, however, is that the required number of occurrences of a signature needs to be manually chosen. To address this limitation, we formalize the problem of selecting the best signature using the minimum description length (MDL) principle. To this end, we propose an encoding for signature models and for any data stream given such a signature model. As finding the MDL-optimal solution is unfeasible, we propose a novel algorithm that is an instance of *widening*, i.e., a diversified beam search that heuristically explores promising parts of the search space. Finally, we demonstrate the effectiveness of the problem formalization and the algorithm on a real-world retail dataset, and show that our approach yields relevant signatures.

Keywords: Signature discovery · Minimum description length · Widening

1 Introduction

When analyzing (human) activity logs, it is especially important to discover recurrent behavior. Recurrent behavior can indicate, for example, personal preferences or habits, and can be useful in contexts such as personalized marketing. Some types of behavior are elusive to traditional data mining methods: for example, behavior that has some temporal regularity but not strong enough to be periodic, and which does not form simple itemsets or sequences in the log. A prime example is the set of products that is essential to a retail customer: all of these products are bought regularly, but often not periodically due to different

C. Gautrais—This work has received funding from the European Research Council (ERC) under the European Union's Horizon 2020 research and innovation programme (grant agreement No [694980] SYNTH: Synthesising Inductive Data Models).

M. R. Berthold et al. (Eds.): IDA 2020, LNCS 12080, pp. 197–209, 2020.
https://doi.org/10.1007/978-3-030-44584-3_16

depletion rates, and they are typically bought over several transactions—in any arbitrary order—rather than all at the same time.

To model and detect such behavior, we have proposed *signature patterns* [3]: patterns that identify irregular recurrences in an event sequence by segmenting the sequence (see Fig. 1). We have shown the relevance of signature patterns in the retail context, and demonstrated that they are general enough to be used in other domains, such as political speeches [2]. As a disadvantage, however, signature patterns require the analyst to provide the number of recurrences, i.e., the number of segments in the segmentation. This number of segments influences the signature: fewer segments give a more detailed signature, while more segments result in a simpler signature. Although in some cases domain experts may have some intuition on how to choose the number of segments, it is often difficult to decide on a good trade-off between the number of segments and the complexity of the signature. The main problem that we study in this paper is therefore how to automatically set this parameter in a principled way, based on the data.

Our first main contribution is a problem formalization that defines the best signature for a given dataset, so that the analyst no longer needs to choose the number of segments. By considering the signature corresponding to each possible number of segments as a model, we can naturally formulate the problem of selecting the best signature as a model selection problem. We formalize this problem using the minimum description length (MDL) principle [4], which, informally, states that the best model is the one that compresses the data best. The MDL principle perfectly fits our purposes because (1) it allows to select the simplest model that adequately explains the data, and (2) it has been previously shown to be very effective for the selection of pattern-based models (e.g., [7,11]).

After defining the problem using the MDL principle, the remaining question is how to solve it. As the search space of signatures is extremely large and the MDL-based problem formulation does not offer any properties that could be used to substantially prune the search space, we resort to heuristic search. Also here, the properties of signature patterns lead to technical challenges. In particular, we empirically show that a naïve beam search often gets stuck in suboptimal solutions. Our second main contribution is therefore to propose a diverse beam search algorithm, i.e., an instance of *widening* [9], that ensures that a diverse set of candidate solutions is maintained on each level of the beam search. For this, we define a distance measure for signatures based on their segmentations.

2   Preliminaries

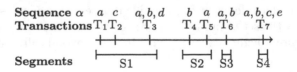

Fig. 1. A sequence of transactions and a 4-segmentation. We have the signature items $\mathcal{R} = \{a, b\}$, the remaining items $\mathcal{E} = \{c, d, e\}$, the set of items $\mathcal{I} = \{a, b, c, d, e\}$, the segmentation $S = \langle [T_1, T_2, T_3], [T_4, T_5], [T_6], [T_7] \rangle$.

Signatures. Let us first recall the definition of a *signature* as presented in [3]. Let $\mathcal{I}$ be the set of all items, and let $\alpha = \langle T_1 \ldots T_n \rangle$, $T_i \subseteq \mathcal{I}$ be a sequence of itemsets. A *k-segmentation* of α, denoted $S(\alpha, k) = \langle S_1 \ldots S_k \rangle$, is a sequence of k non-overlapping consecutive sub-sequences of α, denoted S_i and called *segments*, each consisting of consecutive transactions. An example of a 4-segmentation is given in Fig. 1. Given $S(\alpha, k) = \langle S_1 \ldots S_k \rangle$, a k-segmentation of α, we have $Rec(S(\alpha, k)) = \bigcap_{S_i \in S(\alpha,k)} (\bigcup_{T_j \in S_i} T_j)$: the set of all recurrent items that are present in each segment of $S(\alpha, k)$. For example in Fig. 1, the segmentation $S(\alpha, 4) = \langle S_1, S_2, S_3, S_4 \rangle$ gives $Rec(S(\alpha, 4)) = \{a, b\}$. Given k and α, one can compute $S_{max}(\alpha, k)$, the set of k-segmentation of α yielding the largest sets of recurrent items: $S_{max}(\alpha, k) = \mathrm{argmax}_{S(\alpha,k)} |Rec(S(\alpha, k))|$. For example, in Fig. 4, $\langle S_1, S_2, S_3, S_4 \rangle$ is the only 4-segmentation yielding two recurrent items. As all other 4-segmentations either yield zero or one recurrent item, $S_{max}(\alpha, 4) = \{\langle S_1, S_2, S_3, S_4 \rangle\}$. A k-signature (also named signature when k is clear from context) is then defined as a maximal set of recurrent items in a k-segmentation S, with $S \in S_{max}(\alpha, k)$. As $S_{max}(\alpha, k)$ can contain several segmentations, we define the k-signature set $Sig(\alpha, k)$, which contains all k-signatures: $Sig(\alpha, k) = \{Rec(S_m(\alpha, k)) \mid S_m \in S_{max}(\alpha, k)\}$. k gives the number of recurrences of the recurrent items in sequence α. Given a number of recurrences k, finding a k-*signature* relies on finding a k-segmentation that maximizes the size of the itemset that occurs in each segment of that segmentation. For example, in Fig. 1, given segmentation $S = \langle S_1, S_2, S_3, S_4 \rangle$ and given that $S_{max}(\alpha, 4) = \{S\}$, we have $Sig(\alpha, 4) = \{Rec(S)\} = \{\{a, b\}\}$. For simplicity, the segmentation associated with a k-signature in $Sig(\alpha, k)$ is denoted $S = \langle S_1 \ldots S_k \rangle$, and the signature items are denoted $\mathcal{R} \subseteq \mathcal{I}$. The remaining items are denoted $\mathcal{E}$, i.e., $\mathcal{E} = \mathcal{I} \backslash \mathcal{R}$.

Minimum Description Length (MDL). Let us now briefly introduce the basic notions of the minimum description length (MDL) principle [4] as it is commonly used in compression-based pattern mining [7]. Given a set of models $\mathcal{M}$ and a dataset $\mathcal{D}$, the best model $M \in \mathcal{M}$ is the one that minimizes $L(\mathcal{D}, M) = L(M) + L(\mathcal{D}|M)$, with $L(M)$ the length, in bits, of the encoding of M, and $L(\mathcal{D}|M)$ the length, in bits, of the encoding of the data given M. This is called *two-part MDL* because it separately encodes the model and the data given the model, which results in a natural trade-off between model complexity and data complexity. To fairly compare all models, the encoding has to be *lossless*. To use the MDL principle for model selection, the model class $\mathcal{M}$ has to be defined (in our case, the set of all signatures), as well as how to compute the length of the model and the length of the data given the model. It should be noted that only the *encoded length* of the data is of interest, not the encoded data itself.

3   Problem Definition

To extract recurrent items from a sequence using signatures, one must define the number of segments k. Providing meaningful values for k usually requires expert knowledge and/or many tryouts, as there is no general rule to automatically set

k. Our problem is therefore to devise a method that adjusts k, depending on the data at hand. As this is a typical model selection problem, our approach relies on the minimum description length principle (MDL) to find the best model from a set of candidate models. However, the signature model must be refined into a probabilistic model to use the MDL principle for model selection. Especially, the occurrences of items in α should be defined according to a probability distribution. With no information about these occurrences, the uniform distribution is the most natural choice. Indeed, without information on the transaction in which an item occurs, the best is to assume it can occur uniformly at random in any transaction of the sequence α. Moreover, the choice of the uniform distribution has been shown to minimize the worst case description length [4].

To make the signature model probabilistic, we assume that it generates three different types of occurrences independently and uniformly. As the signature gives the information that there is at least one occurrence of every signature item in every segment, the first type of occurrences correspond to this one occurrence of signature items in every segment. These are generated uniformly over all the transactions of every segment. The second type of occurrences are the remaining signature items occurrences. Here, the information is that these items already have occurrences generated by the previous type of occurrences. As α is a sequence of itemsets, an item can occur at most once in a transaction. Hence, for a given signature item, the second type of occurrences for this item are distributed uniformly over the transactions where this item does not already occur for the first type of occurrences. Finally, the third type are the occurrences of the remaining items: the items that are not part of the signature. There is no information about these items occurrences, hence we assume them to be generated uniformly over all transactions of α.

With these three types of occurrences, the signature model is probabilistic: all occurrences in α are generated according to a probability distribution that takes into account the information provided by the signature specification. Hence, we can now define the problem we are tackling:

Problem 1. Let $\mathbb{S}$ denote the set of signatures for all values of k, $\mathbb{S} = \bigcup_{k=1}^{|\alpha|} Sig(\alpha, k)$. Given a sequence α, it follows from the MDL principle that the best signature $S \in \mathbb{S}$ is the one that minimizes the two-part encoded length of S and α, i.e.,

$$S_{MDL} = \operatorname{argmin}_{S \in \mathbb{S}} L(\alpha, S),$$

where $L(\alpha, S)$ is the two-part encoded length that we present in the next section.

4   An Encoding for Signatures

As typically done in compression-based pattern mining [7], we use a two-part MDL code that leads to decomposing the total encoded length $L(\alpha, S)$ into two parts: $L(S)$ and $L(\alpha|S)$, with the relation $L(\alpha, S) = L(S) + L(\alpha|S)$. In the upcoming subsection we define $L(S)$, i.e., the encoded length of a signature, after which Subsect. 4.2 introduces $L(\alpha|S)$, i.e., the length of the sequence α given a signature S. In the remainder of this paper, all logarithms are in base 2.

4.1 Model Encoding: $L(S)$

A signature is composed of two parts: (1) the signature items, and (2) the signature segmentation. The two parts are detailed below.

Signature Items Encoding. The encoding of the signature items consists of three parts. The signature items are a subset of $\mathcal{I}$, hence we first encode the number of items in $\mathcal{I}$. A common way to encode non-negative integer numbers is to use the universal code for integers [4,8], denoted $L_{\mathbb{N}}$[1]. This yields a code of size $L_{\mathbb{N}}(|\mathcal{I}|)$. Next, we encode the number of items in the signature, using again the universal code for integers, with length $L_{\mathbb{N}}(|\mathcal{R}|)$. Finally, we encode the items of the signature. As the order of signature items is irrelevant, we can use an $|\mathcal{R}|$-combination of $|\mathcal{I}|$ elements without replacement. This yields a length of $\log(\binom{|\mathcal{I}|}{|\mathcal{R}|})$. From $\mathcal{R}$ and $\mathcal{I}$, we can deduce $\mathcal{E}$.

Segmentation Encoding. We now present the encoding of the second part of the signature: the signature segmentation. To encode the segmentation, we encode the segment boundaries. These boundaries are indexed on the size of the sequence, hence we first need to encode the number of transactions n. This can be done using again the universal code for integers, which is of size $L_{\mathbb{N}}(n)$. Then, we need to encode the number of segments $|S|$, which is of length $L_{\mathbb{N}}(|S|)$. To encode the segments, we only have to encode the boundaries between two consecutive segments. As there are $|S| - 1$ such boundaries, a naive encoded length would be $(|S|-1)*\log(n)$. An improved encoding takes into account the previous segments. For example, when encoding the second boundary, we know that its value will not be higher than $n - |S_1|$. Hence, we can encode it in $\log(n - |S_1|)$ instead of $\log(n)$ bits. This principle can be applied to encode all boundaries. Another way to further reduce the encoded length is to use the fact that we know that each signature segment contains at least one transaction. We can therefore subtract the number of remaining segments to encode the boundary of the segment we are encoding. This yields an encoded length of $\sum_{i=1}^{|S|-1} \log(n - (|S| - i) - \sum_{j=1}^{i-1} |S_j|)$.

Putting Everything Together. The total encoded length of a signature S is

$$L(S) = L_{\mathbb{N}}(|\mathcal{I}|) + L_{\mathbb{N}}(|\mathcal{R}|) + \log\left(\binom{|\mathcal{I}|}{|\mathcal{R}|}\right) +$$

$$L_{\mathbb{N}}(n) + L_{\mathbb{N}}(|S|) + \sum_{i=1}^{|S|-1} \log(n - (|S| - i) - \sum_{j=1}^{i-1} |S_j|).$$

---

[1] $L_{\mathbb{N}} = \log^*(n) + \log(2.865064)$, with $\log^*(n) = \log(n) + \log(\log(n)) + \dots$.

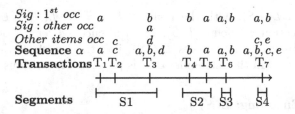

Fig. 2. A sequence of transactions and its encoding scheme. We have $\mathcal{R} = \{a, b\}$, $\mathcal{E} = \{c, d, e\}$ and $\mathcal{I} = \{a, b, c, d, e\}$. The first occurrence of each signature item in each segment is encoded in the red stream, the remaining signature items occurrences in the orange stream, and the items from $\mathcal{E}$ in the blue stream. (Color figure online)

4.2   Data Encoding: $L(\alpha|S)$

We now present the encoding of the sequence given the model: $L(\alpha|S)$. This encoding relies on the refinement of the signature model into a probabilistic model presented in Sect. 3. To summarize, we have three separate encoding streams that encode the three different types of occurrences presented in Sect. 3: (1) one that encodes one occurrence of every signature item in every segment, (2) one that encodes the rest of the signature items occurrences, and (3) one that encodes the remaining items occurrences. An example illustrating the three different encoding streams is presented in Fig. 2.

Encoding One Occurrence of Each Signature Item in Each Segment. As stated in Sect. 3, the signature says that in each segment, there is at least one occurrence of each signature item. The size of each segment is known (from the encoding of the model, in Subsect. 4.1), hence we encode one occurrence of each signature item in segment S_i by encoding the index of the transaction, within segment S_i, that contains this occurrence. From Sect. 3, this occurrence is uniformly distributed over the transactions in S_i. As encoding an index over $|S_i|$ equiprobable possibilities costs $\log(|S_i|)$ bits and as in each segment, $|\mathcal{R}|$ occurrences are encoded this way, we encode each segment in $|\mathcal{R}| * \log(|S_i|)$ bits.

Encoding the Remaining Signature Items' Occurrences. As presented in Fig. 2, we now encode remaining signature items occurrences to guarantee a lossless encoding. Again, this encoding relies on encoding transactions where signature items occur. For each item a, we encode its occurrences $occ(a) = \sum_{T_i \in \alpha} \sum_{p \in T_i} \mathbf{1}_{a=p}$ by encoding to which transaction it belongs. As S occurrences have already been encoded using the previous stream, there are $occ(a) - |S|$ remaining occurrences to encode. These occurrences can be in any of the $n - |S|$ remaining transactions. From Sect. 3, we use a uniform distribution to encode them. More precisely, the first occurrence of item a can belong to any of the $n - |S|$ transactions where a does not already occur. For the second occurrence of a, there are now only $n - |S| - 1$ transactions where a can occur. By applying this principle, we encode all the remaining occurrences of a as $\sum_{i=0}^{occ(a) - |S| - 1} \log(n - |S| - i)$. For each item, we also use $L_{\mathbb{N}}(occ(a) - |S|)$ bits to encode the number of occurrences. This yields a total length of $\sum_{a \in \mathcal{R}} L_{\mathbb{N}}(occ(a) - |S|) + \sum_{i=0}^{occ(a)-|S|-1} \log(n - |S| - i)$.

Remaining Items Occurrences Encoding. Finally, we encode the remaining items occurrences, i.e., the occurrences of items in $\mathcal{E}$. The encoding technique is identical to the one used to encode additional signature items occurrences, with the exception that the remaining items occurrences can initially be present in any of the n transactions. This yields a total code of $\sum_{a \in \mathcal{E}} L_{\mathbb{N}}(occ(a)) + \sum_{i=0}^{occ(a)} \log(n - i)$.

Putting Everything Together. The total encoded length of the data given the model is given by: $L(\alpha|S) = \sum_{S_i \in S} |\mathcal{R}| * \log(|S_i|) + \sum_{a \in \mathcal{R}} L_{\mathbb{N}}(occ(a) - |S|) + \sum_{i=0}^{occ(a)-|S|-1} \log(n - |S| - i) + \sum_{a \in \mathcal{E}} L_{\mathbb{N}}(occ(a)) + \sum_{i=0}^{occ(a)} \log(n - i)$.

5   Algorithms

The previous section presented how a sequence is encoded, completing our problem formalization. The remaining problem is to find the signature minimizing the code length, that is, finding S_{MDL} such that $S_{MDL} = \operatorname{argmin}_{S \in \mathbb{S}} L(\alpha, S)$.

Naive Algorithm. A naive approach would be to directly mine the whole set of signatures $\mathbb{S}$ and find the signature that minimizes the code length. However, mining a signature with k segments has time complexity $O(n^2 k)$. Mining the whole set of signatures requires k to vary from 1 to n, resulting in a total complexity of $O(n^4)$. The quartic complexity does not allow us to quickly mine the complete set of possible signatures on large datasets, hence we have to rely on heuristic approaches.

To quickly search for the signature in $\mathbb{S}$ that minimizes the code length, we initially rely on a top-down greedy algorithm. We start with one segment containing the whole sequence, and then search for the segment boundary that minimizes the encoded length. Then, we recursively search for a new single segment boundary that minimizes the encoded length. We stop when no segment can be added, i.e., when the number of segments is equal to the number of transactions. During this process, we record the signature with the best encoded length. However, this algorithm can perform early segment splits that seem promising initially, but that eventually impair the search for the best signature.

5.1   Widening for Signatures

To solve this issue, a solution is to keep the w signatures with the lowest code length at each step instead of keeping only the best one. This technique is called *beam search* and has been used to tackle optimization problems in pattern mining [6]. The beam width w is the number of solutions to keep at each step of the algorithm. However, the beam search technique suffers from having many of the best w signatures that tend to be similar and correspond to slight variations of one signature. Here, this means that most signatures in the beam would

Algorithm 1. Widening algorithm for signature code length minimization.

1: **function** SIGNATURE MINING($\alpha = \langle T_1, \ldots, T_n \rangle$, β, w)
2:     BestKSign $= \emptyset$, BestSign $= \emptyset$
3:     **for** $k = 1 \rightarrow n$ **do**
4:         AllKSign = Split1Segment(BestKSign)
5:         $S_{opt} = \text{argmin}_{S \in AllKSign} L(\alpha, S)$
6:         BestSign = BestSign $\bigcup \{S_{opt}\}$
7:         BestKSign $= \{S_{opt}\}$
8:         $\theta = threshold(\beta, w, \text{AllKSign})$
9:         **while** $S_{opt} \neq \emptyset$ and $|\text{BestKSign}| < w$ **do**
10:             $S_{opt} = \text{argmin}_{S \in AllKSign} L(\alpha, S), \nexists S_i \in BestKSign, d(S_i, S) \leq \theta$
11:             BestKSign = BestKSign $\bigcup \{S_{opt}\}$
12:     **return** $\text{argmin}_{S \in BestSign} L(\alpha, S)$

Algorithm 2. Distance threshold computation.

1: **function** THRESHOLD(β, w, $AllSign$)
2:     KBest $= \beta * |AllSign|$
3:     BestS = GetBestSign(AllSign, KBest)
4:     **return** $\text{argmin}_\theta \{N(\theta), N(\theta) = |\{S \in BestS, d(S, BestS[0]) < \theta\}|, N(\theta) \geq |BestS|/w\}$

---

have segmentations that are very similar. The widening technique [9] solves this issue by adding a diversity constraint into the beam. Different constraints exist [5,6,9], but a common solution is to add a distance constraint between each pair of elements in the beam: all pairwise distances between the signatures in the beam have to be larger than a given threshold θ. As this threshold is dependent on the data and the beam width, we propose a method to automatically set its value.

Algorithm 1 presents the proposed widening algorithm. Line 3 iterates over the number of segments. Line 4 computes all signatures having k segments that are considered to enter the beam. More specifically, function *Split1Segment* computes the direct refinements of each of all signatures in *BestKSign*. A direct refinement of a signature corresponds to splitting one segment in the segmentation associated with that signature. Line 5 selects the refinement having the smallest code length. If several refinements yield the smallest code length, one of these refinements is chosen at random. Lines 8 to 11 perform the widening step by adding new signatures to the beam while respecting the pairwise distance constraint. Line 8 computes the distance threshold (θ) depending on the diversity parameter (β), the beam width (w), and the current refinements. Algorithm 2 presents the details of the threshold computation. With this threshold, we recursively add a new element in the beam, until either the beam is full or no new element can be added (line 9). Lines 10 and 11 add the signature having the smallest code length and being at a distance of at least θ to any current element of the beam. Line 12 returns the best overall signature we have encountered.

Distance Between Signatures. We now define the distance measure for signatures (used in line 10 of Algorithm 1). As the purpose of the signature distance is to ensure diversity in the beam, we will use the segmentation to define the distance between two elements of the beam, i.e., between two signatures. Terzi et al. [10] presented several distance measures for segmentations. The *disagreement distance* is particularly appealing for our purposes as it compares how transactions belonging to the same segment in one segmentation are allocated to the other segmentation. Let $S_a = \langle S_{a1} \ldots S_{ak} \rangle$ and $S_b = \langle S_{b1} \ldots S_{bk} \rangle$ be two k-segmentations of a sequence α. We denote by $d(S_a, S_b)$ the disagreement distance between segmentation a and segmentation b. The disagreement distance corresponds to the number of transaction pairs that belong to the same segment in one segmentation, but that are not in the same segment in the other segmentation. Techniques on how to efficiently compute this distance are presented in [10].

Defining a Distance Threshold. Algorithm 1 uses a distance threshold θ between two signatures, that controls the diversity constraint in the beam. If θ is equal to 0, there is no diversity constraint, as any distance between two different signatures is greater than 0. Higher values of θ enforce more diversity in the beam: good signatures will not be included in the beam if they are too close to signatures already in the beam. However, setting the θ threshold is not easy. For example θ depends on the beam width w. Indeed, with large beam widths, θ should be low enough to allow many good signatures to enter the beam.

To this end, we introduce a method that automatically sets the θ parameter, depending on the beam width and on a new parameter β that is easier to interpret. The β parameter ranges from 0 to 1 and controls the strength of the diversity constraint. The intuition behind β is that its value will approximately correspond to the relative rank of the worst signature in the beam. For example, if β is set to 0.2, it means that signatures in the beam are in the top-20% in ascending order of code length. Algorithm 2 details how θ is derived from β and w; this algorithm is called by the *threshold* function in line 8 of Algorithm 1.

Knowing the set of all candidate signatures that are considered to enter the beam, we retain only the proportion β of the best signatures (line 3 of Algorithm 2). Then, in line 4 we extract the best signature. Finally, we look for the distance threshold θ such that the number of signatures within a distance of θ from the best signature is equal to the number of considered signatures divided by the beam width w (line 5). The rationale behind this threshold is that since we are adding w signatures to the beam and we want to use the proportion β of the best signatures, the distance threshold should approximately discard $1/w$ of the proportion β of the best signatures around each signature of the beam.

6   Experiments

This section, analyzes runtimes and code lengths of variants of our algorithm on a real retail dataset[2]. We show that our method runs significantly faster than the naive baseline, and give advice on how to choose the w and β parameters. Next, we illustrate the usefulness of the encoding to analyze retail customers.

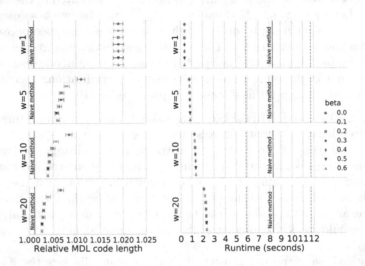

Fig. 3. Left: Mean relative code length for different instances of the widening algorithm. For each customer, the relative code length is computed with regard to the smallest code length found for this customer. Averaging these lengths across all customers gives the mean relative code length. The β parameter sets the diversity constraint and w the beam width. The solid black line shows the mean code length of the naive algorithm. Bootstrapped 95% confidence intervals [1] are displayed. **Right**: Mean runtime in seconds for different instances of the widening algorithm. The dotted black lines shows a bootstrapped 95% confidence interval of the naive algorithm's mean runtime.

6.1   Algorithm Runtime and Code Length Analysis

We here analyze the runtimes and code lengths obtained by variants of Algorithm 1. 3000 customers having more than 40 baskets in the Instacart 2017 dataset are randomly selected[3]. Customers having few purchases are less relevant, as we are looking for purchase regularities. These 3000 customers are analyzed individually, hence the algorithm is evaluated on different sequences.

---

[2] Code is available at https://bitbucket.org/clement_gautrais/mdl_signature_ida 2020/.

[3] The Instacart Online Grocery Shopping Dataset 2017, Accessed from https://www.instacart.com/datasets/grocery-shopping-2017on05/04/2018.

Code Length Analysis. To assess the performance of the different algorithms, we analyze the code length yielded by each algorithm on each of these 3000 customers. We evaluate different instances of the widening algorithm with different beam widths w and diversity constraints β. The resulting relative mean code lengths per algorithm instance are presented in Fig. 3 left. When increasing the beam width, the code length always decreases for a fixed β value. This is expected, as increasing the beam size allows the widening algorithm to explore more solutions. As increasing the beam size improves the search, we recommend setting it as high as your computational budget allows you to do.

Increasing the β parameter usually leads to better code lengths. However, for $w = 5$, higher β values give slightly worse results. Indeed, if β is too high, good signatures might not be included in the beam, if they are too close to existing solutions. Therefore, we recommend setting the β value to a moderate value, for example between 0.3 and 0.5. A strong point of our method is that it is not too sensitive to different β values. Hence, setting this parameter to its optimal value is not critical. The enforced diversity is highly relevant, as a fixed beam size with some diversity finds code lengths that are similar to the ones found by a larger beam size with no diversity. For example, with $w = 5$ and $\beta = 0.3$, the code lengths are better than with $w = 10$ and $\beta = 0$. As using a beam size of 5 with $\beta = 0.3$ is faster than using a beam size of 10 with $\beta = 0$, it shows that using diversity is highly suited to decrease runtime while yielding smaller code lengths.

Runtime Analysis. We now present runtimes of different widening instances in Fig. 3 right. The beam width mostly influences the runtime, whereas the β value has a smaller influence. Overall, increasing β slightly increases computation time, while yielding a noticeable improvement in the resulting code length, especially for small beam sizes. Our method also runs 5 to 10 times faster than the naive method. In this experiment, customers have a limited number of baskets (at most 100), thus the $O(n^4)$ complexity of the naive approach exhibits reasonable runtimes. However in settings with more transactions (retail data over a longer period for example), the naive approach will require hours to run, and the performance gain of our widening approach will be a necessity. Another important thing is that the naive method has a high variability in runtimes. Confidence intervals are narrow for the widening algorithm (they are barely noticeable on the plot), whereas it spans over 5 s for the naive algorithm.

6.2 Qualitative Analysis

Figure 4 presents two signatures of a customer, to illustrate that signatures are of practical use to analyze retail customers, and that finding signatures with smaller code lengths is of interest. We use the widening algorithm to get a variety of good signatures according to our MDL encoding. The top signature in Fig. 4 is the best signature found: it has the smallest code length. This signature seems to correctly capture the regular behavior of this customer, as it contains 7 products that are regularly bought throughout the whole purchase sequence.

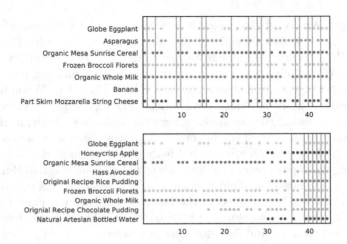

Fig. 4. Example of two signatures found by our algorithms. Gray vertical lines are segments boundaries and each dot represents an item occurrence in a purchase sequence. **Top**: best signature (code length of 5221.33 bits) found by the widening algorithm, with $w = 20$ and $\beta = 0.5$. **Bottom**: signature found by the beam search algorithm: $w = 1$ and $\beta = 0$, with a code length of 5338.46 bits (the worst code length).

Knowing these 7 favorite products, a retailer could target its offers. The segments also give some information regarding the temporal behavior of this customer. For example, because segments tend to be smaller and more frequent towards the end of the sequence, one could guess that this customer is becoming a regular.

On the other hand, the bottom signature is significantly worse than the top one. It is clear that it mostly contains products that are bought only at the end of the purchase sequence of this customer. This phenomenon occurs because the beam search algorithm, with $w = 1$, only picks the best solution at each step of the algorithm. Hence, it can quickly get stuck in a local minimum. This example shows that considering larger beams and adding diversity is an effective approach to optimize code length. Indeed, having a large and diverse beam is necessary to have the algorithm explore different segmentations, yielding better signatures.

7   Conclusions

We tackled the problem of automatically finding the best number of segments for signature patterns. To this end, we defined a model selection problem for signatures based on the minimum description length principle. Then, we introduced a novel algorithm that is an instance of widening. We evaluated the relevance and effectiveness of both the problem formalization and the algorithm on a retail dataset. We have shown that the widening-based algorithm outperforms the beam search approach as well as a naive baseline. Finally, we illustrated the practical usefulness of the signature on a retail use case. As part of future work, we would like to study our optimization techniques on larger databases (thousands of transactions), like online news feeds. We would also like to work on model selection for *sets* of interesting signatures, to highlight diverse recurrences.

References

1. Davison, A.C., Hinkley, D.V., et al.: Bootstrap Methods and Their Application, vol. 1. Cambridge University Press, Cambridge (1997)
2. Gautrais, C., Cellier, P., Quiniou, R., Termier, A.: Topic signatures in political campaign speeches. In: Proceedings of EMNLP 2017, pp. 2342–2347 (2017)
3. Gautrais, C., Quiniou, R., Cellier, P., Guyet, T., Termier, A.: Purchase signatures of retail customers. In: Kim, J., Shim, K., Cao, L., Lee, J.-G., Lin, X., Moon, Y.-S. (eds.) PAKDD 2017. LNCS (LNAI), vol. 10234, pp. 110–121. Springer, Cham (2017). https://doi.org/10.1007/978-3-319-57454-7_9
4. Grünwald, P.D.: The Minimum Description Length Principle. MIT Press, Cambridge (2007)
5. Ivanova, V.N., Berthold, M.R.: Diversity-driven widening. In: Tucker, A., Höppner, F., Siebes, A., Swift, S. (eds.) IDA 2013. LNCS, vol. 8207, pp. 223–236. Springer, Heidelberg (2013). https://doi.org/10.1007/978-3-642-41398-8_20
6. van Leeuwen, M., Knobbe, A.: Diverse subgroup set discovery. Data Min. Knowl. Disc. **25**(2), 208–242 (2012)
7. van Leeuwen, M., Vreeken, J.: Mining and using sets of patterns through compression. In: Aggarwal, C., Han, J. (eds.) Frequent Pattern Mining, pp. 165–198. Springer, Cham (2014). https://doi.org/10.1007/978-3-319-07821-2_8
8. Rissanen, J.: A universal prior for integers and estimation by minimum description length. Ann. Stat. **11**, 416–431 (1983)
9. Shell, P., Rubio, J.A.H., Barro, G.Q.: Improving search through diversity. In: Proceedings of the AAAI National Conference on Artificial Intelligence, pp. 1323–1328. AAAI Press (1994)
10. Terzi, E.: Problems and algorithms for sequence segmentations. Ph.D. thesis (2006)
11. Vreeken, J., van Leeuwen, M., Siebes, A.: KRIMP: mining itemsets that compress. Data Min. Knowl. Disc. **23**(1), 169–214 (2011). https://doi.org/10.1007/s10618-010-0202-x

Addressing the Resolution Limit and the Field of View Limit in Community Mining

Shiva Zamani Gharaghooshi[1]([✉]), Osmar R. Zaïane[1]([✉]), Christine Largeron[2],
Mohammadmahdi Zafarmand[1], and Chang Liu[1]

[1] Alberta Machine Intelligence Institute, University of Alberta,
Edmonton, AB, Canada
{zamanigh,zaiane,zafarman,chang6}@ualberta.ca
[2] Laboratoire Hubert Curien, Université de Lyon, Saint-Etienne, France
Christine.Largeron@univ-st-etienne.fr

Abstract. We introduce a novel efficient approach for community detection based on a formal definition of the notion of community. We name the links that run between communities weak links and links being inside communities strong links. We put forward a new objective function, called SIWO (Strong Inside, Weak Outside) which encourages adding strong links to the communities while avoiding weak links. This process allows us to effectively discover communities in social networks without the resolution and field of view limit problems some popular approaches suffer from. The time complexity of this new method is linear in the number of edges. We demonstrate the effectiveness of our approach on various real and artificial datasets with large and small communities.

Keywords: Community detection · Social network analysis

1 Introduction

Community detection is an important task in social network analysis and can be used in different domains where entities and their relations are presented as graphs. It allows us to find linked nodes that we call communities inside graphs. There are community detection methods that partition the graph into subgroups of nodes such as the spectral bisection method [4] or the Kernighan-Lin algorithm [27]. There are also hierarchical methods such as the divisive algorithms based on edge betweenness of Girwan et al. [18] or agglomerative algorithms based on dynamical process such as Walktrap [20], Infomap [24] or Label propagation [22]. We do not detail them and refer the interested reader to [7,10,12], but we come back on another class of hierarchical algorithms that aim at maximizing Q-modularity introduced by Newman et al. [18]. After the greedy agglomerative algorithm initially introduced by Newman [19], Blondel et al. [5] proposed Louvain, one of the fastest algorithms to optimize Q-modularity and to solve the community detection task. However, Fortunato et al. [11] showed that Q-Modularity suffers from the resolution limit which means by optimizing

© The Author(s) 2020
M. R. Berthold et al. (Eds.): IDA 2020, LNCS 12080, pp. 210–222, 2020.
https://doi.org/10.1007/978-3-030-44584-3_17

Q-modularity, communities that are smaller than a scale cannot be resolved. The field of view limit [25] is in contrast to the resolution limit leads to overpartitioning the communities with a large diameter.

To overcome the resolution limit of Q-modularity, several proposals have been made, notably by [2,17,23], who introduced variants of this criterion allowing the detection of community structures at different levels of granularity. However, these revised criteria make the method time-consuming since they require to tune a parameter. Therefore, we retain the greedy approach of Louvain for its efficiency and ability to handle very large networks, but we introduce SIWO because it relies on the notions of strong and weak links defined in Sect. 2.

We consider that a community corresponds to a subgraph sparsely connected to the rest of the graph. Contrary to the majority of methods which do not formally define what is a community and simply consider that it corresponds to a subset of nodes densely connected internally, we define the conditions a subgraph should meet to be considered as a community in Sect. 2. In Sect. 3, we present the generic community detection algorithm. We can apply this general process regardless of the objective function to improve other community detection methods as our experiments show.

Finally, the extensive experiments described in Sects. 4 and 5, confirm that our objective function is less sensitive to the resolution and the field of view limit compared to the objective functions mentioned earlier. Also, our algorithm has consistently good performance regardless of the size of communities in a network and is efficient on large size networks having up to a million edges.

2   Notations and Definitions

2.1   Strong and Weak Links

A community is oftentimes defined as a subgraph in which nodes are densely connected while sparsely connected to the rest of the graph. One way to find such subgraphs is to divide the network into parts so that the number of links lying inside that part is maximized. However, if there is no prior information about the number of communities or their sizes, one can maximize the number of links within communities by putting all the nodes in one community, but the final result will not be the true communities. To avoid this approach, we penalize the missing links within the communities and we introduce the notions of strong and weak links.

   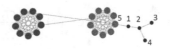

Fig. 1. A network with two communities; each consists of a clique of size 5.   **Fig. 2.** A network with 2 communities and 4 dangling nodes (1, 2, 3, and 4).

Weak links lie between communities, while strong links are inside them. We develop our criterion so that it encourages adding strong links to the communities while avoiding weak ones instead of penalizing the missing links. As these different types of links play different roles in graph connectivity; removing a weak link may divide the graph into disconnected subgraphs, whereas removing a random link would not. Let us focus on the link between nodes i and j in Fig. 1 and also the link between nodes j and k in this graph. Node j is connected to all the neighbors of node k, whereas node i and j have no common neighbors. As generally, nodes in the same community are more likely to have common neighbors, (i, j) can be considered as a weak link whereas (j, k) as a strong link and it is exactly what we want to capture through weights assigned to the links.

2.2   Edge Strength

Given a graph $G = (V, E)$ where V is the set of nodes and E the set of edges, we propose to assign a weight in the range of $(-1, 1)$ to each edge; such that strong links have larger weights. As nodes in the same community tend to have more common neighbors compared to nodes in different communities, if $S_{xy} > S_{xy'}$ then e_{xy} is more likely to be a strong link compared to $e_{xy'}$ with S_{xy} defined by:

$$S_{xy} = |\{k \in V : (x, k) \in E, (y, k) \in E\}| \tag{1}$$

We can compare two links according to S only if they share a node. Thus, if we consider nodes x and y that have 5 and 20 links incident to them, then S can be in range of $[0, 4]$ and $[0, 19]$ for x and y respectively. Consequently, for comparisons, we have to scale down S values to $(-1, 1)$. If S_{xy} has the maximum value of S_x^{max} ($S_x^{max} = \max_{y:(x,y)\in E} S_{xy}$) for a particular node x. We divide the range $[-1, 1]$ into $S_x^{max} + 1$ equal length segments. Each S value in the range of $[0, S_x^{max}]$ is then mapped to the center of $(n + 1)^{th}$ segment using equation:

$$w_{xy}^x = S_{xy} \frac{2}{S_x^{max} + 1} + \frac{1}{S_x^{max} + 1} - 1 \tag{2}$$

where w_{xy}^x is the scaled value of S_{xy} from the viewpoint of node x (min-max normalization could also work). We can also scale S_{xy} from the viewpoint of node y: $w_{xy}^y = S_{xy} \frac{2}{S_y^{max}+1} + \frac{1}{S_y^{max}+1} - 1$ where $S_y^{max} = \max_{x:(y,x)\in E} S_{xy}$. To decide whether we should trust x or y, we need to look at the importance of each one in the network. Local clustering coefficient (CC) [28], given below, is a measure that reflects the importance of nodes and it can be computed even on large graphs, for instance with Mapreduce [15].

$$CC(x) = \frac{|\{e_{ij} : i, j \in N_x, e_{ij} \in E\}|}{\binom{d_x}{2}} \tag{3}$$

where d_x and N_x are respectively the degree and the set of neighbors of node x. CC is in the range of $[0,1]$ with 1 for nodes whose neighbors form cliques, and 0 for nodes whose neighbors are not connected to each other directly. Here, we scale each edge from the viewpoint of the endpoint that is more likely to be in a dense neighborhood characterized by a large CC:

$$w_{xy} = \begin{cases} w_{xy}^x, & \text{if } CC(x) \geq CC(y) \\ w_{xy}^y, & \text{otherwise} \end{cases} \tag{4}$$

2.3 SIWO Measure

The new measure that we propose encourages adding strong links into the communities while keeping the weak links outside of the communities (**S**trong **I**nside, **W**eak **O**utside). This measure is defined as follows:

$$SIWO = \sum_{i,j \in V} \frac{w_{ij}\delta(c_i, c_j)}{2} \tag{5}$$

where c_i is the community of node i and $\delta(x,y)$ is 1 if $x = y$ and 0 otherwise. SIWO is the sum of weights of the edges that reside in the communities. This objective function provides a way to partition the set of nodes but it does not specify the conditions required by a subset of nodes to be a community. These conditions are defined in the following.

2.4 Community Definition

Following [21] we consider that a subgraph C is a community in a weak sense if the following condition is satisfied:

$$\frac{1}{2} \sum_{v \in C} |N_v^C| > \sum_{v \in C} |N_v - N_v^C| \tag{6}$$

where N_v is the set of the neighbors of node v and N_v^C is the set of the neighbors of node v that are also in community C. This condition means that the collective of the nodes in a community have more neighbors within the community than outside. In this paper, we expand this definition by adding one more condition. Given a partition $p = \{C_1, C_2, ..., C_t\}$ of a network, subgraph C_i is considered as a **qualified community** if it satisfies the following conditions:

1. C_i is a community in a weak sense (Eq. 6).
2. The number of links within C_i exceeds the number of links towards any other subgraph C_j $(j \neq i)$ in the partition p taken separately, such that:

$$\frac{1}{2} \sum_{v \in C_i} |N_v^{C_i}| > \sum_{v \in C_i} |N_v^{C_j}|, j \in [1..t], j \neq i \tag{7}$$

3   The SIWO Method

This method has four steps: pre-processing, optimizing SIWO, qualified community identification, and post-processing. They are discussed in detail below.

Step 1. Pre-processing

The first step calculates the edge strength weights (w_{ij}) needed during the SIWO optimization. Moreover, to reduce the computational time, we remove the dangling nodes temporally. Node x is a dangling node if there exists node y such that by removing e_{xy}, the network would be divided into two disconnected parts with $part_x$ (the part containing node x) being a tree. Since $part_x$ has a tree structure, it cannot form a community on its own. So all the nodes in $part_x$ belong to the same community as node y. In Fig. 2, nodes 1, 2, 3 and 4 are dangling nodes and they belong to the same community as node 5, unless we consider them outliers. Even though such tree-structured subgraphs attached to the network are very sparse and cannot be considered as communities, they satisfy Eqs. (6) and (7) defined for qualified communities. So we do not need to consider them during the community detection process. To remove them (and the links incident to them), we need to investigate every node of the network in the first time to identify nodes with degree of 1. However, after the first visit, we only need to check the list of the neighbors of the nodes that are removed in the previous time.

Step 2. Optimizing SIWO

We use Louvain's optimization process to maximize SIWO since it has been proven to be very efficient but we replace the modularity by our criterion. This greedy optimization process has two main phases, iteratively performed until a local maximum of the objective function (SIWO measure) is reached. The first phase starts by placing each node of graph G in its community. Then each node is moved to the neighbor community which results in the maximum gain of the SIWO value. If no gain can be achieved, the node stays in its community. In the second phase, a new weighted graph G' is created in which each node corresponds to a community in G. Two nodes in G' are connected if there exists at least one edge lying between their corresponding communities in G. Finally, we assign each edge e_{xy} in G' a weight equal to the sum of the weights of edges between the communities that match with x and y. These two phases are repeated until no further improvement in the SIWO objective function can be achieved.

Step 3. Qualified Community Identification

This step determines qualified communities complying with Eqs. (6) and (7) for the dense subgraphs discovered in the previous step. However, there may exist communities consisting of one node weakly connected to all of its neighbors $(S_x^{max} = 0)$ and that have links with non-positive weight incident to it, we call them Lone communities. Since the decision about the communities of such nodes can not be made on edge strength, we let the majority of their neighbors decide about their communities but, to reduce the computational time, like for dangling nodes, we temporarily remove these nodes in this step and bring them back in the final step. Then, we identify the unqualified communities which do not satisfy Eqs. (6) or (7). We keep merging each unqualified community with one of its neighboring communities (qualified or not) until no more unqualified community exists. For that, first, we assign a weight equal to 1 to each edge. Then, we repeat the two phases of Louvain. In phase 1, we create a new graph G^* in which each node corresponds to a community identified in step 3 for the first iteration of in phase 2 for the next ones and where each edge e_{xy} is assigned a weight equal to the sum of the weights of edges between the communities that correspond to x and y. We also add a self-loop to each node that has a weight equal to the sum of the weights of the edges that reside in its corresponding community. In phase 2, we visit all nodes in G^*. If a node x has a self-loop with a weight that is larger than (1) half of sum of the weights of the edges incident to it and (2) weight of any edge connecting x to another node in G^*, it means the community assigned to x satisfies both the conditions in Eqs. (6) and (7), we let x stay in its community. Otherwise, we move node x to the neighboring community that results in the maximum decrease in the sum of the weights of the edges that lie between communities of G^*.

Step 4. Post-processing
Finally, each lone community that was temporarily removed is sequentially added back to the network and merged with the community in which it has the most neighbors. If two or more communities tie and they have more than one connection to the node, then one is chosen at random. Otherwise, we choose the community of the most important neighbor, based on the largest degree of centrality within its community. Since we add lone nodes one after the other, the community that a former node is assigned to, might not be the best for that node. To resolve this issue, once all lone nodes are added to the network, we repeat moving each one of them to the community of the majority of its neighbors until no further movement can be made. Dangling nodes are also added to the network in the reverse order that they were removed and they are assigned to the community of their unique neighbor.

4   The Resolution Limit of SIWO

Fortunato and Barthélemy [11] used two sample networks, shown in Fig. 3, to demonstrate how Q-modularity is affected by the resolution limit. The first example is a ring of cliques where each clique is connected to its adjacent cliques through a single link. If the number of cliques is larger than about $\sqrt{m}$ with m being the total number of edges in the network, then optimizing Q-modularity results in merging the adjacent cliques into groups of two or more, despite that each clique corresponds to a community. The second example is a network containing 4 cliques: 2 of size k and 2 of size p. If $k \gg p$, Q-modularity similarly fails to find the correct communities and the cliques of size p will be merged.

To prove how SIWO resolves the resolution limit of Q-modularity, the exact structure of the network should be known; which is not possible. So, we analyze whether SIWO is affected by the resolution limit on these networks Given the definition of SIWO, let us consider the edge e_{xy} between two adjacent cliques

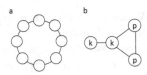

Fig. 3. Schematic examples (a) a ring of cliques; adjacent cliques are connected through a single link (b) a network with 2 cliques of size k and 2 cliques of size p.

in the first network. Since x and y do not have any common neighbors, the edge between them has a non-positive weight. Therefore, by maximizing SIWO measure in our algorithm, the adjacent cliques will not be merged. For the edge e_{xy} between the cliques of size p in the second network, since x and y have at most one common neighbor, the edge between them has a non-positive weight. Therefore, the cliques in the second network will not be merged either.

5   Experimental Results

We compared the performance of our method with the most widely used and efficient algorithms, as pointed out in several recent state of art studies [8,29], on both real and synthetic networks. The algorithms are: 1- Fastgreedy [6]; 2- Infomap; 3- Infomap+ which is Infomap to which we added the third step of our algorithm (to relieve its sensitivity to the field of view limit and demonstrate that our framework can be used to improve other algorithms); 4- Label Propagation [22]; 5- Louvain[1] [5]; 6- Walktrap[2] [20]. It should be noted that Infomap is the only algorithm that suffers from the filed of view limit among these algorithms.

The results are evaluated according to the Adjusted Rand Index (ARI) [14] and Normalized Mutual Information (NMI) [26]. As both ARI and NMI show similar results, we only present ARI results for lack of space. We also compared the results of different methods according to the ratio of the number of detected communities over the true number of communities in the ground-truth to observe how a method is affected by the resolution and the field of view limits.

5.1   Real Networks

We used 5 real networks and the ground-truth communities are available for 4 of them. Table 1 presents the properties of these networks.

We compared SIWO and Louvain on Eurosis network [9] which represents scientific web pages from 12 European countries and the hyperlinks between them without known ground-truth communities. However, since each European country has its own language, web pages in different countries are sparsely connected to each other. Moreover, as reported in [9], some of the countries can be

---

[1] https://github.com/taynaud/python-louvain.
[2] https://www-complexnetworks.lip6.fr/~latapy/PP/walktrap.html.

Table 1. Properties of real networks

Network	#nodes	#edges	#C	Network	#nodes	#edges	#C
Karate [30]	34	78	2	Eurosis [9]	1218	5999	-
Polbooks^a	105	441	3	Polblogs [1]	1222	16717	2
Football [13]	115	613	12				

^{a}http://www.orgnet.com

divided into smaller components e.g. Montenegro network includes three components: 1- Telecom and Engineering, 2- Faculties and 3- High Schools. Louvain detects 13 communities whereas SIWO detects 16 communities in this network. Louvain assigns all nodes in Montenegro network to one giant community. However, SIWO puts Faculties and High Schools in one community and Telecom and Engineering web pages in another community. These two communities are connected to each other with only 7 links. However, Louvain cannot separate them due to its resolution limit.

Table 2. Comparison of 7 algorithms according to ARI and the ratio of the number of detected communities over the true number of communities in the ground-truth on real networks. Tables shows the average results and standard deviation computed on 10 iterations of the algorithms on each network.

		Karate	Polbooks	Football	Polblogs
SIWO	ARI	**1 ± 0**	**0.67 ± 0**	0.79 ± 0	0.77 ± 0
	$\overline{C}/C_r$	**1 ± 0**	1.3 ± 0	**1 ± 0**	**1.5 ± 0**
Fastgreedy	ARI	0.68 ± 0	0.63 ± 0	0.47 ± 0	0.78 ± 0
	$\overline{C}/C_r$	1.5 ± 0	1.3 ± 0	0.5 ± 0	5 ± 0
Infomap	ARI	0.7 ± 0	0.64 ± 0	0.84 ± 0	0.68 ± 0
	$\overline{C}/C_r$	1.5 ± 0	1.6 ± 0	0.9 ± 0	17.5 ± 0
Infomap+	ARI	0.70 ± 0	0.66 ± 0	**0.84 ± 0**	0.76 ± 0
	$\overline{C}/C_r$	1.5 ± 0	1.3 ± 0	0.9 ± 0	1.5 ± 0
Label_prop	ARI	0.66 ± 0.3	0.66 ± 0	0.73 ± 0	**0.8 ± 0**
	$\overline{C}/C_r$	1.2 ± 0.35	**1.1 ± 0.1**	0.8 ± 0.1	2.1 ± 0
Louvain	ARI	0.46 ± 0	0.55 ± 0	0.8 ± 0	0.77 ± 0
	$\overline{C}/C_r$	2 ± 0	1.3 ± 0	0.8 ± 0	4.5 ± 0
Walktrap	ARI	0.32 ± 0	0.65 ± 0	0.81 ± 0	0.76 ± 0
	$\overline{C}/C_r$	3 ± 0	1.3 ± 0	0.8 ± 0	5.5 ± 0

Table 2 presents the comparison with respect to ARI and $\overline{C}/C_r$, the ratio of the number of detected communities over the true number of communities (both ARI and $\overline{C}/C_r$ should be as close to 1 as possible) in the ground-truth, on real networks with ground-truth communities. It shows that SIWO performs better on Karate and Polbooks based on ARI. It also outperforms the others methods on Karate, Football, and Polblogs networks according to $\overline{C}/C_r$ measure (SIWO could detect the exact communities with respect to the ground-truth on these networks). Infomap detects a considerably larger number of communities in Polblogs network which indicates this algorithm is sensitive to the field of view limit [25]. However, Infomap+ is much less sensitive to this limit which implies the third step of SIWO, added to Infomap+, is effective in resolving the field of view limit. Considering results for all networks, SIWO is the top performer among these algorithms on a variety of networks.

5.2   Synthetic Networks

To analyze the effect of the resolution and field of view limit, it is important to test how community detection algorithms perform on networks with small/large communities. Therefore, in this work we generated two sets of networks using LFR [16] to test the different algorithms: one with large communities and one with small communities. The first set is in favor of algorithms that suffer from resolution limit such as Louvain and the second set is in favor of algorithms with field of view limit such as Infomap. Each set includes networks with a varying number of nodes and mixing parameter. The mixing parameter controls the fraction of edges that lie between communities. We do not generate networks with mixing parameter ≥ 0.5 since beyond this point and including 0.5, the communities in the ground truth no longer satisfy the definition of community. The input parameters used to generate these two sets are presented in Table 3. Figures 4 and 5 present respectively ARI or the ratio of the number of detected communities over the true number of communities ($\overline{C}/C_r$). Panels correspond to networks with a specific number of nodes (1000 to 100000) and they are divided into two parts; the lower (respectively upper) part illustrates the average ARI (or $\overline{C}/C_r$) (respectively standard deviation) computed over 20 graphs (10 small and 10 large communities) as a function of the mixing parameter.

Table 3. Input parameters of LFR benchmark: Set 1 contains networks with large communities and Set 2 contains networks with small communities. For each combination of parameters we generated 10 networks.

	Set 1	Set 2
#nodes (N)	$[1, 10, 50, 100] \times 10^3$	$[1, 10, 50, 100] \times 10^3$
Average and max degrees	20 - $N/10$	20 - $\sqrt{N}$
Mixing parameter	$[1, \dots, 7] \times 0.1$	$[1, \dots, 7] \times 0.1$
Min and max community sizes	$N/20$ - $N/10$	Default - by default $\sqrt{N}$

Figure 4 shows the performance of Fastgreedy decreases as the mixing parameter increases. Louvain and Walktrap perform well on the smallest networks in the set; however, its performance drops when we apply it to the networks with sizes 50000 and larger. Label propagation, Infomap and Infomap+ perform well up to when the mixing parameter reaches 0.3. However, a larger mixing parameter causes a rapid decrease in the ARI value when applying these algorithms to the two largest networks in the set. These three algorithms have a large standard deviation and their outputs are not stable on these networks. SIWO correctly detects the communities when the mixing parameter is less than or equal to 0.3 (ARI $\simeq$ 1) regardless of size of the network and has the best performance overall.

Figure 5 clearly shows the resolution limit of Louvain and Fastgreedy as they underestimate the number of communities. SIWO is the best performer in terms of the number of communities and it has a very small standard deviation whereas, Infomap+ and Label propagation have a large standard deviation and fail to find the correct number of communities when the mixing parameter exceeds 0.3.

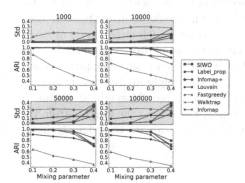

Fig. 4. Evaluation according to ARI on synthetic networks generated with LFR.

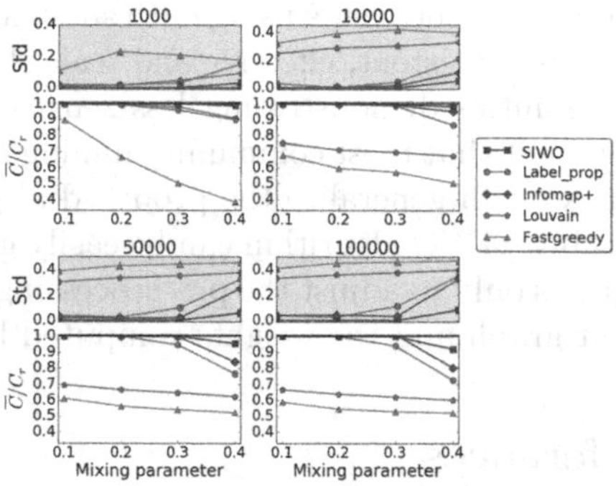

Fig. 5. Evaluation of SIWO, Label propagation, Infomap+, Louvain and Fastgreedy according to $\overline{C}/C_r$ on synthetic networks generated with LFR.

6   Scalability

We analyze how the computational cost of SIWO varies with the size of the network. The pre-processing step has two phases: removing dangling nodes which requires a time of the order of n where n is the number of nodes, and calculating the edge strength weights which requires a time of the order of $nd^2 = 2md$ where m is the number of edges and d is the average degree. In many real networks d is much smaller than n and it does not grow with n [10]. The second and third step follows the same greedy process as Louvain does. Louvain is theoretically cubic but was demonstrated experimentally to be quasi-linear [3] and has been applied with success to handle large size networks having several million nodes, and 100 million links. The time complexity of the post-processing step depends on the number of Lone communities and if all the nodes are in Lone communities, it requires a time $O(nd^2)$. Overall, the time complexity of SIWO is $O(n + md)$, which is similar to Louvain due to the fact that d is small and $n = 2m/d$. SIWO can detect communities in a networks with 100000 nodes and 1 million edges, in about 1 min on a commodity i7 and 8GB RAM laptop. The current implementation of SIWO is in Python[3], derived from python-louvain.

7   Conclusion

This paper introduces SIWO, a novel objective function based on edge strength for community detection, and a formal definition of community, that we use to lead the community detection process after optimizing the objective function. This framework can also be applied to other community detection methods to remedy their inability that causes the resolution or the field of view limit. Our extensive experiments using both small and large networks confirm that our algorithm is consistent, effective and scalable for networks with either large or small communities demonstrating less sensitivity to the resolution limit and field of view limit that most community mining algorithms suffer from. As a future direction, we will generalize the proposed algorithm for weighted/directed networks. Notably, SIWO algorithm can be easily generalized to handle weighted graphs. It requires only to adjust the pre-processing step by combining the weights from the input graph and the weights computed by SIWO to evaluate the edge strength.

References

1. Adamic, L.A., Glance, N.: The political blogosphere and the 2004 U.S. election. In: Proceedings of the 3rd International Workshop on Link Discovery, pp. 36–43 (2005)
2. Arenas, A., Fernandez, A., Gomez, S.: Analysis of the structure of complex networks at different resolution levels. New J. Phys. **10**(5), 053039 (2008)
3. Aynaud, T., Blondel, V.D., Guillaume, J.L., Lambiotte, R.: Multilevel Local Optimization of Modularity, pp. 315–345. Wiley, Hoboken (2013)
4. Barnes, E.R.: An algorithm for partitioning the nodes of a graph. SIAM J. Alg. Discr. Meth. **3**(4), 541–550 (1982)
5. Blondel, V.D., Guillaume, J.L., Lambiotte, R., Lefebvre, E.: Fast unfolding of communities in large networks. J. Stat. Mech. Theor. Exp. **2008**(10), P10008 (2008)
6. Clauset, A., Newman, M.E.J., Moore, C.: Finding community structure in very large networks. Phys. Rev. E **70**, 066111 (2004)
7. Coscia, M., Giannotti, F., Pedreschi, D.: A classification for community discovery methods in complex networks. Stat. Anal. Data Min. **4**(5), 512–546 (2013)
8. Emmons, S., Kobourov, S., Gallant, M., Börner, K.: Analysis of network clustering algorithms and cluster quality metrics at scale. PLoS One **11**(7), 1–18 (2016)

---

[3] SIWO Code and datasets available at https://www.dropbox.com/sh/eehjt5qblll0yvg/AACW2XjHJjHX2Q876Vbk0e4Ya?dl=0 .

9. EUROSIS Final Report: Webmapping of science and society actors in Europe, final report. www.eurosfaire.prd.fr/7pc/documents/1274371553_finalreporteurosis3_1. doc. Accessed 01 June 2018

10. Fortunato, S.: Community detection in graphs. Phy. Rep. **486**(3–5), 75–174 (2010)

11. Fortunato, S., Barthélemy, M.: Resolution limit in community detection. PNAS **104**(1), 36–41 (2007)

12. Fortunato, S., Hric, D.: Community detection in networks: a user guide. Phys. Rep. **659**, 1–44 (2016)

13. Girvan, M., Newman, M.E.J.: Community structure in social and biological networks. PNAS **99**(12), 7821–7826 (2002)

14. Hubert, L., Arabie, P.: Comparing partitions. J. Classif. **2**(1), 193–218 (1985)

15. Kolda, T.G., Pinar, A., Plantenga, T.D., Seshadhri, C., Task, C.: Counting triangles in massive graphs with MapReduce. SIAM J. Sci. Comput. **36**(5), S48–S77 (2014)

16. Lancichinetti, A., Fortunato, S., Radicchi, F.: Benchmark graphs for testing community detection algorithms. Phys. Rev. E **78**(4), 1–5 (2008)

17. Li, Z., Zhang, S., Wang, R.S., Zhang, X.S., Chen, L.: Quantitative function for community detection. Phys. Rev. E **77**(3), 36109 (2008)

18. Newman, M., Girvan, M.: Finding and evaluating community structure in networks. Phys. Rev. E **69**, 026113 (2004)

19. Newman, M.: Fast algorithm for detecting community structure in networks. Phys. Rev. E - Stat. Nonlinear Soft Matter Phys. **69**, 066133 (2004)

20. Pons, P., Latapy, M.: Computing communities in large networks using random walks. In: Yolum, I., Güngör, T., Gürgen, F., Özturan, C. (eds.) ISCIS 2005. LNCS, vol. 3733, pp. 284–293. Springer, Heidelberg (2005). https://doi.org/10. 1007/11569596_31

21. Radicchi, F., Castellano, C., Cecconi, F., Loreto, V., Parisi, D.: Defining and identifying communities in networks. PNAS **101**(9), 2658–63 (2004)

22. Raghavan, N., Albert, R., Kumara, S.: Near linear time algorithm to detect community structures in large-scale networks. Phys. Rev. E - Stat. Nonlinear Soft Matter Phys. **76**, 036106 (2007)

23. Reichardt, J., Bornholdt, S.: Statistical mechanics of community detection. Phys. Rev. E **74**, 16110 (2006)

24. Rosvall, M., Bergstrom, C.: Maps of random walks on complex network reveal community structure. PNAS **105**(4), 1118–1123 (2008)

25. Schaub, M.T., Delvenne, J.C., Yaliraki, S.N., Barahona, M.: Markov dynamics as a zooming lens for multiscale community detection: non clique-like communities and the field-of-view limit. PLoS One **7**, e32210 (2012)

26. Strehl, A., Ghosh, J.: Cluster ensembles – a knowledge reuse framework for combining multiple partitions. J. Mach. Learn. Res. **3**, 583–617 (2003)

27. Kernighan, B.W., Lin, S.: An efficient heuristic procedure for partitioning graphs. Bell Syst. Tech. J. **49**(2), 291–307 (1970)

28. Watts, D.J., Strogatz, S.H.: Collective dynamics of 'small-world' networks. Nature **393**(6684), 440–442 (1998)

29. Yang, Z., Algesheimer, R., Tessone, C.J.: A comparative analysis of community detection algorithms on artificial networks. Sci. Rep. **6**, 30750 (2016)

30. Zachary, W.: An information flow model for conflict and fission in small groups. J. Anthropol. Res. **33**, 452–473 (1977)

Estimating Uncertainty in Deep Learning for Reporting Confidence: An Application on Cell Type Prediction in Testes Based on Proteomics

Biraja Ghoshal[1]([⊠]), Cecilia Lindskog[2], and Allan Tucker[1]

[1] Brunel University London, Uxbridge UB8 3PH, UK
biraja.ghoshal@brunel.ac.uk
[2] Department of Immunology, Genetics and Pathology, Rudbeck Laboratory,
Uppsala University, 75185 Uppsala, Sweden
https://www.brunel.ac.uk/computer-science

Abstract. Multi-label classification in deep learning is a practical yet challenging task, because class overlaps in the feature space means that each instance is associated with multiple class labels. This requires a prediction of more than one class category for each input instance. To the best of our knowledge, this is the first deep learning study which quantifies uncertainty and model interpretability in multi-label classification; as well as applying it to the problem of recognising proteins expressed in cell types in testes based on immunohistochemically stained images. Multi-label classification is achieved by thresholding the class probabilities, with the optimal thresholds adaptively determined by a grid search scheme based on Matthews correlation coefficients. We adopt MC-Dropweights to approximate Bayesian Inference in multi-label classification to evaluate the usefulness of estimating uncertainty with predictive score to avoid overconfident, incorrect predictions in decision making. Our experimental results show that the MC-Dropweights visibly improve the performance to estimate uncertainty compared to state of the art approaches.

Keywords: Uncertainty estimation · Multi-label classification · Cell type prediction · Human Protein Atlas · Proteomics

1 Introduction

Proteins are the essential building blocks of life, and resolving the spatial distribution of all human proteins at an organ, tissue, cellular, and subcellular level greatly improves our understanding of human biology in health and disease. The testes is one of the most complex organs in the human body [15]. The spermatogenesis process results in the testes containing the most tissue-specific genes than elsewhere in the human body. Based on an integrated 'omics' approach using transcriptomics and antibody-based proteomics, more than 500 proteins with distinct testicular protein expression patterns have previously been identified [10], and transcriptomics data suggests that over 2,000 genes are elevated

© The Author(s) 2020
M. R. Berthold et al. (Eds.): IDA 2020, LNCS 12080, pp. 223–234, 2020.
https://doi.org/10.1007/978-3-030-44584-3_18

in testes compared to other organs. The function of a large proportion of these proteins are however largely unknown, and all genes involved in the complex process of spermatogenesis are yet to be characterized. Manual annotation provides the standard for scoring immunohistochemical staining pattern in different cell types. However, it is tedious, time-consuming and expensive as well as subject to human error as it is sometimes challenging to separate cell types by the human eye. It would be extremely valuable to develop an automated algorithm that can recognise the various cell types in testes based on antibody-based proteomics images while providing information on which proteins are expressed by that cell type [10]. This is, therefore, a multi-label image classification problem.

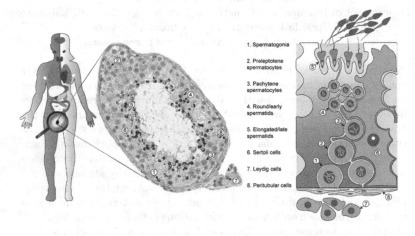

Fig. 1. Schematic overview: cell type-specific expression of testis elevated genes [10]

Exact Bayesian inference with deep neural networks is computationally intractable. There are many methods proposed for quantifying uncertainty or confidence estimates. Recently Gal [5] proved that a dropout neural network, a well-known regularisation technique [13], is equivalent to a specific variational approximation in Bayesian neural networks. Uncertainty estimates can be obtained by training a network with dropout and then taking Monte Carlo (MC) samples of the prediction using dropout during test time. Following Gal [5], Ghoshal et al. [7] also showed similar results for neural networks with Drop-weights and Teye [14] with batch normalisation layers in training (Fig. 1).

In this paper, we aim to:

1. Present the first approach in multi-label pattern recognition that can recognise various cell types-specific protein expression patterns in testes based on antibody-based proteomics images and provide information on which cell types express the protein with estimated uncertainty.
2. Show Multi-Label Classification (MLC) is achieved by thresholding the class probabilities, with the Optimal Thresholds adaptively determined by a grid search scheme based on Matthews correlation coefficient.

3. Demonstrate through extensive experimental results that a Deep Learning Model with MC-Dropweights [7] is significantly better than a wide spectrum of MLC algorithms such as Binary Relevance (BR), Classifier Chain (CC), Probabilistic Classifier Chain (PCC) and Condensed Filter Tree (CFT), Cost-sensitive Label Embedding with Multidimensional Scaling (CLEMS) and state-of-the-art MC-Dropout [5] algorithms across various cell types.
4. Develop Saliency Maps in order to increase model interpretability visualizing descriptive regions and highlighting pixels from different areas in the input image. Deep learning models are often accused of being "black boxes", so they need to be precise, interpretable, and uncertainty in predictions must be well understood.

Our objective is not to achieve state-of-the-art performance on these problems, but rather to evaluate the usefulness of estimating uncertainty leveraging MC-Dropweights with predictive score in multi-label classification to avoid over-confident, incorrect predictions for decision making.

2   Multi-label Cell-Type Recognition and Localization with Estimated Uncertainty

2.1   Problem Definition

Given a set of training data D, where $X = \{x_1, x_2 \ldots x_N\}$ is the set of N images and the corresponding labels $Y = \{y_1, y_2 \ldots y_N\}$ is the cell-type information. The vector $y_i = \{y_{i,1}, y_{i,2} \ldots y_{i,M}\}$ is a binary vector, where $y_{i,j} = 1$ indicates that the ith image belongs to the jth cell-type. Note that an image may belong to multiple cell-types, i.e., $1 <= \sum_j y_{i,j} <= M$. Based on $D(X, Y)$, we constructed a Bayesian Deep Learning model giving an output of the predictive probability with estimated uncertainty of a given image x_i belonging to each cell category. That is, the constructed model acts as a function such that $f : X \to Y$ using weights of neural net parameters ω where $(0 <= \hat{y}_{x,j} <= 1)$ as close as possible to the original function that has generated the outputs Y, output the estimated value $(\hat{y}_{i,1}, \hat{y}_{i,2}, \ldots, \hat{y}_{i,M})$ as close to the actual value $(y_{i,1}, y_{i,2}, \ldots, y_{i,M})$.

2.2   Solution Approach

We tailored Deep Convolutional Neural Network (DCNN) architectures for cell type detection and localisation by considering a large image capacity, binary-cross entropy loss, sigmoid activation, along with Dropweights in the fully connected layer and Batch Normalization formulation of propagating uncertainty in deep learning to estimate meaningful model uncertainty.

Multi-label Setup: There are multiple approaches to transform the multi-label classification into multiple single-label problems with the associated loss function [8]. In this study, we used immunohistochemically stained testes tissue consisting of 8 cell types corresponding to 512 testis elevated genes.

Therefore, we define a 8-dimensional class label vector $Y = \{y_1, y_2 \ldots y_N\}$; $Y \in \{0, 1\}$, given 8 cell types. y_c indicates the presence with respect to according cell type expressing the protein in the image while an all-zero vector [0; 0; 0; 0; 0; 0; 0; 0] represents the "Absence" (no cell type expresses the protein in the scope of any of 8 categories).

Multi-label Classification Cost Function: The cost function for Multi-label Classification has to be different considering the fact that a prediction for a class is not mutually exclusive. So we selected the sigmoid function with the addition of binary cross-entropy.

Data Augmentation: We used Keras' image pre-processing package to apply affine transformations to the images, such as rotation, scaling, shearing, and translation during training and inference. This reduces the epistemic uncertainty during training, captures heteroscedastic aleatoric uncertainty during inference and overall improves the performance of models.

Multi-label Classification Algorithm: In Bayesian classification, the mean of the predictive posterior corresponds to the parameter point estimates, and the width of the posterior reflects the confidence of the predictions. The output of the network is an M-dimensional probability vector, where each dimension indicates how likely each cell type in a given image expresses the protein. The number of cell types that simultaneously express the protein in an image varies. One method to solve this multi-label classification problem is placing thresholds on each dimension. However different dimensions may be associated with different thresholds. If the value of the ith dimension of $\hat{y}$ is greater than a threshold, we can say that the i-th cell-type is expressed in the given tissue. The main problem is defining the threshold for each class label.

A threshold based on Matthews Correlation Coefficient (MCC) is used on the model outcome to determine the predicted class to improve the accuracy of the models.

We adopted a grid search scheme based on Matthews Correlation Coefficients (MCC) to estimate the optimal thresholds for each cell type-specific protein expression [2]. Details of the optimal threshold finding algorithm is shown in Algorithm 1.

The idea is to estimate the threshold for each cell category in an image separately. We convert the predicted probability vector with the estimated threshold into binary and calculate the Matthews correlation coefficient (MCC) between the threshold value and the actual value. The Matthews correlation coefficient for all thresholds are stored in the vector ω, from which we find the index of threshold that causes the largest correlation. The Optimal Threshold for the ith dimension is then determined by the corresponding value. We then leveraged Bias-Corrected Uncertainty quantification method [6] using Deep Convolutional Neural Network (DCNN) architectures with Dropweights [7].

Input: Ground Truth Vector: $\{y_{i,1}, y_{i,2}, \ldots, y_{i,M}\}$;
Estimated Probability Vector: $\{\hat{y}_{i,1}, \hat{y}_{i,2}, \ldots, \hat{y}_{i,M}\}$;
Upper Bound for threshold $= \Omega$, and Threshold Stride $=$ S
Result: The Optimal Thresholds $T = (ot_1, ot_2, \ldots, ot_M)$
Initialization: The set of threshold $T = (ot_1 = 0, ot_2 = 0, \ldots, ot_M = 0)$;
for $i \leftarrow 1$ **to** M **do**
    $j \leftarrow 0$;
    $\omega \leftarrow 0$;
    $\pi \leftarrow 0$;
    for $j < \Omega$ **do**
      Initialize M-dimensional binary vector $\mathbf{v} \leftarrow (v_1 = 0, v_2 = 0, \ldots, v_M = 0)$
      ;
      if $\hat{y}_i > j$ **then**
        | $v_i \leftarrow 1$;
      end
      else
        | $v_i \leftarrow 0$;
      end
      $\omega \leftarrow \omega.append(MCC(\mathbf{y}[1:i], v))$;
      $\pi = \pi.append(j)$;
      $j = j + S$
    end
    $\hat{m} \leftarrow argmax_m \omega = (\omega_1, \omega_2, \ldots, \omega_m, \ldots)$;
    $ot_i = \pi[\hat{m}]$
end

Algorithm 1. Find Optimal Threshold

Network Architecture: Our models are trained and evaluated using Keras with Tensorflow backend. For the DNN architecture, we used a generic building block containing the following model structure: Conv-Relu-BatchNorm-MaxPool-Conv-Relu-BatchNorm-MaxPool-Dense-Relu-Dropweights and Dense-Relu-Dropweights-Dense-Sigmoid, with 32 convolution kernels, 3×3 kernel size, 2×2 pooling, dense layer with 512 units, 128 units, and 8 feed-forward Dropweights probabilities 0.3. We optimised the model using Adam optimizer with the default learning rate of 0.001. The training process was conducted in 1000 epochs, with mini-batch size 32. We repeated our experiments three times for an algorithm and calculated a mean of the results.

3 Estimating Bias-Corrected Uncertainty Using Jackknife Resampling Method

3.1 Bayesian Deep Learning and Estimating Uncertainty

There are many measures to estimate uncertainty such as softmax variance, expected entropy, mutual information, predictive entropy and averaging predictions over multiple models. In supervised learning, information gain, i.e. mutual information between the input data and the model parameters is considered as the most relevant measure of the epistemic uncertainty [4,12]. Estimation of entropy from the finite set of data suffers from a severe downward bias when the data is under-sampled. Even small biases can result in significant inaccuracies when estimating entropy [9]. We leveraged Jackknife resampling method to calculate bias-corrected entropy [11].

Given a set of training data D, where $\mathbf{X} = \{x_1, x_2 \ldots x_N\}$ is the set of N images and the corresponding labels $\mathbf{Y} = \{y_1, y_2 \ldots y_N\}$, a BNN is defined in terms of a prior $p(\omega)$ on the weights, as well as the likelihood $p(D|\omega)$. Consider class probabilities $p(y_{x_i} = c \mid x_i, \omega_t, D)$ with $\omega_t \sim q(\omega \mid D)$ with $\mathcal{W} = (\omega_t)_{t=1}^T$, a set of independent and identically distributed (i.i.d.) samples draws from $q(\omega \mid, D)$. The below procedure computes the Monte Carlo (MC) estimate of the posterior predictive distribution, its Entropy and Mutual Information(MI):

$$\sum_{i=1}^N \mathbb{I}_{\mathrm{MC}}(y_i; \omega \mid x_i, D) = \mathbb{H}\big(\hat{p}(y_i \mid x_i, D)\big) - \frac{1}{|\mathcal{W}|} \sum_{\omega \in \mathcal{W}} \mathbb{H}\big(p(y_i \mid x_i, \omega, D)\big). \quad (1)$$

where

$$\hat{p}(y_i \mid x_i, D) = \frac{1}{|\mathcal{W}|} \sum_{\omega \in \mathcal{W}} p(y_i \mid x_i, \omega, D). \quad (2)$$

The stochastic predictive entropy is $H[y \mid x, \omega] = \mathbb{H}(\hat{p}) = -\sum_c \hat{p}_c \log(\hat{p}_c)$, where $\hat{p}_c = \frac{1}{T} \sum_t p_{tc}$ is the entire sample maximum likelihood estimator of probabilities.

The first term in the MC estimate of the mutual information is called the plug-in estimator of the entropy. It has long been known that the plug-in estimator underestimates the true entropy and plug-in estimate is biased [11,17].

A classic method for correcting the bias is the Jackknife resampling method [3]. In order to solve the bias problem, we propose a Jackknife estimator to estimate the epistemic uncertainty to improve an entropy-based estimation model. Unlike MC-Dropout, it does not assume constant variance. If $\mathcal{D}(X, Y)$ is the observed random sample, the ith Jackknife sample, x_i, is the subset of the sample that leaves-one-out observation $x_i : x_{(i)} = (x_1, \ldots x_{i-1}, x_{i+1} \ldots x_n)$. For sample size N, the Jackknife standard error $\hat{\sigma}$ is defined as: $\sqrt{\frac{(N-1)}{N} \sum_{i=1}^N (\hat{\sigma}_i - \hat{\sigma}_{(\odot)})^2}$, where $\hat{\sigma}_{(\odot)}$ is the empirical average of the Jackknife replicates: $\frac{1}{N} \sum_{i=1}^N \hat{\sigma}_{(i)}$. Here, the Jackknife estimator is an unbiased estimator of the variance of the sample mean. The Jackknife correction of a plug-in estimator $\mathbb{H}(\cdot)$ is computed according to the method below [3]:

Given a sample $(p_t)_{t=1}^T$ with p_t discrete distribution on $1 \ldots C$ classes, T corresponds to the total number of MC-Dropweights forward passes during the test.

1. for each $t = 1 \ldots T$
   - calculate the leave-one-out estimator: $\hat{p}_c^{-t} = \frac{1}{T-1} \sum_{j \neq i} p_{jc}$
   - calculate the plug-in entropy estimate: $\hat{H}_{-t} = \mathbb{H}(\hat{p}^{-t})$
2. calculate the bias-corrected entropy $\hat{H}_J = T\hat{H} + \frac{(T-1)}{T} \sum_{t=1}^T \hat{H}_{(-i)}$, where $\hat{H}_{(-i)}$ is the observed entropy based on a sub-sample in which the ith individual is removed.

We leveraged the following relation:

$$\mu_{-i} = \frac{1}{T-1} \sum_{j \neq i} x_j = \mu + \frac{\mu - x_i}{T-1}.$$

while resolving the i-th data point out of the sample mean $\mu = \frac{1}{T} \sum_i x_i$ and recompute the mean μ_{-i}. This makes it possible to quickly calculate leave-one-out estimators of a discrete probability distribution.

The epistemic uncertainty can be obtained as the difference between the approximate predictive posterior entropy (or total entropy) and the average uncertainty in predictions (i.e: aleatoric entropy):

$$I(\mathbf{y} : \omega) = H_e(\mathbf{y}|\mathbf{x}) = \hat{H}_J(\mathbf{y}|\mathbf{x}) - H_a(\mathbf{y}|\mathbf{x}) = \hat{H}_J(\mathbf{y}|\mathbf{x}) - \mathbb{E}_{q(\omega|\mathbf{D})}[\hat{H}_J(\mathbf{y}|\mathbf{x}, \omega)]$$

Therefore, the mutual information $I(\mathbf{y} : \omega)$ i.e. as a measure of bias-corrected epistemic uncertainty, represents the variability in the predictions made by the neural network weight configurations drawn from approximate posteriors. It derives an estimate of the finite sample bias from the leave-one-out estimators of the entropy and reduces bias considerably down to $O(n^{-2})$ [3].

The bias-corrected uncertainty estimation model explains regions of ambiguous data space or difficult to classify, as data distribution with noise in the inputs or model, which was trained with different domain data. Consequently, these inputs should be assigned a higher aleatoric uncertainty. As a result, we can expect high model uncertainty in these regions.

Following Gal [5], we define the stochastic versions of Bayesian uncertainty using MC-Dropweights, where the class probabilities $p(y_{x_i} = c \mid x_i, \omega_t, D)$ with $\omega_t \sim q(\omega \mid D)$ and $W = (\omega_t)_{t=1}^T$ along with a set of independent and identically distributed (i.i.d.) samples drawn from $q(\omega \mid, D)$, can be approximated by the average over the MC-Dropweights forward pass.

We trained the multi-label classification network with all eight classes. We dichotomised the network outputs using optimal threshold with Algorithm 1 for each cell type, with a 1000 MC-Dropweights forward passes at test time. In these detection tasks, $p(y_{x_i} >= 0; OptimalThreshold_i \mid x_i, \omega_t, D)$, where 1 marks the presence of cell type, is sufficient to indicate the most likely decision along with estimated uncertainty.

3.2   Dataset

Our main dataset is taken from The Human Protein Atlas project, that maps the distribution of all human proteins in human tissues and organs [15]. Here, we used high-resolution digital images of immunohistochemically stained testes tissue consisting of 8 cell types: spermatogonia, preleptotene spermatocytes, pachytene spermatocytes, round/early spermatids, elongated/late spermatids, sertoli cells, leydig cells, and peritubular cells, publicly available on the Human Protein Atlas version 18 (v18.proteinatlas.org), as shown in Fig. 2:

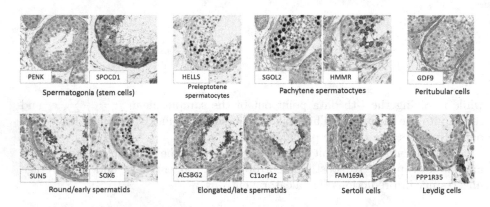

Fig. 2. Examples of proteins expressed only in one cell-type [10]

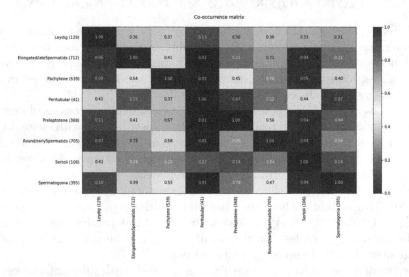

Fig. 3. Annotated heatmap of a correlation matrix between cell types

A relationship was observed between spermatogonia and preleptotene spermatocytes cell types and between round/early spermatids and elongated/late spermatids cell types along with Pachytene spermatocytes cells. Figure 3 illustrates the correlation coefficients between cell types. The observable pattern is that very few cell types are strongly correlated with each other.

3.3   Results and Discussions

We conducted the experiments on Human Protein Atlas datasets to validate the proposed algorithm, MC-Dropweights in Multi-Label Classification.

Multi-label Classification Model Performance: Model evaluation metrics for multi-label classification are different from those used in multi-class (or binary) classification. The performance metrics of multi-label classifiers can be classified as label-based (i.e.: it is assumed that labels are mutually exclusive) and example-based [16]. In this work, example-based measures (Accuracy score, Hamming-loss, F1-Score) and Rank-Loss are used to evaluate the performance of the classifiers.

Table 1. Performance metrics

%Metrics	BR	CC	PCC	CFT	CLEMS	MC-Dropout	MC-Dropweights
Hamming loss	0.2445	0.2420	0.2420	0.2375	0.2370	0.207	0.1925
Rank loss	3.6700	3.5740	3.1580	3.2920	3.1120	2.862	2.626
F1 score	0.5038	0.5184	0.5733	0.5373	0.5902	0.6306	0.6627
Avg. accuracy score	0.4236	0.4389	0.4643	0.4573	0.5052	0.6150	0.7067

In the first experiment, we compared the MC-Dropweights neural network-based method with five machine learning MLC algorithms introduced in Sect. 1: binary relevance (BR), Classifier Chain (CC), Probabilistic Classifier Chain (PCC) and Condensed Filter Tree (CFT), Cost-Sensitive Label Embedding with Multi-dimensional Scaling (CLEMS) and the MC-Dropout neural network model. Table 1 shows that MC-Dropweights exhibits considerably better performance overall the algorithms, which demonstrates the importance of considering the Dropweights in the neural network.

Cell Type-Specific Predictive Uncertainty: The relationship between uncertainty and predictive accuracy grouped by correct and incorrect predictions is shown in Fig. 4. It is interesting to note that, on average, the highest uncertainty is associated with Elongated/late Spermatids and Round/early Spermatids. This indicates that there is some feature which contributes greater uncertainty to the Spermatids class types than to the other cell types.

Cell Type Localization: Estimated uncertainty with Saliency Mapping is a simple technique to uncover discriminative image regions that strongly influence the network prediction in identifying a specific class label in the image. It highlights the most influential features in the image space that affect the predictions of the model [1] and visualises the contributions of individual pixels to epistemic and aleatoric uncertainties separately. We calculated the class activation maps (CAM) [18] using the activations of the fully connected layer and the weights from the prediction layer as shown in Fig. 5.

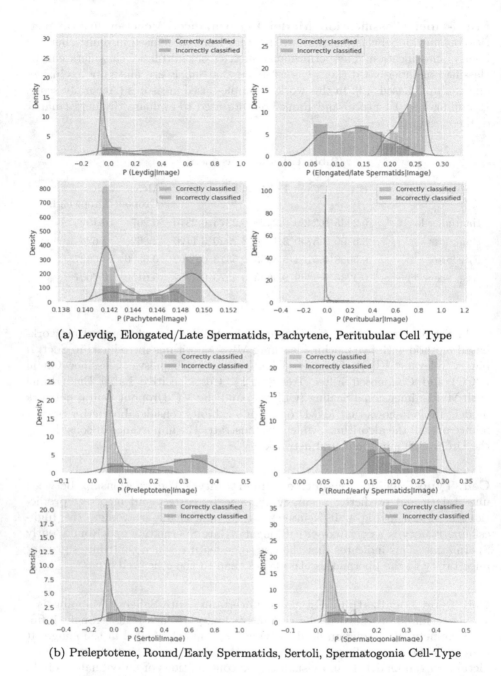

(a) Leydig, Elongated/Late Spermatids, Pachytene, Peritubular Cell Type

(b) Preleptotene, Round/Early Spermatids, Sertoli, Spermatogonia Cell-Type

Fig. 4. Distribution of uncertainty values for all protein images, grouped by correct and incorrect predictions. Label assignment was based on optimal thresholding (Algorithm 1). For an incorrect prediction, there is a strong likelihood that the predictive uncertainty is also high in all cases except for Spermatids.

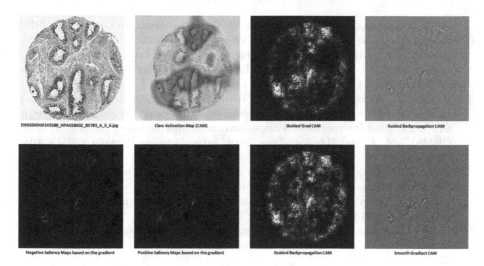

Fig. 5. Saliency maps for some common methods towards model explanation

4   Conclusion and Discussion

In this study, a multi-label classification method was developed using deep learning architecture with Dropweights for the purposes of predicting cell types-specific protein expression with estimated uncertainty, which can increase the ability to interpret, with confidence and make models based on deep learning more applicable in practice. The results show that a Deep Learning Model with MC-Dropweights yields the best performance among all popular classifiers.

Building truly large-scale, fully-automated, high precision, very high dimensional, image analysis system that can recognise various cell type-specific protein expression, specifically for Elongated/Late Spermatids and Round/early Spermatids remains a strenuous task. The properties in the dataset such as label correlations, label cardinality can strongly affect the uncertainty quantification in predictive probability performance of a Bayesian Deep learning algorithm in multi-label settings. There is no systematic study on how and why the performance varies over different data properties; any such study would be of great benefit in progressing multi-label algorithms.

References

1. Adebayo, J., Gilmer, J., Muelly, M., Goodfellow, I., Hardt, M., Kim, B.: Sanity checks for saliency maps. In: Advances in Neural Information Processing Systems, pp. 9505–9515 (2018)
2. Chu, W.T., Guo, H.J.: Movie genre classification based on poster images with deep neural networks. In: Proceedings of the Workshop on Multimodal Understanding of Social, Affective and Subjective Attributes, pp. 39–45. ACM (2017)
3. DasGupta, A.: Asymptotic Theory of Statistics and Probability. Springer, New York (2008). https://doi.org/10.1007/978-0-387-75971-5
4. Depeweg, S., Hernández-Lobato, J.M., Doshi-Velez, F., Udluft, S.: Decomposition of uncertainty in Bayesian deep learning for efficient and risk-sensitive learning. arXiv preprint arXiv:1710.07283 (2017)

5. Gal, Y.: Uncertainty in deep learning. Ph.D. thesis, University of Cambridge (2016)
6. Ghoshal, B., Tucker, A., Sanghera, B., Wong, W.: Estimating uncertainty in deep learning for reporting confidence to clinicians in medical image segmentation and diseases detection. In: Computational Intelligence - Special Issue on Foundations of Biomedical (Big) Data Science, vol. 1 (2019)
7. Ghoshal, B., Tucker, A., Sanghera, B., Wong, W.: Estimating uncertainty in deep learning for reporting confidence to clinicians when segmenting nuclei image data. 2019 IEEE 32nd International Symposium on Computer-Based Medical Systems (CBMS), vol. 1, pp. 318–324, June 2019. https://doi.org/10.1109/CBMS.2019.00072
8. Huang, K.H., Lin, H.T.: Cost-sensitive label embedding for multi-label classification. Mach. Learn. **106**(9–10), 1725–1746 (2017)
9. Macke, J., Murray, I., Latham, P.: Estimation bias in maximum entropy models. Entropy **15**(8), 3109–3129 (2013)
10. Pineau, C., et al.: Cell type-specific expression of testis elevated genes based on transcriptomics and antibody-based proteomics. J. Proteome Res. **18**, 4215–4230 (2019)
11. Quenouille, M.H.: Notes on bias in estimation. Biometrika **43**(3/4), 353–360 (1956)
12. Shannon, C.E.: A mathematical theory of communication. Bell Syst. Tech. J. **27**(3), 379–423 (1948)
13. Srivastava, N., Hinton, G., Krizhevsky, A., Sutskever, I., Salakhutdinov, R.: Dropout: a simple way to prevent neural networks from overfitting. Journal Mach. Learn. Res. **15**(1), 1929–1958 (2014)
14. Teye, M., Azizpour, H., Smith, K.: Bayesian uncertainty estimation for batch normalized deep networks. arXiv preprint arXiv:1802.06455 (2018)
15. Uhlén, M., et al.: Tissue-based map of the human proteome. Science **347**(6220), 1260419 (2015)
16. Wu, X.Z., Zhou, Z.H.: A unified view of multi-label performance measures. In: Proceedings of the 34th International Conference on Machine Learning, vol. 70, pp. 3780–3788. JMLR. org (2017)
17. Yeung, R.W.: A new outlook on Shannon's information measures. IEEE Trans. Inf. Theory **37**(3), 466–474 (1991)
18. Zhou, B., Khosla, A., Lapedriza, A., Oliva, A., Torralba, A.: Learning deep features for discriminative localization. In: CVPR (2016)

Adversarial Attacks Hidden in Plain Sight

Jan Philip Göpfert[1(✉)], André Artelt[1], Heiko Wersing[2], and Barbara Hammer[1]

[1] Bielefeld University, Bielefeld, Germany
jgoepfert@techfak.uni-bielfeld.de
[2] Honda Research Institute Europe GmbH, Offenbach, Germany

Abstract. Convolutional neural networks have been used to achieve a string of successes during recent years, but their lack of interpretability remains a serious issue. Adversarial examples are designed to deliberately fool neural networks into making any desired incorrect classification, potentially with very high certainty. Several defensive approaches increase robustness against adversarial attacks, demanding attacks of greater magnitude, which lead to visible artifacts. By considering human visual perception, we compose a technique that allows to hide such adversarial attacks in regions of high complexity, such that they are imperceptible even to an astute observer. We carry out a user study on classifying adversarially modified images to validate the perceptual quality of our approach and find significant evidence for its concealment with regards to human visual perception.

1 Introduction

The use of convolutional neural networks has led to tremendous achievements since Krizhevsky et al. [1] presented AlexNet in 2012. Despite efforts to understand the inner workings of such neural networks, they mostly remain black boxes that are hard to interpret or explain. The issue was exaggerated in 2013 when Szegedy et al. [2] showed that "adversarial examples" – images perturbed in such a way that they fool a neural network – prove that neural networks do not simply generalize correctly the way one might naïvely expect. Typically, such adversarial attacks change an input only slightly, but in an adversarial manner, such that humans do not regard the difference of the inputs relevant, but machines do. There are various types of attacks, such as one pixel attacks, attacks that work in the physical world, and attacks that produce inputs fooling several different neural networks without explicit knowledge of those networks [3–5].

Adversarial attacks are not strictly limited to convolutional neural networks. Even the simplest binary classifier partitions the entire input space into labeled regions, and where there are no training samples close by, the respective label can only be nonsensical with regards to the training data, in particular near decision boundaries. One explanation of the "problem" that convolutional neural networks have is that they perform extraordinarily well in high-dimensional settings, where the training data only covers a very thin manifold, leaving a lot of "empty space" with ragged class regions. This creates a lot of room for an

M. R. Berthold et al. (Eds.): IDA 2020, LNCS 12080, pp. 235–247, 2020.
https://doi.org/10.1007/978-3-030-44584-3_19

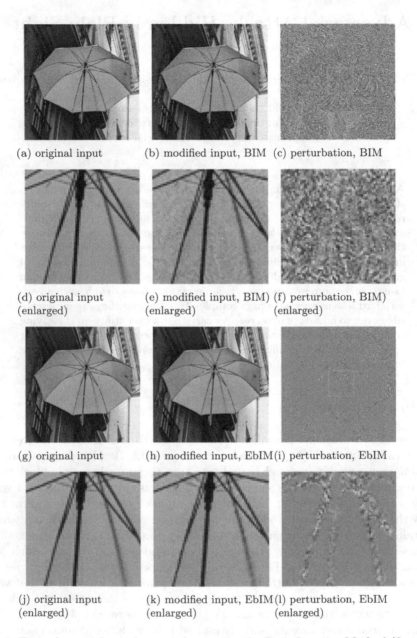

(a) original input

(b) modified input, BIM

(c) perturbation, BIM

(d) original input
(enlarged)

(e) modified input, BIM)
(enlarged)

(f) perturbation, BIM)
(enlarged)

(g) original input

(h) modified input, EbIM

(i) perturbation, EbIM

(j) original input
(enlarged)

(k) modified input, EbIM
(enlarged)

(l) perturbation, EbIM
(enlarged)

Fig. 1. Two adversarial attacks carried out using the Basic Iterative Method (first two rows) and our Entropy-based Iterative Method (last two rows). The original image (a) (and (g)) is correctly classified as *umbrella* but the modified images (b) and (h) are classified as *slug* with a certainty greater than 99 %. Note the visible artifacts caused by the perturbation (c), shown here with maximized contrast. The perturbation (i) does not lead to such artifacts. (d), (e), (f), (j), (k), and (l) are enlarged versions of the marked regions in (a), (b), (c), (g), (h), and (i), respectively.

attacker to modify an input sample and move it away from the manifold on which the network can make meaningful predictions, into regions with nonsensical labels. Due to this, even adversarial attacks that simply blur an image, without any specific target, can be successful [6]. There are further attempts at explaining the origin of the phenomenon of adversarial examples, but so far, no conclusive consensus has been established [7–10].

A number of defenses against adversarial attacks have been put forward, such as defensive distillation of trained networks [11], adversarial training [12], specific regularization [9], and statistical detection [13–16]. However, no defense succeeds in universally preventing adversarial attacks [17,18], and it is possible that the existence of such attacks is inherent in high-dimensional learning problems [6]. Still, some of these defenses do result in more robust networks, where an adversary needs to apply larger modifications to inputs in order to successfully create adversarial examples, which begs the question how robust a network can become and whether robustness is a property that needs to be balanced with other desirable properties, such as the ability to generalize well [19] or a reasonable complexity of the network [20].

Strictly speaking, it is not entirely clear what defines an adversarial example as opposed to an incorrectly classified sample. Adversarial attacks are devised to change a given input minimally such that it is classified incorrectly – in the eyes of a human. While astonishing parallels between human visual information processing and deep learning exist, as highlighted e. g. by Yamins and DiCarlo [21] and Rajalingham et al. [22], they disagree when presented with an adversarial example. Experimental evidence has indicated that specific types of adversarial attacks can be constructed that also deteriorate the decisions of humans, when they are allowed only limited time for their decision making [23]. Still, human vision relies on a number of fundamentally different principles when compared to deep neural networks: while machines process image information in parallel, humans actively explore scenes via saccadic moves, displaying unrivaled abilities for structure perception and grouping in visual scenes as formalized e. g. in the form of the Gestalt laws [24–27]. As a consequence, some attacks are perceptible by humans, as displayed in Fig. 1. Here, humans can detect a clear difference between the original image and the modified one; in particular in very homogeneous regions, attacks lead to structures and patterns which a human observer can recognize. We propose a simple method to address this issue and answer the following questions. How can we attack images using standard attack strategies, such that a human observer does not recognize a clear difference between the modified image and the original? How can we make use of the fundamentals of human visual perception to "hide" attacks such that an observer does not notice the changes?

Several different strategies for performing adversarial attacks exist. For a multiclass classifier, the attack's objective can be to have the classifier predict *any* label other than the correct one, in which case the attack is referred to as *untargeted*, or *some specifically chosen* label, in which case the attack is called *targeted*. The former corresponds to minimizing the likelihood of the original label being assigned; the latter to maximizing that of the target label. Moreover, the classifier can be fooled into classifying the modified input with extremely high confidence, depending on the method employed. This, in particular, can however lead to visible artifacts in the resulting images (see Fig. 1). After looking at a number of examples, one can quickly learn to make out typical patterns that depend on the classifying neural network. In this work, we propose a method for changing this procedure such that this effect is avoided.

For this purpose, we extend known techniques for adversarial attacks. A particularly simple and fast method for attacking convolutional neural networks is the aptly named Fast Gradient Sign Method (FGSM) [4,7]. This method, in its original form, modifies an input image x along a linear approximation of the objective of the network. It is fast but limited to untargeted attacks. An extension of FGSM, referred to as the Basic Iterative Method (BIM) [28], repeatedly adds small perturbations and allows targeted attacks. Moosavi-Dezfooli et al. [29] linearize the classifier and compute smaller (with regards to the ℓ_p norm) perturbations that result in untargeted attacks. Using more computationally demanding optimizations, Carlini and Wagner [17] minimize the ℓ_0, ℓ_2, or ℓ_∞ norm of a perturbation to achieve targeted attacks that are still harder to detect. Su et al. [3] carry out attacks that change only a single pixel, but these attacks are only possible for some input images and target labels. Further methods exist that do not result in obvious artifacts, e. g. the Contrast Reduction Attack [30], but these are again limited to untargeted attacks – the input images are merely corrupted such that the classification changes. None of the methods mentioned here regard human perception directly, even though they all strive to find imperceptibly small perturbations. Schönherr et al. [31] successfully do this within the domain of acoustics.

We rely on BIM as the method of choice for attacks based on images, because it allows robust targeted attacks with results that are classified with arbitrarily high certainty, even though it is easy to implement and efficient to execute. Its drawbacks are the aforementioned visible artifacts. To remedy this issue, we will take a step back and consider human perception directly as part of the attack. In this work, we propose a straightforward, very effective modification to BIM that ensures targeted attacks are visually imperceptible, based on the observation that attacks do not need to be applied homogeneously across the input image and that humans struggle to notice artifacts in image regions of high local complexity. We hypothesize that such attacks, in particular, do not change saccades as severely as generic attacks, and so humans perceive the original image and the modified one as very similar – we confirm this hypothesis in Sect. 3 as part of a user study.

2   Adversarial Attacks

Recall the objective of a targeted adversarial attack. Given a classifying convolutional neural network f, we want to modify an input x, such that the network assigns a different label $f(x')$ to the modified input x' than to the original x, where the target label $f(x')$ can be chosen at will. At the same time, x' should be as similar to x as possible, i. e. we want the modification to be small. This results in the optimization problem:

$$\min \|x' - x\| \quad \text{such that} \quad f(x') = y \neq f(x), \tag{1}$$

where $y = f(x')$ is the target label of the attack. BIM finds such a small perturbation $x' - x$ by iteratively adapting the input according to the update rule

$$x \leftarrow x - \epsilon \cdot \text{sign}[\nabla_x J(x, y)] \tag{2}$$

until f assigns the label y to the modified input with the desired certainty, where the certainty is typically computed via the softmax over the activations of all class-wise outputs. $\text{sign}[\nabla_x J(x, y)]$ denotes the sign of the gradient of the objective function $J(x, y)$, and is computed efficiently via backpropagation; ϵ is the step size. The norm of the perturbation is not considered explicitly, but because in each iteration the change is distributed evenly over all pixels/features in x, its ℓ_∞-norm is minimized.

2.1 Localized Attacks

The main technical observation, based on which we hide attacks, is the fact that one can weigh and apply attacks locally in a precise sense: During prediction, a convolutional neural network extracts features from an input image, condenses the information contained therein, and conflates it, in order to obtain its best guess for classification. Where exactly in an image a certain feature is located is of minor consequence compared to how strongly it is expressed [32,33]. As a result, we find that during BIM's update, it is not strictly necessary to apply the computed perturbation evenly across the entire image. Instead, one may choose to leave parts of the image unchanged, or perturb some pixels more or less than others, i. e. one may localize the attack. This can be directly incorporated into Eq. (2) by setting an individual value for ϵ for every pixel.

For an input image $x \in [0, 1]^{w \times h \times c}$ of width w and height h with c color channels, we formalize this by setting a strength map $\mathcal{E} \in [0, 1]^{w \times h}$ that holds an update magnitude for each pixel. Such a strength map can be interpreted as a grayscale image where the brightness of a pixel corresponds to how strongly the respective pixel in the input image is modified. The adaptation rule (2) of BIM is changed to the update rule

$$x_{ijk} \leftarrow x_{ijk} - \epsilon \cdot \mathcal{E}_{ijk} \cdot \text{sign}[\nabla_x J(x, y)] \tag{3}$$

for all pixel values (i, j, k). In order to be able to express the overall strength of an attack, for a given strength map $\mathcal{E}$ of size w by h, we call

$$\kappa(\mathcal{E}) = \frac{\sum_{i,j \in \overline{w} \times \overline{h}} \mathcal{E}_{i,j}}{w \cdot h} \tag{4}$$

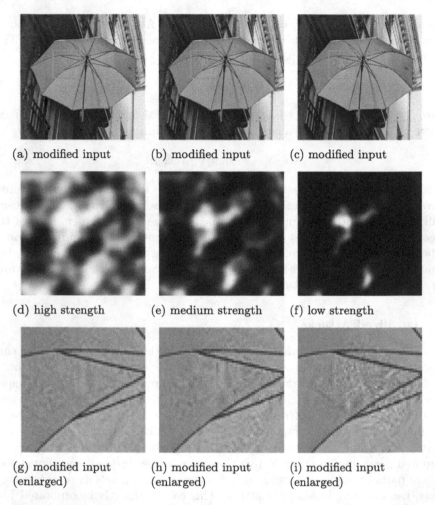

Fig. 2. Localized attacks with different relative total strengths. The strength maps (d), (e), and (f), which are based on Perlin noise, scaled such that the relative total strength is 0.43, 0.14, and 0.04, are used to create the adversarial examples in (a), (b), and (c), respectively. In each case, the attacked image is classified as *slug* with a certainty greater than 99 %. The attacks took 14, 17, and 86 iterations. (g), (h), and (i) are enlarged versions of the marked regions in (a), (b), and (c).

the *relative total strength* of $\mathcal{E}$, where for $n \in \mathbb{N}$ we let $\overline{n} = \{1, \ldots, n\}$ denote the set of natural numbers from 1 to n. In the special case where $\mathcal{E}$ only contains either black or white pixels, $\kappa(\mathcal{E})$ is the ratio of white pixels, i.e. the number of attacked pixels over the total number of pixels in the attacked image.

As long as the scope of the attack, i.e. $\kappa(\mathcal{E})$, remains large enough, adversarial attacks can still be carried out successfully – if not as easily – with more iterations required until the desired certainty is reached. This leads to the attacked pixels being perturbed more, which in turn leads to even more pronounced artifacts. Given a strength map $\mathcal{E}$, it can be modified to increase or decrease $\kappa(\mathcal{E})$ by adjusting its brightness or by applying appropriate morphological operations. See Fig. 2 for a demonstration that uses pseudo-random noise as a strength map.

2.2 Entropy-Based Attacks

The crucial component necessary for "hiding" adversarial attacks is choosing a strength map $\mathcal{E}$ that appropriately considers human perceptual biases. The strength map essentially determines which "norm" is chosen in Eq. (1). If it differs from a uniform weighting, the norm considers different regions of the image differently. The choice of the norm is critical when discussing the visibility of adversarial attacks. Methods that explicitly minimize the ℓ_p norm of the perturbation for some p, only "accidentally" lead to perturbations that are hard to detect visually, since the ℓ_p norm does not actually resemble e. g. the human visual focus for the specific image. We propose to instead make use of how humans perceive images and to carefully choose those pixels where the resulting artifacts will not be noticeable.

Instead of trying to hide our attack in the background or "where an observer might not care to look", we instead focus on those regions where there is high local complexity. This choice is based on the rational that humans inspect images in saccadic moves, and a focus mechanism guides how a human can process highly complex natural scenes efficiently in a limited amount of time. *Visual interest* serves as a selection mechanism, singling out relevant details and arriving at an optimized representation of the given stimuli [34]. We rely on the assumption that adversarial attacks remain hidden if they do not change this scheme. In particular, regions which do not attract focus in the original image should not increase their level of interest, while relevant parts can, as long as the adversarial attack is not adding additional relevant details to the original image.

Due to its dependence on semantics, it is hard – if not impossible – to agnostically compute the magnitude of *interest* for specific regions of an image. Hence, we rely on a simple information theoretic proxy, which can be computed based on the visual information in a given image: the entropy in a local region. This simplification relies on the observation that regions of interest such as edges typically have a higher entropy than homogeneous regions and the entropy serves as a measure for how much information is already contained in a region – that is, how much relative difference would be induced by additional changes in the region.

Algorithmically, we compute the *local entropy* at every pixel in the input image as follows: After discarding color, we bin the gray values, i. e. the intensities, in the neighborhood of pixel i, j such that $B_{i,j}$ contains the respective occurrence ratios. The occurrence ratios can be interpreted as estimates of the intensity probability in this neighborhood, hence the local entropy $S_{i,j}$ can be calculated as the Shannon entropy

$$S_{i,j} = -\sum_{p \in B_{i,j}} p \log p. \qquad (5)$$

Through this, we obtain a measure of local complexity for every pixel in the input image, and after adjusting the overall intensity, we use it as suggested above to scale the perturbation pixel-wise during BIM's update. In other words, we set

$$\mathcal{E} = \phi(S) \qquad (6)$$

where ϕ is a nonlinear mapping, which adjusts the brightness. The choice of a strength map based on the local entropy of an image allows us to perform an attack as straightforward as BIM, but localized, in such a way that it does not produce visible artifacts, as we will see in the following experiments.

While we could attach our technique to any attack that relies on gradients, we use BIM because of the aforementioned advantages including simplicity, versatility, and robustness, but also because as the direct successor to FGSM we consider it the most typical attack at present. As a method of performing adversarial attacks, we refer to our method as the *Entropy-based Iterative Method (EbIM)*.

3 A Study of How Humans Perceive Adversarial Examples

It is often claimed that adversarial attacks are imperceptible[1]. While this can be the case, there are many settings in which it does not necessarily hold true – as can be seen in Fig. 1. When robust networks are considered and an attack is expected to reliably and efficiently produce adversarial examples, visible artifacts appear. This motivated us to consider human visual perception directly and thereby our method. To confirm that there are in fact differences in how adversarial examples produced by BIM and EbIM are perceived, we conducted a user study with 35 participants.

---

[1] We do not want to single out any specific source for this claim, and it should not necessarily be considered strictly false, because there is no commonly accepted rigorous definition of what constitutes an adversarial example or an adversarial attack, just as it remains unclear how to best measure adversarial robustness. Whether an adversarial attack results in noticeable artifacts depends on a multitude of factors, such as the attacked model, the underlying data (distribution), the method of attack, and the target certainty.

3.1   Generation of Adversarial Examples

To keep the course of the study manageable, so as not to bore our relatively small number of participants, and still acquire statistically meaningful (i. e. with high statistical power) and comparable results, we randomly selected only 20 labels and 4 samples per label from the validation set of the *ILSVRC 2012 classification challenge* [35], which gave us a total of 80 images. For each of these 80 images we generated a targeted high confidence adversarial example using BIM and another one using EbIM – resulting in a total of 240 images. We set a fixed target class and the target certainty to 0.99. We attacked the pretrained *Inception v3* model [36] as provided by *keras* [37]. We set the parameters of BIM to $\epsilon = 1.0$, *stepsize* $= 0.004$ and *max_iterations* $= 1000$. For EbIM, we binarized the entropy mask with a threshold of 4.2. We chose these parameters such that the algorithms can reliably generate targeted high certainty adversarial examples across all images, without requiring expensive per-sample parameter searches.

3.2   Study Design

For our study, we assembled the images in pairs according to *three different conditions*:

(i) The original image versus itself.
(ii) The original image versus the adversarial example generated by BIM.
(iii) The original image versus the adversarial example generated by EbIM.

This resulted in 240 pairs of images that were to be evaluated during the study.

All image pairs were shown to each participant in a random order – we also randomized the positioning (left and right) of the two images in each pair. For each pair, the participant was asked to determine whether the two images were identical or different. If the participant thought that the images were identical they were to click on a button labeled "Identical" and otherwise on a button labeled "Different" – the ordering of the buttons was fixed for a given participant but randomized when they began the study. To facilitate completion of the study in a reasonable amount of time, each image pair was shown for 5 s only; the participant was, however, able to wait as long as they wanted until clicking on a button, whereby they moved on to the next image pair.

3.3   Hypotheses Tests

Our hypothesis was that it would be more difficult to perceive the changes in the images generated by EbIM than by BIM. We therefore expect our participants to click "Identical" more often when seeing an adversarial example generated by EbIM than when seeing an adversarial generated by BIM.

As a test statistic, we compute *for each participant* and *for each of the three conditions separately*, the percentage of time they clicked on "Identical". The values can be interpreted as a mean if we encode "Identical" as 1 and "Different" as 0. Hereinafter we refer to these mean values as μ_{BIM} and μ_{EbIM}. For each of

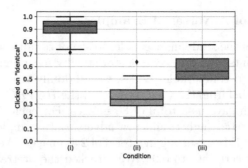

Fig. 3. Percentage of times users clicked on "Identical" when seeing two identical images (condition (i), blue box), a BIM adversarial (condition (ii), orange box), or an EbIM adversarial (condition (iii), green box). (Color figure online)

the three conditions, we provide a boxplot of the test statistics in Fig. 3 – the scores of EbIM are much higher than BIM, which indicates that it is in fact much harder to perceive the modifications introduced by EbIM compared to BIM. Furthermore, users almost always clicked on "Identical" when seeing two identical images.

Finally, we can phrase our belief as a hypothesis test. We determine whether we can reject the following five hypotheses:

(1) H_0: $\mu_{\text{BIM}} \geq \mu_{\text{EbIM}}$, i.e. attacks using BIM are as hard or harder to perceive than EbIM.

(2) H_0: $\mu_{\text{BIM}} \geq 0.5$, i.e. whether attacks using BIM are easier or harder to perceive than a random prediction

(3) H_0: $\mu_{\text{EbIM}} \leq 0.5$, i.e. whether attacks using EbIM are easier or harder to perceive than a random prediction

(4) H_0: $\mu_{\text{BIM}} \geq \mu_{\text{NONE}}$, i.e. whether attacks using BIM are as easy or easier to perceive than identical images.

(5) H_0: $\mu_{\text{EbIM}} \geq \mu_{\text{NONE}}$, i.e. whether attacks using EbIM are as easy or easier to perceive than identical images.

We use a *one-tailed t-test* and the (non-parametric) *Wilcoxon signed rank test* with a significance level $\alpha = 0.05$ in both tests. The cases (1), (4) and (5) are tested as a *paired test* and the other two cases (2) and (3) as *one sample tests*.

Because the t-test assumes that the mean difference is normally distributed, we test for normality[2] by using the *Shapiro-Wilk normality test*. The Shapiro-Wilk normality test computes a p-value of 0.425, therefore we assume that the mean difference follows a normal distribution. The resulting p-values are listed in Table 1 – we can reject all null hypotheses with very low p-values.

---

[2] Because we have 35 participants, we assume that normality approximately holds because of the central limit theorem.

Table 1. p-values of each hypothesis (columns) under each test (rows). We reject all null hypotheses.

Test	Hyp. (1)	Hyp. (2)	Hyp. (3)	Hyp. (4)	Hyp. (5)
t-test	2.20×10^{-16}	1.03×10^{-10}	2.13×10^{-5}	2.20×10^{-16}	2.20×10^{-16}
Wilcoxon	1.28×10^{-7}	9.10×10^{-7}	6.75×10^{-5}	1.28×10^{-7}	1.28×10^{-7}

In order to compute the power of the t-test, we compute the effect size by computing *Cohen's d*. We find that $d \approx 2.29$ which is considered a huge effect size [38]. The power of the one-tailed t-test is then approximately 1.

We have empirically shown that adversarial examples produced by EbIM are significantly harder to perceive than adversarial examples generated by BIM. Furthermore, adversarial examples produced by EbIM are not perceived as differing from their respective originals.

4   Discussion

Adversarial attacks will remain a potential security risk on the one hand and an intriguing phenomenon that leads to insight into neural networks on the other. Their nature is difficult to pinpoint and it is hard to predict whether they constitute a problem that will be solved. To further the understanding of adversarial attacks and robustness against them, we have demonstrated two key points:

– Adversarial attacks against convolutional neural networks can be carried out successfully even when they are localized.
– By reasoning about human visual perception and carefully choosing areas of high complexity for an attack, we can ensure that the adversarial perturbation is barely perceptible, even to an astute observer who has learned to recognize typical patterns found in adversarial examples.

This has allowed us to develop the Entropy-based Iterative Method (EbIM), which performs adversarial attacks against convolutional neural networks that are hard to detect visually even when their magnitude is considerable with regards to an ℓ_p-norm. It remains to be seen how current adversarial defenses perform when confronted with entropy-based attacks, and whether robust networks learn special kinds of features when trained adversarially using EbIM.

Through our user study we have made clear that not all adversarial attacks are imperceptible. We hope that this is only the start of considering human perception explicitly during the investigation of deep neural networks in general and adversarial attacks against them specifically. Ideally, this would lead to a concise definition of what constitutes an adversarial example.

References

1. Krizhevsky, A., et al.: ImageNet classification with deep convolutional neural networks. In: Advances in Neural Information Processing Systems, vol. 25 (2012). https://doi.org/10.1145/3065386
2. Szegedy, C., et al.: Intriguing properties of neural networks (2013)
3. Su, J., et al.: One pixel attack for fooling deep neural networks (2017)
4. Kurakin, A., et al.: Adversarial examples in the physical world (2016)
5. Papernot, N., et al.: Practical black-box attacks against deep learning systems using adversarial examples (2016)
6. Chakraborty, A., et al.: Adversarial attacks and defences: a survey (2018)
7. Goodfellow, I.J., et al.: Explaining and harnessing adversarial examples (2014)
8. Luo, Y., et al.: Foveation-based mechanisms alleviate adversarial examples, 19 November 2015
9. Cisse, M., et al.: Parseval networks: improving robustness to adversarial examples. In: ICML, 28 April 2017
10. Ilyas, A., et al.: Adversarial examples are not bugs, they are features, 6 May 2019
11. Papernot, N., et al.: Distillation as a defense to adversarial perturbations against deep neural networks. In: 2016 IEEE Symposium on Security and Privacy (SP), May 2016. https://doi.org/10.1109/sp.2016.41
12. Madry, A., et al.: Towards deep learning models resistant to adversarial attacks. In: Proceedings of the International Conference on Learning Representations (ICLR) (2018)
13. Crecchi, F., et al.: Detecting adversarial examples through nonlinear dimensionality reduction. In: Proceedings of the European Symposium on Artificial Neural Networks, Computational Intelligence and Machine Learning (ESANN) (2019)
14. Feinman, R., et al.: Detecting Adversarial Samples from Artifacts, 1 March 2017
15. Grosse, K., et al.: On the (statistical) detection of adversarial examples, 21 February 2017
16. Metzen, J.H., et al.: On detecting adversarial perturbations, 14 February 2017
17. Carlini, N., et al.: Towards evaluating the robustness of neural networks. In: 2017 IEEE Symposium on Security and Privacy (SP) (2017)
18. Athalye, A., et al.: Obfuscated gradients give a false sense of security: circumventing defenses to adversarial examples. In: ICML, 1 February 2018
19. Tsipras, D., et al.: Robustness may be at odds with accuracy. In: Proceedings of the International Conference on Learning Representations (ICLR) (2019)
20. Nakkiran, P.: Adversarial robustness may be at odds with simplicity (2019)
21. Yamins, D.L.K., et al.: Using goal-driven deep learning models to understand sensory cortex. Nat. Neurosci. **19**, 356–365 (2016). https://doi.org/10.1038/nn.4244
22. Rajalingham, R., et al.: Large-scale, high-resolution comparison of the core visual object recognition behavior of humans, monkeys, and state-of-the-art deep artificial neural networks. J. Neurosci. **38**(33), 7255–7269 (2018). https://doi.org/10.1523/JNEUROSCI.0388-18.2018. ISSN 0270–6474
23. Elsayed, G., et al.: Adversarial examples that fool both computer vision and time-limited humans. In: Advances in Neural Information Processing Systems, vol. 31, pp. 3910–3920 (2018)
24. Wersing, H., et al.: A competitive-layer model for feature binding and sensory segmentation. Neural Comput. **13**(2), 357–387 (2001). https://doi.org/10.1162/089976601300014574

25. Ibbotson, M., et al.: Visual perception and saccadic eye movements. Curr. Opin. Neurobiol. **21**(4), 553–558 (2011). https://doi.org/10.1016/j.conb.2011.05.012. ISSN 0959–4388. Sensory and Motor Systems
26. Lewicki, M., et al.: Scene analysis in the natural environment. Front. Psychol. **5**, 199 (2014). https://doi.org/10.3389/fpsyg.2014.00199. ISSN 1664–1078
27. Jäkel, F., et al.: An overview of quantitative approaches in Gestalt perception. Vis. Res. **126**, 3–8 (2016). https://doi.org/10.1016/j.visres.2016.06.004. ISSN 0042–6989. Quantitative Approaches in Gestalt Perception
28. Kurakin, A., et al.: Adversarial machine learning at scale (2016)
29. Moosavi-Dezfooli, S.-M., et al.: DeepFool: a simple and accurate method to fool deep neural networks. In: 2016 IEEE Conference on Computer Vision and Pattern Recognition (CVPR), pp. 2574–2582 (2016)
30. Rauber, J., et al.: Foolbox v0.8.0: a Python toolbox to benchmark the robustness of machine learning models (2017)
31. Schönherr, L., et al.: Adversarial attacks against automatic speech recognition systems via psychoacoustic hiding (2018)
32. Sabour, S., et al.: Dynamic routing between capsules. In: Advances in Neural Information Processing Systems (2017)
33. Brown, T.B., et al.: Adversarial Patch, 27 December 2017
34. Carrasco, M.: Visual attention: the past 25 years. Vis. Res. **51**, 1484–1525 (2011)
35. Russakovsky, O., et al.: ImageNet large scale visual recognition challenge. Int. J. Comput. Vis. **115**(3), 211–252 (2015). https://doi.org/10.1007/s11263-015-0816-y
36. Szegedy, C., et al.: Rethinking the inception architecture for computer vision. In: 2016 IEEE Conference on Computer Vision and Pattern Recognition (CVPR), pp. 2818–2826 (2016)
37. Chollet, F., et al.: Keras (2015). https://keras.io
38. Sawilowsky, S.S.: New effect size rules of thumb. J. Mod. Appl. Stat. Methods **8**(2), 597–599 (2009). https://doi.org/10.22237/jmasm/1257035100

Enriched Weisfeiler-Lehman Kernel for Improved Graph Clustering of Source Code

Frank Höppner[(⊠)] and Maximilian Jahnke

Department of Computer Science, Ostfalia University of Applied Sciences,
38302 Wolfenbüttel, Germany
f.hoeppner@ostfalia.de

Abstract. To perform cluster analysis on graphs we utilize graph kernels, Weisfeiler-Lehman kernel in particular, to transform graphs into a vector representation. Despite good results, these kernels have been criticized in the literature for high dimensionality and high sensitivity, so we propose an efficient subtree distance measure that is subsequently used to enrich the vector representations and enables more sensitive distance measurements. We demonstrate the usefulness in an application, where the graphs represent different source code snapshots, and a cluster analysis of these snapshots provides the lecturer an overview about the overall performance of a group of students.

1 Motivation

Graphs are a universal data structure and have become very popular over recent years in various domains with structured data (e.g. protein function prediction, drug toxicity prediction, malware detection, etc.). To apply existing clustering or classification techniques to graphs, either a distance (or similarity) measure is needed, or a transformation into a vector representation for which most clustering and classification algorithms were developed for. In this paper we are concerned about repeatedly clustering graphs to understand the evolution of student's source code. As will be explained in Sect. 2, we settle on Weisfeiler-Lehman (WL) graph kernels [9] to decompose the graph into subtrees and to define a similarity function over the number of common substructures across graphs. It has been criticized, however, that WL subtree kernels produce (a) many different substructures and thus only a few substructures will be common across graphs, which establishes (b) a tendency of *being only similar to itself*. In this paper we propose to include the subtree similarity in an efficient post-processing step to tackle both problems: We exploit the fact that many of the substructures may be formally distinct but actually quite similar. By enriching the vector representations we obtain positive effects for the overall graph similarity.

M. R. Berthold et al. (Eds.): IDA 2020, LNCS 12080, pp. 248–260, 2020.
https://doi.org/10.1007/978-3-030-44584-3_20

Algorithm 1. WLSK(G, l_{i-1})

Require: graph $G = (V, E)$, label function $l_{i-1} : V \to \Sigma^*$
Ensure: returns new label function $l_i : V \to \Sigma^*$
1: **for** $v \in V$ **do**
2:     store node label $l_{i-1}(v)$ in s
3:     **for** $w \in V, (v, w) \in E$ in (some lexicographical) order of $l_{i-1}(w)$ **do**
4:         append $l_{i-1}(w)$ to s
5:     **end for**
6:     compress $s \leftarrow h(s)$ by applying a hash function h
7:     assign new label to node $v : l_i(v) \leftarrow s$
8: **end for**
9: **return** l_i

---

2   Related Work

2.1   Measuring Similarity Directly

A common approach to compare graphs is to calculate the *edit distance* between graphs F and G: the minimal number of steps to transform G to F. For the special case of trees, these steps consists of node deletion, node insertion, and node relabelling. A survey on tree edit distance can be found in [1], an efficient algorithmic $O(n^3)$ solution, n being the maximal number of nodes in F and G, is proposed in [2]. To adapt a tree edit distance to a specific application, there are approaches to learn appropriate cost parameters [6]. With general graphs, the editing process becomes more complicated as additional operations need to be considered (edge insertion and edge deletion). A survey on graph edit distance is given in [3]. Its computation is exponential in the number of nodes and therefore infeasible for large graphs.

2.2   Measuring Similarity Indirectly

Instead of coping with the full graph, one may decompose the graph into a set of smaller entities and compare these sets instead of the graphs. These entities may be frequent subgraphs (e.g. [8]), walks (short paths), graphlets (e.g. [10]) or subtrees (e.g. [9]). Many *graph kernel* approaches explicitly construct a vector representation, where the ith element indicates how often the ith substructure occurs in the graph. From this vector a kernel or similarity matrix may be calculated. Recent approaches, such as `subgraph2vec` [5], use deep learning to translate graphs into such a vector representation.

This section particularly reviews the construction of a WL subtree kernel (following [9]), as it will be foundation of the next section. The subtree kernel transforms a graph into a vector, where a non-zero entry indicates the occurrence of a specific subtree in the graph. The total number of dimensions is determined by all subtrees that have been identified in the full set of graphs.

Given a graph $G = (V, E)$, a label function $l : V \to \Sigma^*$ yields for each node $v \in V$ a label over a finite alphabet Σ. The initial labels $l_0(v)$ are provided together with the graph G (original labels). A new label function l_i is obtained by calling $WLSK(G, l_{i-1})$, which is shown in Algorithm 1: It constructs new labels by concatenating all child labels deterministically (by processing children in some lexicographic order). A series of n WLSK calls provides a sequence of n label functions $l_0, \ldots, l_n$, where a node label $l_i(v)$ takes all children of v up to depth i into account. A label $l_i(v)$ may thus serve as a kind of fingerprint of the neighbourhood of v (hashcode). Let $L_i = \{l_i^1, l_i^2, \ldots, l_i^{k_i}\} = l_i(V)$ be the set of all different l_i-labels in G. The final vector representation of a graph is obtained from

$$\Phi(G) = \left(\#l_0^1, \ldots, \#l_0^{k_0}, \#l_1^1, \ldots, \#l_1^{k_1}, \ldots, \#l_{n-1}^1, \ldots, \#l_{n-1}^{k_{n-1}}\right)$$

where $\#l_i^j$ denotes how many nodes received the label l_i^j. Originally this approach was proposed as a test of isomorphism [11], as isomorphic graphs exhibit identical substructures (labels).

Figure 1 shows an illustrative example. On the top left we have two graphs G_1 and G_2 with nodes v_1–v_7 and v_8–v_{14}, resp. The (numeric) label is written in the node, the node identifiers are shown in gray. The table next to the graphs shows, for each node, how the new label s is constructed from the current node label and its successors. For instance, node v_1 of G_1 has label 0 and successors with labels $2, 0, 1$. Algorithm 1 creates new labels by appending the node label and the successor labels (in sorted order), which yields "$0 : 0, 1, 2$" for v_1. The rightmost table shows a dictionary, where each new label (here: $0 : 0, 1, 2$) gets a fresh ID (here: 3). Algorithm 1 refers to this step as hashing the node label into a new ID (or hashcode) – we use consecutive numbers just for illustrative purposes. Children need to be ordered deterministically to get the same hash for identical subtrees. The new label $l_1(v_1) = 3$ thus encodes a subtree of depth 1 with root 0 and children $0, 1, 2$. Once all new labels are determined (lower half of Fig. 1) the nodes v_1 and v_8 still have the same label: $l_1(v_1) = 3 = l_1(v_8)$, because their subtree of depth 1 was identical. After another WLSK iteration, however, the subtrees of depth 2 are no longer identical for v_1 and v_8, so their l_2-labels are no longer the same: $l_2(v_1) = 11 \neq 17 = l_2(v_8)$. The final vector representation for G_1 and G_2 (after 2 iterations) consists of counts for each label (from all depths):

$$\Phi(G_1) = (4, 1, 2,\ 1, 1, 1, 1, 2, 1, 0, 0,\ 1, 1, 1, 1, 2, 1, 0, 0, 0, 0)$$
$$\Phi(G_2) = (\underbrace{3, 2, 2,}_{L_0-}\ \underbrace{2, 0, 0, 0, 2, 1, 1, 1,}_{L_1-}\ \underbrace{0, 0, 0, 0, 2, 1, 1, 1, 2, 1}_{L_2-\text{label counts}})$$

The vector representation $\Phi(G)$ enables us to construct a kernel matrix or apply standard clustering and classification directly.

2.3 Discussion

Measuring graph similarity indirectly is in general more efficient than direct approaches. Among the kernel approaches it has been pointed out that with some

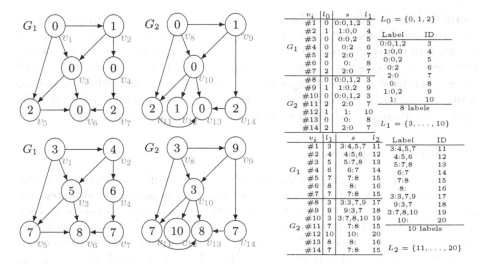

Fig. 1. Illustrative example of 2 WLSK iterations. left: initial labels l_0, middle: l_1, right: l_2

substructures, e.g. short paths (aka walks), many different graphs refer to the same point at the same point in the feature space (cf. [7]). Subtree kernels (and in particular WLSK) have been reported to be efficient[1] and well-performing in subsequent task (e.g. SVM classification). However, from the example in Fig. 1 we can also acknowledge the critique of the approach: Although G_2 has been obtained from G_1 by removing v_4 and adding v_{12} only, the vector representations are very different. Spotting differences early is good when checking for isomorphic graphs, but may be less desirable for similarity assessment (e.g. clustering). Despite the few changes, more than half of the labels occur exclusively in only one of the graphs (13 entries out of 21 that are zero in one of the two graphs). Continuous (rather than integer) features may help, as provided by some deep learning approaches, but deep learning requires a huge amount of training data, which makes them unsuitable for datasets of moderate size.

3   Enriching WL Subtree Kernels

Revisiting Fig. 1, node v_3 of G_1 and node v_{10} of G_2 differ only by a missing node labelled '1'. From the different l_1-hashcodes for both nodes (5 for v_3 and 3 for v_{10}) we cannot conclude what they have in common. Secondly, node v_2 of G_1 and v_9 of G_2 are similar in the sense that nodes labelled 0 and 2 can be reached, only in G_1 there is an intermediate node v_4. If we accept that node pairs (v_2, v_9) and (v_3, v_{10}) are somewhat similar, this should then positively affect the l_2-similarity of v_1 and v_8, too. We want to take this kind of similarity into account without

---

[1] The only necessary data structure is a hash table that collects how often each node label occurred.

sacrificing the efficiency of WLSK. Instead of integer features (subtree counts) we introduce continuous features to better reflect a partial matching of subtrees. We stick to the WLSK construction, but propose a post-processing step, which replaces the zero entries in the vector representation. As many subtrees (with different hashcodes) are in fact similar, we obtain highly correlating dimensions which are safe to remove and thus reduces the dimensionality. We optionally apply dimensionality reduction to arrive at a vector of moderate size.

3.1   Subtree Similarity

Given a graph $G = (V, E)$, let $L_i = l_i(V)$ be the set of all hashcodes for subtrees of depth i (cf. tables on the right of Fig. 1). The hashcodes compress the newly constructed node labels, but no longer contain any information about the subtree. So we track this information in tables: For all occurred hashcodes $h \in L_i$, we denote the root node label by $r_h \in L_{i-1}$ and the multiset of successor labels by $S_h \subseteq L_{i-1}$. (*Example:* For $h = 11 \in L_2$ in Fig. 1 we have $r_h = 3$ and $S_h = \{4, 5, 7\}$.)

Next we define a series of distance functions $d_i : L_i \times L_i \to \mathbb{R}$ to capture the distance between subtree hashcodes of the same depth i. We start with a distance d_0 for the original graph node labels. In absence of any background knowledge we use for the initial level

$$d_0(h, h') := \begin{cases} 0 & \text{if } h = h' \\ 1 & \text{otherwise} \end{cases}, \tag{1}$$

but generally assume that some background information can be provided to arrive at meaningful distances for the initial node labels.

For non-trivial subtrees (that is, $i > 0$) we recursively define distance functions $d_i(h, h')$. It is natural to define the distance as the sum of distances between root and child nodes. This requires to assign child nodes of h uniquely to child nodes of h', which is provided by a bijective function $f : S_h \to S_{h'}$:

$$d_i(h, h') := \underbrace{d_{i-1}(r_h, r_{h'})}_{\text{root node distance}} + \underbrace{\min_{f \in B(S_h, S_{h'})} \sum_{k \in S_h} d_{i-1}(k, f(k))}_{\text{distance of best subtree alignment}} \tag{2}$$

Here $B(S, T)$ denotes the set of bijective functions $f : S \to T$. The first term measures the distance between the root node labels and the second term identifies the minimal distance among all node assignments. Finding the assignment with minimal distance is known as the *assignment problem*, which has well-known solutions and we adopt the Munkres algorithm for this task [4].

We are likely to deal with unbalanced assignments, that is, different numbers of children for h and h'. A bijective assignment requires $|S_h| = |S_{h'}|$, so we add the necessary number of *missing nodes* (denoted by $\perp$) to the smaller multiset.[2]

---

[2] More formally $B(S, T)$ is the set of bijective functions $f : S' \to T'$ where $|S'| = k = |T'|$, $S \subseteq S'$, $T \subseteq T'$, S' has $k - |S|$ (and T' has $k - |T|$) additional $\perp$ elements.

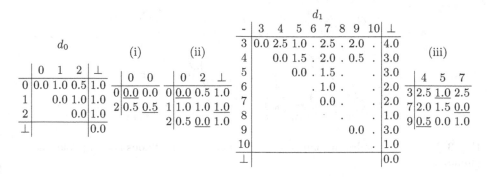

Fig. 2. Left: A priori distances d_0 between labels of L_0. Case (i): Assignment matrix for d_1 distance of $l_1(v_2) = 4$ and $l_1(v_9) = 9$. Case (ii): Assignment matrix for d_1 distance of v_3 ($\{0, 2\}$) and v_{10} ($\{0, 1, 2\}$). Right: Derived d_1-distances from case (i) and (ii). Case (iii): Assignment matrix for d_2 distance of v_0 ($\{4, 5, 7\}$) and v_8 ($\{3, 7, 9\}$) (Color figure online)

We extend the distance d_0 to the case of missing nodes, which corresponds to an additional row/column in the d_0-matrix (see d_0 example matrix in Fig. 1(left)). Again, these $\bot$-distances may be an arbitrary constant or specifically provided for each label $h \in L_0$ using background knowledge. Then Eq. (2) extends naturally to $\bot$-values:

$$d_i(h, \bot) := d_{i-1}(r_h, \bot) + \sum_{k \in S_h} d_{i-1}(k, \bot) \tag{3}$$

Figure 2 shows an example. The leftmost table shows the d_0-distances between original node labels (cf. Fig. 1: $L_0 = \{0, 1, 2\}$), including the case of a missing label $\bot$. For the sake of illustration we assume a distance of $\frac{1}{2}$ for the label pair $(0, 2)$. Consider the comparison of v_2 and v_9 for depth-1 subtrees: $d_1(h, h')$ with $h = l_1(v_2)$, $h' = l_1(v_9)$. Both root nodes are identical ($r_h = r_{h'} = 0$), but the multisets of successors are not ($S_h = \{0, 0\}, S_{h'} = \{0, 2\}$). Matrix (i) shows the distance matrix for the assignment problem: all nodes of h' (rows) have to be assigned to a node of h (columns). As the child nodes represent l_0-hashcodes, we take the distances from the d_0 table. An optimal assignment is marked in red and we obtain a distance $d_1(h, h') = 0 + (0 + \frac{1}{2}) = \frac{1}{2}$. Matrix (ii) shows a second example for the d_1 comparison of v_3 vs v_{10}: As v_{10} has three children but v_3 only two, we introduce one $\bot$-element to obtain a square matrix. The optimal assignment is shown in red, the d_1-distance becomes 1.0. Both examples contribute two values to the d_1-distance (fourth matrix), from which we may then calculate, e.g., $d_2(l_2(v_1), l_2(v_8)) = 0 + (\frac{1}{2} + 1 + 0) = 1.5$ (matrix (iii)).

3.2 Updating Vector Representations

Once the WLSK algorithm has been executed, we determine all d_i-distances from the l_i-labels alone (without revisiting the graphs). Then we update the vector

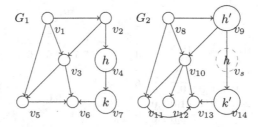

Fig. 3. Insertion of nodes to compensate side-effects of superfluous nodes. (Color figure online)

representations of all graphs, the zero entries in particular. Suppose $\mathbf{x}$ is a vector representation of G and $\mathbf{x}_h = 0$ for some $h \in L_i$, which means that subtree h is not present in G. Among the subtrees that *do occur* in G we can now find the one most similar to $h' \in L_i$ (smallest distance $d_i(h, h')$) and replace $\mathbf{x}_h$ by

$$\mathbf{x}_h \leftarrow k(d_i(h, h')) \cdot x_{h'}$$

where $k : \mathbb{R}^+ \rightarrow [0, 1]$ is a monotonically decreasing function that turns distances into similarities with $k(0) = 1$. The multiplication with $x_{h'}$ accounts for the fact that h' may occur multiple times in G. We used $k(d) = e^{-(d/\delta)^2}$, where δ is a user-defined threshold.

3.3   Compensating Superfluous Nodes

We say v is an *superfluous node* if it is just a *stopover* on the way to yet another node, but does not contribute to the graph structure itself, that is, if the in- and out-degree of v is 1. In Fig. 1 the node v_4 in G_1 is such a superfluous node. In some applications nodes with certain labels may occur occasionally, but do not carry any important information. Their existence/absence should therefore affect the graph similarity not too much.

The discussed distance measure can cope with such differences when comparing, e.g., the subtree of v_2 with that of v_9. But if we consider v_4 as an superfluous intermediate node, it brings another undesired effect: It may introduce completely new subtrees which are not present in other graphs. In the example of Fig. 1 the node v_4 introduces subtrees with hashcodes 6 (at depth 1) and 14 (at depth 2), which are not present in G_2. When measuring the similarity of G_1 and G_2, such subtrees make the graphs appear less similar.

We address such cases by considering the insertion of a superfluous node in our distance calculation. Figure 3 shows the situation once more: To enrich the vector representation of G_2 we seek a closest match for label h. According to Sect. 3.1 we consider, amongst others, the node v_9 with label h' as a candidate. With both nodes having a single child only, finding the optimal bijective assignment f is trivial ($f(k) = k'$) and Eq. (2) boils down to $d_{i-1}(r_h, r_{h'}) + d_{i-1}(k, k')$. Now we additionally consider the *insertion* of a superfluous node v_s with the same label as v_4, as shown in Fig. 3 (red). Note that a hashcode $l_i(v_s)$ for the newly inserted node was not necessarily generated earlier. How would the distance between a node v_4 and v_s evaluate? According to (2) we have

$$d_i(l_i(v), l_i(v_s)) = d_{i-1}(l_{i-1}(v), l_{i-1}(v_s)) + d_{i-1}(k, k')$$

The second part consists of a single term because both nodes have a single child only. Note that it does not depend on v_s. Substituting the first term repeatedly by its definition eventually leads us to

$$d_i(l_i(v), l_i(v_s)) = \underbrace{d_0(l_0(v), l_0(v_s))}_{\text{0 by construction}} + \sum_{j=0}^{i-1} d_j(l_j(k), l_j(l)) \tag{4}$$

The level-0-distance to the newly inserted node is 0 by construction, however, we replace it by a penalty term $d_I(l_0(v))$ to reflect the fact that we had to insert a new node. As with $d_0(\cdot, \cdot)$ we assume that $d_I(\cdot)$ can be derived meaningfully from the application context: If, for instance, nodes with a certain label h are optional, we choose a low insertion distance $d_I(h)$ and may otherwise set $d_I(h) = \infty$ to prevent undesired insertions.

We thus arrive at a distance $d_i^*(h, h')$ for the insertion of a superfluous node

$$\min \begin{cases} \min\{d_I(r_h), d_I(r_{h'})\} + \sum_{j=0}^{i-1}(l_j(k), l_j(k')) & \text{if } S_h = \{k\} \wedge S_{h'} = \{k'\} \\ \infty & \text{otherwise} \end{cases} \tag{5}$$

which yields ∞ if the prerequisites of a superfluous nodes are not given and considers node insertion on both sides (inner min-term). The original distance (2) may then be replaced by $\min\{d_i(h, h'), d_i^*(h, h')\}$ to reflect the occurrence of superfluous nodes appropriately. These changes can be handled during the precalculation of the distance matrices, the vector enrichment remains unchanged.

3.4 Complexity

Enriching the vector representations requires two steps: (1) The calculation of all distance matrices d_i requires to calculate $\sum_i |L_i|^2$ entries. For each entry we have to solve an assignment problem, which is $O(d^2 \log d)$ where d is the maximal node degree. The method is therefore unattractive for highly connected graphs. But many applications with large graphs have a bounded node degree. (2) Secondly, the vector representations $\mathbf{x}$ of all n graphs need to be enriched. This takes $O(m_z \cdot m_{nz})$ for each graph, where m_z (resp. m_{nz}) is the number of entries in $\mathbf{x}$ with zero (resp. non-zero) entries: for each 0-entry in $\mathbf{x}$ we have to find the most similar 1-entry. The number of all labels from all graphs ($m = \sum_i |L_i|$) is much larger than the number of nodes in a single graph, whereas m_{nz} is bounded by the number of nodes in a single graph. With $m_{nz} \ll m_z$ we may consider m_{nz} as a constant (max. no. of nodes) and arrive at $O(n \cdot m)$ for the vector enrichment.

Exercise: Write a function to count the number of entries in an integer array having a 3 at the last digit.

```
public static int count3(int [] x){
    int count=0;
    int i=0;
    int zaehler=0;
    while (i < x.length) {
        zaehler=x[i] % 10;
        if (zaehler == 3) {
            ++count;
} } }
```

Fig. 4. Example of a source code snapshot and its graph representation. The student has not yet finished the solution at this stage/snapshot, the return statement is still missing.

4 Application

We demonstrate the usefulness of the proposed modification in an application from computer science education. The increase in the number of CS students over the last years calls for tools that help lecturers to assess the stage of development of a whole group of students – rather than inspecting the solutions one by one. Our dataset consists of editing streams from the students source code editor (for selected exercises of an introductory programming course using Java). In our preliminary evaluation we have about 30–50 such streams per task. We extract snapshots of the code whenever a student starts to edit a different code line than before. (Many snapshot thus do not represent compileable code.) The goal is to compare editing paths against each other, for instance, to identify the most common paths or outliers. We replace the textual representation of the source snapshot by a graph capturing the abstract syntax tree and the variable usage, as can be seen in the example of Fig. 4. We want to cluster the snapshots and to construct a new graph where nodes correspond to clusters (of code snapshots) and edges indicate editing paths of students. For the experiments we applied some preprocessing (e.g. variable renaming in the graph) and assigned low insertion costs to expression- and declaration-nodes, because students may phrase conditions quite differently. Our use case for superfluous nodes (Sect. 3.3) are code blocks ({ }), which are optional if the code within the block consists of a single statement only (e.g. the ++count in Fig. 4).

4.1 Effect on Distances

To measure the effect of the enriched kernel we have manually subdivided a set of snapshots into *similar* and *dissimilar snapshots*. In a clustering setting we want

Table 1. Effect of vector enrichment on distances.

Kernel depth d	σ	Standard vector	Enriched vector
		f	f
2	$3 \cdot d$	$\frac{8.43-4.76}{2.44} = 1.50$	$\frac{7.03-1.75}{0.97} = 5.44$
3	$3 \cdot d$	$\frac{9.48-6.43}{3.06} = 0.99$	$\frac{11.18-4.24}{2.36} = 2.93$
4	$3 \cdot d$	$\frac{9.85-7.82}{3.46} = 0.58$	$\frac{14.67-6.63}{3.99} = 2.01$
5	$3 \cdot d$	$\frac{9.91-8.52}{3.64} = 0.38$	$\frac{15.07-8.34}{4.54} = 1.48$

the modification to carve out clusters more clearly. We therefore compare the mean distance μ_w (and variance σ_w) *within* the group of similar graphs against the mean distance μ_b (and variance σ_b) *between* both groups. By the factor f we denote the size of the gap between both means in multiples of the within-group standard deviation σ_w, that is, $f = \frac{\mu_b - \mu_s}{\sigma_s}$. The factor f may be considered as a measure of separation between the cluster of similar graphs and the remaining graphs. From Table 1 we find that the enriched representation consistently yields higher values of f for the enriched than for the standard vector representation.

4.2 Dimensionality

New node labels are introduced for every new subtree, which introduces a high dimensional vector representation that has been identified as problematic in the literature (Sect. 2.3). Enriching the vector representation can help to overcome this problem, because labels with minor changes will receive similar (enriched) entries. For instance, a dataset with 718 code snapshot graphs generated as many as 5179 different subtree labels (depth 3). After enrichment we identified the number of attributes that might be removed from the dataset because it contains a highly correlating attribute already. This leads to a substantial reduction in the number of columns: Depending on the Pearson correlation threshold of 0.9/0.95/0.99 as much as 77%/68%/55% of the attributes can be discarded.

4.3 Code Graph Clustering

To reduce the dimensionality further, a principal component analysis (PCA) may be applied. Figure 5 shows the scatter plot of the principal components (PC) #2 against PC #1, #3 and #4 for the standard representation (top) and the enriched vectors (bottom). The colors indicate cluster memberships from a mean shift clustering over 4 principal components. Note that, by construction of the dataset, we do not expect the source code snapshots to fall apart completely in well separated clusters, because the data represents the evolution towards a final solution, snapshots differ by incremental changes only. In the standard case the data scatters more uniformly and less structured (left; PC1 vs PC2), while the enriched data shows two long-stretched clusters that reflect a somewhat linear code evolution for two different approaches to solve the exercise, which

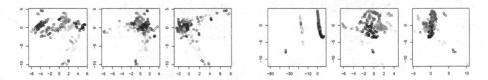

Fig. 5. Principal component #2 versus principal component #1, #3 and #4 for standard (left) and enriched (right) vectors. (Color figure online)

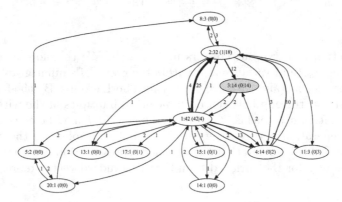

Fig. 6. Snapshot evolution for a group of students: Nodes represent clusters, edges represent snapshot transitions. (Color figure online)

corresponds much better to our expectation. When taking additional component into account (PC3), the scatterplot in the middle (PC2 vs PC3) offers a clearer structure for the enriched data (e.g. the separation of the curved red cluster at the top) than the original data.

Figure 6 shows how the clusters are used in the context of our application. Each cluster (like those in Fig. 5, but for a different exercise) corresponds to a node in this graph. Whenever a student changes the code and thereby moves to a different cluster, a (directed) edge is inserted. The number of students who have followed a path is written nearby the edge. Clusters that have only one incoming and one outgoing edge are not shown for the sake of brevity. The green color indicates the degree of unit-test fulfillment. The node labels $a : b(c|d)$ carry information about the cluster id a, number of students b that came across this node, number of students c (resp. d) who started (resp. ended) in this node. From this example the lecturer can immediately recognize that 42 students start in cluster #1, from where most students (25) transition to cluster #2 and 10 more students reach the same cluster via cluster #4 as an intermediate step. Cluster #2 does not yet correspond to a perfect solution, but only 12 students manage to reach the green cluster #3 from cluster #2. Other clusters and edges have much smaller numbers, they cover exotic solutions or trial-and-error approaches. The graph provides a good overview about the students performance as a group.

5   Conclusions

Weisfeiler-Lehman subtree kernels can be used to transform graphs into a meaningful vector representation, but suffer from high dimensionality and sparsity, such that the similarity assessment is limited. We overcome both problems by taking the subtree distances into account – which are simpler to assess than general tree distance, because only subtrees of equal depth need to be considered. Based on the subtree distance we enrich the zero entries of graph vectors and improve the similarity assessment. A removal of highly correlating attributes reduces the dimensionality considerably. The modifications turned out to be advantageous in a use case of source code snapshot clustering.

References

1. Bille, P.: A survey on tree edit distance and related problems. Theor. Comput. Sci. **337**, 217–239 (2005)
2. Demaine, E.D., Mozes, S., Rossman, B., Weimann, O.: An optimal decomposition algorithm for tree edit distance. In: International Colloquium on Automata, Languages, and Programming, pp. 146–157 (2007)
3. Gao, X., Xiao, B., Tao, D., Li, X.: A survey of graph edit distance. Pattern Anal. Appl. **13**(1), 113–129 (2010)
4. Munkres, M.: Algorithms for the assignment and transportation problems. J. Soc. Ind. Appl. Math. **5**(1), 32–38 (1957)
5. Narayanan, A., Chandramohan, M., Chen, L., Liu, Y., Saminathan, S.: subgraph2vec: learning distributed representations of rooted sub-graphs from large graphs. In: Workshop on Mining and Learning with Graphs (2016)
6. Paassen, B.: Metric learning for structured data. Ph.D. thesis, Bielefeld University (2019)
7. Ramon, J., Gärtner, T.: Expressivity versus efficiency of graph kernels. In: Proceedings of the First International Workshop on Mining Graphs, Trees and Sequences, pp. 65–74 (2003)
8. Seeland, M., Girschick, T., Buchwald, F., Kramer, S.: Online structural graph clustering using frequent subgraph mining. In: Balcázar, J.L., Bonchi, F., Gionis, A., Sebag, M. (eds.) ECML PKDD 2010. LNCS (LNAI), vol. 6323, pp. 213–228. Springer, Heidelberg (2010). https://doi.org/10.1007/978-3-642-15939-8_14
9. Shervashidze, N., Schweitzer, P., van Leeuwen, E.J., Mehlhorn, K., Borgwardt, K.M.: Weisfeiler-lehman graph kernels. J. Mach. Learn. Res. **12**, 2539–2561 (2011)
10. Shervashidze, N., Vishwanathan, S., Petri, T.H., Mehlhorn, K., Borgwardt, K.M.: Efficient graphlet kernels for large graph comparison. In: Proceedings of the International Conference on Artificial Intelligence and Statistics (AISTATS) (2009)
11. Weisfeiler, B., Lehman, A.: A reduction of a graph to a canonical form and an algebra arising during this reduction. Nauchno-Technicheskaya Informatsia **2**(9), 12–16 (1968)

Overlapping Hierarchical Clustering (OHC)

Ian Jeantet[✉], Zoltán Miklós, and David Gross-Amblard

Univ Rennes, CNRS, IRISA, Rennes, France
{ian.jeantet,zoltan.miklos,
david.gross-amblard}@irisa.fr

Abstract. Agglomerative clustering methods have been widely used by many research communities to cluster their data into hierarchical structures. These structures ease data exploration and are understandable even for non-specialists. But these methods necessarily result in a tree, since, at each agglomeration step, two clusters have to be merged. This may bias the data analysis process if, for example, a cluster is almost equally attracted by two others. In this paper we propose a new method that allows clusters to overlap until a strong cluster attraction is reached, based on a density criterion. The resulting hierarchical structure, called a quasi-dendrogram, is represented as a directed acyclic graph and combines the advantages of hierarchies with the precision of a less arbitrary clustering. We validate our work with extensive experiments on real data sets and compare it with existing tree-based methods, using a new measure of similarity between heterogeneous hierarchical structures.

1 Introduction

Agglomerative hierarchical clustering methods are widely used to analyze large amounts of data. These successful methods construct a dendrogram – a tree structure – that enables a natural exploration of data which is very suitable even for non-expert users. Various tools offer intuitive top-down or bottom-up exploration strategies, zoom-in and zoom-out operations, etc.

Let us consider the following real-life scenario: a social science researcher would like to understand the structure of specific scientific domains based on a large corpus of publications, such as dblp or Wiley. A contemporary approach is to construct a word embedding [23] of the key terms in publications, that is, to map terms into a high-dimensional space such that terms frequently used in the same context appear close together in this space (for the sake of simplicity, we omit interesting issues such as preprocessing, polysemy, etc.). Identifying for example the denser regions in this space directly leads to insights on the key terms of Science. Moreover, building a dendrogram of key terms using an agglomerative method is typically used [9,14] to organize terms into hierarchies. This dendogram (Fig. 1a) eases data exploration and is understandable even for non-specialists of data science.

© The Author(s) 2020
M. R. Berthold et al. (Eds.): IDA 2020, LNCS 12080, pp. 261–273, 2020.
https://doi.org/10.1007/978-3-030-44584-3_21

Despite its usefulness, the dendrogram structure might be limiting. Indeed, any embedding of key terms has a limited precision, and key terms proximity is a debatable question. For example, in Fig. 1a, we can see that the *bioinformatics* key term is almost equally attracted by *biology* and *computing*, meaning that these terms appear frequently together, but in different contexts (e.g. different scientific conferences). Unfortunately, with classical agglomerative clustering, a merging decision has to be made, even if the advantage of one cluster on another is very small. Let us suppose that arbitrarily, *biology* and *bioinformatics* are merged. This may suggest to our analyst (not expert in computer science) that *bioinformatics* is part of *biology*, and its link to *computing* may only appear at the root of the dendrogram. Clearly, an interesting part of information is lost in this process.

In this paper, our goal is to combine the advantages of hierarchies while avoiding early cluster merge. Going back to the previous example, we would like to provide two different clusters showing that *bioinformatics* is closed both to *biology* and *computing*. At a larger level of granularity, these clusters will still collapse, showing that these terms belong to a broader community. This way, we deviate from the strict notion of trees, and produce a directed acyclic graph that we call a quasi-dendrogram (Fig. 1b).

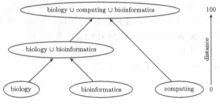 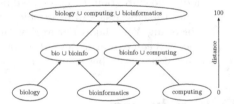

(a) A classical dendrogram, hiding the early relationship between *bioinformatics* and *computing*.

(b) A **quasi-dendrogram**, preserving the relationships of *bioinformatics*.

Fig. 1. Dendrogram and quasi-dendrogram for the structure of Science.

Our contributions are the following:

- We propose an agglomerative clustering method that produces a directed acyclic graph of clusters instead of a tree, called a quasi-dendrogram,
- We define a density-based merging condition to identify these clusters,
- We introduce a new similarity measure to compare our method with other, quasi-dendrogram or tree-based ones,
- We show through extensive experiments on real and synthetic data that we obtain high quality results with respect to classical hierarchical clustering, with reasonable time and space complexity.

The rest of the paper is organized as follows: Sect. 2 describes our proposed overlapping hierarchical clustering framework[1]. Section 3 details our experimen-

---

[1] Source code available at https://gitlab.inria.fr/ijeantet/ohc.

tal evaluation. Section 4 presents the related works, while Sect. 5 concludes the paper.

2   Overlapping Hierarchical Clustering

2.1   Intuition and Basic Definitions

In a nutshell, our method obtains clusters in a gradual agglomerative fashion and in a precise way. At each step, when we increase the neighbourhood of the clusters by including more interconnections, we consider the points that fall in this connected neighbourhood and we take the decision to merge some of them whenever they are connected enough to a cluster using a *density criterion* λ. Taking interconnections into account may lead to overlapping clusters.

More precisely, we consider a set $V = \{X_1, \ldots, X_N\}$ of N points in a n-dimensional space, i.e. $X_i \in V \subset \mathbb{R}^n$ where $n \geq 1$ and $|V| = N$. In order to explore this space in an iterative way, we consider points that are close up to a limit distance $\delta \geq 0$. We define the δ-neighbourhood graph of V as follows:

Definition 1 (δ-neighbourhood graph). *Let $V \subset \mathbb{R}^n$ be a finite set of data points and $E \subset V^2$ a set of pair of elements of V, let d be a metric on $\mathbb{R}^n$ and let $\delta \geq 0$ be a positive number. The δ-neighbourhood graph $G_\delta(V, E)$ is a graph with vertices labelled with the data points in V, and where there is an edge $(X, Y) \in E$ between $X \in V$ and $Y \in V$ if and only if $d(X, Y) \leq \delta$.*

Property 1. If $\delta = 0$ then the δ-neighbourhood graph consists of isolated points while if $\delta = \delta_{max}$, where δ_{max} is the maximum distance between any two nodes in V then $G_\delta(V, E)$ is the complete graph on V.

Varying δ will allow to progressively extend the neighbourhood of the vectors to form bigger and bigger clusters. Clusters will be formed according to the density of a region of the graph.

Definition 2 (Density). *The density [16] dens(G) of a graph $G(V, E)$ is given by the ratio of the number of edges of G to the number of edges of G if it were a complete graph, that is, $dens(G) = \frac{2|E|}{|V|(|V|-1)}$. If $|V| = 1$, $dens(G) = 1$.*

A cluster is simply defined as a subset of the nodes of the graph and its density is defined as the density of the corresponding subgraph.

2.2   Computing Hierarchies with Overlaps

Our algorithm, called OHC, computes a hierarchy of clusters that we can identify in the data. We call the generated structure a quasi-dendrogram and it is defined as follows.

Definition 3 (Quasi-dendrogram). *A quasi-dendrogram is a hierarchical structure, represented as a directed acyclic graph, where the nodes are labelled with a set of data points, the clusters, such as:*

- *The leaves (i.e. the nodes with 0 in-degree) correspond to the singletons, i.e. contain a unique data point. The level of the leaf nodes is 0.*
- *There is only one root node (node with 0 out-degree) that corresponds to the set of all the data points.*
- *Each node (except the root node) has one or more parent nodes. The parent relationship corresponds to inclusion of the corresponding clusters.*
- *The nodes at a level δ represent a set of (potentially overlapping) clusters that is a cover of all the data points. Also, for each pair of points of a given cluster, it exists a path between points of this cluster that have a distance less than δ. In other terms, a node contains a part of a connected subgraph of the δ-neighbourhood graph.*

The OHC method works as presented in Algorithm 1. We first compute the distance matrix of the data points (I3). We chose the cosine distance, widely use in NLP. Then we construct and maintain the δ-neighbourhood graph $G_\delta(V, E)$, starting from $δ = 0$ (I4).

We also initialize the set of clusters, i.e. the leaves of our quasi-dendrogram, with the individual data points (I4). At each iteration, we increase δ (I6) and consider the new added links to the graph (I8) and the impacted clusters (I9). We extend these clusters by integrating the most linked neighbour vertices if the density does not change more than a given threshold λ (I10–15). We remove all the clusters included in these extended clusters (I16) and add the new set of clusters to the hierarchy as a new level (I18). We stop when all the points are in the same cluster which means that we reached the root of the quasi-dendrogram.

Also to improve the efficiency of this algorithm we use dynamic programming to avoid to recompute information related to the clusters like their density and the list of their neighbour vertices. It lead to significant improvements in the execution time of the algorithm. We will discuss this further in the Sect. 3.3.

Property 2 (λ = 0). When $λ = 0$, each level $δ_i$ of a quasi-dendrogram contains exactly the cliques (complete subgraphs) of the $δ_i$-neighbourhood graph $G_{δ_i}$.

Property 3 (λ = 1). When $λ = 1$, each level $δ_i$ of a quasi-dendrogram contains exactly the connected subgraphs of the $δ_i$-neighbourhood graph $G_{δ_i}$.

3   Experimental Evaluation

3.1   Experimental Methodology

Tests: The tests we performed were focused on the quality of the hierarchical structures produced by our algorithm. To measure this quality we used the classical hierarchy produced by *SLINK*, an optimal single-linkage clustering algorithm proposed in *Sibson et al.* [28], as a baseline. Our goal was to study the behaviour of the **merging criterion** parameter λ that we introduced, as long as its influence on the **execution time**, to verify if for $λ = 1$ we experimentally obtain the same hierarchy as *SLINK* (Property 3) and hence observe the **conservativeness** of our algorithm. We also compared our method to other agglomerative

---

Algorithm 1. Overlapping Hierarchical Clustering (OHC)

1: Input:
    – $V = \{x_1, \ldots, x_N\}$, N data points.
    – $\lambda \geq 0$, a merging density threshold.
2: Output: quasi-dendrogram H.
3: Preprocessing: obtain $\Delta = (\delta_1, \ldots, \delta_m)$ the distances between data points in increasing order.
4: Initialization:
    – Create the graph $G(V, E_0 = \emptyset)$.
    – Set a list of clusters $C = [\{x_1\}, \ldots, \{x_N\}]$.
    – Add the list of clusters to the level 0 of H.
5: i=1.
6: **while** $\#C > 1$ and $i \leq m$ **do**
7:     **for each** pair $(u, v) \in V^2$ such as $d(u, v) = \delta_i$ **do**
8:         Add (u, v) to $E_{\delta_{i-1}}$.
9:         Determine the impacted clusters C_{imp} of C containing either u or v.
10:     **for each** impacted cluster $C_{imp_j} \in C_{imp}$ **do**
11:         Look for the points $\{p_1, \ldots, p_k\}$ that are the most linked to C_{imp_j} in G_{δ_i}.
12:         Compute the density $dens(S_j)$ of the subgraph $S_j = C_{imp_j} \cup \{p_1, \ldots, p_k\}$.
13:         **if** $S_j \neq C_{imp_j}$ and $|dens(S_j) - dens(C_{imp_j})| \leq \lambda$ **then**
14:             Continue to add the most linked neighbors to S_j the same way if possible.
15:             When S_j stops growing remove C_{imp_j} from the list of clusters C and add S_j to the
list of new clusters C_{new}.
16:     Remove all cluster of C included in one of the clusters of C_{new}.
17:     Concatenate C_{new} to C.
18:     Add the list of clusters to the level δ_i of H.
19:     i=i+1.
20: **return** H

---

methods such as the *Ward* variant [29] and *HDBSCAN** [8]. To compare such structures we needed to create a new similarity measure which is described in Sect. 3.2.

Datasets: To partially see the scalability of our algorithm but also to avoid too long running times we had to limit the size of the datasets to few thousand vectors. To be able to compare the results, we run the tests on datasets of same size that we fixed to **1000 vectors**.

– The first dataset is composed of 1000 **randomly** generated 2-dimensional points.
– To test the algorithm on real data and in our motivating scenario, the second dataset was created from the **Wiley** collection via their API[2]. We extracted the titles and abstracts of the scientific papers and trained a word embedding model on the data of a given period of time by using the classical *SGNS* algorithm from *Mikolov et al.* [22] following the recommendation of *Levy et al.* [20]. We set the vocabulary size to only 1000 key words per year even though this dataset allows us to extract up to 50000 of them. This word embedding algorithm created 1000 300-dimensional vectors for each year over 20 years.

Experimental Setting: All our experiments are done on a Intel Xeon 5 Core 1.4 GHz, running MacOS 10.2 on a SSD hard drive. Our code is developed with

---

[2] https://onlinelibrary.wiley.com/library-info/resources/text-and-datamining.

Python 3.5 and the visualization part was done on a Jupyter NoteBook. We used the *SLINK* and Ward implementations from the scikit-learn python package and the *HDBSCAN** implementation of *McInnes et al.* [21].

3.2 A Hierarchy Similarity Measure

As there is no ground truth on the hierarchy of the data we used, we need a similarity measure to compare the hierarchical structures produced by hierarchical clustering algorithms. The goal is not only to compare the topology but also the content of the nodes of the structure. However up to our knowledge there is very little in the literature about hierarchy comparison especially when the structure is similar to a DAG or a quasi-dendrogram. *Fowlkes and Mallows* [19] defined a similarity measure per level and the new similarity function we propose is based on the same principle. First we construct a similarity between two given levels of the hierarchies, and then we extend it to the global structures by exploring all the existing levels.

Level Similarity: Given two hierarchies h_1 and h_2 and a cardinality i, we assume that it is possible to identify a set l_1 (resp. l_2) of i clusters for a given level of hierarchy h_1 (resp. h_2). Then, to measure the similarity between l_1 and l_2, we take the maximal Jaccard similarity among one cluster of l_1 and every clusters of l_2. The average of these similarities, one for each cluster of l_1, will give us the similarity between the two sets. If we consider the similarity matrix of h_1 and h_2 with a cluster of l_1 for each row, a cluster of l_2 for each column and the Jaccard similarity between each pair of clusters at the respective coordinates in the matrix, we can compute the similarity between l_1 and l_2 by taking the average of the maximal value for each row. Hence, the similarity function between two sets of clusters l_1, l_2 is defined as:

$$sim_l(l_1, l_2) = mean\{max\{J(c_1, c_2) \mid c_2 \in l_2\}|c_1 \in l_1\} \qquad (1)$$

where J is the Jaccard similarity function.

However, taking the maximal value of each row shows how the clusters of the first set are represented in the second. If we take the maximal value of each column we will see the opposite, i.e. how the second set is represented in the first set. Hence with this definition the similarity might not be symmetrical so we propose this corrected similarity measure that shows how both sets are represented in the other one:

$$sim_l^*(l_1, l_2) = mean(sim_l(l_1, l_2), sim_l(l_2, l_1)) \qquad (2)$$

Complete Similarity: Now that we can compare two levels of the hierarchical structures, we can simply average the similarity for each corresponding levels of the same size. For classical dendograms, each level has a distinct number of clusters so identification of levels is easy. Conversely, our quasi-dendrograms may have several distinct levels (pseudo-levels) with the same number of clusters. If so, we need to find the best similarity between these pseudo-levels. For a given level (i.e. number of clusters), we want to build a matching M that maps each pseudo-level $l_1^1, l_1^2, ...$ of h_1 to at least one pseudo-level $l_2^1, l_2^2, ...$ of h_2 and conversely (see Fig. 2). This matching M should maximize the similarity between pseudo-levels while preserving their hierarchical relationship. That is, for a, b, c, d representing the height of pseudo-levels in the hierarchies, if $(l_1^a, l_2^c) \in M$ and $(l_1^b, l_2^d) \in M$, then $(b \geq a \rightarrow d \geq c)$ or $(b < a \rightarrow d < c)$ (no "crossings" in M, such as $((l_1^{231}, l_2^{303})$ with $(l_1^{230}, l_2^{304}))$.

To produce this mapping, our simple algorithm is the following. We initialize M and two pointers with the two highest pseudo-levels $((l_1^{231}, l_2^{304})$, step 1 of Fig. 2). At each step, for each hierarchy, we consider current pointers and their children, and compute all their similarities (step 2). We then add pseudo-levels with maximal similarity to M (here, $(l_1^{230}, l_2^{303}))$. Whenever a child is chosen, the respective pointer advances, and at each step, at least one pointer advances. Once pseudo-levels have been consumed on one side, ending with l, we can finish the process by adding (l^f, l) to M for all remaining pseudo-level l' on the other side (here, $l = l_1^{230}$. On our example, the final matching is $M = \{(l_1^{231}, l_2^{304}), (l_1^{230}, l_2^{303}), (l_1^{230}, l_2^{302}), (l_1^{230}, l_2^{301}), (l_1^{230}, l_2^{300}), (l_1^{230}, l_2^{299})\}$.

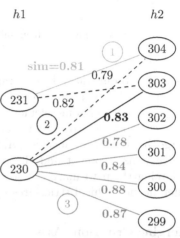

Fig. 2. Computing the similarity between two quasi-dendograms h_1 and h_2 for levels having the same number of clusters.

Finally, from (2) we define the similarity between two hierarchies as

$$sim(h_1, h_2) = mean\{sim_l^*(l_1, l_2) | (l_1, l_2) \in (h_1, h_2) \,\&\, (l_1, l_2) \in M\}. \qquad (3)$$

3.3 Experimental Results

Expressiveness: With this small following example we would like to present the expressiveness of our algorithm compared to classical hierarchical clustering algorithms such as *SLINK*. On the hand-built example shown in Fig. 3a we can clearly distinguish two groups of points, $\{A, B, C, D, E\}$ and $\{G, H, I, J, K\}$ and two points that we can consider as noise, F and L. Due to the chaining effect we expect that the *SLINK* algorithm will regroup the 2 sets of points early in the hierarchy while we would like to prevent it by allowing some cluster overlaps.

Figure 3b shows the dendrogram computed by *SLINK* and we can see as expected that when F merges with the cluster formed by $\{A, B, C, D, E\}$ the next step is to merge this new cluster with $\{G, H, I, J, K\}$.

On the contrary in Fig. 4 that presents the hierarchy built with our method for a specific merging criterion, we can see an example of diamond shape that

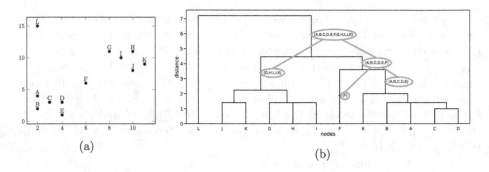

Fig. 3. A hand-built example (a) and its *SLINK* dendrogram (b).

is specific to our quasi-dendrogram. For simplicity the view here slightly differs from the quasi-dendrogram definition as we used dashed arrows to represent the provenance of some elements of a cluster instead of going further down in hierarchy to have a perfect inclusion and respect the lattice-like structure. The merge between the clusters $\{A, B, C, D, E\}$ and $\{G, H, I, J, K\}$ is delayed to the very last moment and the point F will belong to these 2 clusters instead of forcing them to merge. Also depending on the merging criterion we obtain different hierarchical structures by merging earlier of later some clusters.

Merging Criterion: As we can see in Fig. 5b when the merging criterion increases we obtain a hierarchy more and more similar to the one produced by the classical *SLINK* algorithm until we obtain exactly the same for a merging criterion of 1. Knowing this fact it is also normal to have a similarity between OHC and *Ward* (resp. *HDBSCAN**) hierarchies converging to the similarity between *SLINK* and *Ward* (resp. *HDBSCAN**) hierarchies. However we can notice that the OHC and *Ward* hierarchies are the most similar for a merging criterion smaller than 1.

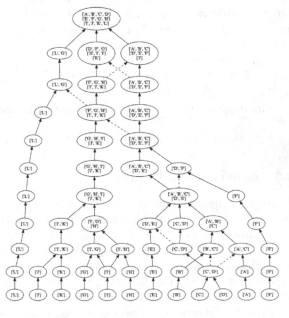

Fig. 4. *OHC* quasi-dendrogram obtained from the hand-built example in Fig. 3a for $\lambda = 0.2$.

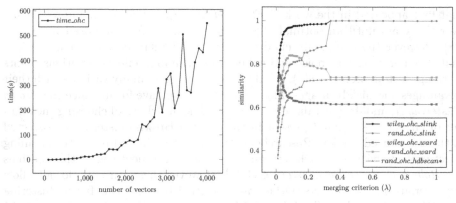

(a) Execution time according to the number of vectors.

(b) Similarity between hierarchical structures according to the merging criterion.

Fig. 5. Study of the merging criterion.

Execution Time: We observe that when the merging criterion increases the execution time decreases. It is due to the fact that when the merging criterion increases we are more likely to completely merge clusters so we reach faster the top of the hierarchy. It means less levels and less overlapping clusters so less computation. However in this case we have the same drawback of chaining effect as the single-linkage clustering that we wanted to avoid. Even if it was not the objective of this work we set $\lambda = 0.1$, as it is an interesting value according to the study of the merging criterion (Fig. 5a), to observe the evolution of the execution time (Fig. 5a). The trend gives a function in $O(n^{2.45})$ so to speed up the process and scale up our algorithm is it possible to precompute a set of possibly overlapping clusters over a given δ-neighbourhood graph with a classical method, for instance CLIQUE, and build the OHC hierarchy on top of that.

4   Related Work

Our goal is to group together data points represented as vectors in $\mathbb{R}^n$. For our motivating application domain of understanding the structure of scientific fields, it is important to construct structures (i) that are hierarchical, (ii) that allow overlaps between the identified groups of vectors and (iii) which groups (clusters) are related to dense areas of the data. There are a number of other application domains where obtaining a structure with these properties is important. In the following, we relate our work to relevant literature.

Hierarchical Clustering: There exist two kinds of hierarchical clustering. Divisive methods follow a top-down strategy while agglomerative techniques compute the hierarchy in a bottom-up fashion. It produces the well known dendrogram structure [1]. One of the oldest methods is the single-linkage clustering that first appeared in the work of *Florek et al.* [18]. It had many improvements over the years until an optimal algorithm named *SLINK* proposed by *Sibson* [28]. However the commonly cited drawback of the single-linkage clustering is that it is not robust to noise and suffers from chaining effects (spurious points merging clusters prematurely). It led to the invention of many variants with their advantages and disadvantages. In the NLP world we have for instance the *Brown clustering* [7] and its generalized version [13]. The drawback of choosing the number of clusters beforehand present in the original Brown clustering is corrected in the generalized version. Researchers also tried to address directly the chaining effect problem with approaches through defining new objective functions such as the *Robust Hierarchical Clustering* [4,11]. However these variants do not allow any overlaps in the clusters. Other variants tried to allow this fuzzy clustering in the hierarchy such as *SOHC* [10], a hierarchical clustering based on a spatial overlapping metric but with a fixed number of clusters, or *HCOSM* [26], that use an overlap similarity measure to merge clusters and then compute a hierarchy from an already determined set of clusters. Generalization of dendrogram to more complex structures like *Pyramidal Clustering* [15] and *Weak Hierarchies* [5] were also proposed. We can find examples to prove that our method produces even more general hierarchical structures that include the weak hierarchies.

Density-Based Clustering: Another important class of work is the density-based clustering. Here, clusters are defined as regions in the data that have a higher density. The data points in the sparse areas that are required to separate clusters are considered as noise or border points. One of the most widely-used algorithms of this category is *DBSCAN* defined by *Ester et al.* [17]. This method connects data points that satisfy a specific density-based criterion: the minimum number of other data points within a given radius must be above a predefined threshold. The main advantage of this method is that it allows detecting clusters of arbitrary shapes. More recently improved versions of *DBSCAN* were proposed such as *HDBSCAN** [8]. This new variant not only improved notions from *DBSCAN* and *OPTICS* [3] but also proposed a procedure to extract a simplified cluster tree from the reachability relation which allows determining a hierarchy of the clusters but again with no overlapping.

Overlapping Clustering: Fuzzy clustering methods [6] allow that certain data points belong to multiple clusters with a different level of confidence. In this way, the boundary of clusters is fuzzy and we can talk about overlaps of these clusters. In our definition it is a different notion, a data point either does or does not belong to a specific cluster and might also belong to multiple clusters. While *HDBSCAN* is closely related to connected components of certain level sets, the clusters do not overlap (since overlap would imply the connectivity).

Community Detection in Networks: A number of algorithmic methods have been proposed to identify communities. The first kind of methods produces a partition where a vertex can belong to one and only one community. Following the *modularity* function of *Newman and Girvan* [24], numerous quality functions have been proposed to evaluate the goodness of a partition with a fundamental drawback, the now proved existence of a resolution limit. The second kind of methods, such as *CLIQUE* [2], *k-clique* [25], *DBLC* [31] or *NMF* [30], aims at finding sets of vertices that respect an edge density criterion which allows overlaps but can lead to incomplete cover of the network. Similarly to *HCOSM*, the method *EAGLE* [27] builds a dendrogram over the set of predetermined clusters, here the maximal cliques of the network so overlaps appear only at the leaf level. Coscia et al. [12] have proposed an algorithm to reconstruct a hierarchical and overlapping community structure of a network, by hierarchically merging local ego neighbourhoods.

5   Conclusion and Future Work

We propose an overlapping hierarchical clustering framework. We construct a quasi-dendrogram hierarchical structure to represent the clusters that is however not necessarily a tree (of specific shape) but a directed acyclic graph. In this way, at each level, we represent a set of possibly overlapping clusters. We experimentally evaluated our method using several datasets and also our new similarity measure that hence proved its usefulness. If the clusters present in the data show no overlaps, the obtained clusters are identical to the clusters we can compute using agglomerative clustering methods. In case of overlapping and nested clusters, however, our method results in a richer representation that can contain relevant information about the structure of the clusters of the underlying dataset. As a future work we plan to identify interesting clusters on the basis of the concept of stability. Such methods give promising results in the context of hierarchical density-based clustering [21], but the presences of overlaps in the clusters requires specific considerations.

References

1. Achtert, E.: Hierarchical subspace clustering. Ph.D. thesis, LMU (2007)
2. Agrawal, R., Gehrke, J., Gunopulos, D., Raghavan, P.: Automatic subspace clustering of high dimensional data. Data Min. Knowl. Disc. **11**(1), 5–33 (2005)
3. Ankerst, M., Breunig, M.M., Kriegel, H.P., Sander, J.: OPTICS: ordering points to identify the clustering structure. In: ACM SIGMOD Record, vol. 28, pp. 49–60. ACM (1999)
4. Balcan, M.F., Liang, Y., Gupta, P.: Robust hierarchical clustering. J. Mach. Learn. Res. **15**(1), 3831–3871 (2014)
5. Bandelt, H.J., Dress, A.W.: Weak hierarchies associated with similarity measures- an additive clustering technique. Bull. Math. Biol. **51**(1), 133–166 (1989). https://doi.org/10.1007/BF02458841
6. Bezdek, James C.: Pattern Recognition with Fuzzy Objective Function Algorithms. Springer, Boston (1981)

7. Brown, P.F., Desouza, P.V., Mercer, R.L., Pietra, V.J.D., Lai, J.C.: Class-based n-gram models of natural language. Comput. Linguist. **18**(4), 467–479 (1992)
8. Campello, R.J., Moulavi, D., Zimek, A., Sander, J.: Hierarchical density estimates for data clustering, visualization, and outlier detection. ACM Trans. Knowl. Discov. Data (TKDD) **10**(1), 5 (2015)
9. Chavalarias, D., Cointet, J.P.: Phylomemetic patterns in science evolution - the rise and fall of scientific fields. PloS One **8**(2), e54847 (2013)
10. Chen, H., Guo, G., Huang, Y., Huang, T.: A spatial overlapping based similarity measure applied to hierarchical clustering. In: Fuzzy Systems and Knowledge Discovery (FSKD 2008), vol. 2, pp. 371–375. IEEE (2008)
11. Cohen-Addad, V., Kanade, V., Mallmann-Trenn, F., Mathieu, C.: Hierarchical clustering: objective functions and algorithms. In: Proceedings of the 29th Annual ACM-SIAM Symposium on Discrete Algorithms, pp. 378–397. SIAM (2018)
12. Coscia, M., Rossetti, G., Giannotti, F., Pedreschi, D.: Uncovering hierarchical and overlapping communities with a local-first approach. ACM Trans. Knowl. Discov. Data **9**(1), 6:1–6:27 (2014)
13. Derczynski, L., Chester, S.: Generalised brown clustering and roll-up feature generation. In: AAAI, pp. 1533–1539 (2016)
14. Dias, L., Gerlach, M., Scharloth, J., Altmann, E.G.: Using text analysis to quantify the similarity and evolution of scientific disciplines. R. Soc. Open Sci. **5**(1), 171545 (2018)
15. Diday, E.: Une représentation visuelle des classes empiétantes: les pyramides (1984)
16. Diestel, R.: Graph Theory. Graduate Texts in Mathematics, vol. 101 (2005)
17. Ester, M., Kriegel, H.P., Sander, J., Xu, X., et al.: A density-based algorithm for discovering clusters in large spatial databases with noise. In: KDD, vol. 96, pp. 226–231 (1996)
18. Florek, K., Łukaszewicz, J., Perkal, J., Steinhaus, H., Zubrzycki, S.: Sur la liaison et la division des points d'un ensemble fini. In: Colloquium Mathematicae, vol. 2, p. 282 (1951)
19. Fowlkes, E.B., Mallows, C.L.: A method for comparing two hierarchical clusterings. J. Am. Stat. Assoc. **78**(383), 553–569 (1983)
20. Levy, O., Goldberg, Y., Dagan, I.: Improving distributional similarity with lessons learned from word embeddings. Trans. Assoc. Comput. Linguist. **3**, 211–225 (2015)
21. McInnes, L., Healy, J.: Accelerated hierarchical density based clustering. In: 2017 IEEE International Conference on Data Mining Workshops (ICDMW), pp. 33–42. IEEE (2017)
22. Mikolov, T., Chen, K., Corrado, G., Dean, J.: Efficient estimation of word representations in vector space. arXiv preprint arXiv:1301.3781 (2013)
23. Mikolov, T., Sutskever, I., Chen, K., Corrado, G.S., Dean, J.: Distributed representations of words and phrases and their compositionality. In: Advances in Neural Information Processing Systems, pp. 3111–3119 (2013)
24. Newman, M.E., Girvan, M.: Finding and evaluating community structure in networks. Phys. Rev. E **69**(2), 026113 (2004)
25. Palla, G., Derényi, I., Farkas, I., Vicsek, T.: Uncovering the overlapping community structure of complex networks in nature and society. Nature **435**(7043), 814 (2005)
26. Qu, J., Jiang, Q., Weng, F., Hong, Z.: A hierarchical clustering based on overlap similarity measure. In: Eighth ACIS International Conference on Software Engineering, Artificial Intelligence, Networking, and Parallel/Distributed Computing (SNPD 2007), vol. 3, pp. 905–910. IEEE (2007)
27. Shen, H., Cheng, X., Cai, K., Hu, M.B.: Detect overlapping and hierarchical community structure in networks. Phys. A: Stat. Mech. Appl. **388**(8), 1706–1712 (2009)

28. Sibson, R.: SLINK: an optimally efficient algorithm for the single-link cluster method. Comput. J. **16**(1), 30–34 (1973)
29. Ward Jr., J.H.: Hierarchical grouping to optimize an objective function. J. Am. Stat. Assoc. **58**(301), 236–244 (1963)
30. Yang, J., Leskovec, J.: Overlapping community detection at scale: a nonnegative matrix factorization approach. In: Proceedings of the Sixth ACM International Conference on Web Search and Data Mining, pp. 587–596. ACM (2013)
31. Zhou, X., Liu, Y., Wang, J., Li, C.: A density based link clustering algorithm for overlapping community detection in networks. Phys. A: Stat. Mech. Appl. **486**, 65–78 (2017)

Digital Footprints of International Migration on Twitter

Jisu Kim[1]($\boxtimes$)(iD), Alina Sîrbu[2]($\boxtimes$)(iD), Fosca Giannotti[3]($\boxtimes$)(iD), and Lorenzo Gabrielli[3]($\boxtimes$)

[1] Scuola Normale Superiore, Pisa, Italy
jisu.kim@sns.it
[2] University of Pisa, Pisa, Italy
alina.sirbu@unipi.it
[3] Istituto di Scienza e Tecnologie dell'Informazione,
National Research Council of Italy, Pisa, Italy
{fosca.giannotti,lorenzo.gabrielli}@isti.cnr.it

Abstract. Studying migration using traditional data has some limitations. To date, there have been several studies proposing innovative methodologies to measure migration stocks and flows from social big data. Nevertheless, a uniform definition of a migrant is difficult to find as it varies from one work to another depending on the purpose of the study and nature of the dataset used. In this work, a generic methodology is developed to identify migrants within the Twitter population. This describes a migrant as a person who has the current residence different from the nationality. The residence is defined as the location where a user spends most of his/her time in a certain year. The nationality is inferred from linguistic and social connections to a migrant's country of origin. This methodology is validated first with an internal gold standard dataset and second with two official statistics, and shows strong performance scores and correlation coefficients. Our method has the advantage that it can identify both immigrants and emigrants, regardless of the origin/destination countries. The new methodology can be used to study various aspects of migration, including opinions, integration, attachment, stocks and flows, motivations for migration, etc. Here, we exemplify how trending topics across and throughout different migrant communities can be observed.

Keywords: International migration · Emigration · Big data · Twitter

This work was supported by the European Commission through the Horizon2020 European project "SoBigData Research Infrastructure—Big Data and Social Mining Ecosystem" (grant agreement no 654024) and partially by the Horizon2020 European project "HumMingBird – Enhanced migration measures from a multidimensional perspective" (grant agreement no 870661).

M. R. Berthold et al. (Eds.): IDA 2020, LNCS 12080, pp. 274–286, 2020.
https://doi.org/10.1007/978-3-030-44584-3_22

1   Introduction

Understanding where migrants are is an important topic because it touches upon multidimensional aspects of the sending and receiving countries' society. It is not only the demographic fabric of countries but also labour market conditions, as well as economic conditions that may alter due to demographic adjustment. Understanding their allocation is essential for both policy makers and researchers to bring the best of its effects.

Official data such as census, survey and administrative data have been traditionally the main data source to study migration. However, these data have some limitations [12]. They are inconsistent across different nations because countries employ different definitions of a migrant. Moreover, collecting traditional data is costly and time consuming, thus tracking instantaneous stocks of migrants becomes difficult. This becomes even harder when tracking emigrants because of the lack of motivation from citizens to declare their departure.

In recent years, however, we are provided with other alternative data sources for migration. The availability of social big data allows us to study social behaviours both at large scale and at a granular level, and to peek into real-world phenomena. Although known to suffer from other types of issues, such as selection bias, these data could bring complementary value to standard statistics.

Here, we propose a method to identify migrants based on Twitter data, to be used in further analyses. According to the official definition, a migrant[1] is "a person who moves to a country other than that of his or her usual residence for a period of at least a year". In the context of Twitter, we define a migrant as "*a person who has the current residence different from the nationality*".

Following this definition, we performed a two step analysis. First, we estimated the current residence for users by examining location information from tweets. The residence is defined as the country where the user spends most of the time in a year. Second, we estimated nationality, by considering the social network of users. In the international literature, nationality is defined as a relationship between a state and an individual, with rights and duties on both sides [1,6]. Related concepts are ethnicity - in terms of cultural features - and citizenship - in terms of political life. In this paper, we employ the term nationality to define the ensemble of features that make a person feel like they belong to a certain country [2,5]. This could be the country where a person was born, raised and/or lived most of their lives. By comparing labels of residence and nationality of a user, we were able to understand whether the person has moved from their home country to a host country, and thus if they are a migrant. We validated our estimation internally, from the data itself, and externally, with two official datasets (Italian register and Eurostat data).

One of the advantages of our methodology is that it is generic enough to allow for identification of both immigrants and emigrants. We also overcome one of the limitations of traditional data by setting up a uniform definition of

---

[1] Recommendations on Statistics of International Migration, Revision 1 (p. 113). United Nations, 1998.

a migrant across different countries. Furthermore, our definition of a migrant is very close to the official definition. We establish the fact that a person has spent a significant period at the current location. Also, we eliminate visitors or short-term stays that do not follow the definition of a migrant. This is also validated by the comparison with official datasets. Another advantage of our method is the fact that it uses only very basic features from the Twitter data: location, language and network information. This is useful since the settings of the freely available Twitter API change constantly. Some of the user attributes that the existing literature use to estimate nationality are no longer available. In addition, we make use of unknown locations of tweets by examining whether they intersect with identified locations. By doing so, we do not neglect any information provided by the tweets from unknown locations which later provide useful information on trending topics of Italian emigrants overseas.

One of the issues with our method is that the migrants that we observed are selected from the Twitter population, and not from the general world population, and it is known that some demographic groups are missing. Nevertheless, we believe that studying the Twitter migrant population can provide important insight into migration phenomena, even if some findings may not apply to the other demographic groups that are not represented in the data.

It is important to note that tracking individual migrants is not the objective of our study, but it is only an intermediate stage to enable further analyses. We simply perform user classification to identify migrants among users in our data, and then aggregate the findings. Further studies we envision are aimed at devising new population-level indices useful to evaluate and improve the quality of life of migrants, through targeted evidence-based policy making. No individual personal information nor migration status is released at any stage during the current analysis, nor in any population-level analysis, which is performed following the highest ethical and privacy standards.

The rest of the paper is organised as follows. In the next section we describe related work that studies migration using big data. In Sect. 3, we provide details of the experimental setting for data collection as well as data pre-processing. We then explain our identification strategy for both residence and nationality in Sect. 4. In Sect. 5, we evaluate our estimation using both internal and external data. Section 6 covers a possible application of our method on studying trending topics among Italian emigrants, while Sect. 7 concludes the paper.

2 Related Work

In the past few years, there have been several works on migration studies using social big data. Most of these employed Twitter data but Facebook, Skype, Email as well as Call Detail Record (CDR) data have also been used to study both international and internal migration [3,9,10,14,16]. Here, we focus on studies that have employed freely available data. The definition of a migrant varied from one work to another depending on the purpose of the study and the nature of the dataset. Thus, the definitions provided fit under different types of migration such as refugees, internal migrants, seasonal migrants or even visitors.

One example of using Twitter to observe migration flows is [15]. They defined residence as the country where the tweets were most frequently sent out for periods of four months. If one's residence changed in the following four months period, it was considered that the person has moved. In a more recent work, [11] measure migration flows from Venezuela to neighbouring countries between 2015 and 2019. They look at the bounding boxes and country labels provided by the tweets and identified the most common country of tweets posted monthly. Their definition of a migrant was "any individual leaving Venezuela during the time window of observation" which was observed when an identified Venezuelan resident appeared for the first time in a different country. Our definition of residence is somewhat similar to these works. However, unlike them, we are measuring stocks of migrants, and not flows. Thus, we take into account the aspect of duration of stay. This naturally eliminates short-term trips and visits.

Apart from geo-tagged tweets, there is other information provided by the Twitter API that can help us infer whether a person is a migrant or not. Although [8] did not directly study migrants, but looked at foreigners present in Qatar, it provides important insights to which of the features provided by Twitter is useful in identifying nationality of users. They gathered features from both profile and tweets of users. For features providing information on profile pictures and name, they performed facial recognition and name ethnicity detection. Their final results showed that ethnicity of name, race, language of tweet, language of mention, location of followers and friends are the first six features that are useful. In this paper, we purely employ data provided by Twitter for the analysis and therefore, we do not have name, ethnicity and race features. Nevertheless, our work also shows that locations of users and friends are the useful features. The difference here is that we propose to use the social network of users as one of the main features in identifying nationality, which is more flexible than having to perform ethnicity detection on names and profile pictures.

3   Experimental Setting for Data Collection

We began with a Twitter dataset collected by the SoBigData.eu Laboratory [4]. We started from a three months period of geo-tagged tweets from August to October 2015. Due to our focus on Italy, we selected from these data the users that tweeted from Italy, obtaining thus 34,160 users. We then crawled the network of geo-enabled friends of these 34,160 users, using the Twitter API. Friends are people that the individual users are following. We focused on friends because we believe that for a user, the information on whom they follow is more informative when it comes to nationality, than who they are followed by. We concentrated on geo-enabled friends because geo-location is necessary for our analysis. By collecting friends, the list of users crossed our initial geographic boundary, i.e., Italy. At this stage, the number of unique users grew to over 250,000. For all users we also scraped the profile information and the 200 most recent tweets using the Twitter API. During this process, we were able to collect all 200 recent tweets for 97% of users and at least 55 tweets for 99% of users. Our final user network consists of 258,455 nodes and 1,205,133 edges which includes both our initial 34,160 users and their geo-tagged friends.

For the process of identifying migration status, we focus on the core users, i.e., 34,160 users. We assign a residence and a nationality to each user, based on the geo-locations included in the data, the language of tweets and profile information. The final dataset includes 237 unique countries from where individuals have sent out their tweets, including 'undefined' location. Even if a user enables geo-tags on their tweets, not all tweets are geo-tagged. As a result, 21% of our tweets are 'undefined'. As for the languages, there are 66 unique languages and 12% of our tweets are in English.

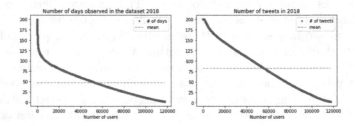

Fig. 1. Distribution of the number of days (left) and the number of tweets (right) observed in the data per user: on average, our users have tweeted 47 days and 82 tweets in 2018.

As for the profile features, we observe that 40% of the users have filled out location description. In addition, most of users have set their profile language to English. The number of unique profile languages detected in our data is 58 which is smaller than the languages used, indicating that some users are using languages different from their profile language when tweeting.

In order to assign a place of residence to users, we needed to restrict the observation time period. We have chosen to look at one year length of tweets from 2018, in order to assign the residence label for the 2018 solar year. We selected users that have tweeted in 2018, identifying 128,305 users. To remove bots, we looked at whether a user is tweeting too many times a day. We considered that tweeting more than 50 tweets on average in a single day was excessive and we have eliminated in this way 39 users. In addition, we removed users that were not very active in 2018. If the number of tweets was less than 20, we checked whether the tweeted days were spread out during the year. If the days were not well spread out, we filtered out the user. On the other hand, if it was well spread out, it meant that the user was regularly tweeting, so the user was kept. During this process, we removed 10,764 users. After removing bots and inactive users, we have 117,502 users. For these, we show the distribution of the number of tweets and number of days in which they tweeted in Fig. 1. On average we see 47 days and 82 tweets.

In addition to the Twitter data, we also collected a list of official and spoken languages for countries identified in our data[2].

4   Identifying Migrants

A migrant is a person that has the residence different from the nationality. We thus consider our core 34,160 Twitter users and assign a residence and nationality based on the information included in our dataset. The difference between the two labels will allow us to detect individuals who have migrated and are currently living in a place different from their home country. The methodology we propose is based on a series of hypotheses: a person that has moved away from their home country stays in contact with their friends back in the home country and may keep using their mother tongue.

4.1   Assigning Residence

In order for a place to be called residence, a person has to spend a considerable amount of time at the location. Our definition of residence is based on the amount of time in which a Twitter user is observed in a country for a given solar year. More precisely, a residence for each user is the country with the longest length of stay which is calculated by taking into account both the number of days in which a user tweets from a country but also the period between consecutive tweets in the same country. In this work we compute residences based on 2018 data.

To compute the residence, we first compute the number of days in which we see tweets for each country for each user. If the top location is not 'undefined', then that is the location chosen as residence. Otherwise, we check whether any tweet sent from 'undefined' country was sent on a same day as tweets sent from the second top country. In case at least one date matched between the two locations, we substitute second country as the user's place of residence. On average, 5 dates matched. This is done under the assumption that a user cannot tweet from two different countries in a day. Although this is not always the case if a user travels, in most of the days of the year this should be true. This approach allowed us to assign a residence in 2018 to 57,180 users.

For the remaining 60,322 users, a slightly different approach was implemented. We computed the length of stay in days by adding together the duration between consecutive tweets in the same country. We selected the country with the largest length of stay. In case the top country was 'undefined', we checked whether 'undefined' locations were in between segments of the second top country, in which case the second country was chosen. In this way, an additional 11,046 users were assigned a place of residence. The remaining 49,276 users were neglected because we considered that we did not have enough information to assign a residence.

---

[2] Retrieved from http://www.geonames.org and https://www.worlddata.info.

4.2   Assigning Nationality

In order to estimate nationalities for Twitter users, we took into account two types of information included in our Twitter data. The first type relates to the users themselves, and includes the countries from which tweets are sent and the languages in which users tweet. For each user u we define two dictionaries loc^u and $lang^u$ where we include, for each country and language the proportion of user tweets in that country/language.

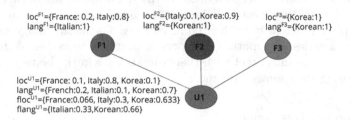

Fig. 2. Example of calculation of the $floc$ and $flang$ values for a user. The calculation of $floc^{U1}$ and $flang^{U1}$ is based of the $floc$ and $flang$ values for the three friends, showing the distribution of tweets in various countries/languages for each.

The second type of information used is related to the user's friends. Again, we look at the languages spoken by friends, and locations from which friends tweet. Specifically, starting from the loc and $lang$ dictionaries of all friends of a user, we define two further dictionaries $floc$ and $flang$. The first stores all countries from where friends tweet, together with the average fraction of tweets in that country, computed over all friends:

$$floc^u[C] = \frac{1}{|F(u)|} \sum_{f \in F(u)} loc^f[C] \tag{1}$$

where $F(u)$ is the set of friends of user u. Similarly, the $flang$ dictionary stores all languages spoken by friends, with the average fraction of tweets in each language l:

$$flang^u[l] = \frac{1}{|F(u)|} \sum_{f \in F(u)} lang^f[l] \tag{2}$$

Figure 2 shows an example of a (fictitious) user with their friends, and the four resulting dictionaries.

The four dictionaries defined above are then used to assign a nationality score to each country C for each user u:

$$N_C^u = w_{loc} loc^u[C] + w_{lang} \sum_{l \in languages(C)} lang^u[l] + \tag{3}$$

$$w_{floc} floc^u[C] + w_{flang} \sum_{l \in languages(C)} flang^u[l] \tag{4}$$

where $languages(C)$ are the set of languages spoken in country C, while w_{loc}, w_{lang}, w_{floc} and w_{flang} are parameters of our model which need to be estimated from the data (one global value estimated for all users). Each of the w value gives a weight to the corresponding user attribute in the calculation of the nationality. To select the nationality for each user we simply select the country C with maximum N_C: $N^u = \text{argmax}_C N_C^u$.

5 Evaluation

To evaluate our strategy for identifying migrants we first propose an internal validation procedure. This defines gold standard datasets for residence and nationality and computes the classification performance of our two strategies to identify the two user attributes. The gold standard datasets are produced using profile information as they are provided by the users themselves. We then perform an external validation where we compare the migrant percentages obtained in our data with those from official statistics.

5.1 Internal Validation: Gold Standards Derived from Our Data

Residence. To devise a gold standard dataset for residence we consider profile locations set by users. We assume that if users declare a location in their profile, then that is most probably their residence. Very few users actually declare a location, and not all of them provide a valid one, thus we only selected profile locations that were identifiable to country level. Among the user accounts for which we could estimate the residence, 3,065 accounts had a valid country in their profile location. Using these accounts as our validation data, we computed the F1 score to measure the performance of our residence calculation. Table 1 shows overall results, and also scores for the most common countries individually. The weighted average of the F1 score is 86%, with individual countries reaching up to 94%, demonstrating the validity of our residence estimation procedure.

Nationality. In order to build a gold standard for nationality, we take into account the profile language declared by the users. The assumption is that profile languages can provide a hint of one's nationality [13]. However, many users might not set their profile language, but use the default English setting. For this reason, we do not include into the gold standard users that have English as their profile language.

Table 1. Average precision, recall and F1 scores, together with scores for the top 7 residences in terms of support size.

	Weighted Avg	Macro avg	Micro avg	IT	KW	US	ID	SG	AU
F1-score	0.858	0.716	0.856	0.928	0.839	0.703	0.945	0.83	0.891
Precision	0.879	0.745	0.856	0.935	0.989	0.572	0.949	0.946	0.883
Recall	0.856	0.727	0.856	0.921	0.728	0.91	0.941	0.739	0.899
Support	3065	3065	3065	343	125	122	119	119	109

Table 2. Average precision, recall and F1 scores for top 8 nationalities in terms of support numbers

	Weighted avg	Macro avg	Micro avg	IT	ES	TR	RU	FR	BR	DE	AR
F1-score	0.99	0.98	0.72	0.99	0.96	0.98	0.95	0.94	0.95	0.92	0.97
Precision	0.99	0.98	0.73	1	0.94	0.98	0.98	0.9	0.96	0.91	0.98
Recall	0.98	0.98	0.75	0.99	0.97	0.99	0.93	0.98	0.94	0.93	0.95
Support	12223	12223	12223	10781	302	173	146	118	113	86	59

The profile language, however, does not immediately translate into nationality. While for some languages the correspondence to a country is immediate, for many others it is not. For instance, Spanish is spoken in Spain and most American countries, so one needs to select the correct one. For this, we look at tweet locations. We consider all countries that match with the profile language and, among these, we select the one with the largest number of tweets, but only if the number of tweets from that country is at least 10% of the total number of tweets of that user. This allows to select the most probable country, also for users who reside outside their native country. If no location satisfies this criterion the user is not included in the gold standard. We were able to identify nationalities of 12,223 users. Due to the fact that during data collection we focused on geo-tags in Italy, the dataset contains a significant number of Italians.

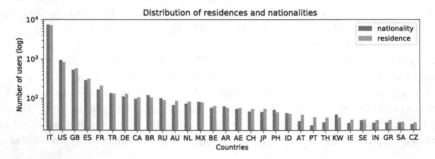

Fig. 3. Distribution of residences and nationalities of top 30 countries, for all users that possess both residence and nationality labels.

We employed this gold standard dataset in two ways. First, we needed to select suitable values for the w weights from Eqs. 3–4. These show the importance of the four components used for nationality computation: own language and location, friends' language and location. We performed a simple grid search and obtained the best accuracy on the gold standard using values 0 for languages and 2 and 1.5 for own and friends' location, respectively. Thus we can conclude that it is the locations that are most important in defining nationality for twitter users, with a slightly stronger weight on the individual's location rather than the friends. The final F1-score, both overall and for top individual nationalities, are included in Table 2, showing a very good performance in all cases.

To assign final residences and nationalities to our core users, we combined the predictions with the gold standards (we predicted only if the gold standard was not present). Figure 3 shows the final distribution of residences and nationalities of top 30 countries for all users that have both the residence and nationality labels. The difference in the residence and nationality can be interpreted as either immigrants or emigrants.

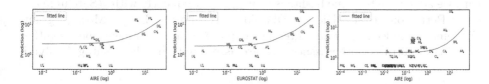

Fig. 4. Comparison between the true and predicted data; the first two plots show predicted versus AIRE/EUROSTAT data on European countries. The last plot shows predicted versus AIRE data on non-European countries.

5.2   External Validations: Validation with Ground Truth Data

In order to validate our results with ground truth data, we study users labelled with Italian nationality and non-Italian residence, i.e. Italian emigrants. We computed the normalised percentage of Italian emigrants resulting from our data for all countries, and compared with two official datasets: AIRE (Anagrafe Italiani residenti all'estero), containing Italian register data, and Eurostat, the European Union statistical office. For comparison we use Spearman correlation coefficients, which allow for quantifying the monotonic relationship between the ground truth data and our estimation by taking ranks of variables into consideration.

Figure 4 displays the various values obtained, compared with official data. A first interesting remark is that even between the official datasets themselves, the numbers do not match completely. The correlation between the two datasets is 0.91. Secondly we observed good agreement between our predictions and the official data for European countries. The correlation with AIRE is 0.753, while with Eurostat it is 0.711 when considering Europe. For non-European countries, however the correlation with AIRE data drops to 0.626. We believe the lower performance is due to several factors related to sampling bias and data quality in the various datasets. This includes bias on Twitter and in our methods, but also errors in the official data, which could be larger in non-EU countries due to less efficient connections in sharing information.

All in all, we believe our method shows good performance and can be successfully used to build population level indices for studying migration. We do not aim to perform nowcasting of immigrant stocks, but rather to identify a population that can be representative enough for further analyses.

6   Case Study: Topics on Twitter

In this section we show that our methodology can be employed to study how trending topics in Italy are also being discussed among Italian emigrants. As an example, we selected one hashtag that has been very popular in the last years: #Salvini. This refers to the Italian politician Matteo Salvini who served as Deputy Prime Minister and Minister of internal affairs in Italy until recently. To this, we added the top nine hashtags that appear frequently with #Salvini in our data: Berlusconi, Conti, Diciott, DiMaio, Facciamorete, Legga, M5S, Migrant, Ottoemezzo. Indeed, they all represent people that are often mentioned together or political parties or other issues that are associated with the hashtag #Salvini.

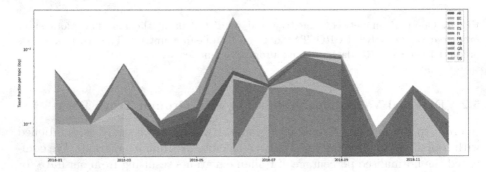

Fig. 5. Stream graph: appearance of hashtags related to #Salvini from Italians across 10 selected residence countries in 2018. The discussion continuously appeared in Italy throughout the year and it became more lively employed by Italians overseas as Salvini gained more political attention.

Figure 5 shows an evolution of the usage of the 10 above mentioned hashtags across different Italian communities both within and abroad Italy. The values shown are the number of tweets from Italian nationals residing in each country that include one of the 10 hashtags, divided by the total number of tweets from Italian nationals from that country. Values are computed monthly. Thus, we show the monthly popularity of the topics in each country. In this way, even the tweets from less represented countries are well shown. As the figure shows, the hashtag was continuously used by Italians in Italy. We observed that the hashtag gradually spread over other residence countries as Salvini received more and more attention. We also observe that most of the attention comes from Italians residing in Europe, with non-European countries less represented.

7   Conclusion and Future Work

We have developed a new methodology to provide a snapshot of migrants within the Twitter population. We considered the length of stay in a country as the key factor to define a user's residence. As for the nationality, connections which migrants maintain with their country of origin provided us with a good indication. In particular, the location of friends seemed to be a strong feature in determining nationality, together with the location of the users themselves. Tweet language, on the other hand, was not considered relevant by our model. This is probably due to the fact that English is the dominating language on Twitter, since a language that is widely understood has to be spoken to get more attention from other users. We have validated our results both with internal and external data. The results show good classification performance scores and good correlation coefficients with official datasets.

The constructed dataset can be applied in different scenarios. We have shown how it can be used to study trending topics on Twitter, and how attention is divided between emigrants and non-migrants of a certain nationality. In the future, we plan to analyse social ties, integration and assimilation of migrants [7]. At the same time, one can investigate the strength of the ties with the community of origin.

References

1. Castillo petruzzi case (1999)
2. Assal, M.A.: Nationality and citizenship questions in Sudan after the Southern Sudan referendum vote. Sudan Report (2011)
3. Blumenstock, J.E.: Inferring patterns of internal migration from mobile phone call records: evidence from Rwanda. Inf. Technol. Dev. **18**(2), 107–125 (2012)
4. Coletto, M., et al.: Perception of social phenomena through the multidimensional analysis of online social networks. Online Soc. Netw. Media **1**, 14–32 (2017)
5. Donner, R.: The Regulation of Nationality in International Law, 2d edn, p. 289. Leiden, Brill Nijhoff (1994). https://brill.com/view/title/14000, ISBN 978-09-41-32077-1
6. Hailbronner, K.: Nationality in public international law and European law. JSTOR (2006)
7. Herdağdelen, et al.: The social ties of immigrant communities in the united states. In: Proceedings of the 8th ACM Conference on Web Science, pp. 78–84. ACM (2016)
8. Huang, W., et al.: Inferring nationalities of Twitter users and studying international linking. In: Proceedings of the 25th ACM Conference on Hypertext and Social Media, pp. 237–242. ACM (2014)
9. Kikas, R., et al.: Explaining international migration in the Skype network: the role of social network features. In: Proceedings of the 1st ACM Workshop on Social Media World Sensors, pp. 17–22. ACM (2015)
10. Lamanna, F., et al.: Immigrant community integration in world cities. PLoS One **13**(3), e0191612 (2018)
11. Mazzoli, M., et al.: Migrant mobility flows characterized with digital data. arXiv preprint arXiv:1908.02540 (2019)
12. Sîrbu, A., et al.: Human migration: the big data perspective. Int. J. Data Sci. Anal. (2020, under review)
13. Stokes, B.: Language: the cornerstone of national identity. Pew Research Center's Global Attitudes Project (2017)

14. Zagheni, E., et al.: Combining social media data and traditional surveys to nowcast migration stocks. In: Annual Meeting of the Population Association of America (2018)
15. Zagheni, E., et al.: Inferring international and internal migration patterns from Twitter data. In: Proceedings of the 23rd International Conference on World Wide Web, pp. 439–444. ACM (2014)
16. Zagheni, E., Weber, I.: You are where you e-mail: using e-mail data to estimate international migration rates. In: Proceedings of the 4th Annual ACM Web Science Conference, pp. 348–351. ACM (2012)

Percolation-Based Detection of Anomalous Subgraphs in Complex Networks

Corentin Larroche[1,2]($\boxtimes$), Johan Mazel[1], and Stephan Clémençon[2]

[1] French National Cybersecurity Agency (ANSSI), Paris, France
{corentin.larroche,johan.mazel}@ssi.gouv.fr
[2] LTCI, Télécom Paris, Institut Polytechnique de Paris, Palaiseau, France
{corentin.larroche,stephan.clemencon}@telecom-paris.fr

Abstract. The ability to detect an unusual concentration of extreme observations in a connected region of a graph is fundamental in a number of use cases, ranging from traffic accident detection in road networks to intrusion detection in computer networks. This task is usually performed using scan statistics-based methods, which require explicitly finding the most anomalous subgraph and thus are computationally intensive.

We propose a more scalable method in the case where the observations are assigned to the edges of a large-scale network. The rationale behind our work is that if an anomalous cluster exists in the graph, then the subgraph induced by the most individually anomalous edges should contain an unexpectedly large connected component. We therefore reformulate our problem as the detection of anomalous sample paths of a percolation process on the graph, and our contribution can be seen as a generalization of previous work on percolation-based cluster detection. We evaluate our method through extensive simulations.

1 Introduction

Detection of a significant connected subgraph in a larger background network is a ubiquitous task: such significant regions can be indicative of fraudulent behavior in social networks [15] or of the propagation of an intruder in a computer network [22], for instance. Therefore, being able to discern them from ambient noise has valuable applications in a number of settings. This anomaly detection problem is, however, remarkably challenging: the large size and complex structure of real-world graphs make the characterization of normal behavior difficult and the search for non-trivial substructures computationally expensive.

The aim of this paper is to propose a scalable method for anomalous connected subgraph detection in a graph with observations attached to its edges. The null distribution of the observations, or an approximation thereof, is assumed to be known. Building upon this knowledge, the degree of abnormality of each individual edge with respect to the model can be measured, and our goal is to detect a significant concentration of anomalous edges in a connected region of

© The Author(s) 2020
M. R. Berthold et al. (Eds.): IDA 2020, LNCS 12080, pp. 287–299, 2020.
https://doi.org/10.1007/978-3-030-44584-3_23

the graph. Usual methods for this task are built around scan statistics [14]. Such methods boil down to maximizing a scoring function over the set of connected regions of the graph, then rejecting the null hypothesis (*i.e.* absence of anomalous subgraph) if the maximum exceeds a certain threshold. This implies solving a combinatorial optimization problem over the class of all connected subgraphs, which is expensive due to the exponentially growing size of the latter.

In contrast, our approach does not require explicitly searching for the best candidate subgraph. Instead, we build on the following idea: under the null hypothesis, the most individually anomalous edges are randomly spread out over the graph. Therefore, removing all but the k most anomalous edges from the graph is equivalent to drawing k edges uniformly at random and extracting the subgraph induced by these edges. In other words, this procedure amounts to bond percolation on a graph. On the other hand, when an anomalous subgraph is present, the location of the individual anomalies is no longer random, and thus the largest connected component of the subgraph induced by the k most anomalous edges should contain an unexpectedly large connected component. This link between anomalous subgraph detection and percolation theory has already been introduced in the context of regular lattices [6,19,20], but to the best of our knowledge, it has not yet been studied for arbitrary graphs.

We argue that our method is more scalable than traditional ones while retaining an acceptable detection power, especially when seeking to detect small anomalous regions in large graphs. We assess this detection performance through numerical experiments on several realistic synthetic graphs.

The rest of this paper is structured as follows. In Sect. 2, we introduce the statistical framework for our problem and present some related work. Section 3 describes our detection method, while Sect. 4 is devoted to its empirical evaluation on simulated data. Finally, we discuss our results and some interesting leads for future work in Sect. 5, then briefly conclude in Sect. 6.

2   Problem Formulation and Related Work

We begin with a thorough formulation of our problem as a case of statistical hypothesis testing, then review the main existing approaches to it.

2.1   Problem Formulation – Statistical Hypothesis Testing

Consider an undirected and connected graph $\mathcal{G} = (\mathcal{V}, \mathcal{E})$, where $\mathcal{V}$ (resp. $\mathcal{E}$) is the set of vertices (resp. edges) of $\mathcal{G}$. Letting $|\mathcal{A}|$ denote the number of elements of a set $\mathcal{A}$, we write $m = |\mathcal{E}|$, and we use $\mathcal{E}$ and $[m] = \{1, \ldots, m\}$ interchangeably to represent the set of edges. We further write $2^{\mathcal{A}}$ for the set of all subsets of $\mathcal{A}$ and $\mathbb{1}\{\cdot\}$ for the indicator function of an event.

Let $\Lambda \subset 2^{\mathcal{E}}$ denote the class of subsets of $\mathcal{E}$ whose induced subgraph in $\mathcal{G}$ is connected. Given a signal $\mathbf{X} = (X_1, \ldots, X_m) \in \mathbb{R}^m$ observed on the edges of $\mathcal{G}$ and a known probability distribution F_0, the null hypothesis is defined as

$H_0 : X_i \overset{iid}{\sim} F_0$. For each $\mathcal{S} \in \Lambda$, we further define the alternative

$$H_\mathcal{S} : \begin{cases} \mathbf{X}_{|\mathcal{S}} \sim F_\mathcal{S} \\ \forall i \notin \mathcal{S}, X_i \sim F_0 \end{cases},$$

where $\mathbf{X}_{|\mathcal{S}}$ is the restriction of $\mathbf{X}$ to $\mathcal{S}$ and $F_\mathcal{S}$ is a joint probability distribution. $F_\mathcal{S}$ is only assumed to be different from $F_0^{\otimes|\mathcal{S}|}$, and it can differ in various ways. In many applications, the observations in $\mathcal{S}$ are simply larger than expected (consider for instance network intrusion detection, where the presence of an intruder results in additional activity in a connected region of the network). The problem considered in this paper can be formulated as

$$H_0 \quad \text{vs.} \quad H_1 = \bigcup_{\mathcal{S} \in \Lambda} H_\mathcal{S}.$$

That is, we want to know whether there exists a connected subgraph of $\mathcal{G}$ inside of which the observations X_i are drawn from an alternative distribution. Note that we only care about detection, leaving the reconstruction of $\mathcal{S}$ aside.

2.2 Related Work – Scan Statistics and Beyond

A lot of existing work deals with a specific instance of the problem defined above, namely elevated mean detection on a graph. In this setting, the observations are independent standard centered normal random variables under the null, while X_i has mean $\mu_\mathcal{S} \mathbb{1}\{i \in \mathcal{S}\}$ under the alternative $H_\mathcal{S}$ (for some $\mu_\mathcal{S} > 0$). Theoretical conditions for detectability in this case are stated in [1]. A closely related problem arises when the observations are associated with vertices rather than edges, and this setting was studied in [3–5]. However, these papers focus on statistical analysis and do not provide computationally tractable tests.

From a more practical perspective, the most common approach to anomalous subgraph detection is based on scan statistics. Broadly speaking, this method consists in defining a scoring function $f : 2^\mathcal{E} \to \mathbb{R}$, computing the test statistic $t = \max_{\mathcal{S} \in \Lambda} f(\mathcal{S})$, then rejecting H_0 if t exceeds a given threshold. This amounts to finding the most anomalous subset $\mathcal{S}^*$ in Λ, and then rejecting the null hypothesis if $\mathcal{S}^*$ is anomalous enough. Defining f requires some hypotheses on the class of alternative distributions $\{F_\mathcal{S}\}$. For instance, when $F_\mathcal{S}$ has a parametric form, $f(\mathcal{S})$ can be defined as the likelihood ratio between $H_\mathcal{S}$ and H_0. In the more general case considered here, however, finding a suitable scoring function is non-trivial. Moreover, computing t implies maximizing f over the combinatorial class Λ, which quickly becomes computationally intensive as the graph grows. Therefore, most related work focuses on making the computation of scan statistics more efficient. Ways to achieve this include the following:

Restriction of the Class Λ. The easiest way to speed up the computation is to simply reduce the size of the search space by considering only a subset of Λ. Such restriction can be based on domain-specific knowledge [17,18,22,25] or more general heuristics [24].

Convex Relaxation. Another classical approach to combinatorial optimization consists in solving a convex relaxation of the problem, and then projecting the solution back onto the original search space. This method was applied to scan statistics [2, 26, 27], using elements of spectral graph theory [9] to find a relaxed form of the connectivity constraint. Similar ideas were also used in a slightly different context [29–31], where the class Λ consists of subgraphs with low cut size rather than connected ones.

Algorithmic Approaches. Finally, efficient optimization algorithms have been used to find exact or approximate values for the scan statistic, including simulated annealing [11, 12], greedy algorithms [28], primal-dual algorithms [28], branch and bound algorithms [32] and dynamic programming algorithms [33].

Despite the popularity of scan statistics, other ideas have also been considered in the literature. We focus on one of these alternative approaches, namely the Largest Open Cluster (LOC) test, which was first studied in the context of object detection in images [19, 20]. The idea of this method is to represent an image as a two-dimensional lattice, each node carrying a random variable standing for the value of the associated pixel. Then, after deleting from the lattice every vertex whose pixel value is lower than a suitable threshold, the largest remaining connected component is expected to be small if there is no object in the image. On the other hand, if an object is present, an unexpectedly large connected component should remain in the thresholded lattice. The theory behind the LOC test has since been extended to lattices of arbitrary dimension [6], but to the best of our knowledge, the underlying idea of using percolation theory to detect anomalous connected subgraphs has not yet been applied to complex, arbitrary-shaped networks.

3   Local Anomaly Detection and Percolation Theory

We now describe our method, first introducing some necessary notions of percolation theory, then highlighting their relevance to our anomaly detection problem. Finally, we provide a detailed description of our testing procedure.

3.1   Some Notions of Percolation Theory

An interesting aspect of the LOC test is that the behavior of its test statistic under the null hypothesis can be described using percolation theory. Therefore, we first review some useful results from this field, which motivate our approach. For more details, see for example [10] and references therein. Since our primary interest is in signals associated with edges, we focus on bond percolation, where edges of a connected graph with n vertices are occupied uniformly at random with probability p or unoccupied with probability $1 - p$.

Let $C(p)$ denote the size of the largest connected component of this graph at occupation probability p. The main focus of percolation theory is to find the limit of $C(p)$ as n becomes large. Extremal values of p yield obvious results: for

$p = 0$, $C(p) = 1$ for any n and for $p = 1$, $\lim_{n\to\infty} C(p) = \infty$. For intermediate values of p, however, there are two possible regimes. If p is small enough, only small connected components are present and $C(p)/n$ converges in probability to 0. On the other hand, larger values of p lead to the emergence of a giant connected component, which contains a constant fraction of the vertices. The transition between the two regimes happens for a critical value of p called the percolation threshold p_c. Note that p_c depends on the graph structure and can be vanishingly small. Although this phase transition is only well-defined in the limit of an infinite graph, a somewhat similar behavior can be observed in the finite case [8,16]. In particular, define the percolation process $\{C(p)\}_{0\le p\le 1}$ as follows: assign to each edge e an independent random variable U_e, uniformly distributed on $[0, 1]$. Then, keeping the U_e fixed, let p vary on $[0, 1]$, deleting e from the graph whenever $U_e > p$. A tightly related process is obtained by considering the imbedded Markov chain $\{\mathcal{G}_k\}_{k\ge 0}$, where $\mathcal{G}_k$ is the subgraph induced by the edges associated with the k smallest random variables. Letting C_k denote the size of the largest connected component of $\mathcal{G}_k$, $\{C_k\}_{k\ge 0}$ can be seen as a discretized version of $\{C(p)\}_{0\le p\le 1}$. Even for finite graphs, sample paths of these two processes do not deviate significantly from the mean trajectory, making them suitable candidates for anomaly detection.

3.2   Application to Anomalous Subgraph Detection

We now motivate the idea of mapping a signal $\mathbf{X}$ onto a sample path of the percolation process. For $i \in [m]$, define $P_i = 1 - F_0(X_i)$ as the upper tail p-value associated with X_i. Define also, for $k \in \{0, \ldots, m\}$, the subgraph $\mathcal{G}_k$ induced by the edges associated with the k smallest p-values, and let S_k denote the size of its largest connected component. Under the null hypothesis, the random variables $\{P_i\}$ are independent and uniformly distributed on $[0, 1]$. Therefore, S_k has the same distribution as C_k for all $k \in \{0, \ldots, m\}$. Under the alternative $H_\mathcal{S}$, however, the distribution of the variables $\{P_i\}_{i\in\mathcal{S}}$ is altered, which induces a deviation in the process $\{S_k\}_{0\le k\le m}$ with respect to the normal percolation process. Our test aims to detect this deviation.

Figure 1 illustrates the normal and anomalous behaviors of the percolation process for three graph models: a two-dimensional square lattice, an Erdős-Rényi random graph [13] and a Barabási-Albert preferential attachment graph [7]. For each model, a graph with 1024 vertices and approximately 2000 edges is generated, and the mean and standard deviation of the fraction of vertices in the largest connected component for each value of p is estimated using 10000 Monte Carlo simulations. Then, for each graph, we generate a subtree $\mathcal{S}$ containing a fraction δ of the vertices, assign to each edge e an independent Gaussian random variable $X_e \sim \mathcal{N}(\mu \mathbb{1}\{e \in \mathcal{S}\}, 1)$ and compute the associated sample path of the percolation process. This experiment was repeated 1000 times for each graph, and the mean sample path for different values of δ and μ is displayed. The two regimes of the percolation process can be observed, and the shape and location of the phase transition both clearly depend on the graph model. While the

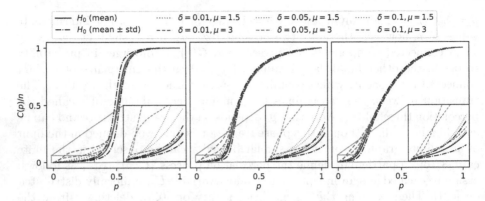

Fig. 1. Evolution of the fraction of vertices in the largest connected component as p varies from 0 to 1, under H_0 and various alternatives, for three kinds of graphs: a two-dimensional square lattice (left), an Erdős-Rényi random graph (center) and a Barabási-Albert preferential attachment graph (right).

separation between the two regimes is quite clear for the lattice and the Erdős-Rényi graph, it is much blurrier for the Barabási-Albert model, which yields more complex structures – most interestingly, heavy-tailed degree distributions. Since such properties are often found in real-world networks, it is important to qualify their impact on the feasibility of percolation-based cluster detection. Figure 1 shows that although the anomalous sample paths become harder to distinguish as the phase transition gets hazier, the normal trajectories are concentrated enough to make even small deviations visible, which motivates our approach.

3.3 Putting It All Together – Description of Our Test

We now proceed with the description of our test. First, define

$$K = \min\left\{k \leq m, \; \mathbb{E}_0[S_k] \geq \sqrt{|\mathcal{V}|}\right\},$$

where $\mathbb{E}_0$ denotes the expected value under H_0. K can be understood as the index corresponding to the onset of the phase transition. Since we aim to detect the appearance of an unexpectedly large connected component in the early steps of the percolation process, the test statistic we use is

$$\chi = \frac{1}{|\mathcal{V}| \cdot K} \sum_{k=1}^{K} S_k.$$

This statistic is equivalent to the area under a piecewise constant interpolation of the sequence of points $\{(k, S_k)\}_{0 \leq k \leq K}$, and is therefore expected to be higher than usual in the presence of an anomalous subgraph.

Estimation of K and calibration of the test are both done through Monte Carlo simulation: using the Newman-Ziff algorithm [23], N random sample paths of the imbedded Markov chain are computed. Let $\{S_k^{(i)}\}_{0 \leq k \leq m}$ denote the trajectory of the largest connected component's size for the ith realization of the process. We get the following estimates:

$$\hat{K} = \min\left\{k \leq m, \frac{1}{N}\sum_{i=1}^{N} S_k^{(i)} \geq \sqrt{|\mathcal{V}|}\right\}, \qquad \hat{\chi} = \frac{1}{|\mathcal{V}| \cdot \hat{K}}\sum_{k=1}^{\hat{K}} S_k.$$

Finally, the empirical p-value can be expressed as

$$\hat{p} = \frac{1}{N}\sum_{i=1}^{N}\mathbb{1}\{\hat{\chi} \leq \hat{\chi}^{(i)}\}, \quad \text{where } \hat{\chi}^{(i)} = \frac{1}{|\mathcal{V}| \cdot \hat{K}}\sum_{k=1}^{\hat{K}} S_k^{(i)} \quad \text{for } i \in \{1, \ldots, N\}.$$

4   Experiments

In order to assess the power of our test, we ran it on several synthetic graphs containing random anomalous trees. This section describes the procedure we used to generate the dataset, then presents our results and their interpretation.

4.1   Generation of the Dataset

The dataset is generated using the stochastic Kronecker graph model [21]. Kronecker graphs exhibit similar structural properties as real-world networks, most importantly power law-distributed degrees and small diameter. Hence, this model allows us to evaluate our test in a somewhat realistic setting.

Two parameter matrices are used: $\Theta_1 = [0.9\ 0.5; 0.5\ 0.3]$ (core-periphery network) and $\Theta_2 = [0.9\ 0.2; 0.2\ 0.9]$ (hierarchical network). For a given matrix and for $i \in \{12, 13, 14, 15\}$, we generate an undirected graph through i iterations of the Kronecker product, and only the largest connected component of this graph is kept in order to obtain a connected network with approximately 2^i vertices. Using this procedure, 10 graphs are generated for each pair of parameters (Θ, i). Thus, we evaluate our test on graphs with sizes ranging from a few thousands to a few tens of thousands of vertices, which covers a wide scope of potential use cases. For each synthetic graph, anomalies are then generated as follows: given $\delta \in (0, 1)$, a random subtree $\mathcal{S}$ containing a fraction δ of the vertices is drawn. Then, a random observation $X_e \sim \mathcal{N}(\mu\mathbb{1}\{e \in \mathcal{S}\}, 1)$ is independently drawn for each edge e of the graph (where μ is a fixed signal strength). For a given graph and a pair of parameters (δ, μ), 1000 anomalous signals $\mathbf{X} = (X_1, \ldots, X_m)$ are generated. 1000 signals are also drawn from the null distribution (that is, $\mathbf{X} \sim \mathcal{N}(0, I_m)$, where I_m is the $m \times m$ identity matrix) for comparison. Finally, for each graph, the null distribution of the test statistic is estimated using 10000 random realizations of the percolation process. Using the obtained histogram, the empirical p-values associated with the normal and anomalous samples are derived, and we construct the Receiver Operating Characteristic (ROC) curve for each pair (δ, μ). This procedure exposes the influence of various parameters on the performance of our test, namely the graph size, the generator matrix, the size δ of the anomalous region and the signal strength μ.

4.2   Detectability Conditions – Empirical Study

Our results are displayed in Table 1 and Figs. 2 and 3. Our main interest is in finding out which parameters have the strongest influence on the power of the test, and we provide some key observations and interpretations below.

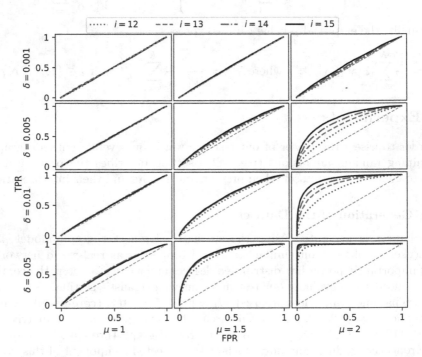

Fig. 2. Aggregated ROC curves of our test for 10 Kronecker graphs with initial matrix $\Theta_1 = [0.9\ 0.5;\ 0.5\ 0.3]$, for several values of the number of Kronecker product iterations i, the proportion δ of vertices in the anomalous tree and the signal strength μ.

Influence of the Graph Size. The first thing we notice in Figs. 2 and 3 is that for a given pair of parameters (δ, μ), the performance of the test consistently improves as the size of the graph increases. One possible explanation for this comes from percolation theory: before the phase transition, the size of the largest connected component is sublinear in the size of the graph. This implies that, for a fixed ratio of vertices in the anomalous component, the difference between the size of the latter and the expected size of the largest component grows with the graph size. Therefore, the anomalous component becomes more visible as the graph grows. Note, however, that some structural properties of our synthetic graphs (*e.g.* density) might not remain identical for different values of i. It is thus difficult to pinpoint the actual influence of the sole number of vertices.

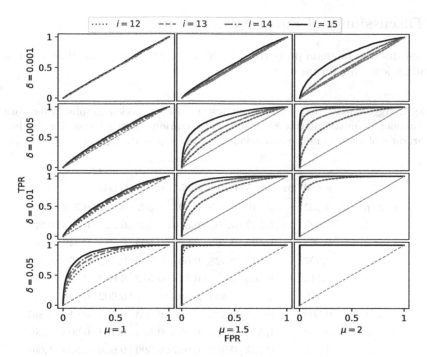

Fig. 3. Aggregated ROC curves of our test for 10 Kronecker graphs with initial matrix $\Theta_2 = [0.9\ 0.2;\ 0.2\ 0.9]$, for several values of the number of Kronecker product iterations i, the proportion δ of vertices in the anomalous tree and the signal strength μ.

Trade-Off Between δ and μ. As could be intuitively expected, our test performs better for higher values of δ and μ. More interestingly, these two parameters are intertwined: what makes an anomalous subgraph detectable is not only the number of vertices it contains (which is controlled by δ), but also the presence of a sufficient fraction of its edges among the most individually anomalous edges of the graph (which is controlled by μ). In terms of experimental results, this translates to poor performance when at least one of these parameters is too low. However, there seems to be a range of values of δ and μ in which a decrease in one can be made up for by an increase in the other. In particular, this implies that even small subgraphs can be detected by our test as long as the signal is strong enough. This is useful in "needle-in-a-haystack" scenarios such as network intrusion detection, where the anomalies one looks for are often localized.

Influence of the Graph Structure. As evidenced by Fig. 1, structural properties of the graph heavily influence the normal behavior of the percolation process, in turn affecting the viability of percolation-based cluster detection. This explains the observable difference in detection power between the two kinds of graphs we consider. Further analysis shows that the generator Θ_1 yields more heavy-tailed degree distributions, which is a plausible cause for the performance gap.

5   Discussion and Future Work

We now discuss the main properties of our test, identifying some limitations and providing leads for future work.

Table 1. Aggregated AUC score of our test for 10 Kronecker graphs, using various combinations of initial matrix Θ, number of iterations of the Kronecker product i, proportion δ of vertices in the anomalous tree and signal strength μ.

		Θ_1				Θ_2			
		$\delta = 0.001$	0.005	0.01	0.05	0.001	0.005	0.01	0.05
$i = 12$	$\mu = 1$	0.502	0.510	0.525	0.591	0.502	0.527	0.582	0.796
	1.5	0.505	0.542	0.603	0.819	0.502	0.626	0.763	0.990
	2	0.503	0.628	0.769	0.981	0.505	0.785	0.949	1.000
$i = 13$	1	0.507	0.513	0.528	0.602	0.505	0.540	0.595	0.838
	1.5	0.513	0.560	0.631	0.847	0.512	0.694	0.848	0.998
	2	0.518	0.699	0.845	0.993	0.531	0.902	0.988	1.000
$i = 14$	1	0.503	0.515	0.525	0.596	0.503	0.550	0.614	0.867
	1.5	0.508	0.570	0.639	0.855	0.524	0.764	0.908	1.000
	2	0.528	0.752	0.887	0.997	0.590	0.969	0.998	1.000
$i = 15$	1	0.500	0.509	0.522	0.586	0.508	0.565	0.634	0.897
	1.5	0.511	0.584	0.645	0.861	0.555	0.840	0.955	1.000
	2	0.551	0.801	0.925	0.999	0.706	0.994	1.000	1.000

Theoretical Guarantees. From a theoretical perspective, our setting is more complex than that of [6]: we consider arbitrary networks instead of regular lattices, and our test statistic depends on the whole sample path of the percolation process rather than the marginal behavior at a given occupation probability. Therefore, the search for theoretical guarantees for our test was left out of the scope of this work, although it would certainly be of great interest.

Computational Cost. The main advantage of our method is its computational efficiency. Indeed, computing the empirical p-value for a given graph and an observed signal only requires $N + 1$ runs of the Newman-Ziff algorithm, which has a very low cost. In contrast, a scan statistic-based test would require $N + 1$ runs of a combinatorial optimization algorithm (one for the observed data and N additional runs to estimate the distribution of the test statistic under the null). Even with a very efficient optimization method, this is significantly more intensive. In terms of complexity, our test requires sorting the observations X_i, running the Newman-Ziff algorithm $N + 1$ times, computing the mean sample path and the index K, and summing the first K values for each of the $N + 1$ trajectories, resulting in $\mathcal{O}(m(\log m + N))$ operations. Note that the algorithm can be further optimized using the fact that the test statistic depends only on the first K steps of the percolation process. Although the exact value of K depends on the graph, we empirically observe that it is generally smaller than the number of vertices $|\mathcal{V}|$. Therefore, early stopping of the Newman-Ziff algorithm and partial sorting can reduce the complexity to $\mathcal{O}(m + |\mathcal{V}|(N + \log|\mathcal{V}|))$.

Detection Power. The expected downside of our method's low computational cost is a loss in detection power. Our simulations show, however, that the proposed test can detect reasonably small anomalous subgraphs in large enough ambient graphs, which is our main goal here. Moreover, it does not rely on prior knowledge of the alternative distribution and can be used with only a rough estimate of F_0, which improves its usability in realistic settings.

Although the influence of some factors on the performance of the test was left out of the scope of this work, a wider analysis would be an interesting topic for future work. These factors include the density of the graph and the shape of the anomalous subgraph. More specifically, we only evaluated our test in the case of random anomalous trees, which provides general results but no insight into the influence of the diameter and the density of the anomalous subgraph.

6   Conclusion

By extending previous work on percolation-based cluster detection to a more general setting, we propose a computationally efficient test to detect an anomalous connected subgraph in an edge-weighted network. The underlying intuition is that it is often possible to find out whether such a subgraph is present without explicitly finding it: instead of enumerating all possible candidates, a much faster method can be obtained by looking for properties of the whole graph which are affected by the apparition of an anomalous cluster. Our work suggests that percolation theory can provide such properties.

Since it scales easily to large graphs and does not rely on extensive knowledge of the null and alternative distributions of the observed signal, we argue that our method is applicable to real-world problems. Moreover, we show through extensive simulations that its detection power remains acceptable, and that it can in particular detect small anomalous regions in large graphs. Therefore, we think the link between cluster detection and percolation theory deserves further exploration, both from a theoretical and applied point of view.

References

1. Addario-Berry, L., Broutin, N., Devroye, L., Lugosi, G., et al.: On combinatorial testing problems. Ann. Stat. **38**(5), 3063–3092 (2010)
2. Aksoylar, C., Orecchia, L., Saligrama, V.: Connected subgraph detection with mirror descent on SDPs. In: ICML (2017)
3. Arias-Castro, E., Candes, E.J., Durand, A., et al.: Detection of an anomalous cluster in a network. Ann. Stat. **39**(1), 278–304 (2011)

4. Arias-Castro, E., Candès, E.J., Helgason, H., Zeitouni, O., et al.: Searching for a trail of evidence in a maze. Ann. Stat. **36**(4), 1726–1757 (2008)
5. Arias-Castro, E., Donoho, D.L., Huo, X., et al.: Near-optimal detection of geometric objects by fast multiscale methods. IEEE Trans. Inf. Theory **51**(7), 2402–2425 (2005)
6. Arias-Castro, E., Grimmett, G.R., et al.: Cluster detection in networks using percolation. Bernoulli **19**(2), 676–719 (2013)
7. Barabási, A.L., Albert, R.: Emergence of scaling in random networks. Science **286**(5439), 509–512 (1999)
8. Callaway, D.S., Newman, M.E., Strogatz, S.H., Watts, D.J.: Network robustness and fragility: percolation on random graphs. Phys. Rev. Lett. **85**(25), 5468 (2000)
9. Chung, F.: Spectral Graph Theory. American Mathematical Society, Providence (1997)
10. Chung, F., Horn, P., Lu, L.: Percolation in general graphs. Internet Math. **6**(3), 331–347 (2009)
11. Duczmal, L., Assuncao, R.: A simulated annealing strategy for the detection of arbitrarily shaped spatial clusters. Comput. Stat. Data Anal. **45**(2), 269–286 (2004)
12. Duczmal, L., Kulldorff, M., Huang, L.: Evaluation of spatial scan statistics for irregularly shaped clusters. J. Comput. Graph. Stat. **15**(2), 428–442 (2006)
13. Erdős, P., Rényi, A.: On the evolution of random graphs. Publ. Math. Inst. Hung. Acad. Sci **5**, 17–60 (1960)
14. Glaz, J., Naus, J., Wallenstein, S.: Scan Statistics. Springer, Berlin (2001)
15. Hooi, B., Song, H.A., Beutel, A., Shah, N., Shin, K., Faloutsos, C.: Fraudar: bounding graph fraud in the face of camouflage. In: KDD (2016)
16. Karrer, B., Newman, M.E., Zdeborová, L.: Percolation on sparse networks. Phys. Rev. Lett. **113**(20), 208702 (2014)
17. Kulldorff, M.: A spatial scan statistic. Commun. Stat. - Theory Methods **26**(6), 1481–1496 (1997)
18. Kulldorff, M., Huang, L., Pickle, L., Duczmal, L.: An elliptic spatial scan statistic. Stat. Med. **25**(22), 3929–3943 (2006)
19. Langovoy, M., Habeck, M., Schölkopf, B.: Spatial statistics, image analysis and percolation theory. arXiv preprint arXiv:1310.8574 (2013)
20. Langovoy, M., Wittich, O.: Robust nonparametric detection of objects in noisy images. J. Nonparametr. Stat. **25**(2), 409–426 (2013)
21. Leskovec, J., Chakrabarti, D., Kleinberg, J., Faloutsos, C., Ghahramani, Z.: Kronecker graphs: an approach to modeling networks. J. Mach. Learn. Res. **11**, 985–1042 (2010)
22. Neil, J., Hash, C., Brugh, A., Fisk, M., Storlie, C.B.: Scan statistics for the online detection of locally anomalous subgraphs. Technometrics **55**(4), 403–414 (2013)
23. Newman, M.E., Ziff, R.M.: Fast Monte Carlo algorithm for site or bond percolation. Phys. Rev. E **64**(1), 016706 (2001)
24. Patil, G., Taillie, C., et al.: Geographic and network surveillance via scan statistics for critical area detection. Stat. Sci. **18**(4), 457–465 (2003)
25. Priebe, C.E., Conroy, J.M., Marchette, D.J., Park, Y.: Scan statistics on enron graphs. Comput. Math. Organ. Theory **11**(3), 229–247 (2005)
26. Qian, J., Saligrama, V.: Efficient minimax signal detection on graphs. In: NeurIPS (2014)
27. Qian, J., Saligrama, V., Chen, Y.: Connected sub-graph detection. In: AISTATS (2014)
28. Rozenshtein, P., Anagnostopoulos, A., Gionis, A., Tatti, N.: Event detection in activity networks. In: KDD (2014)

29. Sharpnack, J., Rinaldo, A., Singh, A.: Detecting anomalous activity on networks with the graph Fourier scan statistic. IEEE Trans. Signal Process. **64**(2), 364–379 (2015)
30. Sharpnack, J., Singh, A., Rinaldo, A.: Changepoint detection over graphs with the spectral scan statistic. In: AISTATS (2013)
31. Sharpnack, J.L., Krishnamurthy, A., Singh, A.: Near-optimal anomaly detection in graphs using Lovasz extended scan statistic. In: NeurIPS (2013)
32. Speakman, S., McFowland III, E., Neill, D.B.: Scalable detection of anomalous patterns with connectivity constraints. J. Comput. Graph. Stat. **24**(4), 1014–1033 (2015)
33. Wu, N., Chen, F., Li, J., Zhou, B., Ramakrishnan, N.: Efficient nonparametric subgraph detection using tree shaped priors. In: AAAI (2016)

A Late-Fusion Approach to Community Detection in Attributed Networks

Chang Liu[1], Christine Largeron[2], Osmar R. Zaïane[1(✉)],
and Shiva Zamani Gharaghooshi[1]

[1] Alberta Machine Intelligence Institute, University of Alberta, Edmonton, Canada
{chang6,zaiane,zamanigh}@ualberta.ca
[2] Laboratoire Hubert Curien, Université de Lyon, Saint-Etienne, France
christine.largeron@univ-st-etienne.fr

Abstract. The majority of research on community detection in attributed networks follows an "early fusion" approach, in which the structural and attribute information about the network are integrated together as the guide to community detection. In this paper, we propose an approach called *late-fusion*, which looks at this problem from a different perspective. We first exploit the network structure and node attributes separately to produce two different partitionings. Later on, we combine these two sets of communities via a fusion algorithm, where we introduce a parameter for weighting the importance given to each type of information: node connections and attribute values. Extensive experiments on various real and synthetic networks show that our late-fusion approach can improve detection accuracy from using only network structure. Moreover, our approach runs significantly faster than other attributed community detection algorithms including early fusion ones.

Keywords: Community detection · Attributed networks · Late fusion

1 Introduction

In many modern applications, data is represented in the form of relationships between nodes forming a *network*, or interchangeably a *graph*. A typical characteristic of these real networks is the *community structure*, where network nodes can be grouped into densely connected modules called communities. Community identification is an important issue because it can help to understand the network structure and leads to many substantial applications [6]. While traditional community detection methods focus on the network topology where communities can be defined as sets of nodes densely connected internally, recently, increasing attention has been paid to the attributes associated with the nodes in order to take into account homophily effects, and several works have been devoted to community detection in attributed networks. The aim of such process is to obtain a partitioning of the nodes where vertices belonging to the same subgroup are densely connected and homogeneous in terms of attribute values.

M. R. Berthold et al. (Eds.): IDA 2020, LNCS 12080, pp. 300–312, 2020.
https://doi.org/10.1007/978-3-030-44584-3_24

In this paper, we propose a new method designed for community detection in attributed networks, called *late fusion*. This is a two-step approach where we first identify two sets of communities based on the network topology and node attributes respectively, then we merge them together to produce the final partitioning of the network that exhibits the homophily effect, according to which linked nodes are more likely to share the same attribute values. The communities based upon the network topology are obtained by simply applying an existing algorithm such like Louvain [2]. For graphs whose node attributes are numeric, we utilize existing clustering algorithms to get the communities (i.e., clusters) based on node attributes. We extend to binary-attributed graphs by generating a virtual graph from the attribute similarities between the nodes, and performing traditional community detection on the virtual graph. Albeit being simple, extensive experiments have shown that our late-fusion method can be competitive in terms of both accuracy and efficiency when compared against other algorithms. We summarize our main contributions in this work are:

1. A new late-fusion approach to community detection in attributed networks, which allows the use of traditional methods as well as the integration of personal preference or prior knowledge.
2. A novel method to identify communities that reflect attribute similarity for networks with binary attributes.
3. Extensive experiments to validate the proposed method in terms of accuracy and efficiency.

The rest of the paper is organized as follows: In Sect. 2, we provide a brief review of community detection algorithms suited for attributed networks, next we present our late fusion approach in Sect. 3. Experiments to illustrate the effectiveness of the proposed method are detailed in Sect. 4. Finally, we summarize our work and point out several future directions in Sect. 5.

2   Related Work

How to incorporate the node attribute information into the process of network community detection has been studied for a long time. One of the early ideas is to transform attribute similarities into edge weights. For example, [13] proposes *matching coefficient* which is the count of shared attributes between two connected nodes in a network; [15] extends the matching coefficient to networks with numeric node attributes; [4] defines edge weights based on self-organizing maps. A drawback of these methods is that new edge weights are only applicable to edges already existed, hence the attribute information is not fully utilized. To overcome this issue, a different approach is to *augment* the original graph by adding virtual edges and/or nodes based on node attribute values. For instance, [14] generates content edges based on the cosine similarity between node attribute vectors, in graphs where nodes are textual documents and the corresponding attribute vector is the TF-IDF vector describing their content. The kNN-enhance algorithm [9] adds directed virtual edges from a node to one of its k-nearest neighbors if their attributes are similar. The SA-Clustering [17] adds both virtual nodes and edges to the original graph, where the virtual nodes represent binary-valued attributes, and the virtual edges connect the real nodes to the virtual nodes representing the attributes that the real nodes own.

Another class of methods is inspired by the modularity measure. These methods incorporate attribute information into an optimization objective like the modularity. [5] injects an attribute based similarity measure into the modularity function; [1] combines the gain in the modularity with multiple common users' attributes as an integrated objective; I-Louvain algorithm [3] proposes inertia-based modularity to describe the similarity between nodes with numeric attributes, and adds the inertia-based modularity to the original modularity formula to form the new optimization objective.

With the wide spreading of deep learning, network representation learning and node embedding (e.g. [8]) motivated new solutions. [12] proposes an embedding based community detection algorithm that applies representation learning of graphs to learn a feature representation of a network structure, which is combined with node attributes to form a cost function. Minimizing it, the optimal community membership matrix is obtained.

Probabilistic models can be used to depict the relationship between node connections, attributes, and community membership. The task of community detection is thus converted to inferring the community assignment of the nodes. A representative of this kind is the CESNA algorithm [16], which builds a generative graphical model for inferring the community memberships.

Whereas the majority of the previous methods exploit simultaneously both types of information, we propose the late-fusion approach that combines two sets of communities obtained separately and independently from the network structure and node attributes via a fusion algorithms.

3   The Late-Fusion Method

Given an attributed network $G = (V, E, A)$, with V being the set of m nodes, E the set of n edges, and A an $m \times r$ attribute matrix describing the attribute values of the nodes with r attributes, the goal is to build a partitioning $\mathcal{P} = \{C_1, ..., C_k\}$ of V into k communities such that nodes in the same community are densely connected and similar in terms of attributes, whereas nodes from distinct communities are loosely connected and different in terms of attribute.

For networks with numeric attributes, we can directly apply a community detection algorithm F_s on G to identify a set of communities based on node connections $\mathcal{P}_s = \{C_1, C_2, ..., C_{k_s}\}$, and a clustering algorithms F_a on A to find a set of clusters based on node attributes $\mathcal{P}_a = \{C_1, C_2, ..., C_{k_a}\}$. When it comes to binary attributed networks, traditional clustering algorithms become inaccessible, we instead build a virtual graph G_a that shares the same node set as G, but there is an edge only when the two nodes are similar enough in terms of attributes. Then we apply F_s on G_a and obtain $\mathcal{P}_a$. Note that we omit categorical attributes since categorical values can be easily converted to the binary case.

The second step is to combine the partitions $\mathcal{P}_s$ and $\mathcal{P}_a$. We first derive the adjacency matrices D_s and D_a from $\mathcal{P}_s$ and $\mathcal{P}_a$ respectively, where $d_{ij} = 1$ when nodes i and j are in the same community in a partitioning $\mathcal{P}$ and $d_{ij} = 0$ otherwise. Next, an integrated adjacency matrix D is given by $D = \alpha D_s + (1 - \alpha)D_a$. Here α is the weighting parameter that leverages the strength between network topology and node attributes. In this way, the information about network topology and node attributes of the original graph G is represented in D. Now G_{int}, derived from the adjacency matrix D, is an integrated, virtual, weighted graph whose edges embody the homophily effect of G. Algorithm 1 shows the steps of our late-fusion approach applied to networks with binary attributes.

Algorithm 1. Late-fusion on networks with binary attributes

Input: $G = (V, E, A), F_s, \alpha$
Output: $\mathcal{P} = \{C_1, C_2, ..., C_k\}$
1 $\mathcal{P}_s = F_s(G_s)$
2 $G_a = $ build_virtual_graph (A)
3 $\mathcal{P}_a = F_s(G_a)$
4 $D_s = $ get_adjacency_matrix$(\mathcal{P}_s)$, $D_a = $ get_adjacency_matrix$(\mathcal{P}_a)$
5 $D = \alpha D_s + (1 - \alpha)D_a$
6 $G_{integrated} = $ from_adjacency_matrix (D)
7 $\mathcal{P} = F_s(G_{integrated})$
8 **return** $\mathcal{P}$

Here we address an important detail: how to build the virtual graph G_a from the node-attribute matrix A? We compute the inner product as the similarity measure between each node pair, and if the inner product exceeds a predetermined threshold, we regard the nodes as similar and add a virtual edge between them. The threshold can be determined heuristically based on the distribution of the node similarities. However, the threshold should be chosen properly so that the resulted G_a would be neither too dense nor too sparse, where both cases could harm the quality of the final communities. Under this guidance, we put forward two thresholding approaches:

1. **Median thresholding (MT):** Suppose S is the $m \times m$ similarity matrix of all nodes in V, we take all the off-diagonal, upper triangular (or lower triangular) entries of S, find the median of these numbers and set it as the threshold. This approach guarantees that we add virtual edges to half of all node pairs who share a similarity value higher than the other half.
2. **Equal-edge thresholding (EET):** We compute $q = 1 - d(G)$ where $d(G)$ is the density of G. Then the qth quantile of the similarity distribution is the chosen threshold. In this approach, we let the original graph G_s be the proxy that decides how we construct the virtual graph G_a.

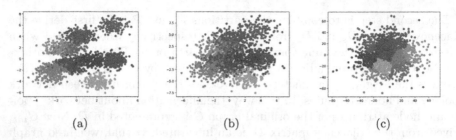

Fig. 1. Node attribute distribution for three groups of experiments. (a) Strong attributes, (b) Medium attributes, (c) Weak attributes. Each color represents a unique community (Color figure online)

4   Experiments

Our proposed method has been evaluated through experiments on multiple synthetic and real networks and results are presented in this section. For networks with numeric attributes, we take advantage of existing clustering algorithms to obtain communities based on attributes (i.e., clusters), and for networks with binary attributes, we employ Algorithm 1 to perform community detection. We have also released our code so that readers can reproduce the results[1].

4.1   Synthetic Networks with Numeric Attributes

Data. We use an attributed graph generator [10] to create three attributed graphs with ground-truth communities, denoted as G_{strong}, G_{medium} and G_{weak}, indicating the corresponding ground-truth partitionings are *strong*, *medium*, and *weak* in terms of modularity Q. To examine the effect of attributes on community detection, for each of G_{strong}, G_{medium} and G_{weak}, we assign three different attribute distributions as shown in Fig. 1, where attributes in Fig. 1a and b are generated from a Gaussian mixture model with a shared standard deviation, and Fig. 1c presents the original attributes generated by [10]. By this way, for each graph having a specific community structure (G_{strong}, G_{medium}, G_{weak}) we have also three types of attributes denoted strong attributes, medium attributes and weak attributes leading in fact to 9 datasets.

Evaluation Measures and Baselines. Normalized Mutual Information (NMI) and Adjusted Rand Index (ARI) and running time are used to evaluate algorithm accuracy and efficiency. Louvain [2] and SIWO [7] have been chosen as baseline algorithms that utilize only the links to identify network communities.

---

[1] https://github.com/changliu94/attributed-community-detection.

Table 1. Properties of synthetic networks

	m	n	k	r	Q
G_{strong}	2000	7430	10	2	0.81
G_{medium}	2000	7445	10	2	0.65
G_{weak}	2000	6988	10	2	0.54

Table 2. Properties of Sina Weibo network

m	n	k	r	Q	I
3490	30282	10	10	0.05	0.04

Note that since the attribute distribution does not affect Louvain and SIWO, the results of Louvain and SIWO are only presented in Table 3. We choose Spectral Clustering (SC) and DBSCAN as two representative clustering algorithms as they both can handle non-flat geometry. We treat the number of clusters as a known input parameter of SC, and the neighborhood size of DBSCAN is set to the average node degree. We adopt default values of the remaining parameters from the *scikit-learn* implementation of these two algorithms. Finally, we take the implementation of the I-Louvain algorithm which exploits links and attribute values as our contender. The code of I-Louvain is available online[2]. Given Louvain, SIWO, SC, and DBSCAN, correspondingly we can have four combinations for our late-fusion method. In all experiments, the α parameter in Algorithm 1 is chosen to be 0.5, i.e., the same weight is allocated to structural and attribute information.

Table 3. Results of strong attributes, time is measured in seconds

	G_{strong}			G_{medium}			G_{weak}		
	NMI	ARI	Time	NMI	ARI	Time	NMI	ARI	Time
Louvain	.795	.797	0.41	.695	.686	0.49	.665	.674	0.64
SIWO	.836	**.850**	0.97	.702	**.705**	1.09	.504	.458	0.98
SC	.802	.713	1.15	.777	.677	0.64	**.768**	.669	0.68
DBSCAN	.469	.103	0.06	.434	.083	0.06	.465	.102	0.24
I-Louvain	.515	.150	39.2	.718	.704	30.0	.608	.503	37.6
Louvain + SC	.824	.704	7.34	.784	.618	5.74	.765	.597	7.14
Louvain + DBSCAN	.818	.813	8.64	.730	.702	8.87	.704	**.690**	10.6
SIWO + SC	**.844**	.738	10.3	**.786**	.636	7.33	.723	.508	6.46
SIWO + DBSCAN	.818	.813	11.7	.730	.702	10.2	.704	**.690**	11.6

---

[2] https://www.dropbox.com/sh/j4aqitujiaifgq4/AAAAH0L3uIPYNWKoLpcAh0TPa.

Table 4. Results of medium attributes, time is measured in seconds

	G_{strong}			G_{medium}			G_{weak}		
	NMI	ARI	Time	NMI	ARI	Time	NMI	ARI	Time
SC	.529	.338	0.83	.522	.322	0.53	.538	.349	0.57
DBSCAN	.096	.012	0.08	.066	.008	0.14	.065	.011	0.09
I-Louvain	.517	.150	36.8	**.707**	**.690**	33.7	.614	.522	33.2
Louvain + SC	.734	.450	5.62	.696	.390	5.96	**.677**	.392	5.66
Louvain + DBSCAN	**.755**	**.726**	9.20	.670	.636	11.9	.641	**.633**	13.6
SIWO + SC	.748	.469	12.7	.699	.402	7.12	.625	.335	7.44
SIWO + DBSCAN	.744	**.726**	8.73	.670	.636	8.98	.641	**.633**	12.4

Results. Table 3, corresponding to strong attributes, shows that late fusion is the best-performing algorithm in terms of NMI on G_{strong} and G_{medium}, and very close to SC on G_{weak} (0.765 against 0.768) whereas it is better in terms of ARI on this last graph. On Tables 4 and 5, corresponding respectively to medium and weak attributes, with the deterioration of the attribute quality, the accuracy of late-fusion degrades, but late fusion still remains at a consistently high level compared to I-Louvain and the clustering algorithms. Moreover, the performance degradation of late-fusion methods is less susceptible to the deterioration of community quality compared to the clustering algorithms, thanks to the complementary structural information. As for the running time, it is expected that classic community detection algorithms Louvain and SIWO are the fastest algorithms, as they do not consider node attributes, but the late-fusion method still outperforms I-Louvain by a remarkable margin.

Table 5. Results of weak attributes, time is measured in seconds

	G_{strong}			G_{medium}			G_{weak}		
	NMI	ARI	Time	NMI	ARI	Time	NMI	ARI	Time
SC	.483	.270	3.31	.514	.307	2.32	.489	.276	2.45
DBSCAN	.000	.000	0.06	.000	.000	0.06	.000	.000	0.14
I-Louvain	.517	.150	35.1	.707	**.690**	34.3	.614	.522	39.5
Louvain + SC	.770	.670	11.8	.705	.613	10.2	**.689**	.564	9.33
Louvain + DBSCAN	.795	**.797**	11.2	.695	.685	10.4	.667	**.674**	12.9
SIWO + SC	**.797**	.703	13.2	**.709**	.635	12.3	.601	.467	11.0
SIWO + DBSCAN	.795	**.797**	11.6	.695	.685	11.3	.667	**.674**	12.6

4.2   Real Network with Numeric Attributes

Data and Baselines. Sina Weibo[3] is the largest online Chinese micro-blog social networking website. Table 2 shows the corresponding properties of the Sina Weibo network built by [9][4]. It includes within-inertia ratio I, a measure of attribute homogeneity of data points that are assigned to the same subgroup. The lower the within-inertia ratio, the more similar the nodes in the same community are. As DBSCAN algorithm performs poorly on the Sina Weibo network and it is costly to infer a good combination of the hyper-parameters of the algorithm, it has been replaced by k-means as a supplement to spectral clustering. The number of clusters required as an input by k-means and SC is inferred from the 'elbow method', which happens to be 10, the actual number of clusters. Moreover, since we have the prior knowledge that the ground truth communities are based on the topics of the forums from which those users are gathered, we reckon that the formation of communities depends more on the attribute values than the structure and set the parameter α at 0.2.

Results. Table 6 presents the results on Sina Weibo network. The two baseline algorithms Louvain and SIWO and the contending algorithm I-Louvain perform poorly on the Sina Weibo network, whereas the clustering algorithms show a high accuracy. Especially, the k-means algorithm together with our four late-fusion methods with the emphasis on attribute information produce results with the best NMI and ARI. This is because modularity of Sina Weibo network is low (0.05 as indicated in Table 2) and the within-inertia ratio is also low (0.04). The results also validate our assumption that communities in this network are mainly determined by the attributes. We will further explore the effect of α in Sect. 4.4.

Table 6. Experimental results on Sina Weibo network

	NMI	ARI	Time
Louvain	.232	.197	1.98
SIWO	.040	.000	3.26
SC	.612	.520	3.16
k-means	**.649**	**.579**	0.25
I-Louvain	.204	.038	261
Louvain+SC	.611	.519	48.9
Louvain+k-means	**.649**	**.579**	42.1
SIWO+SC	.611	.519	37.9
SIWO+k-means	**.649**	**.579**	50.4

Table 7. Properties of Facebook networks

Network ID	m	n	k	r	Q
0	347	5038	24	224	0.179
107	1045	53498	9	576	0.218
348	227	6384	14	161	0.210
414	159	3386	7	105	0.468
686	170	3312	14	63	0.101
698	66	540	13	48	0.239
1684	792	28048	17	319	0.509
1912	755	60050	46	480	0.339
3437	547	9626	32	262	0.026
3980	59	292	17	42	0.242

---

[3] http://www.weibo.com.
[4] This dataset is available online https://github.com/smileyan448/Sinanet.

4.3   Real Network with Binary Attributes

Data. Facebook dataset [11] contains 10 egocentric networks with binary attributes corresponding to anonymous information of the user about the name, work, and education and ground-truth communities. This dataset is available online[5] and Table 7 presents the properties of these networks.

We still treat Louvain and SIWO as our baselines. We use the CESNA algorithm [16], able to handle binary attributes in addition to the links, as our contender[6]. To compare the two thresholding strategies proposed in Section 3, we present experimental results of four late-fusion methods: Louvain + equal-edge thresholding (denoted as Louvain-EET), Louvain + median thresholding (denoted as Louvain-MT), SIWO + equal-edge thresholding (denoted as SIWO-EET), and SIWO + median thresholding (denoted as SIWO-MT). We set α to its default value 0.5.

Table 8. NMI of different community detection results on Facebook network

Network ID	0	107	348	414	686	698	1684	1912	3437	3980	Average
Louvain	.382	.332	.478	**.609**	.284	.281	.047	**.565**	.181	**.729**	.389
SIWO	.390	.363	.375	.586	.215	.259	.053	.557	.174	.605	.358
CESNA	.263	.249	.307	.586	.238	.564	.438	.450	.176	.552	.382
Louvain-EET	**.558**	.355	**.525**	.538	**.463**	**.669**	**.462**	.511	**.310**	.704	**.509**
Louvain-MT	.452	.341	.489	.556	.351	.479	.323	.491	.262	.696	.444
SIWO-EET	.541	**.364**	.452	.531	.406	.630	.460	.509	**.310**	.648	.485
SIWO-MT	.431	.353	.405	.538	.252	.406	.332	.491	.260	.588	.406

Table 9. ARI of different community detection results on Facebook network

Network ID	0	107	348	414	686	698	1684	1912	3437	3980	Average
Louvain	.143	.148	.303	.558	.110	.000	.000	.461	.000	.398	.209
SIWO	.220	.177	.127	.519	.000	.009	.000	.419	.002	.209	.167
CESNA	.073	.097	.156	.480	.001	.202	.310	.361	.014	.067	.176
Louvain-EET	.024	.047	.103	.265	.006	.000	.043	.252	.000	.069	.008
Louvain-MT	.061	.079	.129	.413	.063	.000	.048	.235	.000	.084	.110
SIWO-EET	.043	.045	.124	.252	.003	.000	.057	.235	.000	.095	.009
SIWO-MT	.108	.079	.141	.391	.040	.016	.060	.223	.000	.073	.113

---

[5] http://snap.stanford.edu/data.
[6] The source code of CESNA is available online https://github.com/snap-stanford/snap/tree/master/examples/cesna.

Table 10. Running time of different community detection results on Facebook network, measured in seconds

Network ID	0	107	348	414	686	698	1684	1912	3437	3980	Average
Louvain	0.15	1.83	0.12	0.06	0.09	0.02	0.80	1.28	0.31	0.01	0.47
SIWO	0.34	3.78	0.31	0.16	0.17	0.03	1.46	3.79	0.51	0.02	1.06
CESNA	9.76	103	6.02	2.47	3.12	0.63	38.3	22.9	21.1	0.60	20.8
Louvain-EET	0.72	4.68	0.40	0.25	0.24	0.07	1.95	3.83	0.78	0.03	1.30
Louvain-MT	2.90	20.0	0.82	0.48	0.44	0.08	8.22	9.41	3.28	0.06	4.57
SIWO-EET	1.73	24.4	2.87	0.68	0.76	0.14	5.76	28.5	4.26	0.12	6.92
SIWO-MT	9.45	91.4	5.27	1.73	3.14	0.34	44.9	43.4	13.5	0.17	21.3

Results. Results in terms of NMI, ARI, and running time are respectively presented in Tables 8, 9, and 10. In terms of NMI, results in Table 8 show again that our late-fusion algorithms can significantly improve the community detection accuracy upon Louvain. On average, the late fusion method Louvain+EET outperforms Louvain, SIWO, and CESNA by 30.8%, 42.2%, and 33.2% respectively. The late fusion method Louvain+MT outperforms the three by 14.1%, 24.0%, and 16.2% respectively. However, all of the late-fusion methods perform poorly when evaluated by ARI. This is resulted from the goal of our late-fusion approach. Remember that we aim to find the set of communities such that nodes in the same subgroup are densely connected and similar in terms of attributes, whereas nodes residing in different communities are loosely connected and dissimilar in attributes. This purpose led the late-fusion approach to over-partition communities that are formed by only one of the two sources of information. The over-partitioning greatly hurts the results of ARI. A postprocessing model to resolve the over-partitioning issue with late fusion is left as a future work. The running time results shown in Table 10 again manifests the efficiency advantage of our late-fusion methods over CESNA.

4.4   Effect of Parameter α

In the Sina Weibo experiment, we see the advantage of having a weighting parameter to accordingly leverage the strength of the two sources of information. In this section, we dive deeper into the effect of α on the community detection results. To do so, we devise an experiment where we use the G_{strong} and G_{weak} introduced in Table 1. In reverse, we assign **weak** attributes to G_{strong} and **strong** attributes to G_{weak}. Then we perform our late fusion algorithm on these two graphs with varying α values. In our experiment, we choose SIWO as F_s and k-means as F_a.

Table 11 presents the NMI and ARI of the late fusion with SIWO and k-means when α varies. G_{strong} has communities with a strong structure but weak attributes, so the accuracy score for NMI and ARI goes up as we put more weight on the structure; On the contrary, G_{weak} has weak structural communities but

Table 11. Effect of α

	$\alpha = 0.0$		$\alpha = 0.2$		$\alpha = 0.5$		$\alpha = 0.8$		$\alpha = 1.0$	
	NMI	ARI	NMI	ARI	NMI	ARI	NMI	ARI	NMI	ARI
G_{strong}	0.530	0.359	0.530	0.359	0.756	0.513	0.836	0.850	0.836	0.850
G_{weak}	0.867	0.834	0.867	0.834	0.762	0.470	0.526	0.364	0.526	0.364

strong attributes, hence the accuracy score decreases as α increases. One can also notice that when α is sufficiently high or low, late fusion becomes equivalent to using community detection or clustering only, which is in accordance with our observation done on the Sina Weibo experiment.

In practice, when network communities are mainly determined by the links, α should be greater than 0.5; $\alpha < 0.5$ is recommended if attributes play a more important role in forming the communities; When prior knowledge about network communities is unavailable or both sources of information contribute equally, α should be 0.5.

4.5   Complexity of Late Fusion

It is a known drawback of attributed community detection algorithms that they are very time-consuming due to the need to consider node attributes. Our late-fusion method tries to circumvent this problem by taking advantage of the existing community detection and clustering algorithms that are efficiently optimized, and combining their results by a simple approach. To further show the computational efficiency of our late-fusion method, we compute the running time of the late-fusion method and compare it with other methods.

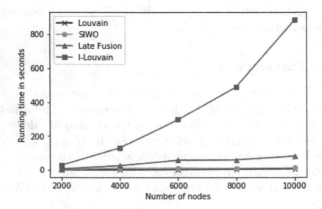

Fig. 2. Running time of Louvain, SIWO, late fusion and I-Louvain on networks of different sizes

We test the running time of four different community detection methods on five graphs with the number of nodes varying from 2000, 4000, 6000, 8000, and 10000. These graphs are also generated by the attributed graph generator [10]. We control the modularity of each graph at the range of 0.64−0.66 and keep other hyperparameters the same. For each size, we randomly sample 10 graphs from the graph generator and plot the average running time of each method. As we can see in Fig. 2, it is expected that our late-fusion method is inevitably slower than the two community detection methods that only utilize node connections. However, our algorithm runs way faster than the I-Louvain algorithm, albeit both being approximately linear in the growth of network sizes.

5   Conclusion and Future Direction

In this paper, we proposed a new approach to the problem of community detection in attributed networks that follows a late-fusion strategy. We showed with extensive experiments that most often, our late-fusion method is not only able to improve the detection accuracy provided by traditional community detection algorithms, but it can also outperform the chosen contenders in terms of both accuracy and efficiency. We learned that combining node connections with attributes to detect communities of a network is not always the best solution, especially when one side of the network properties is strong while the other is weak, using only the best information available can lead to better detection results. It is part of our future work to understand when and how we should use the extra attribute information to help community detection. ARI suffers greatly from over-partitioning issue with our late fusion when applied to networks with binary attributes. A postprocessing model to resolve this issue is desired. We also hope to expand the late-fusion approach to networks with a hybrid of binary and numeric attributes as well as networks with overlapping communities.

References

1. Asim, Y., Ghazal, R., Naeem, W., Majeed, A., Raza, B., Malik, A.K.: Community detection in networks using node attributes and modularity. Int. J. Adv. Comput. Sci. Appl. **8**(1), 382–388 (2017)
2. Blondel, V.D., Guillaume, J.L., Lambiotte, R., Lefebvre, E.: Fast unfolding of communities in large networks. J. Stat. Mech: Theory Exp. **2008**(10), P10008 (2008)
3. Combe, D., Largeron, C., Géry, M., Egyed-Zsigmond, E.: I-Louvain: an attributed graph clustering method. In: Fromont, E., De Bie, T., van Leeuwen, M. (eds.) IDA 2015. LNCS, vol. 9385, pp. 181–192. Springer, Cham (2015). https://doi.org/10.1007/978-3-319-24465-5_16
4. Cruz, J.D., Bothorel, C., Poulet, F.: Semantic clustering of social networks using points of view. In: CORIA, pp. 175–182 (2011)
5. Dang, T., Viennet, E.: Community detection based on structural and attribute similarities. In: International Conference on Digital Society (ICDS), pp. 7–14 (2012)

6. Fortunato, S., Hric, D.: Community detection in networks: a user guide. Phys. Rep. **659**, 1–44 (2016)
7. Gharaghooshi, S.Z., Zaïane, O., Largeron, C., Zafarmand, M., Liu, C.: Addressing the resolution limit and the field of view limit in community mining. In: Berthold, M.R., et al. (eds.) Symposium on Intelligent Data Analysis, IDA 2020. LNCS, vol. 12080, pp. 1–12. Springer, Cham (2020)
8. Grover, A., Leskovec, J.: node2vec: Scalable feature learning for networks. In: Proceedings of the 22nd ACM SIGKDD Conference, pp. 855–864. ACM (2016)
9. Jia, C., Li, Y., Carson, M.B., Wang, X., Yu, J.: Node attribute-enhanced community detection in complex networks. Sci. Rep. **7**, 1–15 (2017)
10. Largeron, C., Mougel, P., Benyahia, O., Zaïane, O.R.: DANCer: dynamic attributed networks with community structure generation. Knowl. Inf. Syst. **53**(1), 109–151 (2017)
11. Leskovec, J., Mcauley, J.J.: Learning to discover social circles in ego networks. In: Advances in Neural Information Processing Systems, pp. 539–547 (2012)
12. Li, Y., Sha, C., Huang, X., Zhang, Y.: Community detection in attributed graphs: An embedding approach. In: AAAI (2018)
13. Neville, J., Adler, M., Jensen, D.: Clustering relational data using attribute and link information. In: Proceedings of the Text Mining and Link Analysis Workshop, 18th International Joint Conference on Artificial Intelligence, pp. 9–15. Morgan Kaufmann Publishers, San Francisco (2003)
14. Ruan, Y., Fuhry, D., Parthasarathy, S.: Efficient community detection in large networks using content and links. In: Proceedings of the 22nd International Conference on World Wide Web, pp. 1089–1098. ACM (2013)
15. Steinhaeuser, K., Chawla, N.V.: Community detection in a large real-world social network. In: Liu, H., Salerno, J.J., Young, M.J. (eds.) Social Computing, Behavioral Modeling, and Prediction, pp. 168–175. Springer, Boston (2008). https://doi.org/10.1007/978-0-387-77672-9_19
16. Yang, J., McAuley, J., Leskovec, J.: Community detection in networks with node attributes. In: ICDM Conference, pp. 1151–1156. IEEE (2013)
17. Zhou, Y., Cheng, H., Yu, J.X.: Graph clustering based on structural/attribute similarities. Proc. VLDB Endow. **2**(1), 718–729 (2009)

Reconciling Predictions in the Regression Setting: An Application to Bus Travel Time Prediction

João Mendes-Moreira[1,2] and Mitra Baratchi[3(✉)]

[1] LIAAD-INESC TEC, Porto, Portugal
jmoreira@fe.up.pt
[2] Faculty of Engineering, University of Porto, Porto, Portugal
[3] LIACS, Leiden University, Leiden, The Netherlands
m.baratchi@liacs.leidenuniv.nl

Abstract. In different application areas, the prediction of values that are hierarchically related is required. As an example, consider predicting the revenue per month and per year of a company where the prediction of the year should be equal to the sum of the predictions of the months of that year. The idea of reconciliation of prediction on grouped time-series has been previously proposed to provide optimal forecasts based on such data. This method in effect, models the time-series collectively rather than providing a separate model for time-series at each level. While originally, the idea of reconciliation is applicable on data of time-series nature, it is not clear if such an approach can also be applicable to regression settings where multi-attribute data is available. In this paper, we address such a problem by proposing Reconciliation for Regression (R4R), a two-step approach for prediction and reconciliation. In order to evaluate this method, we test its applicability in the context of Travel Time Prediction (TTP) of bus trips where two levels of values need to be calculated: (i) travel times of the links between consecutive bus-stops; and (ii) total trip travel time. The results show that R4R can improve the overall results in terms of both link TTP performance and reconciliation between the sum of the link TTPs and the total trip travel time. We compare the results acquired when using group-based reconciliation methods and show that the proposed reconciliation approach in a regression setting can provide better results in some cases. This method can be generalized to other domains as well.

Keywords: Regression · Reconciliation · Bus travel time

1 Introduction

Regression analysis provides a simple framework for predicting numerical target attributes from a set of independent predictive attributes. Addressing any problem using this framework requires designing models that fully capture the relations between predictive and target attributes. This has so far led to many classes of regression models being designed. For instance, multi-target regression models [11] consider predicting

© The Author(s) 2020
M. R. Berthold et al. (Eds.): IDA 2020, LNCS 12080, pp. 313–325, 2020.
https://doi.org/10.1007/978-3-030-44584-3_25

the value of multiple target attributes as opposed to basic regression models that aim at predicting only a single target attribute at a time. In another case, when one target variable is being predicted from a set of hierarchically ordered predictive attributes, the problem is known to be multi-level regression [5].

In this paper, we address the problem of regression for a class of problems where dependent variables are additionally hierarchically organized following different levels of aggregation. An example is the revenue forecasts per month and also per year of a given company. The forecasts for the new year can be the sum of the predictions done for each of the twelve months of the new year or can be done directly for the full new year. However, in many situations, it is important that the sum of the prediction per month is equal to the prediction for the full year. Moreover, relevant questions in this regard can arise. Can we obtain better predictions using both predictions for all months and for the full year? How may we reconcile the sum of the predictions done per month with the prediction done for the full year? Authors of [8] answered these questions for hierarchies of time series, i.e., a sequence of values, typically equally spaced, where this sequence can be aggregated by a given dimension.

This notion of hierarchy can also exist in the regression setting i.e., a problem with a set of n instances $(\mathbf{X_i}, \mathbf{y_i}), \mathbf{i} = \mathbf{1}, ..., \mathbf{n}$. Each $(\mathbf{X_i}, \mathbf{y_i})$ instance has a vector $\mathbf{X_i}$ with p predictive attributes $(x_{i_1}, x_{i_2}, ..., x_{i_p})$ and a quantitative target attribute y_i. The hierarchy can exist in this regression setting when, for instance, two of the p predictive attributes have a 1-to-many relation as referred to in relational databases. Addressing this problem in the regression setting leads to more flexible and robust solutions compared to the time series approach because: (1) any number of observations per time interval can be defined; (2) there are no limitations to the time interval between consecutive observations; and (3) any other type of predictive attribute can be used to better explain the target attribute.

In this work, we present an approach to reconcile predictions in the regression setting. We achieve this by proposing a new method named Reconciling for Regression (R4R). The R4R method is tested for the bus travel time prediction problem. This problem considers that buses run in predefined routes, and each route is composed of several links. Each link is the road stretch between two consecutive bus stops. Reconciling the predictions in this problem aims at reconciling the sum of the predictions done for each link with the prediction done for the full route. According to the authors' knowledge, this is the first work on reconciling predictions in the regression setting. This work is also different from multi-target and multi-level variants being a combination of both (having multiple targets that are hierarchically ordered).

The R4R method can be applied to any other regression problem which exhibits a one-to-many relationship between instances, and also where the aggregated target value (the one) is the sum of the detailed target values (the many). In the previous example: (1) the revenue forecasts for the new year, the many component targets are the revenue forecasts per month, and the one component target is the revenue forecast for the full year; (2) in the bus travel time example, the many component targets are the link predictions while the one component target is the full route prediction. In this paper, we only discuss the sum as aggregation criterion (the one should be equal to the sum of the many), but the proposed method could be easily extended to other aggregation criteria, e.g., the average.

The remainder of this paper is organized as follows. In Sect. 2, we present the previous work on reconciling predictions. Section 3 elaborates the proposed methodology. In Sect. 4, we describe the case study. The results of the case study are presented and discussed in Sect. 5. Finally, the conclusions are presented in Sect. 6.

2 Literature Review

In this section, we review the previous research, both considering (i) the methods for forecasting for hierarchically organized time-series data and (ii) application area of travel time prediction.

Methods for Forecasting Hierarchically Organized Data: Common methods used to reconcile predictions for hierarchically organized time-series data can be further grouped into three categories: bottom-up, top-down and middle-out, based on the level which is predicted first. Bottom-up strategies forecast all the low-level target attributes and use the sum of these predictions as the forecast for the higher-level attribute. On the contrary, top-down approaches predict the top-level attribute and then splits up the predictions for the lower level attributes based on historical proportions that may be estimated. For time-series data with more than two levels of hierarchy, a middle-out approach can be used, combining both bottom-up and top-down approaches [3]. These methods form linear mappings from the initial predictions to reconciled estimates. As a consequence, the sum of the forecasts of the components of a hierarchical time series is equal to the forecast of the whole. However, this is achieved without guaranteeing an optimal solution. Authors of [8] presented a new framework for optimally reconciling forecasts of all series in a hierarchy to ensure they add up. The method first computes the forecast independently for each level of the hierarchy. Afterward, the method provides a means for optimally reconciling the base forecasts so that they aggregate appropriately across the hierarchy. The optimal reconciliation is based on a generalized least squares estimator and requires an estimation of the covariance matrix of the reconciliation errors. Using Australian domestic tourism data, authors of [8] compare their optimal method with bottom-up and conventional top-down forecast approaches. Results show that the optimal combinational approach and the bottom-up approach outperform the top-down method. The same authors extended, in [9], the previous work proposed in [8] to cover non-hierarchical groups of time series, as well as, large groups of time series data with a partial hierarchical structure. A new combinational forecasting approaches is proposed that incorporates the information from a full covariance matrix of forecast errors in obtaining a set of aggregate forecasts. They use a weighted least squares method due to the difficulty of estimating the covariance matrix for large hierarchies.

In [16], an alternative representation that involves inverting only a single matrix of a lower number of dimensions is used. The new combinational forecasting approach incorporates the information from a full covariance matrix of forecast errors in obtaining a set of aggregate consistent forecasts. The approach minimizes the mean squared error of the aggregate consistent forecasts across the entire collection of time series.

A game-theoretically optimal reconciliation method is proposed in [6]. The authors address the problem in two independent steps, by first computing the best possible forecasts for the time series without taking into account the hierarchical structure and next to a game-theoretic reconciliation procedure to make the forecasts aggregate consistent.

The previously mentioned methods are limited by the nature of the time-series approach they take. It is often impossible to take any advantage of additional features and attributes accompanying data with such an approach. Furthermore, many prevalent data imperfection problems such as missing data, lead to imperfect time-series. This fact reduces the applicability of time-series models that require equally distanced samples.

In our work, we take advantage of additional features and the structure of the grouped data to improve and reconcile predictions. Instead of forecasting each time series independently and then combine the predictions, in a regression setting, we can reconcile future events using only some past events. This leads to a solution suitable for online applications.

Application Area of Travel Time Prediction: There exists a considerable amount of research papers that address the problem of travel time prediction for transport applications. Accurate travel time information is essential as it attracts more commuters and increases commuter's satisfaction [1]. The majority of these works are on short-term travel time prediction [19], aimed at applications in advanced traveler information systems. There are also works on long-term travel time prediction [13], which can be used as a planning tool for public transport companies or even for freight transports.

Link travel time prediction can be used for route guidance [17], for bus bunching detection [14], or to predict the bus arrival time at the next station [18] which can promote information services about it. More recently, Global Positioning System (GPS) data is becoming more and more available, allowing its use to predict travel times from GPS trajectories. These trajectories can be used to construct origin-destination matrices of travel times or traffic flows, an important tool for mobility purposes [2].

Using both link travel time predictions and the full trip travel time prediction in order to improve all those predictions is a contribution of this paper for the transportation field.

3   The R4R Method

3.1   Problem Definition

Consider a dataset $D = \langle \mathbf{X}, \mathbf{L}, \mathbf{r} \rangle$. Note that $\mathbf{X}$ in this tuple denotes the **set of predictive attributes** and is a matrix of size $N \times Q$ representing a set of N number of instances each composed of Q number of predictive attributes. Furthermore, $\mathbf{L}$ is the **set of the many component targets** and is a matrix of size $N \times K$ with K being the number of elements of the many component target. $\mathbf{r}$ representes **the set of one component target** and is a vector of length N. Elements of $\{r_n \in \mathbf{r}\}$ represent the target attributes of the one component and each $\{l_{n,k} \in \mathbf{L}\}$ is the kth target attribute of the many component. Also, consider $r_n = \sum_{k=1}^{K} l_{n,k}$ denoting the sum of all the many component targets being equal to a corresponding one component target.

Defining the prediction of each $l_{n,k}$ as $p_{n,k}$, we are looking for a model that ensures that the sum of the predictions of the many component target are as close as possible to the r_n. In other words, after making predictions, we want the following equation to hold:

$$\{\sum_{k=1}^{K} p_{n,k} \approx r_n | n \leq N\} \tag{1}$$

3.2 Methodology

In this section, we elaborate on our proposed method, Reconciling for Regression (R4R), to address the above-mentioned problem. R4R method is composed of two steps. In the first step, it learns models for prediction of the **many component targets**, separately. In the second step, it reconciles the many predictions with the one component.

In order to improve the individual $p_{n,k}$ predictions such that Eq. 1 holds, our proposed framework uses a modified version of the least squares optimization method to compute a set of corrective coefficients (see Eq. 4), that are used to update the individual $p_{n,k}$ predictions.

Step 1, Learning the Predictive Models: at the first step, the predictions of the many component targets are calculated using a specific base learning method. K different models are trained, one for each of the K elements of the many target component. It is possible to select a different learning method for each element to ensure higher accuracy. The resulting predictions for each of the K elements are referred to as $p_{m,k}$, where m is the instance number, and k identifies elements of the many component targets. Algorithm 1 depicts these steps. As a result, this algorithm creates an output $\mathbf{P}$, a matrix of size $M \times K$ composed of predictions $p_{m,k}$. $\mathbf{P}$ is used in the second stage for reconciliation.

Algorithm 1. Learning the predictive model

Input: $\mathbf{D}$ (dataset matrix of size $N \times (Q + K)$), Me (base learning method),γ (a percentage value)
Output: $\mathbf{P}$ (Predictions matrix of size $M \times K$)
1: Split Dataset D into $Train$ set of size $(1 - \gamma)N$ and $Test$ set of size $M = \gamma \times N$;
2: **for** $k = 1$ to K **do**
3:    Train $model_k$ using Me to predict the kth element of the many component target;
4:    **for** $m = 1$ to M **do**
5:       $p_{m,k} :=$ Predict the value of mth instance of kth target in $Test$ using $model_k$; $//$ $p_{m,k}$
         denotes elements of $\mathbf{P}$
6:    **end for**
7: **end for**
8: **return** $\mathbf{P}$;

Step 2, Reconciling Predictions: In the second step, the framework updates the value of predictions resulted from the initial models used in Algorithm 1. This is achieved by estimating a corrective coefficient (θ_k) for each element of the many target component ($p_{m,k}$). This coefficient needs to be multiplied with the model predictions to ensure minimized error from the actual one component target (r_m) and many component target

$(l_{m,k})$. We achieve this goal using a least-squares method on the current training dataset and using the objective functions given by Eqs. 2 and 3 to estimate $\theta = (\theta_1, ..., \theta_K)$.

$$\underset{lb<\theta<ub}{\arg\min}(\sum_{k=1}^{K}(\theta_k p_{m,k}) - r_m)^2, m \leq M \tag{2}$$

$$\underset{lb<\theta<ub}{\arg\min}\sum_{k=1}^{K}(\theta_k p_{m,k} - l_{m,k})^2, m \leq M \tag{3}$$

The first objective function presented in Eq. 2 is attempting to optimize reconciliation based on the value of one component target. The second objective function presented in Eq. 3 aims at minimizing the error of the predictions based on the value of each element of the many component targets, separately. Both of these objective functions can be combined and expanded to Eq. 4. In Eq. 4, the first M rows are representing the objective function presented in Eq. 2. The remaining $M \times (KM)$ rows represent the second objective function as provided in Eq. 3.

$$
\begin{bmatrix}
p_{1,1} & p_{1,2} & \cdots & p_{1,k} & \cdots & p_{1,K} \\
p_{2,1} & p_{2,2} & \cdots & p_{2,k} & \cdots & p_{2,K} \\
\vdots & & & \vdots & & \vdots \\
p_{m,1} & p_{m,2} & \cdots & p_{m,k} & \cdots & p_{m,K} \\
p_{1,1} & 0 & \cdots & 0 & \cdots & 0 \\
0 & p_{1,2} & \cdots & 0 & \cdots & 0 \\
\vdots & & \ddots & \vdots & \ddots & \vdots \\
0 & 0 & \cdots & 0 & \cdots & p_{1,K} \\
\vdots & & \ddots & \vdots & \ddots & \vdots \\
p_{m,1} & 0 & \cdots & 0 & \cdots & 0 \\
0 & p_{m,2} & \cdots & 0 & \cdots & 0 \\
\vdots & & \ddots & \vdots & \ddots & \vdots \\
0 & 0 & \cdots & p_{m,k} & \cdots & 0 \\
\vdots & & \ddots & \vdots & \ddots & \vdots \\
0 & 0 & \cdots & 0 & \cdots & p_{m,K}
\end{bmatrix}
\begin{bmatrix}
\theta_1 \\ \theta_2 \\ \vdots \\ \theta_K
\end{bmatrix}
=
\begin{bmatrix}
r_1 \\ r_2 \\ \vdots \\ r_m \\ l_{1,1} \\ l_{1,2} \\ \vdots \\ l_{1,K} \\ \vdots \\ l_{m,1} \\ l_{m,2} \\ \vdots \\ l_{m,k} \\ \vdots \\ l_{m,K}
\end{bmatrix}
\tag{4}
$$

As seen in Eqs. 2 and 3 we have defined a constraint on the values of θ. The aim is to regularize the modifications to the predictions done for each element of the many component targets in a sensible manner (e.g. negative factors cannot be allowed when negative predictions are not meaningful). Therefore, we assume, without loss of generality, that all values of θ are positive, with lower (lb) and upper (ub) bound constraints, $0 < lb < \theta_k < ub$. Both lb and ub are free input parameters. We reduce the number of free parameters to one (α) by defining a symmetric bound region as $(lb, up) = (1 - \alpha, 1 + \alpha)$.

The process of reconciliation on predictions is explained in Algorithm 2. In the final step of this algorithm, the prediction matrix for all elements of the many component targets is updated using the corrective coefficients θ. A Least Squares method is used to calculate corrective coefficients. To allow robustness against outliers, we suggest using a nk number of nearest neighbors for estimating θ. We assume that similar trips from the past have the same behavior, as shown in [12]. The new predictions are defined as $\mathbf{P}_{new}$. The algorithm takes into account the information of the predictions for both the

---

Algorithm 2. Reconciling predictions

Input: $\mathbf{P}$ (Predictions matrix of size $M \times K$), nk (number of nearest neighbors), lb, up (lower and upper bounds for θ_ks)

Output: $\mathbf{P}_{new}$ (new predictions matrix of size $M \times K$), $\boldsymbol{\theta}$ (vector of corrective coefficients)

1: **for** $k = 1$ to K **do**
2:    get nk nearest neighbor for each prediction;
3:    Calculate θ using the Least Squares method with Bounds (lb,up) according to Eq. 4;
4:    $\mathbf{P}_{new} = \mathbf{P} \cdot \theta$
5: **end for**
6: **return** $\mathbf{P}_{new}, \theta$

---

many component elements and the one component predicted from similar instances in $\mathbf{P}_{m,k}$, in order to verify Eq. 1 on reconciliation.

Table 1. Characteristics of tested STCP bus routes

Bus line	Origin – Destiny	#Stops	#Trips
L200	Bolhão – Castelo do Queijo	30	2526
L201	Viso – Aliados	26	2453
L305	Cordoaria – Hospital S. João	22	3126
L401	Bolhão – S. Roque	26	4476
L502	Bolhão – Matosinhos	32	5966
L900	Trindade – S. Odivio	34	219

4 Case Study

To test the methodology explained in Sect. 3.2, we conduct a series of experiments using a real dataset that has our desired hierarchical organization of target values. Measuring travel time in public transport systems can produce such a dataset. Being able to perform accurate Travel Time Prediction (TTP) is an important goal for public transport companies. On the one hand, *travel time prediction of the link between two consecutive stops* (the many component targets in our model) allows timely informing the roadside users about the arrival of buses at bus stops (in the rest of this paper we refer to this value as link TTP). On the other hand, *total trip travel time prediction* (the one component in our model) is useful to better schedule drivers' duty services (in the rest of this paper, we refer to this value as total TTP) [4].

The dataset used in this section is provided by the Sociedade de Transportes Colectivos do Porto (STCP), the main mass public transportation company in Porto, Portugal.

The experiments described in the following sections are based on the data collected during a period from January 1st to March 30th of 2010 from six bus routes (shown in Table 1). All the six selected bus routes operate between 5:30 a.m. to 2:00 a.m. However, we have considered only bus trips starting after 6 a.m.

The collected dataset has multiple nominal and ordinal attributes that make it suitable for defining a regression problem. We have selected five features that characterize each bus trip: (1) WEEKDAY: the day of the week {Monday, Tuesday, Wednesday, Thursday, Friday, Saturday, Sunday}; (2) DAYTYPE: the type of the day {holiday, normal, non-working day, weekend holiday}; (3) Bus Day Month: {1,...,31}; (4) Shift ID; (5) Travel ID.

We have implemented R4R using the R Software [15] and the *lsq_linear* routine from *Scipy* Python library [10]. For the first stage of R4R, as depicted in Algorithm 1, we use a simple multivariate linear regression as a base learning method. We refer to this base learning method as (Bas). We further split data according to the following format. A 30 days window length is used for selecting training samples, and a 60 days window length is considered for selecting test samples.

In our experiments, the parameter α used for determining the lower and upper bound for the parameter for estimating θ varies from 0.01 to 0.04, which corresponds to 0.96–1.04, minimum and maximum values that θ can take, respectively.

5 Comparative Study

5.1 Can Reconciliation Be Achieved Using R4R?

Firstly, using the proposed R4R method, we try to answer the following question: is it possible to use the total trip travel time to improve the link TTPs guaranteeing a better reconciliation between the sum of the link TTPs and the total TTP simultaneously? To answer this question we measure the relative performance improvement achieved by R4R compared to a multivariate linear regression as the base learning method (denoted by Bas).

We evaluate the performance in predicting the following metrics (i) link travel time prediction (LP), the sum of link travel time predictions (SFP), and full trip time prediction (FP). Methods are compared based on Root Mean Square Error (RMSE) as defined in Eq. 5.

$$RMSE = \sqrt{\frac{1}{N_{test}} \sum_{i=1}^{N_{test}} (\hat{y}_i - y_i)^2} \qquad (5)$$

where y_i and $\hat{y}_i$ represent the target and predicted bus arrival times, for the ith example in the test set, respectively. N_{test} is the total number of test samples. For link travel time prediction indicator, LP, the mean of the RMSE of each bus link is considered.

Results of the comparison of R4R and Bas are presented in Fig. 1. Please note that relative gains are presented for the sake of readability of graphs. The duration of travel-times varies widely. This fact leads to unreadable graphs when actual data is presented.

As seem, R4R outperforms the base multivariate regression model in all cases. This comparison answers the question posed earlier. R4R improves predictions of the base regression learning method, guaranteeing a better reconciliation between the sum of the link TTPs and the total trip travel time, simultaneously.

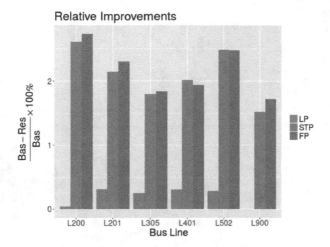

Fig. 1. Relative improvement of R4R (Res) relative to Baseline (Bas) for mean LP - Link Prediction (red), sum of the link travel time predictions (green) and the full trip time prediction (blue). (Color figure online)

5.2 How Does R4R Perform Against Baselines Made for Time Series Data?

We continue our experiments by comparing our proposed methodology R4R with the methods proposed by Hyndman et al. in the recent related works [8,9,16] denoted by (H2011, W2015, and H2016). To compare with these works, we used the available implementation in the R package [7]. It should be considered that these baseline models are designed for time-series data. Therefore, in order to perform comparisons with these approaches, we also define a time series problem using this dataset. This is achieved by representing data in the form of a time series with a resolution of a one-hour interval. We compute the mean link travel time for each hour between 6:00 a.m. to 2:00 a.m the next day, i.e. 20 data points in total for each "bus day". In the majority of the cases, each interval has more than one link travel time. For this reason we averaged the link travel times for each hour. Because the dataset has a considerable amount of missing values, interpolation was used to fill the missing links' travel times. However, the results presented in the paper do not take into account the predictions done for intervals with no data.

The above-mentioned pre-processing tasks that were necessary in order to use the approaches proposed by Hyndman et al. already suggest that it is viable to propose methods such as R4R that perform in a more general and flexible regression setting. Indeed, the discretization of data into a time-series format implies the need to make predictions for intervals instead of point-wise predictions as done in the regression setting. Discretization also implies the necessity of filling missing data when the intervals have no data instances. This problem can be prevented by considering larger intervals. However, larger intervals imply loss of details. Moreover, the regression setting deals naturally with additional attributes that can partially explain the value of the target attribute.

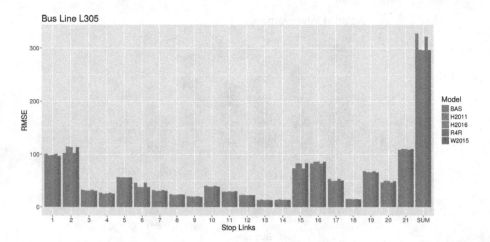

Fig. 2. RMSE for each of the Link Travel Time Predictions of R4R against the methods proposed in H2011 [8], W2015 [16], H2016 [9] applied to bus route L305. SUM is the RMSE of the sum of the LTT prediction for the entire trip against the full trip time. This plot shows the results before the bus starts its journey.

Table 2. Overall mean RMSE for each model, H2011 [8], W2015 [16], H2016 [9] and the new proposed approach R4R. BL - the Bus Line, LP - mean of the RMSE of Link Predictions, STP - RMSE of the sum of the LTT prediction for the entire trip against the full trip time.

BL	MODEL	LP	STP	BL	MODEL	LP	STP	BL	MODEL	LP	STP
L200	BAS	277.88	318.11	L201	BAS	41.56	354.11	L305	BAS	48.37	327.69
L200	R4R	277.76	**309.82**	L201	R4R	**41.43**	346.52	L305	R4R	**48.25**	321.81
L200	H2011	51.50	865.38	L201	H2011	42.01	321.80	L305	H2011	48.59	297.27
L200	H2016	44.40	496.71	L201	H2016	41.69	314.09	L305	H2016	48.49	**295.97**
L200	W2015	**37.38**	319.22	L201	W2015	41.64	**308.85**	L305	W2015	48.40	296.41
L401	BAS	29.26	239.11	L502	BAS	42.62	385.34	L900	BAS	58.60	401.89
L401	R4R	29.17	234.29	L502	R4R	**42.50**	375.75	L900	R4R	58.60	**395.79**
L401	H2011	26.87	193.29	L502	H2011	47.27	**264.14**	L900	H2011	48.20	432.25
L401	H2016	26.80	192.65	L502	H2016	46.69	270.02	L900	H2016	48.24	420.58
L401	W2015	**26.76**	**192.38**	L502	W2015	46.82	287.71	L900	W2015	**48.08**	403.34

Figure 2 presents the results of predictions for bus route L305. It should be mentioned that we have chosen to show only results for $\alpha = 0.01$, the parameter that consistently gave us the best performance in all the experiments we did. Indeed, the errors increase with increasing values of α in all experiments we did. The results show very small differences between the methods under study.

The data provided is not homogeneous. This can adversely affect the performance of the least-squares method when outlying data is used to find the corrective coefficients θ. To avoid such problems, in our proposed framework, we select the nk number of nearest neighbors for each bus trip (also presented in Algorithm 2). Thus, after each link travel time prediction, it is necessary to recompute the whole process, i.e., to select a new set of similar bus trips and further find the coefficients using the least-squares method and update the predictions. Comparing with Hyndman et al. works, this process leads to a more computationally expensive solution. It is also important to find a suitable value for nk. During our experiments, we observed that the best results are achieved for $nk = 3$. Therefore, all results presented in this paper are based on $nk = 3$.

Table 2 shows the general results of predictions using this approach for all bus routes tested using multivariate linear regression as the base learning method (Bas). The results show that R4R outperforms Bas in all cases. There are a number of cases where a version of the time series model proposed by Hyndman. et al. perform better than R4R. These differences can be explained when considering the simple linear regression algorithm we used as a base learner in Algorithm 1. A linear model cannot find non-linear relations between features. Technically, the performance of R4R can be improved further as it allows using any other regression method. Furthermore, using extra features, such as weather conditions, could possibly improve the performance of R4R even further. However, the methods proposed by Hyndman et al. cannot benefit from using extra features.

6   Conclusion

In this paper, we study the problem of the reconciliation of predictions in a regression setting. We presented a two-stage prediction framework for prediction and reconciliation. In order to evaluate the performance and applicability of this method, we conduct a set of experiments using a real dataset collected from buses in Porto, Portugal. The results demonstrate that R4R improves the predictions of the base learning method. R4R is also able to further improve the reconciliation of the link TTPs after each iteration in an online manner. However, this is not shown due to space constraints. We also compare the results achieved in a regression setting with that of a time-series approach. In the case study discussed in this paper, R4R is able to reduce the error of link TTPs and increase reconciliation. An important advantage of the R4R method compared to time series variants is that it provides a flexible framework that can take advantage of any regression model and additional features accompanying data. Furthermore, R4R is not affected by data imperfection problems such as missing data, that reduce the applicability of time-series models that require equally distanced samples.

Acknowledgments. This work is financed by the Portuguese funding agency, FCT - Fundação para a Ciência e a Tecnologia, through national funds, and co-funded by the FEDER, where applicable.

We also thank STCP - Sociedade de Transportes Colectivos do Porto, SA, for providing the data used in this work.

References

1. Amita, J., Jain, S., Garg, P.: Prediction of bus travel time using ann: a case study in Delhi. Transp. Res. Procedia **17**, 263–272 (2016). International Conference on Transportation Planning and Implementation Methodologies for Developing Countries (12th TPMDC) Selected Proceedings, IIT Bombay, Mumbai, India, 10–12 December 2014

2. Bhanu, M., Priya, S., Dandapat, S.K., Chandra, J., Mendes-Moreira, J.: Forecasting traffic flow in big cities using modified tucker decomposition. In: Gan, G., Li, B., Li, X., Wang, S. (eds.) ADMA 2018. LNCS (LNAI), vol. 11323, pp. 119–128. Springer, Cham (2018). https://doi.org/10.1007/978-3-030-05090-0_10
3. Borges, C.E., Penya, Y.K., Fernandez, I.: Evaluating combined load forecasting in large power systems and smart grids. IEEE Trans. Ind. Inf. **9**(3), 1570–1577 (2013)
4. Chen, G., Yang, X., An, J., Zhang, D.: Bus-arrival-time prediction models: link-based and section-based. J. Transp. Eng. **138**(1), 60–66 (2011)
5. De Leeuw, J., Meijer, E., Goldstein, H.: Handbook of Multilevel Analysis. Springer, New York (2008). https://doi.org/10.1007/978-0-387-73186-5
6. van Erven, T., Cugliari, J.: Game-theoretically optimal reconciliation of contemporaneous hierarchical time series forecasts. In: Antoniadis, A., Poggi, J.-M., Brossat, X. (eds.) Modeling and Stochastic Learning for Forecasting in High Dimensions. LNS, vol. 217, pp. 297–317. Springer, Cham (2015). https://doi.org/10.1007/978-3-319-18732-7_15
7. Hyndman, R., Lee, A., Wang, E.: hts: Hierarchical and Grouped Time Series (2017). https://CRAN.R-project.org/package=hts, r package version 5.1.4
8. Hyndman, R.J., Ahmed, R.A., Athanasopoulos, G., Shang, H.L.: Optimal combination forecasts for hierarchical time series. Comput. Stat. Data Anal. **55**(9), 2579–2589 (2011)
9. Hyndman, R.J., Lee, A.J., Wang, E.: Fast computation of reconciled forecasts for hierarchical and grouped time series. Comput. Stat. Data Anal. **97**, 16–32 (2016)
10. Jones, E., Oliphant, T., Peterson, P., et al.: SciPy: open source scientific tools for Python (2001). http://www.scipy.org/. Accessed 10 Jan 2018
11. Kocev, D., Džeroski, S., White, M.D., Newell, G.R., Griffioen, P.: Using single-and multi-target regression trees and ensembles to model a compound index of vegetation condition. Ecol. Model. **220**(8), 1159–1168 (2009)
12. Mendes-Moreira, J.: Travel time prediction for the planning of mass transit companies: a machine learning approach. University of Porto, Porto, Portugal, phD thesis (2008)
13. Mendes-Moreira, J., Jorge, A.M., Freire de Sousa, J., Soares, C.: Comparing state-of-the-art regression methods for long term travel time prediction. Intell. Data Anal. **16**(3), 427–449 (2012)
14. Moreira-Matias, L., Gama, J., Mendes-Moreira, J., Freire de Sousa, J.: An incremental probabilistic model to predict bus bunching in real-time. In: Blockeel, H., van Leeuwen, M., Vinciotti, V. (eds.) IDA 2014. LNCS, vol. 8819, pp. 227–238. Springer, Cham (2014). https://doi.org/10.1007/978-3-319-12571-8_20
15. R Development Core Team: R: A Language and Environment for Statistical Computing. R Foundation for Statistical Computing, Vienna, Austria (2008). http://www.R-project.org, ISBN 3-900051-07-0
16. Wickramasuriya, S.L., Athanasopoulos, G., Hyndman, R.: Forecasting hierarchical and grouped time series through trace minimization. Monash Econometrics and Business Statistics Working Papers 15/15, Monash University, Department of Econometrics and Business Statistics (2015)
17. Wunderlich, K.E., Kaufman, D.E., Smith, R.L.: Link travel time prediction for decentralized route guidance architectures. IEEE Trans. Intell. Transp. Syst. **1**(1), 4–14 (2000)
18. Yu, B., Yang, Z.Z., Chen, K., Yu, B.: Hybrid model for prediction of bus arrival times at next station. J. Adv. Transp. **44**(3), 193–204 (2010)
19. Zhang, X., Rice, J.A.: Short-term travel time prediction. Transp. Res. Part C Emerg. Technol. **11**(3), 187–210 (2003). Traffic Detection and Estimation

A Distribution Dependent and Independent Complexity Analysis of Manifold Regularization

Alexander Mey[1]([✉]) [ID], Tom Julian Viering[1] [ID], and Marco Loog[1,2] [ID]

[1] Delft University of Technology, Delft, The Netherlands
{a.mey,t.j.viering}@tudelft.nl
[2] University of Copenhagen, Copenhagen, Denmark
m.loog@tudelft.nl

Abstract. Manifold regularization is a commonly used technique in semi-supervised learning. It enforces the classification rule to be smooth with respect to the data-manifold. Here, we derive sample complexity bounds based on pseudo-dimension for models that add a convex data dependent regularization term to a supervised learning process, as is in particular done in Manifold regularization. We then compare the bound for those semi-supervised methods to purely supervised methods, and discuss a setting in which the semi-supervised method can only have a constant improvement, ignoring logarithmic terms. By viewing Manifold regularization as a kernel method we then derive Rademacher bounds which allow for a distribution *dependent* analysis. Finally we illustrate that these bounds may be useful for choosing an appropriate manifold regularization parameter in situations with very sparsely labeled data.

Keywords: Semi-supervised learning · Learning theory · Manifold regularization

1 Introduction

In many applications, as for example image or text classification, gathering unlabeled data is easier than gathering labeled data. Semi-supervised methods try to extract information from the unlabeled data to get improved classification results over purely supervised methods. A well-known technique to incorporate unlabeled data into a learning process is manifold regularization (MR) [7,18]. This procedure adds a data-dependent penalty term to the loss function that penalizes classification rules that behave non-smooth with respect to the data distribution. This paper presents a sample complexity and a Rademacher complexity analysis for this procedure. In addition it illustrates how our Rademacher complexity bounds may be used for choosing a suitable Manifold regularization parameter.

We organize this paper as follows. In Sects. 2 and 3 we discuss related work and introduce the semi-supervised setting. In Sect. 4 we formalize the idea of

M. R. Berthold et al. (Eds.): IDA 2020, LNCS 12080, pp. 326–338, 2020.
https://doi.org/10.1007/978-3-030-44584-3_26

adding a distribution-dependent penalty term to a loss function. Algorithms such as manifold, entropy or co-regularization [7,14,21] follow this idea. Section 5 generalizes a bound from [4] to derive sample complexity bounds for the proposed framework, and thus in particular for MR. For the specific case of regression, we furthermore adapt a sample complexity bound from [1], which is essentially tighter than the first bound, to the semi-supervised case. In the same section we sketch a setting in which we show that if our hypothesis set has finite pseudo-dimension, and we ignore logarithmic factors, any semi-supervised learner (SSL) that falls in our framework has at most a constant improvement in terms of sample complexity. In Sect. 6 we show how one can obtain distribution *dependent* complexity bounds for MR. We review a kernel formulation of MR [20] and show how this can be used to estimate Rademacher complexities for *specific* datasets. In Sect. 7 we illustrate on an artificial dataset how the distribution dependent bounds could be used for choosing the regularization parameter of MR. This is particularly useful as the analysis does not need an additional labeled validation set. The practicality of this approach requires further empirical investigation. In Sect. 8 we discuss our results and speculate about possible extensions.

2   Related Work

In [13] we find an investigation of a setting where distributions on the input space $\mathcal{X}$ are restricted to ones that correspond to unions of irreducible algebraic sets of a fixed size $k \in \mathbb{N}$, and each algebraic set is either labeled 0 or 1. A SSL that knows the true distribution on $\mathcal{X}$ can identify the algebraic sets and reduce the hypothesis space to all 2^k possible label combinations on those sets. As we are left with finitely many hypotheses we can learn them efficiently, while they show that every supervised learner is left with a hypothesis space of infinite VC dimension.

The work in [18] considers manifolds that arise as embeddings from a circle, where the labeling over the circle is (up to the decision boundary) smooth. They then show that a learner that has knowledge of the manifold can learn efficiently while for every fully supervised learner one can find an embedding and a distribution for which this is not possible.

The relation to our paper is as follows. They provide specific examples where the sample complexity between a semi-supervised and a supervised learner are infinitely large, while we explore general sample complexity bounds of MR and sketch a setting in which MR can not essentially improve over supervised methods.

3   The Semi-supervised Setting

We work in the statistical learning framework: we assume we are given a feature domain $\mathcal{X}$ and an output space $\mathcal{Y}$ together with an unknown probability distribution P over $\mathcal{X} \times \mathcal{Y}$. In binary classification we usually have that $\mathcal{Y} = \{-1, 1\}$, while for regression $\mathcal{Y} = \mathbb{R}$. We use a loss function $\phi : \mathbb{R} \times \mathcal{Y} \to \mathbb{R}$, which is convex in the first argument and in practice usually a surrogate for the 0–1 loss in classification, and the squared loss in regression tasks. A hypothesis f is a function $f : \mathcal{X} \to \mathbb{R}$. We set (X, Y) to be a random variable distributed according to P, while small x and y are elements of $\mathcal{X}$ and $\mathcal{Y}$ respectively. Our goal is to find a hypothesis f, within a restricted class $\mathcal{F}$, such that the expected loss $Q(f) := \mathbb{E}[\phi(f(X), Y)]$ is small. In the standard supervised setting we choose a hypothesis f based on an i.i.d. sample $S_n = \{(x_i, y_i)\}_{i \in \{1,..,n\}}$ drawn from P. With that we define the empirical risk of a model $f \in \mathcal{F}$ with respect to ϕ and measured on the sample S_n as $\hat{Q}(f, S_n) = \frac{1}{n} \sum_{i=1}^{n} \phi(f(x_i), y_i)$. For ease of notation we sometimes omit S_n and just write $\hat{Q}(f)$. Given a learning problem defined by $(P, \mathcal{F}, \phi)$ and a labeled sample S_n, one way to choose a hypothesis is by the empirical risk minimization principle

$$f_{\text{sup}} = \arg \min_{f \in \mathcal{F}} \hat{Q}(f, S_n). \tag{1}$$

We refer to f_{sup} as the *supervised solution*. In SSL we additionally have samples with unknown labels. So we assume to have $n + m$ samples $(x_i, y_i)_{i \in \{1,..,n+m\}}$ independently drawn according to P, where y_i has not been observed for the last m samples. We furthermore set $U = \{x_1, ..., x_{x_n+m}\}$, so U is the set that contains all our available information about the feature distribution.

Finally we denote by $m^L(\epsilon, \delta)$ the sample complexity of an algorithm L. That means that for all $n \geq m^L(\epsilon, \delta)$ and all possible distributions P the following holds. If L outputs a hypothesis f_L after seeing an n-sample, we have with probability of at least $1 - \delta$ over the n-sample S_n that $Q(f_L) - \min_{f \in \mathcal{F}} Q(f) \leq \epsilon$.

4   A Framework for Semi-supervised Learning

We follow the work of [4] and introduce a second convex loss function $\psi : \mathcal{F} \times \mathcal{X} \to \mathbb{R}_+$ that only depends on the input feature and a hypothesis. We refer to ψ as the *unsupervised loss* as it does not depend on any labels. We propose to *add* the unlabeled data through the loss function ψ and add it as a penalty term to the supervised loss to obtain the semi-supervised solution

$$f_{\text{semi}} = \arg \min_{f \in \mathcal{F}} \frac{1}{n} \sum_{i=1}^{n} \phi(f(x_i), y_i) + \lambda \frac{1}{n+m} \sum_{j=1}^{n+m} \psi(f, x_j), \tag{2}$$

where $\lambda > 0$ controls the trade-off between the supervised and the unsupervised loss. This is in contrast to [4], as they use the unsupervised loss to restrict the hypothesis space directly. In the following section we recall the important insight that those two formulations are equivalent in some scenarios and we can use [4] to generate sample complexity bounds for the here presented SSL framework.

For ease of notation we set $\hat{R}(f, U) = \frac{1}{n+m} \sum_{j=1}^{n+m} \psi(f, x_j)$ and $R(f) = \mathbb{E}[\psi(f, X)]$. We do not claim any novelty for the idea of adding an unsupervised loss for regularization. A different framework can be found in [11, Chapter 10]. We are, however, not aware of a deeper analysis of this particular formulation, as done for example by the sample complexity analysis in this paper. As we are in particular interested in the class of MR schemes we first show that this method fits our framework.

Example: Manifold Regularization. Overloading the notation we write now $P(X)$ for the distribution P restricted to $\mathcal{X}$. In MR one assumes that the input distribution $P(X)$ has support on a compact manifold $M \subset \mathcal{X}$ and that the predictor $f \in \mathcal{F}$ varies smoothly in the geometry of M [7]. There are several regularization terms that can enforce this smoothness, one of which is $\int_M \|\nabla_M f(x)\|^2 dP(x)$, where $\nabla_M f$ is the gradient of f along M. We know that $\int_M \|\nabla_M f(x)\|^2 dP(x)$ may be approximated with a finite sample of $\mathcal{X}$ drawn from $P(X)$ [6]. Given such a sample $U = \{x_1, ..., x_{n+m}\}$ one defines first a weight matrix W, where $W_{ij} = e^{-\|x_i - x_j\|^2/\sigma}$. We set L then as the Laplacian matrix $L = D - W$, where D is a diagonal matrix with $D_{ii} = \sum_{j=1}^{n+m} W_{ij}$. Let furthermore $f_U = (f(x_1), ..., f(x_{n+m}))^t$ be the evaluation vector of f on U. The expression $\frac{1}{(n+m)^2} f_U^t L f_U = \frac{1}{(n+m)^2} \sum_{i,j} (f(x_i) - f(x_j))^2 W_{ij}$ converges to $\int_M \|\nabla_M f\|^2 dP(x)$ under certain conditions [6]. This motivates us to set the unsupervised loss as $\psi(f, (x_i, x_j)) = (f(x_i) - f(x_j))^2 W_{ij}$. Note that $f_U^t L f_U$ is indeed a convex function in f: As L is a Laplacian matrix it is positive definite and thus $f_U^t L f_U$ defines a norm in f. Convexity follows then from the triangle inequality.

5   Analysis of the Framework

In this section we analyze the properties of the solution f_{semi} found in Equation (2). We derive sample complexity bounds for this procedure, using results from [4], and compare them to sample complexities for the supervised case. In [4] the unsupervised loss is used to restrict the hypothesis space directly, while we use it as a regularization term in the empirical risk minimization as usually done in practice. To switch between the views of a constrained optimization formulation and our formulation (2) we use the following classical result from convex optimization [15, Theorem 1].

Lemma 1. *Let $\phi(f(x), y)$ and $\psi(f, x)$ be functions convex in f for all x, y. Then the following two optimization problems are equivalent:*

$$\min_{f \in \mathcal{F}} \frac{1}{n} \sum_{i=1}^{n} \phi(f(x_i), y_i) + \lambda \frac{1}{n+m} \sum_{i=1}^{n+m} \psi(f, x_i) \tag{3}$$

$$\min_{f \in \mathcal{F}} \frac{1}{n} \sum_{i=1}^{n} \phi(f(x_i), y_i) \quad \text{subject to} \quad \sum_{i=1}^{n+m} \frac{1}{n+m} \psi(f, x_i) \leq \tau \tag{4}$$

Where equivalence means that for each λ we can find a τ such that both problems have the same solution and vice versa.

For our later results we will need the conditions of this lemma are true, which we believe to be not a strong restriction. In our sample complexity analysis we stick as close as possible to the actual formulation and implementation of MR, which is usually a convex optimization problem. We first turn to our sample complexity bounds.

5.1   Sample Complexity Bounds

Sample complexity bounds for supervised learning use typically a notion of complexity of the hypothesis space to bound the worst case difference between the estimated and the true risk. As our hypothesis class allows for real-valued functions, we will use the notion of pseudo-dimension $\text{Pdim}(\mathcal{F}, \phi)$, an extension of the VC-dimension to real valued loss functions ϕ and hypotheses classes $\mathcal{F}$ [17,22]. Informally speaking, the pseudo-dimension is the VC-dimension of the set of functions that arise when we threshold real-valued functions to define binary functions. Note that sometimes the pseudo-dimension will have as input the loss function, and sometimes not. This is because some results use the concatenation of loss function and hypotheses to determine the capacity, while others only use the hypotheses class. This lets us state our first main result, which is a generalization of [4, Theorem 10] to bounded loss functions and real valued function spaces.

Theorem 1. *Let $\mathcal{F}_\tau^\psi := \{f \in \mathcal{F} \mid \mathbb{E}[\psi(f,x)] \leq \tau\}$. Assume that ϕ, ψ are measurable loss functions such that there exists constants $B_1, B_2 > 0$ with $\psi(f,x) \leq B_1$ and $\phi(f(x), y) \leq B_2$ for all x, y and $f \in \mathcal{F}$ and let P be a distribution. Furthermore let $f_\tau^* = \arg\min_{f \in \mathcal{F}_\tau^\psi} Q(f)$. Then an unlabeled sample U of size*

$$m \geq \frac{8B_1^2}{\epsilon^2} \left[\ln \frac{16}{\delta} + 2\,\text{Pdim}(\mathcal{F}, \psi) \ln \frac{4B_1}{\epsilon} + 1 \right] \tag{5}$$

and a labeled sample S_n of size

$$n \geq \max \left(\frac{8B_2^2}{\epsilon^2} \left[\ln \frac{8}{\delta} + 2\,\text{Pdim}(\mathcal{F}_{\tau+\frac{\epsilon}{2}}^\psi, \phi) \ln \frac{4B_2}{\epsilon} + 1 \right], \frac{h}{4} \right) \tag{6}$$

is sufficient to ensure that with probability at least $1 - \delta$ the classifier $g \in \mathcal{F}$ that minimizes $\hat{Q}(\cdot, S_n)$ subject to $\hat{R}(\cdot, U) \leq \tau + \frac{\epsilon}{2}$ satisfies

$$Q(g) \leq Q(f_\tau^*) + \epsilon. \tag{7}$$

Sketch Proof: The idea is to combine three partial results with a union bound. For the first part we use Theorem 5.1 from [22] with $h = \text{Pdim}(\mathcal{F}, \psi)$ to show that an unlabeled sample size of

$$m \geq \frac{8B_1^2}{\epsilon^2} \left[\ln \frac{16}{\delta} + 2h \ln \frac{4B_1}{\epsilon} + 1 \right] \tag{8}$$

is sufficient to guarantee $\hat{R}(f) - R(f) < \frac{\epsilon}{2}$ for all $f \in \mathcal{F}$ with probability at least $1 - \frac{\delta}{4}$. In particular choosing $f = f_\tau^*$ and noting that by definition $R(f_\tau^*) \leq \tau$ we conclude that with the same probability

$$\hat{R}(f_\tau^*) \leq \tau + \frac{\epsilon}{2}. \tag{9}$$

For the second part we use Hoeffding's inequality to show that the labeled sample size is big enough that with probability at least $1 - \frac{\delta}{4}$ it holds that

$$\hat{Q}(f_\tau^*) \leq Q(f_\tau^*) + B_2 \sqrt{\ln(\frac{4}{\delta}) \frac{1}{2n}}. \tag{10}$$

The third part again uses Th. 5.1 from [22] with $h = \text{Pdim}(\mathcal{F}_\tau^\psi, \phi)$ to show that $n \geq \frac{8B_2^2}{\epsilon^2} \left[\ln \frac{8}{\delta} + 2h \ln \frac{4B_2}{\epsilon} + 1\right]$ is sufficient to guarantee $Q(f) \leq \hat{Q}(f) + \frac{\epsilon}{2}$ with probability at least $1 - \frac{\delta}{2}$.

Putting everything together with the union bound we get that with probability $1 - \delta$ the classifier g that minimizes $\hat{Q}(\cdot, X, Y)$ subject to $\hat{R}(\cdot, U) \leq \tau + \frac{\epsilon}{2}$ satisfies

$$Q(g) \leq \hat{Q}(g) + \frac{\epsilon}{2} \leq \hat{Q}(f_\tau^*) + \frac{\epsilon}{2} \leq Q(f_\tau^*) + \frac{\epsilon}{2} + B_2 \sqrt{\frac{\ln(\frac{4}{\delta})}{2n}}. \tag{11}$$

Finally the labeled sample size is big enough to bound the last rhs term by $\frac{\epsilon}{2}$. $\square$

The next subsection uses this theorem to derive sample complexity bounds for MR. First, however, a remark about the assumption that the loss function ϕ is globally bounded. If we assume that $\mathcal{F}$ is a reproducing kernel Hilbert space there exists an $M > 0$ such that for all $f \in \mathcal{F}$ and $x \in \mathcal{X}$ it holds that $|f(x)| \leq M\|f\|_\mathcal{F}$. If we restrict the norm of f by introducing a regularization term with respect to the norm $\|.\|_\mathcal{F}$, we know that the image of $\mathcal{F}$ is globally bounded. If the image is also closed it will be compact, and thus ϕ will be globally bounded in many cases, as most loss functions are continuous. This can also be seen as a justification to also use an intrinsic regularization for the norm of f in addition to the regularization by the unsupervised loss, as only then the guarantees of Theorem 1 apply. Using this bound together with Lemma 1 we can state the following corollary to give a PAC-style guarantee for our proposed framework.

Corollary 1. *Let ϕ and ψ be convex supervised and an unsupervised loss function that fulfill the assumptions of Theorem 1. Then f_{semi} (2) satisfies the guarantees given in Theorem 1, when we replace for it g in Inequality (7).*

Recall that in the MR setting $\hat{R}(f) = \frac{1}{(n+m)^2} \sum_{i=1}^{n+m} W_{ij}(f(x_i) - f(x_j))^2$. So we gather unlabeled samples from $\mathcal{X} \times \mathcal{X}$ instead of $\mathcal{X}$. Collecting m samples from $\mathcal{X}$ equates $m^2 - 1$ samples from $\mathcal{X} \times \mathcal{X}$ and thus we only need $\sqrt{m}$ instead of m unlabeled samples for the same bound.

5.2   Comparison to the Supervised Solution

In the SSL community it is well-known that using SSL does not come without a risk [11, Chapter 4]. Thus it is of particular interest how those methods compare to purely supervised schemes. There are, however, many potential supervised methods we can think of. In many works this problem is avoided by comparing to all possible supervised schemes [8,12,13]. The framework introduced in this paper allows for a more fine-grained analysis as the semi-supervision happens on top of an already existing supervised methods. Thus, for our framework, it is natural to compare the sample complexities of f_{sup} with the sample complexity of f_{semi}. To compare the supervised and semi-supervised solution we will restrict ourselves to the square loss. This allows us to draw from [1, Chapter 20], where one can find lower and upper sample complexity bounds for the regression setting. The main insight from [1, Chapter 20] is that the sample complexity depends in this setting on whether the hypothesis class is (closure) convex or not. As we anyway need convexity of the space, which is stronger than closure convexity, to use Lemma 1, we can adapt Theorem 20.7 from [1] to our semi-supervised setting.

Theorem 2. *Assume that $\mathcal{F}^{\psi}_{\tau+\epsilon}$ is a closure convex class with functions mapping to $[0,1]^{1}$, that $\psi(f,x) \leq B_1$ for all $x \in \mathcal{X}$ and $f \in \mathcal{F}$ and that $\phi(f(x),y) = (f(x) - y)^2$. Assume further that there is a $B_2 > 0$ such that $(f(x) - y)^2 < B_2$ almost surely for all $(x,y) \in \mathcal{X} \times \mathcal{Y}$ and $f \in \mathcal{F}^{\psi}_{\tau+\epsilon}$. Then an unlabeled sample size of*

$$m \geq \frac{2B_1^{\,2}}{\epsilon^2}\left[\ln\frac{8}{\delta} + 2\,\mathrm{Pdim}(\mathcal{F},\psi)\ln\frac{2B_1}{\epsilon} + 2\right] \tag{12}$$

and a labeled sample size of

$$n \geq \mathcal{O}\left(\frac{B_2^2}{\epsilon}\left(\mathrm{Pdim}(\mathcal{F}^{\psi}_{\tau+\epsilon})\ln\frac{\sqrt{B_2}}{\epsilon} + \ln\frac{2}{\delta}\right)\right) \tag{13}$$

is sufficient *to guarantee that with probability at least $1 - \delta$ the classifier g that minimizes $\hat{Q}(\cdot)$ w.r.t $\hat{R}(f) \leq \tau + \epsilon$ satisfies*

$$Q(g) \leq \min_{f \in \mathcal{F}^{\psi}_{\tau}} Q(f) + \epsilon. \tag{14}$$

Proof: As in the proof of Theorem 1 the unlabeled sample size is sufficient to guarantee with probability at least $1 - \frac{\delta}{2}$ that $R(f^*_{\tau}) \leq \tau + \epsilon$. The labeled sample size is big enough to guarantee with at least $1 - \frac{\delta}{2}$ that $Q(g) \leq Q(f^*_{\tau+\epsilon}) + \epsilon$ [1, Theorem 20.7]. Using the union bound we have with probability of at least $1 - \delta$ that $Q(g) \leq Q(f^*_{\tau+\epsilon}) + \epsilon \leq Q(f^*_{\tau}) + \epsilon$.   □

Note that the previous theorem of course implies the same learning rate in the supervised case, as the only difference will be the pseudo-dimension term. As in specific scenarios this is also the best possible learning rate, we obtain the following negative result for SSL.

Corollary 2. *Assume that ϕ is the square loss, $\mathcal{F}$ maps to the interval $[0,1]$ and $\mathcal{Y} = [1 - B, B]$ for a $B \geq 2$. If $\mathcal{F}$ and $\mathcal{F}^{\psi}_{\tau}$ are both closure convex, then for sufficiently small $\epsilon, \delta > 0$ it holds that $m^{sup}(\epsilon,\delta) = \tilde{\mathcal{O}}(m^{semi}(\epsilon,\delta))$, where*

---

[1] In the remarks after Theorem 1 we argue that in many cases $|f(x)|$ is bounded, and in those cases we can always map to [0,1] by re-scaling.

$\tilde{O}$ *suppresses logarithmic factors, and* m^{semi}, m^{sup} *denote the sample complexity of the semi-supervised and the supervised learner respectively. In other words, the semi-supervised method can improve the learning rate by at most a constant which may depend on the pseudo-dimensions, ignoring logarithmic factors. Note that this holds in particular for the manifold regularization algorithm.*

Proof: The assumptions made in the theorem allow is to invoke Equation (19.5) from [1] which states that $m^{semi} = \Omega(\frac{1}{\epsilon} + \text{Pdim}(\mathcal{F}_r^\psi))$.[2] Using Inequality (13) as an upper bound for the supervised method and comparing this to Eq. (19.5) from [1] we observe that all differences are either constant or logarithmic in ϵ and δ.   □

5.3   The Limits of Manifold Regularization

We now relate our result to the conjectures published in [19]: A SSL cannot learn faster by more than a constant (which may depend on the hypothesis class $\mathcal{F}$ and the loss ϕ) than the supervised learner. Theorem 1 from [12] showed that this conjecture is true up to a logarithmic factor, much like our result, for classes with finite VC-dimension, and SSL that do *not* make any distributional assumptions. Corollary 2 shows that this statement also holds in some scenarios for all SSL that fall in our proposed framework. This is somewhat surprising, as our result holds explicitly for SSLs that *do* make assumptions about the distribution: MR assumes the labeling function behaves smoothly w.r.t. the underlying manifold.

6   Rademacher Complexity of Manifold Regularization

In order to find out in which scenarios semi-supervised learning can help it is useful to also look at distribution *dependent* complexity measures. For this we derive computational feasible upper and lower bounds on the Rademacher complexity of MR. We first review the work of [20]: they create a kernel such that the inner product in the corresponding kernel Hilbert space contains automatically the regularization term from MR. Having this kernel we can use standard upper and lower bounds of the Rademacher complexity for RKHS, as found for example in [10]. The analysis is thus similar to [21]. They consider a co-regularization setting. In particular [20, p. 1] show the following, here informally stated, theorem.

Theorem 3 ([20, **Propositions 2.1, 2.2**]). *Let H be a RKHS with inner product $\langle \cdot, \cdot \rangle_H$. Let $U = \{x_1, ..., x_{n+m}\}$, $f, g \in H$ and $f_U = (f(x_1), ..., f(x_{n+m}))^t$. Furthermore let $\langle \cdot, \cdot \rangle_{\mathbb{R}^n}$ be any inner product in $\mathbb{R}^n$. Let $\tilde{H}$ be the same space of functions as H, but with a newly defined inner product by $\langle f, g \rangle_{\tilde{H}} = \langle f, g \rangle_H + \langle f_U, g_U \rangle_{\mathbb{R}^n}$. Then $\tilde{H}$ is a RKHS.*

---

[2] Note that the original formulation is in terms of the fat-shattering dimension, but this is always bounded by the pseudo-dimension.

Assume now that L is a positive definite n-dimensional matrix and we set the inner product $\langle f_U, g_U \rangle_{\mathbb{R}^n} = f_U^t L g_U$. By setting L as the Laplacian matrix (Sect. 4) we note that the norm of $\tilde{H}$ automatically regularizes w.r.t. the data manifold given by $\{x_1, ..., x_{n+m}\}$. We furthermore know the exact form of the kernel of $\tilde{H}$.

Theorem 4 ([20, **Proposition 2.2**]). *Let $k(x, y)$ be the kernel of H, K be the gram matrix given by $K_{ij} = k(x_i, x_j)$ and $k_x = (k(x_1, x), ..., k(x_{n+m}, x))^t$. Finally let I be the $n + m$ dimensional identity matrix. The kernel of $\tilde{H}$ is then given by $\tilde{k}(x, y) = k(x, y) - k_x^t (I + LK)^{-1} L k_y$.*

This interpretation of MR is useful to derive computationally feasible upper and lower bounds of the empirical Rademacher complexity, giving distribution *dependent* complexity bounds. With $\sigma = (\sigma_1, ..., \sigma_n)$ i.i.d Rademacher random variables (i.e. $P(\sigma_i = 1) = P(\sigma_i = -1) = \frac{1}{2}$.), recall that the empirical Rademacher complexity of the hypothesis class H and measured on the sample labeled input features $\{x_1, ..., x_n\}$ is defined as

$$\text{Rad}_n(H) = \frac{1}{n} \mathbb{E}_\sigma \sup_{f \in H} \sum_{i=1}^n \sigma_i f(x_i). \tag{15}$$

Theorem 5 ([10, **p. 333**]). *Let H be a RKHS with kernel k and $H_r = \{f \in H \mid \|f\|_H \leq r\}$. Given an n sample $\{x_1, ..., x_n\}$ we can bound the empirical Rademacher complexity of H_r by*

$$\frac{r}{n\sqrt{2}} \sqrt{\sum_{i=1}^n k(x_i, x_i)} \leq \text{Rad}_n(H_r) \leq \frac{r}{n} \sqrt{\sum_{i=1}^n k(x_i, x_i)}. \tag{16}$$

The previous two theorems lead to upper bounds on the complexity of MR, in particular we can bound the maximal reduction over supervised learning.

Corollary 3. *Let H be a RKHS and for $f, g \in H$ define the inner product $\langle f, g \rangle_{\tilde{H}} = \langle f, g \rangle_H + f_U(\mu L) g_U^t$, where L is a positive definite matrix and $\mu \in \mathbb{R}$ is a regularization parameter. Let $\tilde{H}_r$ be defined as before, then*

$$\text{Rad}_n(\tilde{H}_r) \leq \frac{r}{n} \sqrt{\sum_{i=1}^n k(x_i, x_i) - k_{x_i}^t (\frac{1}{\mu} I + LK)^{-1} L k_{x_i}}. \tag{17}$$

Similarly we can obtain a lower bound in line with Inequality (16).

The corollary shows in particular that the difference of the Rademacher complexity of the supervised and the semi-supervised method is given by the term

$k_{x_i}^t(\frac{1}{\mu}I_{n+m} + LK)^{-1}Lk_{x_i}$. This can be used for example to compute generalization bounds [17, Chapter 3]. We can also use the kernel to compute local Rademacher complexities which may yield tighter generalization bounds [5]. Here we illustrate the use of our bounds for choosing the regularization parameter μ without the need for an additional labeled validation set.

7 Experiment: Concentric Circles

We illustrate the use of Eq. (17) for model selection. In particular, it can be used to get an initial idea of how to choose the regularization parameter μ. The idea is to plot the Rademacher complexity versus the parameter μ as in Fig. 1. We propose to use an heuristic which is often used in clustering, the so called elbow criteria [9]. We essentially want to find a μ such that increasing the μ will not result in much reduction of the complexity anymore. We test this idea on a dataset which consists out of two concentric circles with 500 datapoints in $\mathbb{R}^2$, 250 per circle, see also Fig. 2. We use a Gaussian base kernel with bandwidth set to 0.5. The MR matrix L is the Laplacian matrix, where weights are computed with a Gaussian kernel with bandwidth 0.2. Note that those parameters have to be carefully set in order to capture the structure of the dataset, but this is not the current concern: we assume we already found a reasonable choice for those parameters. We add a small L2-regularization that ensures that the radius r in Inequality (17) is finite. The precise value of r plays a secondary role as the behavior of the curve from Fig. 1 remains the same.

Looking at Fig. 1 we observe that for μ smaller than 0.1 the curve still drops steeply, while after 0.2 it starts to flatten out. We thus plot the resulting kernels for $\mu = 0.02$ and $\mu = 0.2$ in Fig. 2. We plot the isolines of the kernel around the point of class one, the red dot in the figure. We indeed observe that for $\mu = 0.02$ we don't capture that much structure yet, while for $\mu = 0.2$ the two concentric circles are almost completely separated by the kernel. If this procedure indeed elevates to a practical method needs further empirical testing.

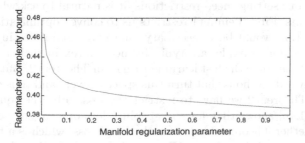

Fig. 1. The behavior of the Rademacher complexity when using manifold regularization on circle dataset with different regularization values μ.

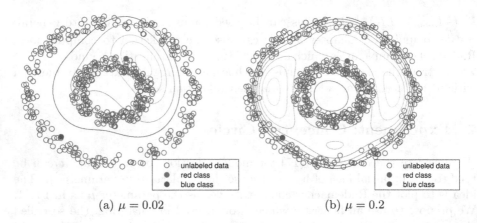

(a) $\mu = 0.02$ (b) $\mu = 0.2$

Fig. 2. The resulting kernel when we use manifold regularization with parameter μ set to 0.02 and 0.2.

8 Discussion and Conclusion

This paper analysed improvements in terms of sample or Rademacher complexity for a certain class of SSL. The performance of such methods depends both on how the approximation error of the class $\mathcal{F}$ compares to that of $\mathcal{F}_\tau^\psi$ and on the reduction of complexity by switching from the first to the latter. In our analysis we discussed the second part. The first part depends on a notion the literature often refers to as a *semi-supervised assumption*. This assumption basically states that we can learn with $\mathcal{F}_\tau^\psi$ as good as with $\mathcal{F}$. Without prior knowledge, it is unclear whether one can test efficiently if the assumption is true or not. Or is it possible to treat just this as a model selection problem? The only two works we know that provide some analysis in this direction are [3], which discusses the sample consumption to test the so-called cluster assumption, and [2], which analyzes the overhead of cross-validating the hyper-parameter coming from their proposed semi-supervised approach.

As some of our settings need restrictions, it is natural to ask whether we can extend the results. First, Lemma 1 restricts us to convex optimization problems. If that assumption would be unnecessary, one may get interesting extensions. Neural networks, for example, are typically not convex in their function space and we cannot guarantee the fast learning rate from Theorem 2. But maybe there are semi-supervised methods that turn this space convex, and thus could achieve fast rates. In Theorem 2 we have to restrict the loss to be the square loss, and [1, Example 21.16] shows that for the absolute loss one cannot achieve such a result. But whether Theorem 2 holds for the hinge loss, which is a typical choice in classification, is unknown to us. We speculate that this is indeed true, as at least the related classification tasks, that use the 0–1 loss, cannot achieve a rate faster than $\frac{1}{\epsilon}$ [19, Theorem 6.8].

Corollary 2 sketches a scenario in which sample complexity improvements of MR can be at most a constant over their supervised counterparts. This may sound like a negative result, as other methods with similar assumptions can achieve exponentially fast learning rates [16, Chapter 6]. But constant improvement can still have significant effects, if this constant can be arbitrarily large. If we set the regularization parameter μ in the concentric circles example high enough, the only possible classification functions will be the one that classifies each circle uniformly to one class. At the same time the pseudo-dimension of the supervised model can be arbitrarily high, and thus also the constant in Corollary 2. In conclusion, one should realize the significant influence constant factors in finite sample settings can have.

References

1. Anthony, M., Bartlett, P.L.: Neural Network Learning: Theoretical Foundations, 1st edn. Cambridge University Press, New York, USA (2009)
2. Azizyan, M., Singh, A., Wasserman, L.A.: Density-sensitive semisupervised inference. Computing Research Repository. abs/1204.1685 (2012)
3. Balcan, M., Blais, E., Blum, A., Yang, L.: Active property testing. In: 53rd Annual IEEE Symposium on Foundations of Computer Science, New Brunswick, NJ, USA, pp. 21–30 (2012)
4. Balcan, M.F., Blum, A.: A discriminative model for semi-supervised learning. J. ACM **57**(3), 19:1–19:46 (2010)
5. Bartlett, P.L., Bousquet, O., Mendelson, S.: Local Rademacher complexities. Ann. Stat. **33**(4), 1497–1537 (2005)
6. Belkin, M., Niyogi, P.: Towards a theoretical foundation for Laplacian-based manifold methods. J. Comput. Syst. Sci. **74**(8), 1289–1308 (2008)
7. Belkin, M., Niyogi, P., Sindhwani, V.: Manifold regularization: a geometric framework for learning from labeled and unlabeled examples. JMLR **7**, 2399–2434 (2006)
8. Ben-David, S., Lu, T., Pál, D.: Does unlabeled data provably help? Worst-case analysis of the sample complexity of semi-supervised learning. In: Proceedings of the 21st Annual Conference on Learning Theory, Helsinki, Finland (2008)
9. Bholowalia, P., Kumar, A.: EBK-means: a clustering technique based on elbow method and k-means in WSN. Int. J. Comput. Appl. **105**(9), 17–24 (2014)
10. Boucheron, S., Bousquet, O., Lugosi, G.: Theory of classification: a survey of some recent advances. ESAIM Probab. Stat. **9**, 323–375 (2005)
11. Chapelle, O., Schölkopf, B., Zien, A.: Semi-Supervised Learning. The MIT Press, Cambridge (2006)
12. Darnstädt, M., Simon, H.U., Szörényi, B.: Unlabeled data does provably help. In: STACS, Kiel, Germany, vol. 20, pp. 185–196 (2013)
13. Globerson, A., Livni, R., Shalev-Shwartz, S.: Effective semisupervised learning on manifolds. In: COLT, Amsterdam, The Netherlands, pp. 978–1003 (2017)
14. Grandvalet, Y., Bengio, Y.: Semi-supervised learning by entropy minimization. In: NeuRIPS, Vancouver, BC, Canada, pp. 529–536 (2004)
15. Kloft, M., Brefeld, U., Laskov, P., Müller, K.R., Zien, A., Sonnenburg, S.: Efficient and accurate Lp-norm multiple kernel learning. In: NeuRIPS, Vancouver, BC, Canada, pp. 997–1005 (2009)
16. Mey, A., Loog, M.: Improvability through semi-supervised learning: a survey of theoretical results. Computing Research Repository. abs/1908.09574 (2019)
17. Mohri, M., Rostamizadeh, A., Talwalkar, A.: Foundations of Machine Learning. The MIT Press, Cambridge (2012)

18. Niyogi, P.: Manifold regularization and semi-supervised learning: some theoretical analyses. JMLR **14**(1), 1229–1250 (2013)
19. Shalev-Shwartz, S., Ben-David, S.: Understanding Machine Learning: From Theory to Algorithms. Cambridge University Press, New York (2014)
20. Sindhwani, V., Niyogi, P., Belkin, M.: Beyond the point cloud: from transductive to semi-supervised learning. In: ICML, Bonn, Germany, pp. 824–831 (2005)
21. Sindhwani, V., Rosenberg, D.S.: An RKHS for multi-view learning and manifold co-regularization. In: ICML, Helsinki, Finland, pp. 976–983 (2008)
22. Vapnik, V.N.: Statistical Learning Theory. Wiley, Hoboken (1998)

Actionable Subgroup Discovery
and Urban Farm Optimization

Alexandre Millot[1], Romain Mathonat[1,2], Rémy Cazabet[3],
and Jean-François Boulicaut[1(✉)]

[1] Univ de Lyon, CNRS, INSA Lyon, LIRIS, UMR5205, 69621 Villeurbanne, France
{alexandre.millot,romain.mathonat,jean-francois.boulicaut}@insa-lyon.fr
[2] Atos, 69100 Villeurbanne, France
[3] Univ de Lyon, CNRS, Université Lyon 1, LIRIS, UMR5205,
69622 Villeurbanne, France
remy.cazabet@univ-lyon1.fr

Abstract. Designing, selling and/or exploiting connected vertical urban farms is now receiving a lot of attention. In such farms, plants grow in controlled environments according to recipes that specify the different growth stages and instructions concerning many parameters (e.g., temperature, humidity, CO_2, light). During the whole process, automated systems collect measures of such parameters and, at the end, we can get some global indicator about the used recipe, e.g., its yield. Looking for innovative ideas to optimize recipes, we investigate the use of a new optimal subgroup discovery method from purely numerical data. It concerns here the computation of subsets of recipes whose labels (e.g., the yield) show an interesting distribution according to a quality measure. When considering optimization, e.g., maximizing the yield, our virtuous circle optimization framework iteratively improves recipes by sampling the discovered optimal subgroup description subspace. We provide our preliminary results about the added-value of this framework thanks to a plant growth simulator that enables inexpensive experiments.

Keywords: Subgroup discovery · Virtuous circle · Urban farms

1 Introduction

Conventional farming methods have to face many challenges like, for instance, soil erosion and/or an overuse of pesticides. The crucial problems related to climate change also stimulate the design of new production systems. The concept of urban farms (see, e.g., AeroFarms, FUL, Infarm[1]) could be part of a solution. It enables the growth of plants in fully controlled environments close to the place where consumers are [8]. Most of the crop protection chemical products can be removed while being able to optimize both the quantity and the quality of plants (e.g., improving the flavor [9] or their chemical proportions [20]).

---

[1] https://aerofarms.com/, http://www.fermeful.com/, https://infarm.com/.

© The Author(s) 2020
M. R. Berthold et al. (Eds.): IDA 2020, LNCS 12080, pp. 339–351, 2020.
https://doi.org/10.1007/978-3-030-44584-3_27

Urban farms can generate large amounts of data that can be pushed towards a cloud environment such that various machine learning and data mining methods can be used. We may then provide new insights about the plant growth process itself (discovering knowledge about not yet identified/understood phenomena) but also offer new services to farm owners. We focus here on services that rely on the optimization of a given target variable, e.g., the yield. The number of parameters influencing plant growth can be relatively large (e.g., temperature, hygrometry, water pH level, nutrient concentration, LED lighting intensity, CO_2 concentration). There are numerous ways of measuring the crop end-product (e.g., energy cost, plant mass and size, flavor and chemical properties). In general, for a given type of plants, expert knowledge exists that concerns the available sub-systems (e.g., to model the impact of nutrient on growth, the effect of LED lighting on photosynthesis, the energy consumption w.r.t. the temperature instruction) but we are far from a global understanding of the interaction between the various underlying phenomena. In other terms, setting the optimal instructions for the diverse set of parameters given an optimization task remains an open problem.

We want to address such an issue by means of data mining techniques. Plant growth recipes are made of instructions in time and space for many numerical attributes. Once a recipe is completed, collections of measures have been collected and we assume that at least one numerical target label value is available, e.g., the yield. Can we learn from available recipe records to suggest new ones that should provide better results w.r.t. the selected target attribute? For that purpose, we investigate the use of subgroup discovery [12,21]. It aims at discovering subsets of objects - called subgroups - with high quality according to a quality measure calculated on the target label. Such a quality measure has to capture deviations in the target label distribution when we consider the overall data set or the considered subset of objects. When addressing only subgroup discovery from numerical data, a few approaches for numerical attributes [6,15] and numerical target labels [14] have been described. To the best of our knowledge, the reference algorithm for subgroup discovery in purely numerical data is SD-Map* [14]. However, like other methods, it uses discretization and leads to loss of information and sub-optimal results.

Our first contribution concerns the proposal of a simple branch and bound algorithm called MinIntChange4SD that exploits the exhaustive enumeration strategy from [11] to achieve a guaranteed optimal subgroup discovery in numerical data without any discretization. Discussing details about this algorithm is out of the scope of this paper and we recently designed a significantly optimized version of MinIntChange4SD in [17]. Our main contribution concerns a new methodology for plant growth recipe optimization that (i) uses MinIntChange4SD to find the optimal subgroup of recipes and (ii) exploits the subgroup description to design better recipes which can in turn be analyzed with subgroup discovery, and so on.

The paper is organized as follows. Section 2 formalizes the problem. In Sect. 3, we discuss related works and their limitations. In Sect. 4, we introduce our new

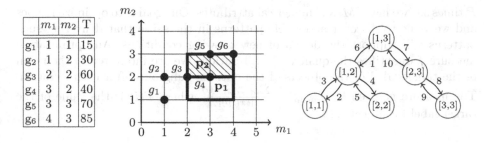

Fig. 1. (**left**) Purely numerical dataset. (**center**) Non-closed ($p_1 = \langle [2,4], [1,3] \rangle$, non-hatched) and closed ($p_2 = \langle [2,4], [2,3] \rangle$, hatched) interval patterns. (**right**) Depth-first traversal of m_2 using minimal changes.

optimal subgroup discovery algorithm and we detail our framework for plant growth recipe optimization. An empirical evaluation of our method is in Sect. 5. Section 6 briefly concludes.

2   Problem Definition

Numerical Dataset. A numerical dataset (G, M, T) is given by a set of objects G, a set of numerical attributes M and a numerical target label T. In a given dataset, the domain of any attribute $m \in M$ (resp. label T) is a finite ordered set denoted D_m (resp. D_T). Figure 1 (left) provides a numerical dataset made of two attributes $M = \{m_1, m_2\}$ and a target label T. A subgroup p is defined by a pattern, i.e., its intent or description, and the set of objects from the dataset where it appears, i.e., its extent, denoted $ext(p)$. For instance, in Fig. 1, the domain of m_1 is $\{1, 2, 3, 4\}$ and the intent $\langle [2, 4], [1, 3] \rangle$ (see the definition of interval patterns later) denotes a subgroup whose extent is $\{g_3, g_4, g_5, g_6\}$.

Quality Measure, Optimal Subgroup. The interestingness of a subgroup in a numerical dataset is measured by a numerical value. We consider here the quality measure based on the mean introduced in [14]. Let p be a subgroup. The quality of p is given by: $q_{mean}^a(p) = |ext(p)|^a \times (\mu_{ext(p)} - \mu_{ext(\emptyset)}), a \in [0, 1]$. $|ext(p)|$ denotes the cardinality of $ext(p)$, $\mu_{ext(p)}$ is the mean of the target label in the extent of p, $\mu_{ext(\emptyset)}$ is the mean of the target label in the overall dataset, and a is a parameter that controls the number of objects of the subgroups. Let (G, M, T) be a numerical dataset, q a quality measure and P the set of all subgroups of (G, M, T). A subgroup $p \in P$ is said to be optimal iff $\forall p' \in P : q(p') \leq q(p)$.

Plant Growth Recipe and Optimization Measure. A plant growth recipe (M, P, T) is given by a set of numerical parameters M specifying the growing conditions thanks to intervals on numerical values, a numerical value P representing the number of stages of the growth cycle, and a numerical target label T to quantify the recipe quality. In a given recipe, each parameter of M is repeated

P times s.t. we have $|M| \times P$ numerical attributes. Our goal is to optimize recipes and we want to discover actionable patterns in the sense that delivering such patterns will support the design of new growing conditions. An optimization measure f quantifies the quality of an iteration. We are interested in the mean of the target label of the objects of the optimal subgroup after each iteration. The measure is given by $f_{mean} = \frac{\sum_{i \in ext(p)} T(i)}{|ext(p)|}$ where $T(i)$ is the value of the target label for object i.

3   Related Work

Designing recipes that optimize a given target attribute (e.g., the mass, the energy cost) is often tackled by domain experts who exploit the scientific literature. However, in our setting, it has two major drawbacks. First, most of the literature remains oriented towards conventional growing conditions and farming methods. In urban farms, there are more parameters that can be controlled. Secondly, the amount of knowledge about plants is unbalanced from one plant to another. Therefore, relying only on expert knowledge for plant recipe optimization is not sufficient. We have an optimization problem and the need for a limited number of iterations. Indeed, experimenting with plant growth recipes is time consuming (i.e., asking for weeks or months). Therefore, we have to minimize the number of experiments that are needed to optimize a given recipe. There are two main families of methods addressing the problem of optimizing a function over numerical variables: *direct* and *model-based* [18]. For *direct* methods, the common idea is to apply various strategies to sequentially evaluate solutions in the search space of recipes. However such methods do not address the problem of minimizing the number of experiments. For *model-based* methods, the idea is to build a model simulating the ground truth using available data and then to use it to guide the search process. For instance, [9] introduced a solution for recipe optimization using this type of method with the goal of optimizing the flavor of plants. Their framework is based on using a surrogate model, in this case a Symbolic Regression [13]. It considers recipe optimization by means of a promising virtuous circle. However, it suffers from several shortcomings: there is no guarantee on the quality of the generated models (i.e., they may not be able to model correctly the ground truth), the number of tested parameters is small (only 3), and the ratio between the number of objects and the number of parameters in the data needs to be at least ten for Symbolic Regression [10]. Clearly, it would restrict the search to only a few parameters.

Heuristic [2,15] and exhaustive [1,5] solutions have been proposed for subgroup discovery. Usually, these approaches consider a set of nominal attributes with a binary label. To work with numerical data, prior discretization of the attributes is then required (see, e.g., [3]) and it leads to loss of information and suboptimal results. A major issue with exhaustive pattern mining is the size of the search space. Fortunately, optimistic estimates can be used to prune the search space and provide tractability in practice [7,21]. [14] introduces a large panel of quality measures and corresponding optimistic estimates for an exhaustive subgroup mining given numerical target labels. They describe SD-Map*, the reference algorithm for subgroup discovery in numerical data. Notice however that for [14] or others [6,15], discretization techniques over the numerical attributes have to be performed. When looking for an exhaustive search of frequent patterns - not subgroups - in numerical data without discretization, we find the MinIntChange algorithm [11]. Using closure operators (see, e.g., [4]) has become a popular solution to reduce the size of the search space. We indeed exploit most of these ideas to design our optimal subgroup discovery algorithm.

4   Optimization with Subgroup Discovery

4.1   An Efficient Algorithm for Optimal Subgroup Discovery

Let us first introduce MinIntChange4SD, our branch and bound algorithm for the optimal subgroup discovery in purely numerical data. It exploits smart concepts about interval patterns from [11].

Interval Patterns, Extent and Closure. In a numerical dataset (G, M, T), an interval pattern p is a vector of intervals $p = \langle [a_i, b_i] \rangle_{i \in \{1,...,|M|\}}$ with $a_i, b_i \in D_{mi}$, where each interval is a restriction on an attribute of M, and $|M|$ is the number of attributes. Let $g \in G$ be an object. g is in the extent of an interval pattern $p = \langle [a_i, b_i] \rangle_{i \in \{1,...,|M|\}}$ iff $\forall i \in \{1, ..., |M|\}, m_i(g) \in [a_i, b_i]$. Let p_1 and p_2 be two interval patterns. $p_1 \subseteq p_2$ means that p_2 encloses p_1, i.e., the hyperrectangle of p_1 is included in that of p_2. It is said that p_1 is a specialization of p_2. Let p be an interval pattern and $ext(p)$ its extent. p is defined as *closed* if and only if it is the most restrictive pattern (i.e., the smallest hyper-rectangle) that contains $ext(p)$. Figure 1 (center) depicts the dataset of Fig. 1 (left) in a cartesian plane as well as examples of interval patterns that are closed (p_2) or not (p_1).

Traversing the Search Space with Minimal Changes. To guarantee the optimal subgroup discovery, we proceed to the so-called minimal changes introduced in MinIntChange. It enables an exhaustive enumeration within the interval pattern search space. A left minimal change consists in replacing the left bound of an interval by the current value closest higher value in the domain of the corresponding attribute. Similarly, a right minimal change consists in replacing the right bound by the current value closest lower value. The search starts with the computation of the minimal interval pattern that covers all the objects of the dataset. The premise is to apply consecutive right or left minimal changes until obtaining an interval whose left and right bounds have the same value for each interval of the minimal interval pattern. In that case, the algorithm backtracks until it finds a pattern on which a minimal change can be applied. Figure 1 (right) depicts the depth-first traversal of attribute m_2 from the dataset of Fig. 1 (left) using minimal changes.

Compressing and Pruning the Search Space. We leverage the concept of closure to significantly reduce the number of candidate interval patterns. After a minimal change and instead of evaluating the resulting interval pattern, we compute its corresponding closed interval pattern. We exploit advanced pruning techniques to reduce the size of the search space thanks to the use of a tight optimistic estimate. We also exploit a combination of *forward checking* and *branch reordering*. Given an interval pattern, the set of all its direct specializations (application of a right or left minimal change on each interval) are computed - forward checking - and those whose optimistic estimate is higher than the best subgroup quality are stored. Branch reordering by descending order of the optimistic estimate value is then carried out which enables to explore the most promising parts of the search space first. It also enables a more efficient pruning by raising the minimal quality early. In fact, providing details about the algorithm is out of the scope of this paper though its source code is available at https://bit.ly/3bA87NE. The important outcome is that it guarantees the discovery of optimal subgroups for a given quality measure. Indeed, provided that it remains tractable, the runtime efficiency is not here an issue given that we want to use the algorithm at some steps of quite slow vegetable growth processes.

4.2   Leveraging Subgroups to Optimize Recipes

A Virtuous Circle. Our optimization framework can be seen as a virtuous circle, where each new iteration uses information previously gathered to iteratively improve the targeted process. First, a set of recipe experiments - which can be created with or without the use of expert knowledge - is created. With the use of expert knowledge, values or domain of values are defined for each attribute and then recipes are produced using these values. When generating recipes without prior knowledge, we create recipes by randomly sampling the values of each attribute. Secondly, we use subgroup discovery to find the best subgroup of recipes according to the chosen quality measure (e.g., the subgroup of recipes with the best average yield). Then, we exploit the subgroup description - i.e., we apply new restrictions on the range of each parameter according to the description - to generate new, better, recipe experiments. Finally these recipes are in turn processed to find the best subgroup for the new recipes, and so on until recipes cannot be improved anymore. This way, we sample recipes in a space which gets smaller after each iteration and where the ratio between good and bad solutions gets larger and larger. Figure 2 depicts a step-by-step example of the process behind the framework. Our framework makes use of several hyperparameters that affect runtime efficiency, the number of iterations and the quality of the results.

Convergence. The first hyperparameter is the parameter a used in the q_{mean}^a quality measure. In standard subgroup discovery, it controls the number of objects in the returned subgroups. A higher value of a means larger subgroups. For us, a larger subgroup means a larger search space to sample. By extension, a higher value of a means more iterations to be able to reach smaller subspaces of

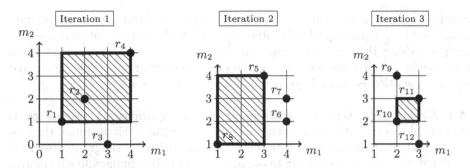

Fig. 2. Example of execution of the optimization framework in 3 iterations. We consider a two-dimensional space (i.e., 2 attributes m_1 and m_2) where 4 recipes are generated during each iteration using our first sampling method. The best subgroup (optimizing the yield) of each iteration (hatched) serves as the next iteration sampling space.

the search space. For that reason, we rename the parameter as the *convergence rate*. The second hyperparameter is called the *minimal improvement* (*minImp*). It defines the minimal improvement of the *Optimization measure* - f_{mean} in our setting - needed from one iteration to another for the framework to keep running. After each iteration, we check whether the following statement is true or false.

$$\frac{f_{mean_{it}} - f_{mean_{it-1}}}{f_{mean_{it-1}}} \geq minImp$$

If it is true, then the optimization framework keeps running, else we consider that the recipes cannot be improved any further. This parameter has a direct effect on the number of iterations needed for the algorithm to converge. A higher value for *minImp* means a lower number of iterations and vice versa. We can also forget *minImp* and set the number of iterations by means of another parameter that would denote a budget.

Sampling the Subspace. After each iteration, to generate new recipes to experiment with, we need to sample the subspace corresponding to the description of the best subgroup. Three sampling methods are currently available and this defines again a new hyperparameter. The first method consists in sampling recipes using the original set of values of each attribute (i.e., in the first iteration) minus the excluded values due to the new restrictions applied on the subspace. Let D_m^1 be the domain of values of attribute m at Iteration 1 and $[a_m^i, b_m^i]$ be the interval of attribute m at Iteration i according to the description of the best subgroup of Iteration $i-1$. Then, $\forall v \in D_m^1, v \in D_m^i \Leftrightarrow b_m^i \geq v \geq a_m^i$. Using this method, the number of values available for sampling for each attribute gets smaller after each iteration, meaning that each iteration is faster than the previous one. The second consists in discretizing the search space through the discretization of each attribute in k intervals of equal length. Parameter k is set before launching the framework. Recipes are then sampled using the discretized domain of values for each attribute. Finally, we can use *Latin Hypercube*

Sampling [16] as a third method. In *Latin Hypercube Sampling*, each attribute is divided in S equally probable intervals, with S the number of samples (i.e., recipes). Using this method, recipes are sampled such that each recipe is the only one in each hyperspace that contains it. The number of samples generated for each iteration is also a hyperparameter of the framework.

An Explainable Generic Framework. Our optimization framework is explainable contrary to black box optimization algorithms. Each step of the process is easily understandable due to the descriptive nature of subgroup discovery. Although we have been referring to our algorithm `MinIntChange4SD` when introducing the optimization framework, other subgroup discovery algorithms can be used, including [14] and [17]. Notice however that the better the quality of the provided subgroup, the better the results returned by our framework will be. Finally, our method can be applied to quite many application domains where we want to optimize a numerical target given collections of numerical features (e.g., hyperparameter optimization in machine learning).

5   Experiments

We work on urban farm recipe optimization while we do not have access to real farming data yet. One of our partners in the FUI DUF 4.0 project (2018–2021) is designing new types of urban farms. We found a way to support the empirical study of our recipe optimizing framework thanks to inexpensive experiments enabled by a simulator. In an urban farm, plants grow in a controlled environment. In the absence of failure, recipe instructions are followed and we can investigate the optimization of the plant yield at the end of the growth cycle. We simulate recipe experiments by using the PCSE[2] simulation environment by setting the characteristics (e.g., the climate) of the different growth stages. We focus on 3 variables that set the amount of solar irradiation (range $[0, 25000]$), wind (range $[0, 30]$) and rain (range $[0, 40]$). The plant growth is split into 3 stages of equal length such that we finally get 9 attributes. In real life, we can control most of the parameters of an urban farm (e.g., providing more or less light) and a recipe optimization iteration needs for new insights about the promising parameter values. This is what we can emulate using the crop simulator: given the description of the optimal subgroup, we get insights to support the design of the next simulations, say experiments, as if we were controlling the growth environment. At the end of the growth cycle, we retrieve the total mass of plants harvested using a given recipe. Note that in the following experiments, unless stated otherwise, no assumption is made on the values of parameters (i.e., no restriction is applied on the range of values defined above and expert knowledge is not taken into account). Table 1 features examples of plant growth recipes. The source code and datasets used in our evaluation are available at https://bit.ly/3bA87NE.

---

[2] https://pcse.readthedocs.io/en/stable/index.html.

Table 1. Examples of growth recipes split in 3 stages (P1, P2, P3), 3 attributes, and a target label (Yield).

R	Rain^{P1}	Irrad^{P1}	Wind^{P1}	Rain^{P2}	Irrad^{P2}	Wind^{P2}	Rain^{P3}	Irrad^{P3}	Wind^{P3}	Yield
r_1	10	23250	5	10	23250	5	15	21000	10	22000
r_2	35	10000	14	5	25000	10	16	19500	30	20500
r_3	15	17500	26	22	15000	18	30	4000	3	8600
r_4	18	22800	17	38	17000	25	38	12000	19	14200

Table 2. Comparison between descriptions of the overall dataset (DS), the optimal subgroup returned by `MinIntChange4SD` (MIC4SD), the optimal subgroup returned by `SD-Map*`. "–" means no restriction on the attribute compared to DS, Q and S respectively the quality and size of the subgroup.

Subgroup	Rain^{P1}	Irrad^{P1}	Wind^{P1}	Rain^{P2}	Irrad^{P2}	Wind^{P2}	Rain^{P3}	Irrad^{P3}	Wind^{P3}	Q	S
DS	[0, 39]	[1170, 23471]	[2, 29]	[0, 37]	[111, 24111]	[0, 29]	[2, 40]	[964, 24197]	[1, 30]	0	30
MIC4SD	[16, 37]	[1170, 22085]	[2, 24]	[7, 37]	[18309, 23584]	[2, 24]	[15, 37]	[12626, 24197]	[1, 25]	33874	7
SD-Map*	[21, 39]	–	–	–	[14455, 24111]	–	–	[12760, 24197]	–	30662	5

5.1 `MinIntChange4SD` vs `SD-Map*`

We study the description of the best subgroup returned by `MinIntChange4SD` and `SD-Map*`, the state-of-the art algorithm for subgroup discovery in numerical data. Table 2 depicts the descriptions for a dataset comprised of 30 recipes generated randomly with the simulator. Besides the higher quality of the subgroup returned by `MinIntChange4SD`, the optimal subgroup description also enables to extract information that is missing from the description obtained with `SD-Map*`. In fact, where `SD-Map*` only offers a strong restriction on 3 attributes, `MinIntChange4SD` provides actionable information on all the considered attributes, i.e., the 9 attributes. This confirms its qualitative superiority over `SD-Map*` which has to proceed to attribute discretizations.

5.2 Empirical Evaluation of the Model Hyperparameters

Our optimization framework involves several hyperparameters whose values need to be studied to define proper ranges or values that will lead to optimized results with a minimized number of recipe experiments. We choose to apply a random search on discretized hyperparameters. Note that in this setting, grid search is a bad solution due to the combinatorial number of hyperparameter values and the high time cost of the optimization process itself. We discretize each hyperparameter in several values (the convergence rate is split into 10 values ranging from 0.1 to 1, the minimal improvement parameter is split into 12 values between 0 and 0.05, the sampling parameter is split between the 3 available methods, and the number of recipes for each iteration is either 20 or 30). We run 100 iterations of random search, with each iteration - read set of parameter values - being tested 10 times and averaged to account for randomness of the

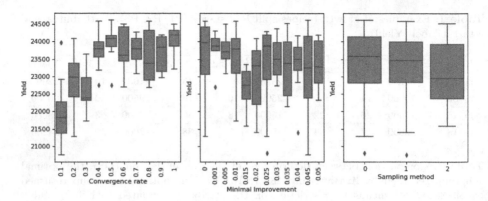

Fig. 3. Yield of the best recipe depending on the value of different hyperparameters using 100 sample recipes for each hyperparameter.

recipes generated. After each iteration of random search, we store the set of hyperparameter values and the corresponding best recipe found. Figure 3 depicts results of the experiments. Optimal values for convergence rate seem to be around 0.5, between 0.001 and 0.01 for minimal improvement, and the best sampling method is tied between the first and second one. Generating 30 recipes for each iteration yields better results than 20 (average yield of 23857 for 30 recipes against 22829 for 20 recipes). To compare our method against other methods, we run our framework with the following parameters: 30 recipes times 5 iterations (for a total of 150 recipes), 0.5 convergence rate, using the second sampling method with $k = 15$. To address the variance in the yield due to randomness in the recipe generation process, we run the framework 10 times, we store the best recipe found at each iteration and then compute the average of the stored recipes. We report the results in Table 3.

5.3   Comparison with Alternative Methods

Good hyperparameter values have been defined for our optimization framework and we can now compare our method with other ones. Let us consider the use of expert knowledge and random search. First, we want to create a model using expert knowledge. With the help of an agricultural engineer, we defined a priori good values for each parameter using expert knowledge and we generated a recipe that can serve as a baseline for our experiments. We then choose to compare our method against a random search model without expert knowledge. We set the number of recipes to 150 for all methods to provide a fair comparison with our own model where the number of recipes is set to 150. To account for randomness in the recipe generation, we run 10 iterations of the random search model, we store the value of the best recipe found in each iteration, and we compute their average yield. Results of the experiments and a description of the best recipe for each method are available in Table 3. Random search and expert knowledge find recipes with almost equal yields, while our framework find recipes with higher

Table 3. Comparison of the description and the yield of the best recipe returned by each method. EK = Expert Knowledge, RS = Random Search, SM = Surrogate Modeling, VC = Virtuous Circle (our framework).

Method	Rain^{P1}	Irrad^{P1}	Wind^{P1}	Rain^{P2}	Irrad^{P2}	Wind^{P2}	Rain^{P3}	Irrad^{P3}	Wind^{P3}	Yield
EK	10	0	5	10	25000	5	10	25000	5	23472
RS	17	23447	8	31	22222	23	39	22385	7	23561
SM	20	44	0	20	24981	0	40	31	30	10170
VC	19	16121	18	25	24052	28	14	21126	7	24336

yield. Note that in industrial settings, an improved yield of 3% to 4% has a significant impact on revenues.

Let us now compare our framework to the Surrogate Modeling method presented in [9]. To be fair, we give the same number of data points to build the Symbolic Regression surrogate model as we used in previous experiments, i.e., 150 for training the model (we evaluated the RMSE of the model on a test set of 38 other samples). We use gplearn [19], with default parameters, except for the number of generations and the number of models evaluated for each generations, which are respectively of 1000 and 2000, as in [9]. Note that the model obtained has a RMSE of 2112, and it is composed of more than 2000 terms (including mathematical operators), therefore the argument of interpretability is questionable. A grid search is finally done on this model and we select the best recipe and obtain their true yield using the PCSE simulation environment. The number of steps for each attribute for the grid search has to be defined. We set it to 5. As we have 9 parameters, it means that the model needs to be evaluated on nearly 9 million potential recipes. Also, the model is composed of hundreds of terms such that experiments are computationally expensive. The best recipe found so far is given in Table 3. The surrogate model predicts a yield value of 21137. Compared to the ground truth of 10170, the model has a strong bias. It illustrates that using a surrogate model for this kind of problem will give good recipes only if it is reliable enough. Interestingly, the RMSE seems to be quite good at first glance, but this does not guarantee that the model will behave correctly on all elements of the search space: on the best recipe found, it largely overestimates the yield, leading to a non-interesting recipe. It seems that this method performs poorly on recipes with more attributes than in [9]. Further studies are here needed.

6 Conclusion

We investigated the optimization of plant growth recipes in controlled environments, a key process in connected urban farms. We motivated the reasons why existing methods fall short of real life constraints, including the necessity to minimize the number of experiments needed to provide good results. We detailed a new optimization framework that leverages subgroup discovery to iteratively find better growth recipes through the use of a virtuous circle. We also introduced an efficient algorithm for the optimal subgroup discovery in purely numerical datasets. It has been recently improved much further in [17]. We avoid discretization and it provides a qualitative added-value (i.e., more interesting optimal subgroups). Future work includes extending our framework to deal with multiple target labels at the same time (e.g., optimizing the yield while keeping the energy cost as low as possible).

Acknowledgment. Our research is partially funded by the French FUI programme (project DUF 4.0, 2018–2021).

References

1. Atzmueller, M., Puppe, F.: SD-Map – a fast algorithm for exhaustive subgroup discovery. In: Fürnkranz, J., Scheffer, T., Spiliopoulou, M. (eds.) PKDD 2006. LNCS (LNAI), vol. 4213, pp. 6–17. Springer, Heidelberg (2006). https://doi.org/10.1007/11871637_6
2. Bosc, G., Boulicaut, J.F., Raïssi, C., Kaytoue, M.: Anytime discovery of a diverse set of patterns with Monte Carlo tree search. Data Min. Knowl. Discov. **32**, 604–650 (2018). https://doi.org/10.1007/s10618-017-0547-5
3. Fayyad, U.M., Irani, K.B.: Multi-interval discretization of continuous-valued attributes for classification learning. In: Proceedings IJCAI, pp. 1022–1029 (1993)
4. Garriga, G.C., Kralj, P., Lavrač, N.: Closed sets for labeled data. J. Mach. Learn. Res. **9**, 559–580 (2008)
5. Grosskreutz, H., Paurat, D.: Fast and memory-efficient discovery of the top-k relevant subgroups in a reduced candidate space. In: Gunopulos, D., Hofmann, T., Malerba, D., Vazirgiannis, M. (eds.) ECML PKDD 2011. LNCS (LNAI), vol. 6911, pp. 533–548. Springer, Heidelberg (2011). https://doi.org/10.1007/978-3-642-23780-5_44
6. Grosskreutz, H., Rüping, S.: On subgroup discovery in numerical domains. Data Min. Knowl. Discov. **19**(2), 210–226 (2009). https://doi.org/10.1007/s10618-009-0136-3
7. Grosskreutz, H., Rüping, S., Wrobel, S.: Tight optimistic estimates for fast subgroup discovery. In: Daelemans, W., Goethals, B., Morik, K. (eds.) ECML PKDD 2008. LNCS (LNAI), vol. 5211, pp. 440–456. Springer, Heidelberg (2008). https://doi.org/10.1007/978-3-540-87479-9_47
8. Harper, C., Siller, M.: OpenAG: a globally distributed network of food computing. IEEE Pervasive Comput. **14**, 24–27 (2015)
9. Johnson, A., Meyerson, E., Parra, J., Savas, T., Miikkulainen, R., Harper, C.: Flavor-cyber-agriculture: optimization of plant metabolites in an open-source control environment through surrogate modeling. PLoS ONE **14**, e0213918 (2019)
10. Jones, D.R., Schonlau, M., Welch, W.J.: Efficient global optimization of expensive black-box functions. J. Global Optim. **13**(4), 455–492 (1998)
11. Kaytoue, M., Kuznetsov, S.O., Napoli, A.: Revisiting numerical pattern mining with formal concept analysis. In: Proceedings IJCAI, pp. 1342–1347 (2011)
12. Klösgen, W.: Explora: a multipattern and multistrategy discovery assistant. In: Advances in Knowledge Discovery and Data Mining, pp. 249–271 (1996)
13. Koza, J.R.: Genetic Programming: On the Programming of Computers by Means of Natural Selection, pp. 162–169. MIT Press, Cambridge (1992)

14. Lemmerich, F., Atzmueller, M., Puppe, F.: Fast exhaustive subgroup discovery with numerical target concepts. Data Min. Knowl. Discov. **30**(3), 711–762 (2015). https://doi.org/10.1007/s10618-015-0436-8
15. Mampaey, M., Nijssen, S., Feelders, A., Knobbe, A.: Efficient algorithms for finding richer subgroup descriptions in numeric and nominal data. In: Proceedings ICDM, pp. 499–508 (2012)
16. McKay, M.D., Beckman, R.J., Conover, W.J.: A comparison of three methods for selecting values of input variables in the analysis of output from a computer code. Technometrics **21**(2), 239–245 (1979)
17. Millot, A., Cazabet, R., Boulicaut, J.F.: Optimal subgroup discovery in purely numerical data. In: Proceedings PaKDD, pp. 1–12 (2020, in press)
18. Rios, L.M., Sahinidis, N.V.: Derivative-free optimization: a review of algorithms and comparison of software implementations. J. Global Optim. **56**(3), 1247–1293 (2013). https://doi.org/10.1007/s10898-012-9951-y
19. Stephens, T.: gplearn (2013). https://github.com/trevorstephens/gplearn
20. Wojciechowska, R., Długosz-Grochowska, O., Kołton, A., Żupnik, M.: Effects of LED supplemental lighting on yield and some quality parameters of lamb's lettuce grown in two winter cycles. Sci. Hortic. **187**, 80–86 (2015)
21. Wrobel, S.: An algorithm for multi-relational discovery of subgroups. In: Komorowski, J., Zytkow, J. (eds.) PKDD 1997. LNCS, vol. 1263, pp. 78–87. Springer, Heidelberg (1997). https://doi.org/10.1007/3-540-63223-9_108

AVATAR - Machine Learning Pipeline Evaluation Using Surrogate Model

Tien-Dung Nguyen[1]([✉]), Tomasz Maszczyk[1], Katarzyna Musial[1],
Marc-André Zöller[2], and Bogdan Gabrys[1]

[1] University of Technology Sydney, Sydney, Australia
TienDung.Nguyen-2@student.uts.edu.au,
{Tomasz.Maszczyk,Katarzyna.Musial-Gabrys,Bogdan.Gabrys}@uts.edu.au
[2] USU Software AG, Karlsruhe, Germany
m.zoeller@usu.de

Abstract. The evaluation of machine learning (ML) pipelines is essential during automatic ML pipeline composition and optimisation. The previous methods such as Bayesian-based and genetic-based optimisation, which are implemented in Auto-Weka, Auto-sklearn and TPOT, evaluate pipelines by executing them. Therefore, the pipeline composition and optimisation of these methods requires a tremendous amount of time that prevents them from exploring complex pipelines to find better predictive models. To further explore this research challenge, we have conducted experiments showing that many of the generated pipelines are invalid, and it is unnecessary to execute them to find out whether they are good pipelines. To address this issue, we propose a novel method to evaluate the validity of ML pipelines using a surrogate model (AVATAR). The AVATAR enables to accelerate automatic ML pipeline composition and optimisation by quickly ignoring invalid pipelines. Our experiments show that the AVATAR is more efficient in evaluating complex pipelines in comparison with the traditional evaluation approaches requiring their execution.

1 Introduction

Automatic machine learning (AutoML) has been studied to automate the process of data analytics to collect and integrate data, compose and optimise ML pipelines, and deploy and maintain predictive models [1–3]. Although many existing studies proposed methods to tackle the problem of pipeline composition and optimisation [2,4–9], these methods have two main drawbacks. Firstly, the pipelines' structures, which define the executed order of the pipeline components, use fixed templates [2,5]. Although using fixed structures can reduce the number of invalid pipelines during the composition and optimisation, these approaches limit the exploration of promising pipelines which may have a variety of structures. Secondly, while evolutionary algorithms based methods [4] enable the randomness of the pipelines' structure using the concept of evolution, this randomness tends to construct more invalid pipelines than valid ones.

© The Author(s) 2020
M. R. Berthold et al. (Eds.): IDA 2020, LNCS 12080, pp. 352–365, 2020.
https://doi.org/10.1007/978-3-030-44584-3_28

Besides, the search spaces of the pipelines' structures and hyperparameters of the pipelines' components expand significantly. Therefore, the existing approaches tend to be inefficient as they often attempt to evaluate invalid pipelines. There are several attempts to reduce the randomness of pipeline construction by using context-free grammars [8,9] or AI planning to guide the construction of pipelines [6,7]. Nevertheless, all of these methods evaluate the validity of a pipeline by executing them (T-method). After executing a pipeline, if the result is a predictive model, the T-method evaluates the pipeline to be valid; otherwise it is invalid. If a pipeline is complex, the complexity of preprocessing/predictor components within the pipeline is high, or the size of the dataset is large, the evaluation of the pipeline is expensive. Consequently, the optimisation will require a significant time budget to find well-performing pipelines.

To address this issue, we propose the AVATAR to evaluate ML pipelines using their surrogate models. The AVATAR transforms a pipeline to its surrogate model and evaluates it instead of executing the original pipeline. We use the business process model and notation (BPMN) [10] to represent ML pipelines. BPMN was invented for the purposes of a graphical representation of business processes, as well as a description of resources for process execution. In addition, BPMN simplifies the understanding of business activities and interpretation of behaviours of ML pipelines. The ML pipelines' components use the Weka libraries[1] for ML algorithms. The evaluation of the surrogate models requires a knowledge base which is generated from many synthetic datasets. To this end, this paper has two main contributions:

- We conduct experiments on current state-of-the-art AutoML tools to show that the construction of invalid pipelines during the pipeline composition and optimisation may lead to bad performance.
- We propose the AVATAR to accelerate the automatic pipeline composition and optimisation by evaluating pipelines using a surrogate model.

This paper is divided into five sections. After the Introduction, Sect. 2 reviews previous approaches to representing and evaluating ML pipelines in the context of AutoML. Section 3 presents the AVATAR to evaluate ML pipelines. Section 4 presents experiments to motivate our research and prove the efficiency of the proposed method. Finally, Sect. 5 concludes this study.

2   Related Work

Salvador et al. [2] proposed an automatic pipeline composition and optimisation method of multicomponent predictive systems (MCPS) to deal with the problem of combined algorithm selection and hyperparameter optimisation (CASH). This proposed method is implemented in the tool AutoWeka4MCPS [2] developed on top of Auto-Weka 0.5 [11]. The pipelines, which are generated by

---

[1] https://www.cs.waikato.ac.nz/ml/weka/.

AutoWeka4MCPS, are represented using Petri nets [12]. A Petri net is a mathematical modelling language used to represent pipelines [2] as well as data service compositions [13]. The main idea of Petri nets is to represent transitions of states of a system. Although it is not clearly mentioned in these previous works [4–7], directed acyclic graph (DAG) is often used to model sequential pipelines in the methods/tools such as AutoWeka4MCPS [14], ML-Plan [6], P4ML [7], TPOT [4] and Auto-sklearn [5]. DAG is a type of graph that has connected vertexes, and the connections of vertexes have only one direction [15]. In addition, a DAG does not allow any directed loop. It means that it is a topological ordering. ML-Plan generates sequential workflows consisting of ML components. Thus, the workflows are a type of DAG. The final output of P4ML is a pipeline which is constructed by making an ensemble of other pipelines. Auto-sklearn generates fixed-length sequential pipelines consisting of scikit-learn components. TPOT construct pipelines consisting of multiple preprocessing sub-pipelines. The authors claim that the representation of the pipelines is a tree-based structure. However, a tree-based structure always starts with a root node and ends with many leaf nodes, but the output of a TPOT's pipeline is a single predictive model. Therefore, the representation of TPOT pipeline is more like a DAG. P4ML uses a tree-based structure to make a multi-layer ensemble. This tree-based structure can be specialised into a DAG. The reason is that the execution of these pipelines will start from leaf nodes and end at root nodes where the construction of the ensembles are completed. It means that the control flows of these pipelines have one direction, or they are topologically ordered. Using a DAG to model an ML pipeline makes it easy to understand by humans as DAGs facilitate visualisation and interpretation of the control flow. However, DAGs do not model inputs/outputs (i.e. possibly datasets, output predictive models, parameters and hyperparameters of components) between vertexes. Therefore, the existing studies use ad-hoc approaches and make assumptions about data inputs/outputs of the pipelines' components.

Although AutoWeka4MCPS, ML-Plan, P4ML, TPOT and Auto-sklearn evaluate pipelines by executing them, these methods have strategies to limit the generation of invalid pipelines. Auto-sklearn uses a fixed pipeline template including preprocessing, predictor and ensemble components. AutoWeka4MCPS also uses a fixed pipeline template consisting of six components. TPOT, ML-Plan and P4ML use grammars/primitive catalogues, which are designed manually, to guide the construction of pipelines. Although these approaches can reduce the number of invalid pipelines, our experiments showed that the wasted time used to evaluate the invalid pipelines is significant. Moreover, using fixed templates, grammars and primitive catalogues reduce search spaces of potential pipelines, which is a drawback during pipeline composition and optimisation.

3 Evaluation of ML Pipelines Using Surrogate Models

Because the evaluation of ML pipelines is expensive in certain cases (i.e., complex pipelines, high complexity pipeline's components and large datasets) in the context of AutoML, we propose the AVATAR[2] to speed up the process by evaluating their surrogate pipelines. The main idea of the AVATAR is to expand the purpose and representation of MCPS introduced in [12]. The AVATAR uses a surrogate model in the form of a Petri net. This surrogate pipeline keeps the structure of the original pipeline, replaces the datasets in the form of data matrices (i.e., components' input/output simplified mappings) by the matrices of transformed-features, and the ML algorithms by transition functions to calculate the output from the input tokens (i.e., the matrices of transformed-features). Because of the simplicity of the surrogate pipelines in terms of the size of the tokens and the simplicity of the transition functions, the evaluation of these pipelines is substantially less expensive than the original ones.

3.1   The AVATAR Knowledge Base

We define transformed-features as the features, which represent dataset's characteristics. These characteristics can be changed because of the transformations of this dataset by ML algorithms. Table 1 describes the transformed-features used

Table 1. Descriptions of the transformed-features of a dataset.

Transformed-feature	Description
BINARY_CLASS	A dataset has binary classes
NUMERIC_CLASS	A dataset has numeric classes
DATE_CLASS	A dataset has date classes
MISSING_CLASS_VALUES	A dataset has missing values in classes
NOMINAL_CLASS	A dataset has nominal classes
SYMBOLIC_CLASS	A dataset has symbolic data in classes
STRING_CLASS	A dataset has string classes
UNARY_CLASS	A dataset has unary classes
BINARY_ATTRIBUTES	A dataset has binary attributes
DATE_ATTRIBUTES	A dataset has date attributes
EMPTY_NOMINAL_ATTRIBUTES	A dataset has an empty column
MISSING_VALUES	A dataset has missing values in attributes
NOMINAL_ATTRIBUTES	A dataset has nominal attributes
NUMERIC_ATTRIBUTES	A dataset has numeric attributes
UNARY_ATTRIBUTES	A dataset has unary attributes
PREDICTIVE_MODEL	A predictive model generated by a predictor

---

[2] https://github.com/UTS-AAi/AVATAR.

for the knowledge base. We select these transformed-features because the capabilities of a ML algorithm to work with a dataset depend on these transformed-features. These transformed-features are extended from the capabilities of Weka algorithms[3].

The purpose of the AVATAR knowledge base is for describing the logic of transition functions of the surrogate pipelines. The logic includes the capabilities and effects of ML algorithms (i.e., pipeline components).

The capabilities are used to verify whether an algorithm is compatible to work with a dataset or not. For example, whether the linear regression algorithm can work with missing value and numeric attributes or not? The capabilities have a list of transformed-features. The value of each capability-related transformed-feature is either 0 (i.e., the algorithm can not work with the dataset which has this transformed-feature) or 1 (i.e., the algorithm can work with the dataset which has this transformed-feature). Based on the capabilities, we can determine which components of a pipeline (i.e., ML algorithms) are not able to process specific transformed-features of a dataset.

The effects describe data transformations. Similar to the capabilities, the effects have a list of transformed-features. Each effect-related transformed-feature can have three values, 0 (i.e., do not transform this transformed-feature), 1 (i.e., transform one or more attributes/classes to this transformed-feature), or −1 (i.e., disable the effect of this transformed-feature on one or more attributes/classes).

To generate the AVATAR knowledge base[4], we have to use synthetic datasets[5] to minimise the number of active transformed-features in each dataset to evaluate which and how transformed-features impact on the capabilities and effects of ML algorithms[6]. Real-world datasets usually have many active transformed-features that make them not suitable for our purpose. We minimise the number of available transformed-features in each synthetic dataset so that the knowledge base can be applicable in a variety of pipelines and datasets. Figure 1 presents the algorithm to generate the AVATAR knowledge base. This algorithm has four main stages:

1. Initialisation: The first stage initialises all transformed-features in the capabilities and effects to 0.
2. Execution: Run ML algorithms with every synthetic dataset and get outputs (i.e., output datasets or predictive models).
3. Find capabilities: If the execution is successful, we set the active transformed-features of the input dataset for the ones in the capabilities.
4. Find effects: If an algorithm is a predictor/transformed-predictor, we set *PREDICTIVE_MODEL* for its effects. If the algorithm is a filter and its

---

[3] http://weka.sourceforge.net/doc.dev/weka/core/Capabilities.html.

[4] https://github.com/UTS-AAi/AVATAR/blob/master/avatar-knowledge-base/avatar_knowledge_base.json.

[5] https://github.com/UTS-AAi/AVATAR/tree/master/synthetic-datasets.

[6] https://github.com/UTS-AAi/AVATAR/blob/master/supplementary-documents/avatar_algorithms.txt.

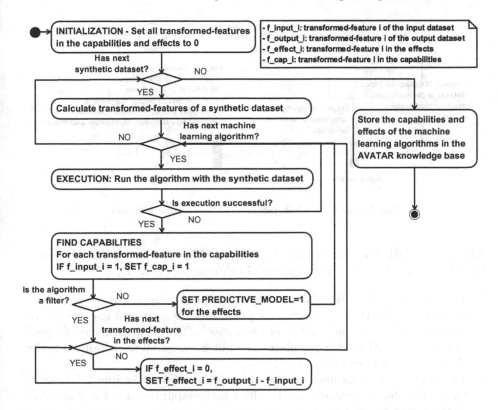

Fig. 1. Algorithm to generate the knowledge base for evaluating surrogate pipelines.

current value is a default value, we set this effect-related transformed-feature equal the difference of the values of this transformed-feature of the output and input dataset.

3.2   Evaluation of ML Pipelines

The AVATAR evaluates a ML pipeline by mapping it to its surrogate pipeline and evaluating this surrogate pipeline. BPMN is the most promising method to represent an ML pipeline. The reasons are that a BPMN-based ML pipeline can be executable, has a better interpretation of the pipeline in terms of control, data flows and resources for execution, as well as integrates into existing business processes as a subprocess. Moreover, we claim that a Petri net is the most promising method to represent a surrogate pipeline. The reason is that it is fast to verify the validity of a Petri net based simplified ML pipeline.

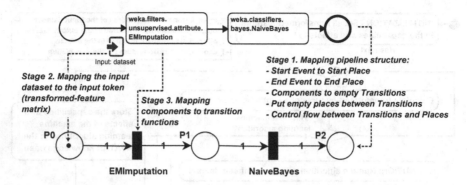

Fig. 2. Mapping a ML pipeline to its surrogate model.

Mapping a ML Pipeline to Its Surrogate Model. The AVATAR maps a BPMN pipeline to a Petri net pipeline via three stages (Fig. 2).

1. The structure of the BPMN-based ML pipeline is mapped to the respective structure of the Petri net surrogate pipeline. The start and end events are mapped to the start and end places respectively. The components are mapped to empty transitions. Empty places are put between all transitions. Finally, all flows are mapped to arcs.
2. The values of transformed-features are calculated from the input dataset to form a transformed-feature matrix which is the input token in the start place of the surrogate pipeline.
3. The transition functions are mapped from the components. In this stage, only the corresponding algorithm information is mapped to the transition function.

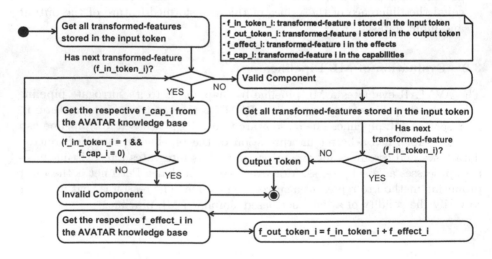

Fig. 3. Algorithm for firing a transition of the surrogate model.

Evaluating a Surrogate Model. The evaluation of a surrogate model will execute a Petri net pipeline. This execution starts by firing each transition of the Petri net pipeline and transforming the input token. As shown in Fig. 3, firing a transition consists of two tasks: (i) the evaluation of the capabilities of each component; and (ii) the calculation of the output token. The first task verifies the validity of the component using the following rules. If the value of a transformed-feature stored in the input token ($f\_in\_token\_i$) is 1 and the corresponding transformed-feature in the component's capabilities (f_cap_i) is 0, this component is invalid. Otherwise, this component is always valid. If a component is invalid, the surrogate pipeline is evaluated as invalid. The second task calculates each transformed-feature stored in the output token ($f\_out\_token\_i$) in the next place from the input token by adding the value of a transformed-feature stored in the input token ($f\_in\_token\_i$) and the respective transformed-feature in the component's effects (f_effect_i).

4 Experiments

To investigate the impact of invalid pipelines on ML pipeline composition and optimisation, we have first conducted a series of experiments with current state-of-the-art AutoML tools. After that, we have conducted the experiments to compare the performance of the AVATAR and the existing methods.

4.1 Experimental Settings

Table 2 summarises characteristics of datasets[7] used for experiments. We use these datasets because they were used in previous studies [2,4,5]. The AutoML tools used for the experiments are AutoWeka4MCPS [2] and Auto-sklearn [5]. These tools are selected because their abilities to construct and optimise hyper-parameters of complex ML pipelines have been empirically proven to be effective in a number of previous studies [2,5,16]. However, these previous experiments

Table 2. Summary of datasets' characteristics: the number of numeric attributes, nominal attributes, distinct classes, instances in training and testing sets.

Dataset	Numeric	Nominal	No. of distinct classes	Training	Testing
abalone	7	1	26	2,924	1,253
car	0	6	4	1,210	518
convex	784	0	2	8,000	50,000
gcredit	7	13	2	700	300
wineqw	11	0	7	3,429	1,469

---

[7] https://archive.ics.uci.edu.

had not investigated the negative impact of the evaluation of invalid pipelines on the quality of the pipeline composition and optimisation yet. This is the goal of our first set of experiments. In the second set of experiments, we show that the AVATAR can significantly reduce the evaluation time of ML pipelines.

4.2   Experiments to Investigate the Impact of Invalid Pipelines

To investigate the impact of invalid pipelines, we use five iterations (Iter) for the first set of experiments. We run these experiments on AWS EC2 $t3a.small$ virtual machines which have 2 vCPU and 2 GB memory. Each iteration uses a different seed number. We set the time budget to 1 h and the memory to 1 GB. We evaluate the pipelines produced by the AutoML tools using three criteria: (1) the number of invalid/valid pipelines, (2) the total evaluation time of invalid/valid pipelines (seconds), and (3) the wasted evaluation time (%). The wasted evaluation time is calculated by the percentage of the total evaluation time of invalid pipelines over the total runtime of the pipeline composition and optimisation. The wasted evaluation time represents the degree of negative impacts of invalid pipelines.

Tables 3 and 4 present negative impacts of invalid pipelines in ML pipeline composition and optimisation of AutoWeka4MCPS and Auto-sklearn using the above criteria. These tables show that not all of constructed pipelines are valid. Because AutoWeka4MCPS can compose pipelines which have up to six components, it is more likely to generate invalid pipelines and the evaluation time

Table 3. Negative impacts of invalid pipelines in pipeline composition and optimisation of AutoWeka4MCPS. (1): the number of invalid/valid pipelines, (2): the total evaluation time of invalid/valid pipelines (s), (3): the wasted evaluation time (%).

Dataset	Criteria	Iter 1	Iter 2	Iter 3	Iter 4	Iter 5
abalone	(1)	16/26	90/79	69/88	34/29	53/80
	(2)	3607.7/1322.5	2007.1/1236.4	4512.9/2172.3	3615.4/277.6	23.2/3509.0
	(3)	73.18	61.88	67.51	92.87	0.66
car	(1)	205/152	108/70	197/313	139/156	85/64
	(2)	3818.1/291.8	3498.5/113.0	4523.6/532.6	5232.2/251.3	4365.1/90.1
	(3)	92.90	96.87	89.47	95.42	97.98
convex	(1)	18/20	2/0	17/11	crashed	crashed
	(2)	76.3/3588.1	3475.2/0.0	1324.7/2331.8		
	(3)	2.08	100.00	36.23		
gcredit	(1)	112/195	229/364	208/166	12/54	30/54
	(2)	2821.0/2260.1	3829.8/285.6	3933.8/184.0	3667.6/34.1	3634.8/64.7
	(3)	55.52	93.06	95.53	99.08	98.25
wineqw	(1)	203/213	121/139	crashed	201/302	36/54
	(2)	4880.6/1052.9	4183.4/1078.6		2418.5/1132.2	1639.2/862.2
	(3)	82.26	79.50		68.11	65.53

Table 4. Negative impacts of invalid pipelines in pipeline composition and optimisation of Auto-sklearn. (1): the number of invalid/valid pipelines, (2): the total evaluation time of invalid/valid pipelines (s), (3): the wasted evaluation time (%).

Dataset	Criteria	Iter 1	Iter 2	Iter 3	Iter 4	Iter 5
abalone		crashed	crashed	crashed	crashed	crashed
car		crashed	crashed	crashed	crashed	crashed
convex	(1)	2/13	2/6	2/8	2/6	2/8
	(2)	560.8/2981.8	537.7/629.2	584.1/1537.5	558.1/977.1	560.0/1655.9
	(3)	15.76	15.07	16.39	15.66	15.72
gcredit		crashed	crashed	crashed	crashed	crashed
wineqw	(1)	0/42	0/22	0/42	0/32	0/32
	(2)	0.0/3523.4	0.0/909.7	0.0/3197.4	0.0/3054.0	0.0/3163.5
	(3)	0.00	0.00	0.00	0.00	0.00

of these invalid pipelines are significant. For example, the wasted evaluation time is 97.98% in the case of using the dataset car and Iter 5. We can see that changing the different random iterations has a strong impact on the wasted evaluation time in the case of AutoWeka4MCPS. For example, the experiments with the dataset abalone show that the wasted evaluation time is in the range between 0.66% and 92.87%. The reason is that Weka libraries them-self can evaluate the compatibility of a single component pipeline without execution. If the initialisation of the pipeline composition and optimisation with a specific seed number results in pipelines consisting of only one predictor, and these pipelines are well-performing, it tends to exploit similar ML pipelines. As a result, the wasted evaluation time is low. However, this impact is negligible in the case of Auto-sklearn. The reason is that Auto-sklearn uses meta-learning to initialise with promising ML pipelines. The experiments with the datasets abalone, car and gcredit show that Auto-sklearn limits the generation of invalid pipelines by making assumption about cleaned input datasets, because the experiments crash if the input datasets have multiple attribute types. It means that Auto-sklearn can not handle invalid pipelines effectively.

4.3 Experiments to Compare the Performance of AVATAR and the Existing Methods

In order to demonstrate the efficiency of the AVATAR, we have conducted a second set of experiments. We run these experiments on a machine with an Intel core i7-8650U CPU and 16 GB memory. We compare the performance of the AVATAR and the T-method that requires the executions of pipelines. The T-method is used to evaluate the validity of pipelines in the pipeline composition and optimisation of AutoWeka4MCPS and Auto-sklearn. We randomly generate ML pipelines which have up to six components (i.e., these component types are missing value handling, dimensionality reduction, outlier removal, data transformation, data sampling and predictor). The predictor is put at the end

Table 5. Comparison of the performance of the AVATAR and T-method

Dataset		abalone	car	convex	gcredit	winequality
T-method	Invalid/valid pipelines	683/ 1,097	4,387/ 6,817	252/ 428	4,557/ 7,208	1,276/ 1,951
	Total evaluation time of invalid/valid pipelines (s)	27,711.9/ 15,484.1	18,627.9/ 24,459.4	5,818.3/ 37,765.1	19,597.9/ 23,452.5	10,830.1/ 32,326.9
AVATAR	Invalid/valid pipelines	663/ 1,117	4,387/ 6,817	250/ 430	4,552/ 7,213	1,262/ 1,965
	Total evaluation time of invalid/valid pipelines (s)	3.5/4.9	43.1/64.8	19.6/131.1	57.0/89.2	17.1/25.4
Pipelines have different/similar evaluated results		20/1,760	0/11,204	2/678	5/11,760	14/3,213
The percentage of pipelines that the AVATAR can validate accurately (%)		98.88	100.00	99.71	99.96	99.57

of the pipelines because a valid pipeline always has a predictor at the end. Each pipeline is evaluated by the AVATAR and the T-method. We set the time budget to 12 h per dataset. We use the following criteria to compare the performance: the number of invalid/valid pipelines, the total evaluation time of invalid/valid pipelines (seconds), the number of pipelines that have the same evaluated results between the AVATAR and the T-method, and the percentage of the pipelines that the AVATAR can validate accurately (%) in comparison to the T-method.

Table 5 compares the performance of the AVATAR and the T-method using the above criteria. We can see that the total evaluation time of invalid/valid pipelines of the AVATAR is significantly lower than the T-method. While the evaluation time of pipelines of the AVATAR is quite stable, the evaluation time of pipelines of the T-method is much higher and depends on the size of the datasets. It means that the AVATAR is faster than the T-method in evaluating both invalid and valid pipelines regardless of the size of datasets. Moreover, we can see that the accuracy of the AVATAR is approximately 99% in comparison with the T-method. We have carefully reviewed the pipelines which have different evaluated results between the AVATAR and the T-method. Interestingly, the AVATAR evaluates all of these pipelines to be valid and vice versa in the case of the T-method. The reason is that executions of these pipelines cause the out of memory problem. In other words, the AVATAR does not consider the allocated memory as an impact on the validity of a pipeline. A promising solution is to reduce the size of an input dataset by adding a sampling component with appropriate hyperparameters. If the sampling size is too small, we may miss important features. If the sampling size is large, we may continue to run into the problem of out of memory. We cannot conclude that if we allocate more memory, whether the executions of these pipelines would be successful or not. It proves that the validity of a pipeline also depends on its execution environment such as memory. These factors have not been considered yet in the AVATAR. This is an interesting research gap that should be addressed in the future.

Table 6. Five invalid pipelines with the longest evaluation time using the T-method on the gcredit dataset.

Pipeline	#1	#2	#3	#4	#5
T-method (s)	11.092	11.068	11.067	11.067	11.066
AVATAR (s)	0.014	0.012	0.011	0.011	0.011

Finally, we take a detailed look at the invalid pipelines with the longest evaluation time using the T-method on the gcredit dataset, as shown in Table 6. Pipeline #1 (11.092 s) has the structure *ReplaceMissingValues → PeriodicSampling → NumericToNominal → PrincipalComponents → SMOreg*. This pipeline is invalid because *SMOreg* does not work with nominal classes, and there is no component transforming the nominal to numeric data. We can see that the AVATAR is able to evaluate the validity of this pipeline without executing it in just 0.014 s.

5 Conclusion

We empirically demonstrate the problem of generation of invalid pipelines during pipeline composition and optimisation. We propose the AVATAR which is a pipeline evaluation method using a surrogate model. The AVATAR can be used to accelerate pipeline composition and optimisation methods by quickly ignoring invalid pipelines to improve the effectiveness of the AutoML optimisation process. In future, we will improve the AVATAR to evaluate pipelines' quality besides their validity. Moreover, we will investigate how to employ the AVATAR to reduce search spaces dynamically.

Acknowledgment. This research is sponsored by AAi, University of Technology Sydney (UTS).

References

1. Kadlec, P., Gabrys, B.: Architecture for development of adaptive on-line prediction models. Memetic Computing **1** (2009). https://doi.org/10.1007/s12293-009-0017-8. Article number. 241

2. Salvador, M.M., Budka, M., Gabrys, B.: Automatic composition and optimization of multicomponent predictive systems with an extended auto-WEKA. IEEE Trans. Autom. Sci. Eng. **16**(2), 946–959 (2019)
3. Zöller, M.A., Huber, M.F.: Survey on automated machine learning. arXiv preprint arXiv:1904.12054 (2019)
4. Olson, R.S., Moore, J.H.: TPOT: a tree-based pipeline optimization tool for automating machine learning. In: Workshop on Automatic Machine Learning, pp. 66–74 (2016)
5. Feurer, M., Klein, A., Eggensperger, K., Springenberg, J., Blum, M., Hutter, F.: Efficient and robust automated machine learning. In: Advances in Neural Information Processing Systems, pp. 2962–2970 (2015)
6. Mohr, F., Wever, M., Hüllermeier, E.: ML-Plan: automated machine learning via hierarchical planning. Mach. Learn. **107**, 1495–1515 (2018). https://doi.org/10.1007/s10994-018-5735-z
7. Gil, Y., et al.: P4ML: a phased performance-based pipeline planner for automated machine learning. In: AutoML Workshop at ICML (2018)
8. de Sá, A.G.C., Pinto, W.J.G.S., Oliveira, L.O.V.B., Pappa, G.L.: RECIPE: a grammar-based framework for automatically evolving classification pipelines. In: McDermott, J., Castelli, M., Sekanina, L., Haasdijk, E., García-Sánchez, P. (eds.) EuroGP 2017. LNCS, vol. 10196, pp. 246–261. Springer, Cham (2017). https://doi.org/10.1007/978-3-319-55696-3_16
9. Tsakonas, A., Gabrys, B.: GRADIENT: grammar-driven genetic programming framework for building multi-component, hierarchical predictive systems. Expert Syst. Appl. **39**, 13253–13266 (2012)
10. Chinosi, M., Trombetta, A.: Modeling and validating BPMN diagrams. In: 2009 IEEE Conference on Commerce and Enterprise Computing, pp. 353–360. IEEE (2009)
11. Thornton, C., Hutter, F., Hoos, H.H., Leyton-Brown, K.: Auto-WEKA: combined selection and hyperparameter optimization of classification algorithms. In: Proceedings of the 19th ACM SIGKDD International Conference on Knowledge Discovery and Data Mining, pp. 847–855. ACM (2013)
12. Salvador, M.M., Budka, M., Gabrys, B.: Modelling multi-component predictive systems as Petri nets (2017)
13. Tan, W., Fan, Y., Zhou, M., Tian, Z.: Data-driven service composition in enterprise SOA solutions: a Petri net approach. IEEE Trans. Autom. Sci. Eng. **7**, 686–694 (2010)
14. Martin Salvador, M., Budka, M., Gabrys, B.: Towards automatic composition of multicomponent predictive systems. In: Martínez-Álvarez, F., Troncoso, A., Quintián, H., Corchado, E. (eds.) HAIS 2016. LNCS (LNAI), vol. 9648, pp. 27–39. Springer, Cham (2016). https://doi.org/10.1007/978-3-319-32034-2_3
15. Barker, A., van Hemert, J.: Scientific workflow: a survey and research directions. In: Wyrzykowski, R., Dongarra, J., Karczewski, K., Wasniewski, J. (eds.) PPAM 2007. LNCS, vol. 4967, pp. 746–753. Springer, Heidelberg (2008). https://doi.org/10.1007/978-3-540-68111-3_78
16. Balaji, A., Allen, A.: Benchmarking automatic machine learning frameworks. arXiv preprint arXiv:1808.06492 (2018)

Detection of Derivative Discontinuities in Observational Data

Dimitar Ninevski$^{(\boxtimes)}$ and Paul O'Leary

University of Leoben, 8700 Leoben, Austria
automation@unileoben.ac.at
http://automation.unileoben.ac.at

Abstract. This paper presents a new approach to the detection of discontinuities in the n-th derivative of observational data. This is achieved by performing two polynomial approximations at each interstitial point. The polynomials are coupled by constraining their coefficients to ensure continuity of the model up to the (n − 1)-th derivative; while yielding an estimate for the discontinuity of the n-th derivative. The coefficients of the polynomials correspond directly to the derivatives of the approximations at the interstitial points through the prudent selection of a common coordinate system. The approximation residual and extrapolation errors are investigated as measures for detecting discontinuity. This is necessary since discrete observations of continuous systems are discontinuous at every point. It is proven, using matrix algebra, that positive extrema in the combined approximation-extrapolation error correspond exactly to extrema in the difference of the Taylor coefficients. This provides a relative measure for the severity of the discontinuity in the observational data. The matrix algebraic derivations are provided for all aspects of the methods presented here; this includes a solution for the covariance propagation through the computation. The performance of the method is verified with a Monte Carlo simulation using synthetic piecewise polynomial data with known discontinuities. It is also demonstrated that the discontinuities are suitable as knots for B-spline modelling of data. For completeness, the results of applying the method to sensor data acquired during the monitoring of heavy machinery are presented.

Keywords: Data analysis · Discontinuity detection · Free-knot splines

1 Introduction

In the recent past *physics informed data science* has become a focus of research activities, e.g., [9]. It appears under different names e.g., *physics informed* [12]; *hybrid learning* [13]; *physics-based* [17], etc.; but with the same basic idea of embedding physical principles into the data science algorithms. The goal is to ensure that the results obtained obey the laws of physics and/or are based on physically relevant features. Discontinuities in the observations of continuous

© The Author(s) 2020
M. R. Berthold et al. (Eds.): IDA 2020, LNCS 12080, pp. 366–378, 2020.
https://doi.org/10.1007/978-3-030-44584-3_29

systems violate some very basic physics and for this reason their detection is of fundamental importance. Consider Newton's second law of motion,

$$F(t) = \frac{\mathrm{d}}{\mathrm{d}t}\left\{m(t)\,\frac{\mathrm{d}}{\mathrm{d}t}y(t)\right\} = \dot{m}(t)\,\dot{y}(t) + m(t)\,\ddot{y}(t). \tag{1}$$

Any discontinuities in the observations of $m(t)$, $\dot{m}(t)$, $y(t)$, $\dot{y}(t)$ or $\ddot{y}(t)$ indicate a violation of some basic principle: be it that the observation is incorrect or something unexpected is happening in the system. Consequently, detecting discontinuities is of fundamental importance in physics based data science. A function $s(x)$ is said to be C^n discontinuous, if $s \in C^{n-1}\backslash C^n$, that is if $s(x)$ has continuous derivatives up to and including order $n - 1$, but the n-th derivative is discontinuous. Due to the discrete and finite nature of the observational data, only jump discontinuities in the n-th derivative are considered; asymptotic discontinuities are not considered. Furthermore, in more classical data modelling, C^n jump discontinuities form the basis for the locations of knots in B-Spline models of observational data [15].

1.1   State of the Art

There are numerous approaches in the literature dealing with estimating regression functions that are smooth, except at a finite number of points. Based on the methods, these approaches can be classified into four groups: local polynomial methods, spline-based methods, kernel-based methods and wavelet methods. The approaches vary also with respect to the available a priori knowledge about the number of points of discontinuity or the derivative in which these discontinuities appear. For a good literature review of these methods, see [3]. The method used in this paper is relevant both in terms of local polynomials as well as spline-based methods; however, the new approach requires no a priori knowledge about the data.

In the local polynomial literature, namely in [8] and [14], ideas similar to the ones presented here are investigated. In these papers, local polynomial approximations from the left and the right side of the point in question are used. The major difference is that neither of these methods use constraints to ensure that the local polynomial approximations enforce continuity of the lower derivatives, which is done in this paper. As such, they use different residuals to determine the existence of a change point. Using constrained approximation ensures that the underlying physical properties of the system are taken into consideration, which is one of the main advantages of the approach presented here. Additionally, in the aforementioned papers, it is not clear whether only co-locative points are considered as possible change points, or interstitial points are also considered. This distinction between collocative and interstitial is of great importance. Fundamentally, the method presented here can be applied to discontinuities at either locations. However, it has been assumed that discontinuities only make sense between the sampled (co-locative) points, i.e., the discontinuities are interstitial.

In [11] on the other hand, one polynomial instead of two is used, and the focus is mainly on detecting C^0 and C^1 discontinuities. Additionally, the number of change-points must be known a-priori, so only their location is approximated; the required a-priori knowledge make the method unsuitable in real sensor based system observation.

In the spline-based literature there are heuristic methods (top-down and bottom-up) as well as optimization methods. For a more detailed state of the art on splines, see [2]. Most heuristic methods use a discrete geometric measure to calculate whether a point is a knot, such as: discrete curvature, kink angle, etc, and then use some (mostly arbitrary) threshold to improve the initial knot set. In the method presented here, which falls under the category of bottom-up approaches, the selection criterion is based on calculus and statistics, which allows for incorporation of the fundamental physical laws governing the system, in the model, but also ensures mathematical relevance and rigour.

1.2   The New Approach

This paper presents a new approach to detecting C^n discontinuities in observational data. It uses constrained coupled polynomial approximation to obtain two estimates for the nth Taylor coefficients and their uncertainties, at every interstitial point. These correspond approximating the local function by polynomials, once from the left $f(x, \alpha)$ and once from the right $g(x, \beta)$. The constraints couple the polynomials to ensure that $\alpha_i = \beta_i$ for every $i \in [0 \dots n - 1]$. In this manner the approximations are C^{n-1} continuous at the interstitial points, while delivering an estimate for the difference in the nth Taylor coefficients. All the derivations for the coupled constrained approximations and the numerical implementations are presented. Both the approximation and extrapolation residuals are derived. It is proven that the discontinuities must lie at local positive peaks in the extrapolation error. The new approach is verified with both known synthetic data and on real sensor data obtained from observing the operation of heavy machinery.

2   Detecting C^n Discontinuities

Discrete observations $s(x_i)$ of a continuous system $s(x)$ are, by their very nature, discontinuous at every sample. Consequently, some measure for discontinuity will be required, with uncertainty, which provides the basis for further analysis.

The observations are considered to be the co-locative points, denoted by x_i and collectively by the vector x; however, we wish to estimate the discontinuity at the interstitial points, denoted by ζ_i and collectively as ζ. Using interstitial points, one ensures that each data point is used for only one polynomial approximation at a time. Furthermore, in the case of sensor data, one expects the discontinuities to happen between samples. Consequently the data is segmented at the interstitial points, i.e. between the samples. This requires the use of interpolating functions and in this work we have chosen to use polynomials.

Polynomials have been chosen because of their approximating, interpolating and extrapolating properties when modelling continuous systems: The Weierstrass approximation theorem [16] states that if $f(x)$ is a continuous real-valued function defined on the real interval $x \in [a,b]$, then for every $\varepsilon > 0$, there exists a polynomial $p(x)$ such that for all $x \in [a,b]$, the supremum norm $\|f(x) - p(x)\|_\infty < \varepsilon$. That is *any* function $f(x)$ can be approximated by a polynomial to an arbitrary accuracy ε given a sufficiently high degree.

The basic concept (see Fig. 1) to detect a C^n discontinuity is: to approximate the data to the left of an interstitial point by the polynomial $\mathsf{f}(x, \boldsymbol{\alpha})$ of degree d_L and to the right by $\mathsf{g}(x, \boldsymbol{\beta})$ of degree d_R, while constraining these approximations to be C^{n-1} continuous at the interstitial point. This approximation ensures that,

$$\mathsf{f}^{(k-1)}(\zeta_i) = \mathsf{g}^{(k-1)}(\zeta_i), \qquad \text{for every } k \in [1 \ldots n]. \tag{2}$$

while yielding estimates for $\mathsf{f}^{(n)}(\zeta_i)$ and $\mathsf{g}^{(n)}(\zeta_i)$ together with estimates for their variances $\lambda_{f(\zeta_i)}$ and $\lambda_{g(\zeta_i)}$. This corresponds exactly to estimating the Taylor coefficients of the function twice for each interstitial point, i.e., once from the left and once from the right. It they differ significantly, then the function's n^{th} derivative is discontinuous at this point. The Taylor series of a function $f(x)$ around the point a is defined as,

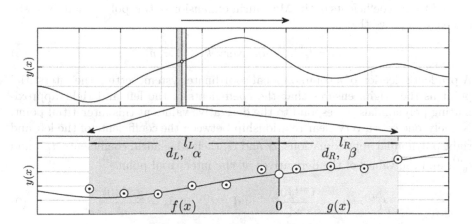

Fig. 1. Schematic of a finite set of discrete observations (dotted circles) of a continuous function. The span of the observation is split into a left and right portion at the interstitial point (circle), with lengths l_L and l_R respectively. The left and right sides are considered to be the functions $f(x)$ and $g(x)$; modelled by the polynomials $\mathsf{f}(x, \boldsymbol{\alpha})$ and $\mathsf{g}(x, \boldsymbol{\beta})$ of degrees d_L and d_R.

$$f(x) = \sum_{k=0}^{\infty} \frac{f^{(k)}(a)}{k!} (x - a)^k \tag{3}$$

for each x for which the infinite series on the right hand side converges. Furthermore, any function which is $n + 1$ times differentiable can be written as

$$f(x) = \tilde{\mathsf{f}}(x) + R(x) \tag{4}$$

where $\tilde{f}(x)$ is an nth degree polynomial approximation of the function $f(x)$,

$$\tilde{f}(x) = \sum_{k=0}^{n} \frac{f^{(k)}(a)}{k!}(x-a)^k \tag{5}$$

and $R(x)$ is the remainder term. The Lagrange form of the remainder $R(x)$ is given by

$$R(x) = \frac{f^{(n+1)}(\xi)}{(n+1)!}(x-a)^{n+1} \tag{6}$$

where ξ is a real number between a and x.

A Taylor expansion around the origin (i.e. $a = 0$ in Eq. 3) is called a Maclaurin expansion; for more details, see [1]. In the rest of this work, the nth Maclaurin coefficient for the function $f(x)$ will be denoted by

$$t_f^{(n)} \triangleq \frac{f^{(n)}(0)}{n!}. \tag{7}$$

The coefficients of a polynomial $f(x, \boldsymbol{\alpha}) = \alpha_n x^n + \ldots + \alpha_1 x + \alpha_0$ are closely related to the coefficients of the Maclaurin expansion of this polynomial. Namely, it's easy to prove that

$$\alpha_k = t_f^{(k)}, \quad \text{for every } k \in [0 \ldots n]. \tag{8}$$

A prudent selection of a common local coordinate system, setting the interstitial point as the origin, ensures that the coefficients of the left and right approximating polynomials correspond to the derivative values at this interstitial point. Namely, one gets a very clear relationship between the coefficients of the left and right polynomial approximations, $\boldsymbol{\alpha}$ and $\boldsymbol{\beta}$, their Maclaurin coefficients, $t_f^{(n)}$ and $t_g^{(n)}$, and the values of the derivatives at the interstitial point

$$t_f^{(n)} = \alpha_n = \frac{f^{(n)}(0)}{n!} \quad \text{and} \quad t_g^{(n)} = \beta_n = \frac{g^{(n)}(0)}{n!}. \tag{9}$$

From Eq. 9 it is clear that performing a left and right polynomial approximation at an interstitial point is sufficient to get the derivative values at that point, as well as their uncertainties.

3   Constrained and Coupled Polynomial Approximation

The goal here is to obtain $\Delta t_{\text{fg}}^{(n)} \triangleq t_f^{(n)} - t_g^{(n)}$ via polynomial approximation. To this end two polynomial approximations are required; whereby, the interstitial point is used as the origin in the common coordinate system, see Fig. 1. The approximations are coupled [6] at the interstitial point by constraining the coefficients such that $\alpha_i = \beta_i$, for every $i \in [0 \ldots n-1]$. This ensures that the two polynomials are C^{n-1} continuous at the interstitial points. This also reduces the degrees of freedom during the approximation and with this the variance of the solution is reduced. For more details on constrained polynomial approximation see [4, 7].

To remain fully general, a local polynomial approximation of degree d_L is performed to the left of the interstitial point with the support length l_L creating $f(x, \alpha)$; similarly to the right d_R, l_R, $g(x, \beta)$. The x coordinates to the left, denoted as x_L are used to form the left Vandermonde matrix V_L, similarly x_R form V_R to the right. This leads to the following formulation of the approximation process,

$$y_L = V_L \, \alpha \quad \text{and} \quad y_R = V_R \, \beta. \tag{10}$$

$$\begin{bmatrix} V_L & 0 \\ 0 & V_R \end{bmatrix} \begin{bmatrix} \alpha \\ \beta \end{bmatrix} = \begin{bmatrix} y_L \\ y_R \end{bmatrix} \tag{11}$$

A C^{n-1} continuity implies $\alpha_i = \beta_i$, for every $i \in [0 \ldots n-1]$ which can be written in matrix form as

$$\begin{bmatrix} 0 & I_{n-1} \, | \, 0 & -I_{n-1} \end{bmatrix} \begin{bmatrix} \alpha \\ \beta \end{bmatrix} = 0 \tag{12}$$

Defining

$$V \triangleq \begin{bmatrix} V_L & 0 \\ 0 & V_R \end{bmatrix}, \gamma \triangleq \begin{bmatrix} \alpha \\ \beta \end{bmatrix}, y \triangleq \begin{bmatrix} y_L \\ y_R \end{bmatrix} \text{ and } C \triangleq \begin{bmatrix} 0 & I_{n-1} \, | \, 0 & -I_{n-1} \end{bmatrix}$$

We obtain the task of least squares minimization with homogeneous linear constraints,

$$\boxed{\begin{aligned} & \min_{\gamma} \quad \|y - V \, \gamma\|_2^2 \\ & \text{Given} \quad C \gamma = 0. \end{aligned}} \tag{13}$$

Clearly γ must lie in the null-space of C; now, given N, an ortho-normal vector basis set for null $\{C\}$, we obtain,

$$\gamma = N \, \delta. \tag{14}$$

Back-substituting into Eq. 13 yields,

$$\min_{\delta} \|y - V \, N \, \delta\|_2^2 \tag{15}$$

The least squares solution to this problem is,

$$\delta = (V \, N)^+ \, y, \tag{16}$$

and consequently,

$$\boxed{\gamma = \begin{bmatrix} \alpha \\ \beta \end{bmatrix} = N \, (V \, N)^+ \, y} \tag{17}$$

Formulating the approximation in the above manner ensures that the difference in the Taylor coefficients can be simply computed as

$$\Delta t_{\text{fg}}^{(n)} = t_{\text{f}}^{(n)} - t_{\text{g}}^{(n)} = \alpha_n = \beta_n. \tag{18}$$

Now defining $d = [1, \mathbf{0}_{d_L-1}, -1, \mathbf{0}_{d_R-1}]^{\text{T}}$, $\Delta t_{\text{fg}}^{(n)}$ is obtained from γ as

$$\Delta t_{\text{fg}}^{(n)} = d^{\text{T}} \gamma = d^{\text{T}} N (V N)^+ y. \tag{19}$$

3.1   Covariance Propagation

Defining, $K = N (V N)^+$, yields, $\gamma = K y$. Then given the covariance of y, i.e., Λ_y, one gets that,

$$\boxed{\Lambda_\gamma = K \Lambda_y K^{\text{T}}.} \tag{20}$$

Additionally, from Eq. 19 one could derive the covariance of the difference in the Taylor coefficients

$$\Lambda_\Delta = d \Lambda_\gamma d^{\text{T}} \tag{22}$$

Keep in mind that, if one uses approximating polynomials of degree n to determine a discontinuity in the nth derivative, as done so far, Λ_Δ is just a scalar and corresponds to the variance of $\Delta t_{\text{fg}}^{(n)}$.

4   Error Analysis

In this paper we consider three measures for error:

1. the norm of the approximation residual;
2. the combined approximation and extrapolation error;
3. the extrapolation error.

4.1   Approximation Error

The residual vector has the form

$$r = y - V\gamma = \begin{bmatrix} y_L - V_L\alpha \\ y_R - V_R\beta \end{bmatrix}.$$

The approximation error is calculated as

$$\begin{aligned}
E_a &= \|r\|_2^2 = \|y_L - V_L\alpha\|_2^2 + \|y_R - V_R\beta\|_2^2 \\
&= (y_L - V_L\alpha)^{\text{T}} (y_L - V_L\alpha) + (y_R - V_R\beta)^{\text{T}} (y_R - V_R\beta) \\
&= y^{\text{T}}y - 2\alpha^{\text{T}}V_L^{\text{T}}y_L + \alpha^{\text{T}}V_L^{\text{T}}V_L\alpha - 2\beta^{\text{T}}V_R^{\text{T}}y_R + \beta^{\text{T}}V_R^{\text{T}}V_R\beta.
\end{aligned}$$

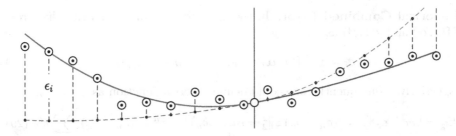

Fig. 2. Schematic of the approximations around the interstitial point. Red: left polynomial approximation $f(x, \alpha)$; dotted red: extrapolation of $f(x, \alpha)$ to the RHS; blue: right polynomial approximation, $g(x, \beta)$; dotted blue: extrapolation of $g(x, \beta)$ to the LHS; ε_i is the vertical distance between the extrapolated value and the observation. The approximation is constrained with the conditions: $f(0, \alpha) = g(0, \beta)$ and $f'(0, \alpha) = g'(0, \beta)$. (Color figure online)

4.2 Combined Error

The basic concept, which can be seen in Fig. 2, is as follows: the left polynomial $f(x, \alpha)$, which approximates over the values x_L, is extended to the right and evaluated at the points x_R. Analogously, the right polynomial $g(x, \beta)$ is evaluated at the points x_L. If there is no C^n discontinuity in the system, the polynomials f and g must be equal and consequently the extrapolated values won't differ significantly from the approximated values.

Analytical Combined Error. The extrapolation error in a continuous case, i.e. between the two polynomial models, can be computed with the following 2-norm,

$$\varepsilon_x = \int_{x_{min}}^{x_{max}} \{f(x, \alpha) - g(x, \beta)\}^2 \, dx. \tag{23}$$

Given, the constraints which ensure that $\alpha_i = \beta_i \, i \in [0, \ldots, n-1]$, we obtain,

$$\varepsilon_x = \int_{x_{min}}^{x_{max}} \{(\alpha_n - \beta_n) x^n\}^2 \, dx. \tag{24}$$

Expanding and performing the integral yields,

$$\varepsilon_x = (\alpha_n - \beta_n)^2 \left\{ \frac{x_{max}^{2n+1} - x_{min}^{2n+1}}{2n+1} \right\} \tag{25}$$

Given fixed values for x_{min} and x_{max} across a single computation implies that the factor,

$$k = \frac{x_{max}^{2n+1} - x_{min}^{2n+1}}{2n+1} \tag{26}$$

is a constant. Consequently, the extrapolation error is directly proportional to the square of the difference in the Taylor coefficients,

$$\varepsilon_x \propto (\alpha_n - \beta_n)^2 \propto \left\{ \Delta t_{fg}^{(n)} \right\}^2. \tag{27}$$

Numerical Combined Error. In the discrete case, one can write the errors of $f(x, \alpha)$ and $g(x, \beta)$ as

$$e_f = y - f(x, \alpha) \quad \text{and} \quad e_g = y - g(x, \beta) \tag{28}$$

respectively. Consequently, one could define an error function as

$$E_{fg} = \|e_f - e_g\|_2^2 = \|(a_n - b_n) z\|_2^2 = (a_n - b_n)^2 z^T z^n = (a_n - b_n)^2 \sum x_i^n \tag{29}$$

where $z \triangleq x.\hat{}\ n$. From these calculations it is clear that in the discrete case the error is also directly proportional to the square of the difference in the Taylor coefficients and that $E_{fg} \propto \varepsilon_x$. This proves that the numerical computation is consistent with the analytical continuous error.

4.3   Extrapolation Error

One could also define a different kind of error, based just on the extrapolative properties of the polynomials. Namely, using the notation from the beginning of Sect. 3, one defines

$$r_{ef} = y_L - g(x_L, \beta) = y_L - V_L \beta \quad \text{and} \quad r_{eg} = y_R - f(x_R, \alpha) = y_R - V_R \alpha$$

and then calculates the error as

$$E_e = r_{ef}^T r_{ef} + r_{eg}^T r_{eg}$$
$$= (y_L - V_L \beta)^T (y_L - V_L \beta) + (y_R - V_R \alpha)^T (y_R - V_R \alpha)$$
$$= y^T y - 2\beta^T V_L^T y_L + \beta^T V_L^T V_L \beta - 2\alpha^T V_R^T y_R + \alpha^T V_R^T V_R \alpha.$$

In the example in Sect. 5, it will be seen that there is no significant numerical difference between these two errors.

5   Numerical Testing

The numerical testing is performed with: synthetic data from a piecewise polynomial, where the locations of the C^n discontinuities are known; and with real sensor data emanating from the monitoring of heavy machinery.

5.1   Synthetic Data

In the literature on splines, functions of the type $y(x) = e^{-x^2}$ are commonly used. However, this function is analytic and C^∞ continuous; consequently it was not considered a suitable function for testing. In Fig. 3 a piecewise polynomial with a similar shape is shown; however, this curve has C^2 discontinuities at known locations. The algorithm was applied to the synthetic data from the piecewise polynomial, with added noise with $\sigma = 0.05$ and the results for a single case can be seen in Fig. 3. Additionally, a Monte Carlo simulation with $m = 10000$ iterations was performed and the results of the algorithm were compared to the true locations of the two known knots. The mean errors in the location of the knots are: $\mu_1 = (5.59 \pm 2.05) \times 10^{-4}$ with 95% confidence, and $\mu_2 = (-4.62 \pm 1.94) \times 10^{-4}$. Errors in the scale of 10^{-4}, in a support with a range $[0, 1]$, and 5% noise amplitude in the curve can be considered a highly satisfactory result.

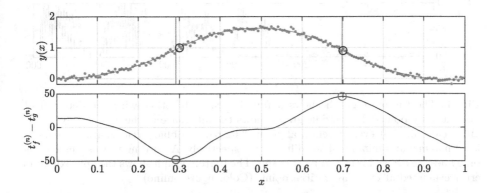

Fig. 3. A piecewise polynomial of degree $d = 2$, created from the knots sequence $x_k = [0, 0.3, 0.7, 1]$ with the corresponding values $y_k = [0, 0.3, 0.7, 1]$. The end points are clamped with $y'(x)_{0,1} = 0$. Gaussian noise is added with $\sigma = 0.05$. Top: the circles mark the known points of C^2 discontinuity; the blue and red lines indicate the detected discontinuities; additionally the data has been approximated by the b-spline (red) using the detected discontinuities as knots. Bottom: shows $\Delta t_{fg}^{(n)} = t_f^{(n)} - t_g^{(n)}$, together with the two identified peaks. (Color figure online)

5.2 Sensor Data

The algorithm was also applied to a set of real-world sensor data[1] emanating from the monitoring of heavy machinery. The original data set can be seen in Fig. 4 (top). It has many local peaks and periods of little or no change, so the algorithm was used to detect discontinuities in the first derivative, in order to determine the peaks and phases. The peaks in the Taylor differences were used in combination with the peaks of the extrapolation error to determine the points of discontinuity. A peak in the Taylor differences means that the Taylor coefficients are significantly different at that interstitial point, compared to other interstitial points in the neighbourhood. However, if there is no peak in the extrapolation errors at the same location, then the peak found by the Taylor differences is deemed insignificant, since one polynomial could model both the left and right values and as such the peak isn't a discontinuity. Additionally, it can be seen in

---

[1] For confidentiality reasons the data has been anonymized.

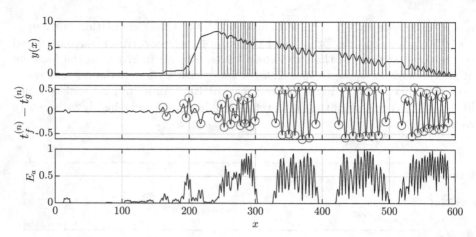

Fig. 4. The top-most graph shows a function $y(x)$, together with the detected C^1 discontinuity points. The middle graph shows the difference in the Taylor polynomials $\Delta t_{\mathrm{fg}}^{(n)}$ calculated at every interstitial point. The red and blue circles mark the relevant local maxima and minima of the difference respectively. According to this, the red and blue lines are drawn in the top-most graph. The bottom graph shows the approximation error evaluated at every interstitial point. (Color figure online)

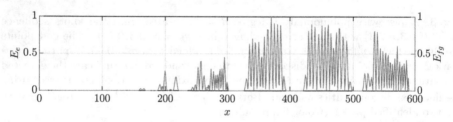

Fig. 5. The two error functions, E_e and E_{fg} as defined in Sect. 4, for the example from Fig. 4. One can see that the location of the peaks doesn't change, and the two errors don't differ significantly.

Fig. 5 that both the extrapolation error and the combined error, as defined in Sect. 4, have peaks at the same locations, and as such the results they provide do not differ significantly.

6   Conclusion and Future Work

It may be concluded, from the results achieved, that the coupled constrained polynomial approximation yield a good method for the detection of C^n discontinuities in discrete observational data of continuous systems. Local peaks in the square of the difference of the Taylor polynomials provide a relative measure as a means of determining the locations of discontinuities.

Current investigations indicate that the method can be implemented directly as a convolutional operator, which will yield a computationally efficient solution.

The use of discrete orthogonal polynomials [5,10] is being tested as a means of improving the sensitivity of the results to numerical perturbations.

Acknowledgements. This work was partially funded by:

1. The COMET program within the K2 Center "Integrated Computational Material, Process and Product Engineering (IC-MPPE)" (Project No 859480). This program is supported by the Austrian Federal Ministries for Transport, Innovation and Technology (BMVIT) and for Digital and Economic Affairs (BMDW), represented by the Austrian research funding association (FFG), and the federal states of Styria, Upper Austria and Tyrol.
2. The European Institute of Innovation and Technology (EIT), a body of the European Union which receives support from the European Union's Horizon 2020 research and innovation programme. This was carried out under Framework Partnership Agreement No. 17031 (MaMMa - Maintained Mine & Machine).

The authors gratefully acknowledge this financial support.

References

1. Burden, R.L., Faires, J.D.: Numerical Analysis, 9th edn. Pacific Grove, Brooks/-Cole (2010)
2. Dung, V.T., Tjahjowidodo, T.: A direct method to solve optimal knots of B-spline curves: an application for non-uniform B-spline curves fitting. PLoS ONE **12**(3), 1–24 (2017). https://doi.org/10.1371/journal.pone.0173857
3. Gijbels, I., Goderniaux, A.C.: Data-driven discontinuity detection in derivatives of a regression function. Commun. Stat.-Theory Methods **33**(4), 851–871 (2005). https://doi.org/10.1081/STA-120028730
4. Klopfenstein, R.W.: Conditional least squares polynomial approximation. Math. Comput. **18**(88), 659–662 (1964). http://www.jstor.org/stable/2002954
5. O'Leary, P., Harker, M.: Discrete polynomial moments and Savitzky-Golay smoothing. Int. J. Comput. Inf. Eng. **4**(12), 1993–1997 (2010). https://publications.waset.org/vol/48
6. O'Leary, P., Harker, M., Zsombor-Murray, P.: Direct and least square fitting of coupled geometric objects for metric vision. IEE Proc. Vis. Image Sig. Process. **152**, 687–694 (2006). https://doi.org/10.1049/ip-vis:20045206
7. O'Leary, P., Ritt, R., Harker, M.: Constrained polynomial approximation for inverse problems in engineering. In: Abdel Wahab, M. (ed.) NME 2018. LNME, pp. 225–244. Springer, Singapore (2019). https://doi.org/10.1007/978-981-13-2273-0_19
8. Orváth, L., Kokoszka, P.: Change-point detection with non-parametric regression. Statistics **36**(1), 9–31 (2002). https://doi.org/10.1080/02331880210930
9. Owhadi, H.: Bayesian numerical homogenization. Multiscale Model. Simul. **13**(3), 812–828 (2015). https://doi.org/10.1137/140974596
10. Persson, P.O., Strang, G.: Smoothing by Savitzky-Golay and Legendre filters. In: Rosenthal, J., Gilliam, D.S. (eds.) Mathematical Systems Theory in Biology, Communications, Computation, and Finance. IMA, vol. 134, pp. 301–315. Springer, New York (2003). https://doi.org/10.1007/978-0-387-21696-6_11

11. Qiu, P., Yandell, B.: Local polynomial jump-detection algorithm in nonparametric regression. Technometrics **40**(2), 141–152 (1998). https://doi.org/10.1080/
00401706.1998.10485196
12. Raissi, M., Perdikaris, P., Karniadakis, G.: Physics-informed neural networks:
a deep learning framework for solving forward and inverse problems involving nonlinear partial differential equations. J. Comput. Phys. **378**, 686–707
(2019). https://doi.org/10.1016/j.jcp.2018.10.045. http://www.sciencedirect.com/
science/article/pii/S0021999118307125
13. Saxena, H., Aponte, O., McConky, K.T.: A hybrid machine learning model for
forecasting a billing period's peak electric load days. Int. J. Forecast. **35**(4), 1288–
1303 (2019). https://doi.org/10.1016/j.ijforecast.2019.03.025
14. Spokoiny, V.: Estimation of a function with discontinuities via local polynomial fit
with an adaptive window choice. Ann. Stat. **26** (1998). https://doi.org/10.1214/
aos/1024691246
15. Wahba, G.: Spline models for observational data. Soc. Ind. Appl. Math. (1990).
https://doi.org/10.1137/1.9781611970128
16. Weierstrass, K.: Über die analytische darstellbarkeit sogenannter willkürlicher
functionen einer reellen veränderlichen. Sitzungsberichte der Königlich Preußis-
chen Akademie der Wissenschaften zu Berlin, **1885**(II), 633–639, 789–805 (1885)
17. Yaman, B., Hosseini, S.A.H., Moeller, S., Ellermann, J., Uğurbil, K., Akçakaya,
M.: Self-supervised physics-based deep learning MRI reconstruction without fully-
sampled data (2019)

Improving Prediction with Causal Probabilistic Variables

Ana Rita Nogueira[1,2]([✉]), João Gama[1]([✉]), and Carlos Abreu Ferreira[1]([✉])

[1] LIAAD - INESC TEC, Rua Dr. Roberto Frias, 4200-465 Porto, Portugal
ana.r.nogueira@inesctec.pt, jgama@fep.up.pt, cgf@isep.ipp.pt
[2] Faculdade de Ciências da Universidade do Porto,
Rua do Campo Alegre 1021/1055, 4169-007 Porto, Portugal

Abstract. The application of feature engineering in classification problems has been commonly used as a means to increase the classification algorithms performance. There are already many methods for constructing features, based on the combination of attributes but, to the best of our knowledge, none of these methods takes into account a particular characteristic found in many problems: causality. In many observational data sets, causal relationships can be found between the variables, meaning that it is possible to extract those relations from the data and use them to create new features. The main goal of this paper is to propose a framework for the creation of new supposed causal probabilistic features, that encode the inferred causal relationships between the target and the other variables. In this case, an improvement in the performance was achieved when applied to the *Random Forest* algorithm.

Keywords: Causality · Causal discovery · Conditional probability · Feature engineering · Causal features

1 Introduction

In regular classification problems, a set of data, classified with a finite set of classes, is used as input so that the chosen classification algorithm can build a model, that represents the behaviour of the learning set. This classifier can have better or worse results, depending on the data and how the algorithm handles it.

Nevertheless, in many problems, applying only machine learning algorithms may not be the answer [4]. Instead, the use of feature engineering can be a way of improving the performance of these algorithms.

Feature engineering is a process by which new information is extracted from the available data, to create new features. These new features are related to the original variables, but also with the target variable, being a better representation of the knowledge embedded in the data, hence helping the algorithms achieve more accurate results [4]. These types of solutions are usually problem-related, being that one solution might work in one particular problem, but not

M. R. Berthold et al. (Eds.): IDA 2020, LNCS 12080, pp. 379–390, 2020.
https://doi.org/10.1007/978-3-030-44584-3_30

in the other. However, there is one particular characteristic common to many classification problems: causality. In observational data, there is the possibility of existing causal relationships between variables, especially in data related to medical problems (among others) [16,17]. This fact should be taken into consideration, for example when selecting or creating new features, since it can give clues to which variables are the most important to the problem.

By definition, causality, more specifically causal discovery, relates to the search of possible cause-effect relationships between variables [13]. The application of causal discovery in the various tasks of machine learning may be challenging, both at the level of the causal process or the sampling process to generate the observed data [9]. Despite this fact, this subject has been the focus of several researchers over the years, given the importance and the potential impact that the discovery of causal relationships between events can have in the problem-solving. In the words of Judea Pearl: *"while probabilities encode our beliefs about a static world, causality tells us whether and how probabilities change when the world changes, be it by intervention or by an act of imagination"* [20]. By discovering causal relationships, it is possible to uncover, not only correlations but also relations that explain how and why the variables behave the way they do.

In this paper, we propose a framework to create new features for discrete data sets (discrete features + discrete target) based on the causal relationships uncovered in the data. These attributes are created through the generation of a causal network, using a modified version of PC [21], and posterior probabilistic analysis of the relations a target variable and the variables considered as relevant. The relevant variables can be chosen by two different methods: parents and children of the target and Markov blanket [19].

This paper is organised as follows: Sect. 2 describes some important definitions. Section 3 describes the proposed framework and Sect. 4 the results obtained in the tests.

2   Background

In this section, we introduce some important notations that are used throughout the document.

2.1   PC

PC is a constraint-based algorithm and was proposed by Spirtes et al. [21]. This algorithm relies on the *faithfulness* assumption (*"If we have a joint probability distribution P of the random variables in some set V and a DAG G =(V,E), (G,P) satisfies the faithfulness condition if, G entails all and only conditional independencies in P"* [18]), meaning that all the independencies in a *DAG* (directed acyclic graph) need to respect the d-separation criterion [8].

This algorithm is divided into two phases. In the first phase, the algorithm starts with a fully connected undirected graph. It removes an edge if the two nodes are independent, *i.e.*, if there is a set of nodes adjacent to both variables in which they are conditionally independent [12]. One of the most applied statistical independence tests is G^2, proposed by Spirtes et al. [21], and then used in non-causal Bayesian networks by Tsamardinos et al. [24].

In the second phase [12], the algorithm orients the edges by first searching for v-structures $(A \rightarrow B \leftarrow C)$ and then by applying a set of rules, to create a completed partially directed acyclic graph $(CPDAG)$, that is equivalent to the original one, where the faithfulness is respected.

2.2   Cochran-Mantel-Haenszel Test

The Cochran-Mantel-Haenszel test, [2] is an independence test that studies the influence of two variables on each other, and takes into account the possible influence of other variables on this dependence, $i.e.$, it searches for causal dependence [11].

There are two different versions of this test: the normal Cochran-Mantel-Haenszel test, which is used in $2 \times 2 \times K$ tables (being K the number of tables created), and the Generalised Cochran-Mantel-Haenszel tests, which is used in $I \times J \times K$ tables (being that I and J represent the number of categories in the studied variables, and K the number of layer categories [6]).

It is important to note that these type of contingency tables (three-way tables) are representations of the association between two variables if the influence of the other covariates is controlled.

Since many causal discovery algorithms (for discrete data) are used in data sets that are composed by a mixture of binary and non-binary discrete variables, the normal Cochran-Mantel-Haenszel test for $2 \times 2 \times K$ contingency tables is not enough. In such cases, the generalised version of this test can be applied instead (Generalised Cochran-Mantel-Haenszel test Eq. (1) [15]).

$$Q_{CMH} = G' Var\{G|H_0\}^{-1} G \qquad G_h = B_h(n_h - m_h)$$
$$G = \sum_h G_h \qquad Var\{G|H_0\} = \sum_h Var\{G_h|H_0\} \qquad B_h = C_h \bigotimes R_h \qquad (1)$$

In the equations presented previously, B_h represents the product of Kronecker between C_h and R_h, Var the co-variance matrix, $(nh - mh)$ the difference between the observed and the expected, C_h and R_h the columns scores and row scores respectively, and H_0[1] the null hypothesis.

3   Framework

In many machine learning problems, the application of only classification algorithms might not be the answer to obtain satisfactory results [4]. The application

---

[1] "For each of the separate levels of the co-variable set $h = 1, 2, ..., q$, the response variable is distributed at random with respect to the sub-populations, i.e. the data in the respective rows of the h_{th} table can be regarded as a successive set of simple random samples of sizes $\{Nhi.\}$ from a fixed population corresponding to the marginal total distribution of the response variable $\{Nh.j\}$." [15].

of feature engineering to the target data can be a way of improving such results. There are already several methods to improve the overall performance of an algorithm through the creation or modification of attributes, but, to the best of our knowledge, none of them explores the potential causal relationships between the target variable and the other variables.

The addition of these new inferred causal attributes may help improve the performance of classification algorithms, since they encode the relationship between the target and the other variables, thus feeding more information about the data set and its behaviour to the model. Moreover, these features may also aid in the generated models interpretability, since they encode the underlying relationships between the variables, thus being possible to explain more easily the decisions made by them.

In this section, we present a new framework to create new features using causal probabilities retrieve from a model that represents the causal associations between variables. This framework can be divided into four different phases:

1. Creation of the causal model (in this approach we suggest the usage of a modified version of PC);
2. Identification of the relevant variables. These variables are directly related to the target variable:
   - They are its parents and children;
   - They belong to its Markov blanket (*i.e.* parents, children and spouses).
3. Inference the probabilities associated with each pair {*target variable, associated variable*};
4. Creation of the new features using this probabilities. The number of features should be: *number of associated variables × number of classes*.

In the first step, the framework starts by creating a full causal model, that represents the causal associations between all the variables. This is done through the application of a modified version of PC [21]. In this modified version, the state of the art independence test (usually X^2 or G^2) is replaced by the Generalised Cochran-Mantel-Haenszel test presented in Sect. 2.2. This test has the advantage (over X^2 and G^2) of adjusting for confounding factors [22].

It is important to note that, in some cases, PC can't direct every edge, hence it creates a CPDAG. In those cases, we apply a method to direct such edges. This method, proposed by Dor and Tarsi [5] searches recursively for possible ways to direct undirected edges.

In the second step, the framework selects the relevant variables. To select these attributes, we propose two different approaches: parents and children and Markov blanket.

In the parents and children (P-C) approach, as the name says, the variables selected are the ones that, in the causal graph, have an edge directed to the target (parents) or from the target (children).

In the Markov blanket (MB) approach, both the parents and children of the target are selected, as well as the nodes that have edges directed to the child nodes (also called spouse nodes). It is important to note that the most common way to select the variables that influence the target is through Markov Blanket

Table 1. Example of probabilities generated by the probability queries

		Attr		
		0	1	2
Target	0	0.63	0.53	0.13
	1	0.34	0.29	0.67
	2	0.14	0.25	0.56

(often used in causal feature selection methods [10]). However, several authors proposed to use only parents and children, as these variables can be considered to be the ones with the most influence in the target within its Markov blanket [1,3,23].

In the third step, the framework infers a set of probabilities that represent the influence of each relevant variable on the classes of the target: posterior probability distribution (Eq. (2)). In these probabilistic queries, the objective is to find what the influence that a evidence (particular values of the relevant variable) has on the value of the target [14]. This is performed for all the values in each variable and the resulting probability matrix is similar to Table 1.

$$P(Target = t | Attr = a) = \frac{occurrences_{t \cap a}}{ococcurrences_a} \tag{2}$$

Finally, in the fourth step, the new features are created and added to the data set. Each new feature represents the probability of the relevant variables influence on a specific class, *i.e.*, if we have, for example, a target variable with two classes ($\{0, 1\}$) and a relevant variable *Attr*, there will be created two new features representing the influence of *Attr* in each class (each instance of the feature represent the, influence the value of *Attr* in that instance on the class represented in that feature).

An overview of the framework can be seen in Fig. 1.

3.1 An Illustrative Example

To explain in more detail how this approach works, we will use as example a data set with 6 discrete variables (A, B, C, D, E and F), with 5000 instances. The values for variables A, B, C, D, and E can be $\{0, 1, 2\}$, while F can have the values $\{0, 1\}$. For this example, we will use variable **B** as the target.

As it was explained in *Step 1*, the approach starts by generating the full network with PC and Generalised Cochran Mantel Haenszel. The generated network can be seen in Fig. 2.

After the creation of the full network, the relevant variables are selected. The selected variables can be parents or children (P-C) of **B** ($\{A, E\}$) or the Markov blanket (MB) of **B** ($\{A, E, F\}$).

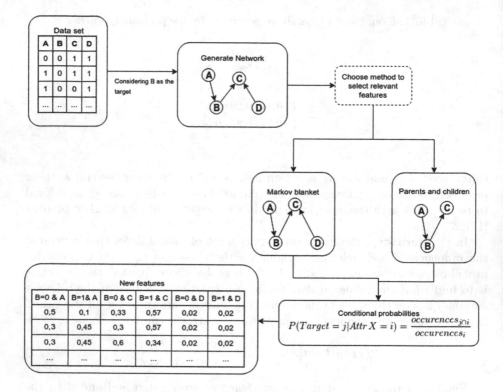

Fig. 1. Example of the operation of the proposed framework

In the third step of the framework generates inference probabilities for the chosen variables (Table 2). Taking A = 0 and B = 0 as an example, the probabilities are obtained for each one of the target values are obtained by by dividing the number of times both $A = 0$ and $B = 0$ occur, by the number of times A = 0 occurs, or in other words $P(B = 0|A = 0) = 0.86$.

These probabilities are then added to the global data set. The resulting data set is similar to Table 3. There is a difference between the number of new features created, since the number of generated features is equal to the product between the number of values in the target and the number of relevant variables. Since the MB approach selects more variables than the P-C approach, in theory, the number of generated features will be higher. So, in the case of P-C features we have 6 new features and in the case of MB we generate 9 new features.

4   Results and Discussion

To evaluate the proposed approaches and make a comparative study, the following configuration of experiments was designed: the performance of Random Forest, using the original data, as well has the versions generated by the two proposed approaches were compared.

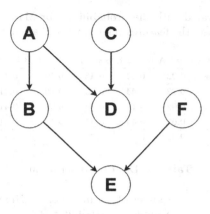

Fig. 2. Example: network generated

Table 2. Probabilities generated for the Markov blanket variables. In parents and children's case, the probabilities for F are not generated.

		A			E			F	
		0	1	2	0	1	2	0	1
Target	0	0.86	0.45	0.11	0.74	0.46	0.15	0.47	0.48
	1	0.03	0.22	0.09	0.08	0.11	0.16	0.11	0.12
	2	0.11	0.32	0.78	0.19	0.44	0.68	0.41	0.41

This comparative analysis was made through 10-fold cross validation, in several public data sets (Table 4). For each fold, the two approaches are applied to the train set and then the resulting conditional probabilities are used to create the new features for both the train and test set (this ensures that no information about the classes in the test set is added to the new features).

To choose the optimal parameters for the approaches presented in the following sections, a sensitivity analysis was performed. This analysis consisted of obtaining the error (*1 - accuracy*) for the presented data sets (by dividing them into 70% train, 30% test). In the case of PC this test was repeated for significance levels 1% and 5%. In these tests we concluded that the error of the algorithms in the data sets did not change much when the parameters were changed. For this reason, for all the data sets we select and present a significance level of 5%.

The performance of this algorithm was compared in terms of error rate (Table 5). This comparison was performed using the *No new features* as a reference. The classification algorithm performance, trained with causal features in each data set were compared to the reference using the Wilcoxon signed ranked-test. The sign $+/-$ indicates that algorithm is significantly better/worse than the reference with a p-value of less than 0.05. Besides this, the algorithms are also compared in terms of average and geometric mean of the errors, average ranks, average error ratio, win/losses, significant win/losses (number of times

Table 3. Features generated with the probabilities for Markov blanket variables. In parents and children's case, the features related with F are not generated.

A	B	C	D	E	F	B=2 A	B=0 A	B=1 A	B =2 E	B=0 E	B= 1 E	B=2 & F	B=0 & F	B=1 & F
1	2	1	0	1	1	0.35	0.44	0.22	0.44	0.45	0.10	0.41	0.48	0.12
1	0	2	0	1	1	0.35	0.44	0.22	0.44	0.45	0.10	0.41	0.48	0.12
0	0	0	0	0	0	0.11	0.87	0.02	0.19	0.73	0.08	0.41	0.47	0.11
0	0	0	0	1	1	0.11	0.87	0.02	0.44	0.45	0.10	0.41	0.48	0.12
0	0	1	2	0	0	0.11	0.87	0.02	0.19	0.73	0.08	0.41	0.47	0.11

Table 4. Data set description

Data set	Number of examples	Number of attributes	Number of classes
breast cancer	286	10	0(70%) 1(30%)
cervical	858	16	0(94%) 1(6%)
corral	160	7	0(56%) 1(44%)
earthquake	10000	5	0(2%) 1(98%)
head injury	3121	11	0(92%) 1(8%)
lucas	2000	12	0(28%) 1(72%)
medpar	1495	9	0(66%) 1(34%)
mifem	1275	10	0(25%) 1(75%)
qualitative bankruptcy	250	7	0(43%) 1(57%)
respiratory	555	5	0(51%) 1(49%)
survey	10000	6	0(56%) 1(28%) 2(16%)
titanic	1316	4	0(62%) 1(38%)
xd6	973	10	0(67%) 1(33%)

that the reference was better or worse than the algorithm, using signed ranked-test) and the Wilcoxon signed ranked-test. For the Wilcoxon signed ranked-test we consider also a p-value of 0.05.

If we analyse Table 5, it is possible to see that, in general, +*Causal features P-C* (the addition of features representing the conditional probability of parents and children features on the target) has a better performance than *No new features*, since the value obtained in the Wilcoxon test is 0.0266 (less then the p-value of 0.05), which means that the difference between the performance is significant. This difference can also be seen in the values of the average and geometric ranks. More specifically, if we look at the average ranks, we can see that +*Causal features P-C* has lower ranks (in average) than *No new features* (1.436 against 2.538).

If we now compare the second approach proposed (+*Causal features MB*) with the reference, we can see that there is a positive difference in the results

Table 5. Error rates of Random Forest for classification with causal features

Data set	No new features	+Causal features P-C	+Causal features MB
breast cancer	28.6 ± 9.88	28.6 ± 7.49	28 ± 8.39
cervical	6.88 ± 1.51	6.65 ± 1.66	6.53 ± 1.49
corral	5.62 ± 5.47	+ 0.01 ± 0.10	+ 0.01 ± 0.10
earthquake	0.26 ± 0.14	0.20 ± 0.14	0.20 ± 0.14
head injury	7.08 ± 1.23	7.43 ± 0.83	7.05 ± 0.69
lucas	15.2 ± 2.02	14.5 ± 2.12	14.5 ± 2.12
medpar	32.70 ± 4.29	33.00 ± 3.91	34.10 ± 3.23
mifem	20.1 ± 4.28	20.00 ± 4.30	19.9 ± 3.63
qualitative bankruptcy	0.40 ± 1.26	0.01 ± 0.10	0.80 ± 2.53
respiratory	40.90 ± 6.79	40.20 ± 6.20	41.2 ± 6.90
survey	44.60 ± 2.26	44.4 ± 2.05	44.4 ± 2.05
titanic	21.4 ± 2.52	20.20 ± 2.19	20.5 ± 1.83
xd6	0.41 ± 0.72	0.10 ± 0.10	0.10 ± 0.10
Average Mean	17.242	16.562	16.715
Geometric Mean	7.161	2.889	4.039
Average Ranks	2.538	1.462	1.538
Average Error Ratio	1	0.764	0.914
Wicoxon test		0.0266	0.1465
Win/Losses		10/2	10/3
Significant win/losses		1/0	1/0

Table 6. AUC for Lucas data set

	AUC
No new features	0.877
+Causal features P-C	0.887
+Causal features MB	0.889

(although not significant). It is possible to see this difference, once again, in the average and geometric mean, as well as in the average rank (1.538).

In Table 6, it is possible to see the AUC values for the three analysed approaches, for lucas data set[2]. The results presented in this table were obtained by dividing this data set in train and test (70%/30%). The model scores were then obtained for the test data (with a 50% cutoff).

In this table it is possible to see that *+Causal features MB* has the highest area, meaning that, in the data set with the causal probabilistic features that represent the relations between the target and its Markov blanket, Random Forest can distinguish better the classes than with the data from the other approaches, thus having a better performance [7]. Although *+Causal features MB* was the

---

[2] http://www.causality.inf.ethz.ch/data/LUCAS.html.

best approach in terms of AUC, the other proposed approach +*Causal features P-C* also obtained an AUC higher than the reference.

Finally, from these results, we can conclude that there is evidence that applying causality to the creation of new features can have a positive impact on the classification algorithms performance.

5   Conclusion

The achievement of satisfactory results in a classification problem not only depends on the chosen classifier but also the data being processed. One possible way to improve the performance of classifiers is to apply feature engineering, or in other words, use the original data to infer new information, creating new attributes and altering others, to obtain more descriptive features. Furthermore, most of the proposed methodologies do not take into account the possible causal relationships in the data. This information can help to create more accurate models, since we are encoding in one variable, information about the interaction between variables, thus reinforcing their importance.

In this paper we proposed a framework that uses causal discovery to create new features based on posterior probabilistic analysis of the relations between a target variable and the variables considered as relevant, being these variables the parents and children of the Markov Blanket of the target.

In the experiments, we compared the approaches with the original data, using Random Forest in public data sets. From these results, we can conclude that there is evidence that the application of causality in the creation of new supposed probabilistic features may have a positive impact on the overall performance of the classification algorithm.

In the future, we intend to study the application of these techniques in other classifiers, as well as in the classification of mixed data (continuous and discrete variables).

Acknowledgments. This research was carried out in the context of the project Fail-Stopper (DSAIPA/DS/0086/2018) and supported by the *Fundação para a Ciência e Tecnologia* (FCT), Portugal for the PhD Grant SFRH/BD/146197/2019.

References

1. Aliferis, C.F., Statnikov, A., Tsamardinos, I., Mani, S., Koutsoukos, X.D.: Local causal and Markov blanket induction for causal discovery and feature selection for classification part I: algorithms and empirical evaluation. J. Mach. Learn. Res. **11**(Jan), 171–234 (2010)
2. Birch, M.: The detection of partial association, I: the 2×2 case. J. Roy. Stat. Soc.: Ser. B (Methodol.) **26**(2), 313–324 (1964)
3. Bühlmann, P., Kalisch, M., Maathuis, M.H.: Variable selection in high-dimensional linear models: partially faithful distributions and the PC-simple algorithm. Biometrika **97**(2), 261–278 (2010)

4. Domingos, P.: A few useful things to know about machine learning. Commun. ACM **55**(10), 78–87 (2012)
5. Dor, D., Tarsi, M.: A simple algorithm to construct a consistent extension of a partially oriented graph. R-185, pp. 1–4, October 1992
6. Everitt, B.S.: The Analysis of Contingency Tables. CRC Press (1992)
7. Gama, J., Carvalho, A.C.P.d.L., Faceli, K., Lorena, A.C., Oliveira, M., et al.: Extração de conhecimento de dados: data mining. Sílabo (2015)
8. Geiger, D., Verma, T., Pearl, J.: d-separation: from theorems to algorithms. In: Machine Intelligence and Pattern Recognition, vol. 10, pp. 139–148. Elsevier (1990)
9. Glymour, C., Zhang, K., Spirtes, P.: Review of causal discovery methods based on graphical models. Front. Genet. **10**, 524 (2019). https://doi.org/10.3389/fgene.2019.00524
10. Guyon, I., Clopinet, C., Elisseeff, A., Aliferis, C.: Causal feature selection. Training **32**, 1–40 (2007)
11. Jin, Z., Li, J., Liu, L., Le, T.D., Sun, B., Wang, R.: Discovery of causal rules using partial association. In: Proceedings - IEEE International Conference on Data Mining, ICDM pp. 309–318 (2012)
12. Kalisch, M., Buehlmann, P.: Estimating high-dimensional directed acyclic graphs with the PC-algorithm. J. Mach. Learn. Res. **8**, 613–636 (2005)
13. Kleinberg, S.: Why: A Guide to Finding and Using Causes. O'Reilly Media Inc., Newton (2015)
14. Koller, D., Friedman, N.: Probabilistic Graphical Models: Principles and Techniques. MIT Press, Cambridge (2009)
15. Landis, J.R., Heyman, E.R., Koch, G.G.: Average partial association in three-way contingency tables: a review and discussion of alternative tests. Int. Stat. Rev./Revue Internationale de Statistique **46**(3), 237 (2006)
16. Listl, S., Jürges, H., Watt, R.G.: Causal inference from observational data. Commun. Dent. Oral Epidemiol. **44**(5), 409–15 (2016)
17. Martin, W.: Making valid causal inferences from observational data. Prev. Vet. Med. **113**(3), 281–297 (2014)
18. Neapolitan, R.E., et al.: Learning Bayesian Networks, vol. 38. Pearson Prentice Hall, Upper Saddle River (2004)
19. Pearl, J.: Probabilistic Reasoning in Intelligent Systems: Networks of Plausible Inference. Elsevier, Amsterdam (2014)
20. Pearl, J., Mackenzie, D.: The Book of Why: The New Science of Cause and Effect. Basic Books, New York (2018)
21. Spirtes, P., Glymour, C., Scheines, R.: Causation, Prediction, and Search, vol. 1, 2nd edn. The MIT Press, Cambridge (2001). https://ideas.repec.org/b/mtp/titles/0262194406.html
22. Tripepi, G., Jager, K.J., Dekker, F.W., Zoccali, C.: Stratification for confounding-part 1: the Mantel-Haenszel formula. Nephron Clin. Pract. **116**(4), c317–c321 (2010)
23. Tsamardinos, I., Aliferis, C.F., Statnikov, A.R., Statnikov, E.: Algorithms for large scale Markov blanket discovery. In: FLAIRS Conference, vol. 2, pp. 376–380 (2003)
24. Tsamardinos, I., Brown, L.E., Aliferis, C.F.: The max-min hill-climbing Bayesian network structure learning algorithm. Mach. Learn. **65**(1), 31–78 (2006). https://doi.org/10.1007/s10994-006-6889-7

390   A. R. Nogueira et al.

DO-U-Net for Segmentation and Counting

Applications to Satellite and Medical Images

Toyah Overton[1,2]([✉]) [iD] and Allan Tucker[1] [iD]

[1] Department of Computer Science, Brunel University London, Uxbridge, UK
{toyah.overton,allan.tucker}@brunel.ac.uk
[2] Alcis Holdings Ltd., Guildford, UK

Abstract. Many image analysis tasks involve the automatic segmentation and counting of objects with specific characteristics. However, we find that current approaches look to either segment objects or count them through bounding boxes, and those methodologies that both segment and count struggle with co-located and overlapping objects. This restricts our capabilities when, for example, we require the area covered by particular objects as well as the number of those objects present, especially when we have a large amount of images to obtain this information for. In this paper, we address this by proposing a Dual-Output U-Net. DO-U-Net is an Encoder-Decoder style, Fully Convolutional Network (FCN) for object segmentation and counting in image processing. Our proposed architecture achieves precision and sensitivity superior to other, similar models by producing two target outputs: a segmentation mask and an edge mask. Two case studies are used to demonstrate the capabilities of DO-U-Net: locating and counting Internally Displaced People (IDP) tents in satellite imagery, and the segmentation and counting of erythrocytes in blood smears. The model was demonstrated to work with a relatively small training dataset, achieving a sensitivity of 98.69% for IDP camps of the fixed resolution, and 94.66% for a scale-invariant IDP model. DO-U-Net achieved a sensitivity of 99.07% on the erythrocytes dataset. DO-U-Net has a reduced memory footprint, allowing for training and deployment on a machine with a lower to mid-range GPU, making it accessible to a wider audience, including non-governmental organisations (NGOs) providing humanitarian aid, as well as health care organisations.

Keywords: Convolutional neural networks · U-Net · Segmentation · Counting · Satellite imagery · Blood smear

1 Introduction

Over recent years, the volumes of data collected across all industries globally have grown dramatically. As a result, we find ourselves in an ever greater need for fully automated analysis techniques. The most common approaches to large scale data analysis rely on the use of supervised and unsupervised Machine Learning, and,

M. R. Berthold et al. (Eds.): IDA 2020, LNCS 12080, pp. 391–403, 2020.
https://doi.org/10.1007/978-3-030-44584-3_31

increasingly, Deep Learning. Using only a small number of human-annotated data samples, we can train models to rapidly analyse vast quantities of data without sacrificing the quality or accuracy compared with a human analyst. In this paper, we focus on images - a rich datatype that often requires rapid and accurate analysis, despite its volumes and complexity. Object classification is one of the most common types of analysis undertaken. In many cases, a further step may be required in which the classified and segmented objects of interest need to be accurately counted. While easily performed by humans, albeit slowly, this task is often non-trivial in Computer Vision, especially in the cases where the objects exist in complex environments or when objects are closely co-located and overlapping. We look at two such case studies: locating and counting Internally Displaced People (IDP) shelters in Western Afghanistan using satellite imagery, and the segmentation and counting of erythrocytes in blood smear images. Both applications have a high impact in the real world and are in a need of new rapid and accurate analysis techniques.

1.1   Searching for Shelter: Locating IDP Tents in Satellite Imagery

Over 40 million individuals were believed to have been internally displaced globally in 2018 [1]. Afghanistan is home to 2,598,000 IDPs displaced by conflict and violence, with the numbers growing by 372,000 in the year 2018 alone. In the same year, an additional 435,000 individuals were displaced due to natural disasters, 371,000 of whom were displaced due to drought.

The internally displaced population receive aid from various non-governmental organisations (NGOs), to prevent IDPs from becoming refugees. The Norwegian Refugee Council (NRC) is one such body, providing humanitarian aid to IDPs across 31 countries, assisting 8.5 million people in 2018 [2]. In Afghanistan, the NRC has been providing IDPs with tents as temporary living spaces. Alcis is assisting the NRC with the analysis of the number, flow, and concentration of these humanitarian shelters, enabling valuable aid to be delivered more effectively.

Existing Methods. In the past, Geographical Information System (GIS) Technicians relied mostly on industry-standard software, such as ArcGIS, for the majority of their analysis. These tools provide the user with a small number of built-in Machine Learning algorithms, such as the popularly used implementation of the Support Vector Machine (SVM) algorithm [3]. These generally involve a time consuming, semi-automated process, with each step being revisited multiple times as manual tuning of the model parameters is required. The methodology does not allow for batch processing as all stages must be repeated with human input for each image. An example of the ArcGIS process[1] used by GIS technicians is:

1. Manually segment the image by selecting a sample of objects exhibiting similar shape, spectral and spatial characteristics.

---

[1] Details of the process have been provided by Alcis.

2. Train the image classifier to identify other examples similar to the marked sample, using a built-in machine learning model (e.g. SVM).
3. Identify any misclassified objects and repeat the above steps.

More recently, many GIS specialists have begun to look towards the latest techniques in Data Science and Big Data analysis to create custom Machine Learning solutions. A review paper by Quinn et al. in 2018 [4] weighed up the merits of Machine Learning approaches used to segment and count both refugee and IDP camps. Their work used a sample of 87,123 structures; a magnitude which was required for training using their approach and was seen as a limitation. Quinn et al. used the popular Mask R-CNN [5] segmentation model to analyse their data; a model using a Region Proposal Network to simultaneously classify and segment images. This yielded an average precision of 75%, improving to 78% by applying a transfer learning approach.

1.2 Counting in Vein: Finding Erythrocytes in Blood Smears

The counting of erythrocytes, or red blood cells, in blood smear images, is another application in which one must count complex objects. This is an important task in exploratory and diagnostic medicine, as well as medical research. An average blood smear imaged using a microscope, contains several hundred erythrocytes of varying size, many of which are overlapping, making an accurate manual count both difficult and time-consuming.

Existing Methods. While only a small number of analyses were able to successfully perform an erythrocyte count, Tran et al. [6] have achieved a counting accuracy of 93.30%. Their technique relied on locating the cells using the Seg-Net [7] network. SegNet is an encoder-decoder style FCN architecture producing segmentation masks as its output. Due to the overlap of erythrocyte cells, they performed a Euclidean Distance Transform on the binary segmentation masks to obtain the location of each cell using a connected component labelling algorithm. A work by Alam and Islam [8] proposes an approach using YOLO9000 [9]; a network using a similar approach to Mask R-CNN, to locate elliptical bounding regions that roughly correspond to the outer contours of the cells. Using 300 images, each containing a small number of erythrocytes, for training, they achieve an accuracy of 96.09%. Bounding boxes acted as ground-truth for Alam and Islam, as opposed to segmentation masks used by Tran et al.

2 Data Description

2.1 Satellite Imagery

Working on the ground, the NRC identified areas within Western Afghanistan with known locations of IDP camps. Through their relationship with Maxar

[10], Alcis has access to satellite imagery covering multiple camps, in a range of different environments. Figure 1 shows a section of a camp in Qala'e'Naw, Badghis.

This work uses imagery collected by the WorldView-2 and WorldView-3 satellites [11], by their operator and owner Maxar. WorldView-2 has a multispectral resolution of 1.85 m, while the multispectral resolution of WorldView-3 is 1.24 m [12], allowing tents of approximately 7.5 m long and 4 m wide to be resolved. The WorldView-2 images were captured on either 05/01/19 (DD/MM/YY) or 03/03/19, with the WorldView-3 images captured on 12/03/19. A further set of images, observed between 08/08/18 and 23/09/19 by WorldView-3, became available for some locations. This dataset can be used to show evolution in the camps during this period, allowing for a better understanding of the changes undergone in IDP camps. Due to the orbital position of the satellite, images observed at different times have varying resolutions, as well as other properties, due to differences in viewing angle and atmospheric effects.

Training Data. We developed DO-U-Net using a limited number of satellite images, obtained over a very limited time, with a nearly identical pixel scale. Each tent found in the training imagery has been marked with a polygon, using a custom Graphical User Interface (GUI) tool developed by Alcis. This has been done for a total of 6 images, covering an area of approximately 15 km^2 and containing 5,178 tents. Incidentally, this makes our training dataset nearly 17 times smaller than that used by Quinn et al. in their analysis.

The second satellite dataset includes imagery of varying quality and resolution, providing an opportunity to develop a scale-invariant version of our model. We used 3 additional training images, distinct from the original dataset, to train our scale-invariant DO-U-Net. These images contained 2,338 additional tents, in an area of around 130 km^2, giving a total of 7,516 tents in over 140 km^2.

2.2   Blood Smear Images

We used blood smear images from the Acute Lymphoblastic Leukemia (ALL) Image Database for Image Processing[2]. These images were captured using an optical laboratory microscope, with magnification ranging from 300–500×, and a Canon PowerShot G5 camera. We used the ALL_IDB1 dataset, comprised of 108 images taken during September 2005 from both ALL and non-ALL patients. An example blood smear image from an ALL patient can be seen in Fig. 2.

Training Data. We selected 10 images from the ALL_IDB1 dataset to be used as training data. These images are representative of the diverse nature of the entire dataset, including the varying microscope magnifications and backgrounds. Of the images used, 3 belong to ALL patients, with the remaining 7

---

[2] Provided by the Department of Information Technology at Università degli Studi di Milano, https://homes.di.unimi.it/scotti/all/.

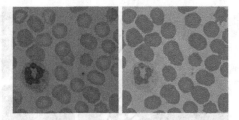

Fig. 1. *Left:* An IDP camp in Badghis, Afghanistan. NRC tents are clearly visible due to their uniform size and light colour. *Right:* The manually created ground-truth annotation for the image.

Fig. 2. *Left:* An image of a blood smear from an Acute Lymphoblastic Leukemia (ALL) patient. *Right:* The manually created ground-truth annotation for the image. The images also contain lymphocytes, which are not marked in the training data.

images coming from non-ALL patients. Similarly to the IDP camp dataset, all erythrocytes in the training data have been manually labelled with a polygon using our custom GUI tool. In the images belonging to ALL patients, a total of 1,300 erythrocytes were marked. A further 3,060 erythrocytes were marked in the images belonging to non-ALL patients, giving a total of 4,360 erythrocytes in the training data.

The training data does not distinguish between typical erythrocytes and those with any forms of morphology - of the 4,360 erythrocytes, just 106 display a clear morphology. The training data also does not contain any annotation for leukocytes. Instead, our focus is on correctly segmenting and counting all erythrocytes in the images.

3   Methodology

Of late, several very advanced and powerful Computer Vision algorithms have been developed, including the popular Mask R-CNN [5] and YOLO [9] architectures. While their performance is undoubtedly impressive, they rely on a large number of images to train their complex networks, as highlighted by Quinn et al. [5]. More recently, many more examples of FCN have been developed, including SegNet [7], DeconvNet [13] and U-Net [14], with the latter emerging as arguably one of the most popular encoder-decoder based architectures. Aimed at achieving a high degree of success even with sparse training datasets and developed to tackle biological image segmentation problems, it is a clear starting block for our architecture.

3.1   U-Net

The classical U-Net, as proposed by Ronneberger et al. has revolutionised the field of biomedical image segmentation. Similarly to other encoder-decoder networks, U-Net is capable of producing highly precise segmentation masks.

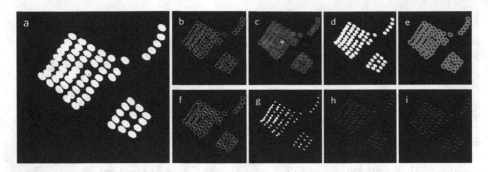

Fig. 3. *a*: Sample segmentation mask in which some tent segmentations are seen to overlap. *b*: Sobel filtered image. *c*: Local entropy filtered image. *d*: Otsu filtered image. *e*: Image with contour ellipses applied. *f*: Image with gradient morphology applied. *g*: Eroded image. *h*: Tophat filtered image. *i*: Blackhat filtered image.

What differentiates it from Mask R-CNN, SegNet and other similar networks is its lack of reliance on large datasets [14]. This is achieved by the introduction of a large number of skip connections, which reintroduce some of the early encoder layers into the much deeper decoder layers. This greatly enriches the information received by the decoder part of the network, and hence reduces the overall size of the dataset required to train the network.

We have deployed the original U-Net on our dataset of satellite images of IDP camps in Western Afghanistan. While we were able to produce segmentation masks that very accurately marked the location of the tents, the segmentation masks contained significant overlaps between tents, as seen in Fig. 3. This overlap prevents us from carrying out an automated count, despite using several post-processing techniques to minimise the impact of these overlaps. The most successful post-processing approaches are shown in Fig. 3. The issues encountered with the classical U-Net have motivated our modifications to the architecture, as described in this work.

3.2   DO-U-Net

Driven by the need to reduce overlap in segmentation masks, we modified the U-Net architecture to produce dual outputs, thus developing the DO-U-Net. The idea of a contour aware network was first demonstrated by the DCAN architecture [15]. Based on a simple FCN, DCAN was trained to use the outer contours of the areas of interest to guide the training of the segmentation masks. This led to improved semantic and instance segmentation of the model, which in their case, looked at non-overlapping features in biomedical imaging.

With the aim of counting closely co-located and overlapping objects, we are predominantly interested in the correct detection of individual objects as opposed to the exact precision of the segmentation mask itself. An examination of the hidden convolutional layers of the classical U-Net showed that the penultimate layer of the network extracts information about the edges of our objects of

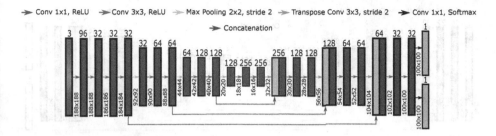

Fig. 4. The DO-U-Net architecture, showing two output layers that target the segmentation and edge masks corresponding to the training images.

interest, without any external stimulus. We introduce a secondary output layer to the network, targeting a mask segmenting the edges of our objects. By subtracting this "edge" mask from the original segmentation mask, we can obtain a "reduced" mask containing only non-overlapping objects.

As our objective was to identify tents of fixed scale in our image dataset, we were able to simplify the model considerably. This reduced the computational requirements in training of the model, allowing not only for much faster development and training but also opening the possibility of deploying the algorithm on a dataset covering a large proportion of the total area of Afganistan, driven by our commercial requirements.

Architecture. Starting with the classical U-Net, we reduce the number of convolutional layers and skip connections in the model. Simultaneously, we minimised the complexity of the model by looking at smaller input regions of the images, thus minimising the memory footprint of the model. We follow the approach of Ronneberger et. al. by using unpadded convolutions throughout the network, resulting in a model with smaller output masks (100×100 px) corresponding to a central region of a larger (188×188 px) input image region. DO-U-Net uses two, independently trained, output layers of identical size. Figure 4 shows our proposed DO-U-Net architecture. The model can also be found in our public online repository[3]. Examples of the output edge and segmentation masks, as well as the final "reduced" mask, can be seen in Figs. 6 and 7. With the reduced memory footprint of our model, we can produce a "reduced" segmentation mask for a single 100×100 px region in 3 ms using `TensorFlow 2.0` with Intel i9-9820X CPU and a single NVIDIA RTX 2080 Ti GPU setup.

Training. The large training images were divided such that no overlap exists between the regions corresponding to the target masks, using zero-padding at the image borders. We train our model against both segmentation and edge masks. The edges of the mark-up polygons, annotated using our custom tool, were used as the "edge" masks during training. Due to the difference in a pixel

---

[3] https://github.com/ToyahJade/DO-U-Net.

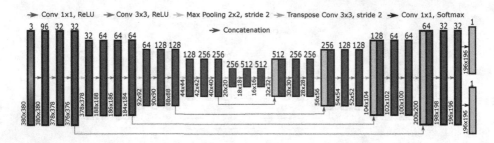

Fig. 5. Scale-Invariant DO-U-Net, redesigned to work with datasets containing objects of variable scale.

size of tents and erythrocytes, the edges were taken to span 2 px and 4 px wide respectively in these case studies.

As our problem deals with segmentation masks covering only a small proportion of the image (<1% in some satellite imagery), the choice of a loss function was a very important factor. We use the Focal Tversky Index, which is suited for training with sparse positive labels compared to the overall area of the training data [16]. Our best result, obtained using the Focal Tversky loss function, gave an improvement of 5% in the Intersect-over-Union (IoU) value compared to the Binary Cross-Entropy loss function, as used by Ronneberger et al. [14]. We found the training to behave most optimally when the combined loss function for the model was heavily weighted toward the edge mask segmentation. Here, we used a 10%/90% split for the object and edge mask segmentation respectively.

Counting. As the resulting "reduced" masks produced by our approach do not contain any overlaps, we can use simple counting techniques, relying on the detection of the bounding polygons for the objects of interest. We apply a threshold to remove all negative values from the image, which may occur due to the subtractions. We then use the Marching Squares Algorithm implemented as part of `Python`'s `skimage.measure` image analysis library [17].

Scale-Invariant DO-U-Net. In addition to the simple DO-U-Net, we propose a scale-invariant version of the architecture with an additional encoder and decoder block. Figure 5 shows the increased depth of the network as is required to capture the generalised model of our objects in the scale varying dataset. The addition of extra blocks resulted in a larger input layer of 380×380 px, corresponding to a segmentation mask of 196×196 px.

4   Results

4.1   IDP Tent Results

Using our DO-U-Net architecture, we were able to achieve a very significant improvement in the counting of IDP tents compared to the popularly used SVM

Fig. 6. *Left:* Segmentation mask produced for NRC tents in a camp near Qala'e'Naw. *Centre:* Edges mask produced for the same image. *Right:* The final output mask.

classifier available in ArcGIS. However, due to the manually intensive nature of the ArcGIS approach[4], we were only able to directly compare a single test camp, located in the Qala'e'Naw region of the Badghis Province. This area contains 921 tents as identified in the ground-truth masks. Using DO-U-Net, we achieved a precision of 99.78% with a sensitivity of 99.46%. Using ArcGIS, we find a precision of 99.86% and a significantly lower sensitivity of 79.37%. Sensitivity, or the true positive rate, measures the probability of detecting an object and is, therefore, the most important metric for us as we aim to locate and count all tents in the region. The scale-invariant DO-U-Net achieved a precision of 98.48% and a sensitivity of 98.37% on the same image.

We also apply DO-U-Net to a larger dataset of five images containing a total of 3,447 tents and find an average precision of 97.01% and an average sensitivity of 98.68%. Similarly, we tested the scale-invariant DO-U-Net using 10 images with varying properties and resolutions containing a total of 5,643 tents. Here, the average precision was reduced to 91.45%, and the average sensitivity dropped to 94.66%. This result is not surprising as, on inspection, we find that without the scale constraints the resulting segmenting masks are contaminated with other structures of similar properties to NRC tents. We also find that, without scale constraints, NRC tents which are partially covered e.g. with tarpaulin may be missed or only partially segmented. Our DO-U-Net and scale-invariant DO-U-Net sensitivities of 98.68% and 94.66% respectively are very strong results when compared to the existing literature.

4.2   Erythrocyte Results

To validate the performance of DO-U-Net at counting erythrocytes, we use 3 randomly selected blood smear images from ALL patients and a further 5 selected images from non-ALL patients. While randomly selected, the images are representative of the entire ALL_IDB1 dataset. On a total of 2,775 erythrocytes, as found in these 8 validation images, DO-U-Net achieved an average precision of 98.31% and an average sensitivity of 99.07%.

---

[4] Results found using ArcGIS methodology can be found at https://storymaps.arcgis.com/stories/d85e5cca27464d97ad4c1bad3da7f140.

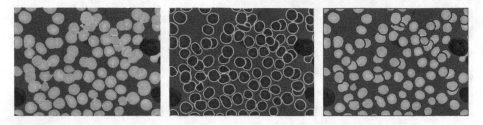

Fig. 7. *Left:* Segmentation mask produced for a blood smear of an ALL patient. *Centre:* Edges mask produced for the same image. *Right:* The final output mask.

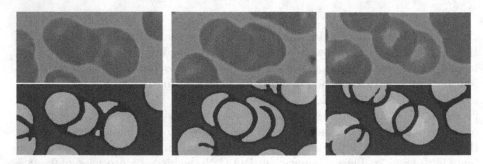

Fig. 8. *Top:* Blood smear images of overlapping cells. *Bottom:* Segmentation masks produced by DO-U-Net. *Left:* An overlapped cell is counted twice when the "edges" from neighbouring cells overlap and break up the cell. *Centre:* A cell is missed due to an incomplete edge mask. *Right:* An uncertainty in the order of the cell overlap leads to the intersect between two cells being counted as an additional cell.

Whilst our proposed DO-U-Net is extremely effective at producing image and edge segmentation masks, as demonstrated in Fig. 7, we do note that the obtained erythrocyte count may not always match the near-perfect segmentation. Figure 8 shows examples of the three most common issues found to occur in the final "reduced" masks. These mistakes arise largely due to the translucent nature of erythrocytes and the difficulty in differentiating between a cell which is overlapping another and a cell which is overlapped. While these cases are rare, this demonstrates that further improvements can be made to the architecture.

4.3   Future Work

Our current model has been designed to segment only one type of object, which is a clear limitation of our solution. As an example, the blood smear images from the ALL-IDP1 dataset contain normal erythrocytes as well as two clear types of morphology: burr cells and dacryocytes. These morphologies may be signs of disease in patients, though burr cells are common artefacts, especially known to occur when the blood sample is aged. It is therefore important to not only count all erythrocytes, but to also differentiate between their various morphologies.

Table 1. Summary of results for DO-U-Net, when tested on our two satellite imagery datasets and the ALL_IDB1 dataset.

Dataset	Number of images	Number of objects	Average precision	Average sensitivity
IDP Camps (Fixed Scale)	5	3,447	97.01%	98.69%
IDP Camps (Scale-Invariant)	10	5,643	91.45%	94.66%
ALL_IDB1	8	2,775	98.31%	99.07%

While our general theory can be applied to identifying different types of object, further modifications to our proposed DO-U-Net would be required.

5   Conclusion

We have proposed a new approach to segmenting and counting closely co-located and overlapping objects in complex image datasets. For this, we developed DO-U-Net: a modified U-Net based architecture, designed to produce both a segmentation and an "edge" mask. By subtracting the latter from the former, we can locate and spatially separate objects of interest before automatically counting them. Our methodology was successful on both of our case studies: locating and counting IDP tents in satellite imagery, and the segmentation and counting of erythrocytes in blood smear images. In the first case study, DO-U-Net increased our sensitivity by approximately 20% compared to a popular ArcGIS based solution, achieving an average sensitivity of 98.69% for a dataset of fixed spatial resolution. Our network went on to achieve a precision of 91.45% and a sensitivity of 94.66% on a set of satellite images with a varying resolution and colour profiles. This is an impressive result when compared to Quinn et al. who achieved a precision of 78%. We also found DO-U-Net to be extremely successful at segmenting and counting erythrocytes in blood smear images, achieving a sensitivity of 99.07% for our test dataset. This is an improvement of 6% over the results found by Tran et al. who used the same training dataset, and 3% over Alam and Islam who used a comparable set of images, giving us a near-perfect sensitivity when counting erythrocytes. The results are summarised in Table 1.

Acknowledgements. We thank Harry Robson, GIS Analyst at Alcis Holdings Ltd., for industry knowledge shared and for performing post-processing in ArcGIS. We would also like to show our gratitude to Richard Brittan, Tim Buckley and the Alcis team for sharing insight with us during the course of this research.

We are also immensely grateful to the Department of Information Technology at Università degli Studi di Milano for providing the ALL_IDB1 dataset from the Acute Lymphoblastic Leukemia Image Database for Image Processing.

References

1. Global Internal Displacement Database, Internal Displacement Monitoring Centre. http://www.internal-displacement.org/database/displacement-data. Accessed 04 Nov 2019
2. We Assisted 8.5 Million People Last Year, Norwegian Refugee Council. https://www.nrc.no/perspectives/2019/we-assisted-8.5-million-people-last-year. Accessed 04 Nov 2019
3. Train Support Vector Machine Classifier, ArcGIS Pro. https://pro.arcgis.com/en/pro-app/tool-reference/spatial-analyst/train-support-vector-machine-classifier.htm. Accessed 10 Nov 2019
4. Quinn, J.A., Nyhan, M.M., Navarro, C., Coluccia, D., Bromley, L., Luengo-Oroz, M.: Humanitarian applications of machine learning with remote-sensing data: review and case study in refugee settlement mapping. Philos. Trans. R. Soc. A Math. Phys. Eng. Sci. **376**, 20170363 (2018). https://doi.org/10.1098/rsta.2017.0363
5. He, K., Gkioxari, G., Dollar, P., Girshick, R.: Mask R-CNN. Facebook AI research (FAIR). Submission Date: 24 January 2018. arXiv:1703.06870
6. Tran, T., Binh Minh, L., Lee, S., Kwon, K.: Blood cell count using deep learning semantic segmentation. Preprints 2019, 2019090075. https://doi.org/10.20944/preprints201909.0075.v1
7. Badrinarayanan, V., Kendall, A., Cipolla, R.: SegNet: a deep convolutional encoder-decoder architecture for image segmentation. IEEE Trans. Pattern Anal. Mach. Intell. **39**(12), 2481–2495 (2017). https://doi.org/10.1109/TPAMI.2016.2644615. Dickinson, S. (ed.)
8. Alam, M.M., Islam, M.T.: Machine learning approach of automatic identification and counting of blood cells. Healthc. Technol. Lett. **6**(4), 103–108 (2019). https://doi.org/10.1049/htl.2018.5098
9. Redmon, J., Farhadi, A.: YOLO9000: better, faster, stronger, Submission Date: 25 December 2016. arXiv: 612.08242
10. Maxar Technologies. https://www.maxar.com/. Accessed 18 Nov 2019
11. WorldView-3, DigitalGlobe Inc. http://worldview3.digitalglobe.com. Accessed 04 Nov 2019
12. The DigitalGlobe Constellation, DigitalGlobe Inc. https://www.digitalglobe.com/company/about-us. Accessed 04 Nov 2019
13. Hyeonwoo, N., Seunghoon, H., Bohyung, H.: Learning deconvolution network for semantic segmentation. In: Proceedings of the IEEE International Conference on Computer Vision, pp. 1520–1528, May 2015
14. Ronneberger, Olaf, Fischer, Philipp, Brox, Thomas: U-Net: Convolutional Networks for Biomedical Image Segmentation. In: Navab, Nassir, Hornegger, Joachim, Wells, William M., Frangi, Alejandro F. (eds.) MICCAI 2015. LNCS, vol. 9351, pp. 234–241. Springer, Cham (2015). https://doi.org/10.1007/978-3-319-24574-4_28
15. Chen, H., Qi, X., Yu, L., Heng, P.: DCAN: deep contour-aware networks for accurate gland segmentation. In: Proceedings of Conference on Computer Vision and Pattern Recognition (CVPR), pp. 2487–2496. IEEE (2016)
16. Abraham, N., Khan, N.M.: A novel focal tversky loss function with improved attention U-Net for lesion segmentation. In: IEEE International Symposium on Biomedical Imaging, ISBI (2019). arXiv:1810.07842
17. Contour Finding, Scikits-image. https://scikit-image.org/docs/0.5/auto_examples/plot_contours.html. Accessed 11 Nov 2019

Enhanced Word Embeddings for Anorexia Nervosa Detection on Social Media

Diana Ramírez-Cifuentes[1]([✉]), Christine Largeron[2], Julien Tissier[2], Ana Freire[1], and Ricardo Baeza-Yates[1]

[1] Universitat Pompeu Fabra, Carrer de Tanger, 122-140, 08018 Barcelona, Spain
{diana.ramirez,ana.freire,ricardo.baeza}@upf.edu
[2] Univ Lyon, UJM-Saint-Etienne, CNRS, Laboratoire Hubert Curien, UMR 5516, 42023 Saint-Etienne, France
{julien.tissier,christine.largeron}@univ-st-etienne.fr

Abstract. Anorexia Nervosa (AN) is a serious mental disorder that has been proved to be traceable on social media through the analysis of users' written posts. Here we present an approach to generate word embeddings enhanced for a classification task dedicated to the detection of Reddit users with AN. Our method extends *Word2vec*'s objective function in order to put closer domain-specific and semantically related words. The approach is evaluated through the calculation of an average similarity measure, and via the usage of the embeddings generated as features for the AN screening task. The results show that our method outperforms the usage of fine-tuned pre-learned word embeddings, related methods dedicated to generate domain adapted embeddings, as well as representations learned on the training set using *Word2vec*. This method can potentially be applied and evaluated on similar tasks that can be formalized as document categorization problems. Regarding our use case, we believe that this approach can contribute to the development of proper automated detection tools to alert and assist clinicians.

Keywords: Social media · Eating disorders · Word embeddings · Anorexia Nervosa · Representation learning

1 Introduction

We present models to identify users with AN based on the texts they post on social media. Word embeddings previously learned in a large corpus, have provided good results on predictive tasks [3]. However, in the case of writings generated by users living with a mental disorder such as AN, we observe specific vocabulary exclusively related with the topic. Terms such as: "*cw*", used to refer to the current weight of a person, or "*ow*" referring to the objective weight,

This work was supported by the University of Lyon - IDEXLYON and the Spanish Ministry of Economy and Competitiveness under the Maria de Maeztu Units of Excellence Program (MDM-2015-0502).

M. R. Berthold et al. (Eds.): IDA 2020, LNCS 12080, pp. 404–417, 2020.
https://doi.org/10.1007/978-3-030-44584-3_32

are elements that are not easily found in large yet general collections extracted from Wikipedia, social media and news websites. Therefore, using pre-learned embeddings may not be the most suitable approach for the task.

We propose a method based on *Dict2vec* [15] to generate word embeddings enhanced for our task domain. The main contributions of our work are the following: (1) a method that modifies *Dict2vec* [15] in order to generate word embeddings enhanced for our classification task, this method has the power to be applied on similar tasks that can be formulated as document categorization problems; (2) different ways to improve the performance of the embeddings generated by our method corresponding to four embeddings variants; and (3) a set of experiments to evaluate the performance of our generated embeddings in comparison to pre-learned embeddings, and other domain adaptation methods.

2 Related Work

In previous work related to detection of mental disorders [8], documents were represented using bag of words (BoW) models, which involve representing words in terms of their frequencies. As these models do not consider contextual information or relations between the terms, other models have been proposed based on word embeddings [3]. These representations are generated considering the distributional hypothesis, which assumes that words appearing in similar contexts are related, and therefore should have close representations [11,13].

Embedding models allow words from a large corpus to be encoded as vectors in a high-dimensional space. The vectors are defined by taking into account the context in which the words appear in the corpus in such a way that two words having the same neighborhood should be close in the vector space.

Among the methods used for generating word embeddings we find *Word2vec* [11], which generates a vector for each word in the corpus considering it as an atomic entity. To build the embeddings, *Word2vec* defines two approaches: one known as continuous bag of words (CBOW) that uses the context to predict a target word; and another one called skip-gram, which uses a word to predict a target context. Another method is *fastText* [2], which takes into account the morphology of words, having each word represented as a bag of character n-grams for training. There is also *GloVe* [13], which proposes a weighted least squares model that does the training on global word-word co-occurrence counts.

In contrast to the previous methods, we can also mention recent methods like Embeddings from Language Models (ELMo) [14] and Bidirectional Encoder Representations from Transformers (BERT) [6] that generate representations which are aware of the context they are being used at. These approaches are useful for tasks where polysemic terms are relevant, and when there are enough sentences to learn these from the context. Regarding our use case, we observe that the vocabulary used by users with AN is very specific and contains almost no polysemic terms, which is why these methods are not addressed in our evaluation framework.

All the methods already mentioned are generally trained over large general purpose corpora. However, for certain domain specific classification tasks we have to work with small corpora. This is the case of mental disorders screening tasks given that the annotation phase is expensive, and requires the intervention of specialists. There are some methods that address this issue by either enhancing the embeddings learned over small corpora with external information, or adapting embeddings learned on large corpora to the task domain.

Among the enhancement methods we find Zhang's et al. [17] work. They made use of word embeddings learned in different health related domains to recognize symptoms in psychiatry. They designed approaches to combine data of the source and target to generate word embeddings, which are considered in our experimental results.

Kuang et al. [9] propose learning weights based on the words' relative importance for the classification task (predictive terms). This method proposes weighting words according to their χ^2 [12] statistics to represent the context. However, this method differs from ours as we generate our embeddings through a different approach, which takes into account the context terms, introduces new domain related vocabulary, considers the predictive terms to be equally important, and moves apart the vectors of terms that are not predictive for the main target class.

Faruqui et al. [7] present an alternative, known as a retrofitting method, which makes use of relational information from semantic lexicons to improve pre-built word vectors. The main disadvantage is that no external new terms representations can be introduced to the enhanced embeddings, and that despite related embeddings are put closer, the embeddings of terms that should not be related (task-wise) cannot be put apart from each other. In our experimental setup, this method is used to define a baseline and to enhance the embeddings generated through our approach.

Our proposal is based on *Dict2vec* [15], which is an extension of the *Word2vec* approach. *Dict2vec* uses the lexical dictionary definitions of words in order to enrich the semantics of the embeddings generated. This approach has proved to perform well on small corpora because in addition to the context defined by *Word2vec*, it introduces a (1) positive sampling, which moves closer the vector of words co-occurring in their mutual dictionary definitions, and a (2) controlled negative sampling which prevents to move apart the vectors of words that appear in the definition of others, as the authors assume that all the words in the definition of a term from a dictionary are semantically related to the word they define.

3   Method Proposed

Our method generates word embeddings enhanced for a classification task dedicated to the detection of users with AN over a small size corpus. In this context, users are represented by documents that contain their writings concatenated, and that are labeled as anorexic (positive) or control (negative) cases. These labels are known as the classes to predict for our task.

Our method is based on *Dict2vec*'s general idea [15]. We extend the *Word2vec* model with both a positive and a negative component, but our method differs from *Dict2vec* because both components are designed to learn vectors for a specific classification task. Within the word embeddings context, we assume that word-level n-grams' vectors, which are predictive for a class, should be placed close to each other given their relation with the class to be predicted. Therefore we first define sets of what we call *predictive pairs* for each class, and use them later for our learning approach.

3.1   Predictive Pairs Definition

Prior to learning our embeddings, we use χ^2 [12] to identify the predictive n-grams. This is a method commonly used for feature reduction, being capable to identify the most predictive features, in this case terms, for a classification task.

Based on the χ^2 scores distribution, we obtain the n terms with the highest scores (most predictive terms) for each of the classes to predict (positive and negative). Later, we identify the most predictive term for the positive class denoted as t_1 or *pivot term*. Depending on the class for which a term is predictive, two types of *predictive pairs* are defined, so that every time a predictive word is found, it will be put close or far from t_1. These predictive pair types are: (1) positive predictive pairs, where each predictive term for the positive class is paired with the term t_1 in order to get its vector representation closer to t_1; and (2) negative predictive pairs, where each term predictive for the negative class is also paired with t_1, but with the goal of putting it apart from t_1.

In order to define the positive predictive terms for our use case, we consider: the predictive terms defined by the χ^2 method, AN related vocabulary (domain-specific) and the k most similar words to t_1 obtained from pre-learned embeddings, according to the cosine similarity. Like this, information coming from external sources that are closely related with the task could be introduced to the training corpus. The terms that were not part of the corpus were appended to it, providing us an alternative to add new vocabulary of semantic significance to the task.

Regarding the negative predictive terms, no further elements are considered asides from the (χ^2) predictive terms of the negative class as for our use case and similar tasks, control cases do not seem to share a vocabulary strictly related to a given topic. In other words, and as observed for the anorexia detection use case, control users are characterized by their discussions on topics unrelated to anorexia.

For the χ^2 method, when having a binary task, the resulting predictive features are the same for both classes (positive and negative). Therefore, we have proceeded to get the top n most predictive terms based on the distribution of the χ^2 scores for all the terms. Later, we decided to take a look at the number of documents containing the selected n terms based on their class (anorexia or control). Given a term t, we calculated the number of documents belonging to the positive class (anorexia) containing t, denoted as PCC; and we also calculated the number of documents belonging to the negative class (control) containing t, named as NCC. Then, for t we calculate the respective ratio of both counts in relation to the total amount of documents belonging to each class: total amount of positive documents (TPD) and total amount of negative documents (TND), obtaining like this a positive class count ratio (PCCR) and a negative class count ratio (NCCR).

For a term to be part of the set of positive predictive terms its PCCR value has to be higher than the NCCR, and the opposite applies for the terms that belong to the set of negative predictive pairs. The positive and negative class count ratios are defined in Eqs. 1a and 1b as:

$$PCCR(t) = \frac{PCC(t)}{TPD} \tag{1a}$$

$$NCCR(t) = \frac{NCC(t)}{TND} \tag{1b}$$

3.2 Learning Embeddings

Once the predictive pairs are defined, the objective function for a target term ω_t (Eq. 2) is defined by the addition of a positive sampling cost (Eq. 3) and a negative sampling cost (Eq. 4a) in addition to *Word2vec*'s usual target, context pair cost given by $\ell(\omega_t, \omega_c)$ where ℓ represents the logistic loss function, and v_t, and v_c are the vectors associated to ω_t and ω_c respectively.

$$J(\omega_t, \omega_c) = \ell(v_t, v_c) + J_{pos}(\omega_t) + J_{neg}(\omega_t) \tag{2}$$

Unlike *Dict2vec*, J_{pos} is computed for each target term where $P(\omega_t)$ is the set of all the words that form a positive predictive pair with the word ω_t, and v_t and v_i are the vectors associated to ω_t and ω_i respectively. β_P is a weight that defines the importance of the positive predictive pairs during the learning phase. Also, as an aspect that differs from *Dict2vec*, the cost given by the predictive pairs is normalized by the size of the predictive pairs set, $|P(\omega_t)|$, considering that all the terms from the predictive pairs set of ω_t are taken into account for the calculations, and therefore when t_1 is found, the impact of trying to move it closer to a big amount of terms is reduced, and it remains as a pivot element to which other predictive terms get close to:

$$J_{pos}(\omega_t) = \beta_P \sum_{\omega_i \in P(\omega_t)} \frac{\ell(v_t \cdot v_i)}{|P(\omega_t)|} \tag{3}$$

On the negative sampling, we modify *Dict2vec*'s approach. We not only make sure that the vectors of the terms forming a positive predictive pair with ω_t are not put apart from it, but we also define a set of words that are predictive for the negative class and define a cost given by the negative predictive pairs. In this case, as explained before, the main goal is to put apart these terms from t_1, so this cost is added to the negative random sampling cost J_{n_r} (Eq. 4b), as detailed in Eq. 4a.

$$J_{neg}(\omega_t) = J_{n\_r}(\omega_t) + \beta_N \sum_{\omega_j \in N(\omega_t)} \frac{\ell(-v_t \cdot v_j)}{|N(\omega_t)|} \tag{4a}$$

$$J_{n\_r}(\omega_t) = \sum_{\substack{\omega_i \in F(\omega_t) \\ \omega_i \notin P(\omega_t)}} \ell(-v_t \cdot v_i) \tag{4b}$$

The negative sampling cost considers, as on *Word2vec*, a set $F(\omega_t)$ of k words selected randomly from the vocabulary. These words are put apart from ω_t as they are likely to not be semantically related. Considering *Dict2vec*'s approach, we make sure as well that any term belonging to the set of positive predictive pairs of ω_t ends up being put apart. In addition to this, we add another negative sampling cost which corresponds to the cost of putting apart from t_1 the most predictive terms from the negative class. In this case, $N(\omega_t)$ represents the set of all the words that form a negative predictive pair with the word ω_t. β_N is a weight to define the importance of the negative predictive pairs during the learning phase.

The global objective function (Eq. 5) is given by the sum of every pair's cost across the whole corpus:

$$J = \sum_{t=1}^{C} \sum_{c=-n}^{n} J(\omega_t, \omega_{t+c}) \tag{5}$$

where C is the corpora size, and n represents the size of the window.

3.3   Enhanced Embeddings Variations

Given a pre-learned embedding which associates for a word ω a pre-learned representation v_{pl}, and an enhanced embedding v obtained through our approach for ω with the same length m as v_{pl}, we generate variations of our embeddings based on existing enhancement methods. First, we denote the embeddings generated exclusively by our approach (predictive pairs) as *Variation 0*, v is an instance of the representation of ω for this variation.

For the next variations, we address ways to combine the vectors of pre-learned embeddings (*i.e.*, v_{pl}) with the ones of our enhanced embeddings (*i.e.*, v). For *Variation 1* we concatenate both representations $v_{pl} + v$, obtaining a $2m$ dimensions vector [16]. *Variation 2* involves concatenating both representations and applying truncated SVD as a dimensionality reduction method to obtain a new representation given by $SVD(v_{pl} + v)$. *Variation 3* uses the values of the pre-learned vector v_{pl} as starting weights to generate a representation using our learning approach. This variation is inspired in a popular transfer learning method that was successfully applied on similar tasks [5]. For these variations (1–3) we take into account the intersection between the vocabularies of both embeddings types (pre-learned and *Variation 0*). Finally, *Variation 4* implies applying Faruqui's retrofitting method [7] over the embeddings of *Variation 0*.

4   Evaluation Framework

4.1   Data Set Description

We used a Reddit data set [10] that consists on posts of users labeled as anorexic and control cases. This data set was defined in the context of an early risk detection shared task, and the training and test sets were provided by the organizers of the eRisk task.[1] Table 1 provides a description of the training and testing data sets statistics. Given the incidence of Anorexia Nervosa, for both sets there is a reduced yet significant amount of AN cases compared to the control cases.

Table 1. Collection description as described on [10].

	Train		Test	
	Anorexia	Control	Anorexia	Control
Users count	20	132	41	279
Writings count	7,452	77,514	17,422	151,364
Avg. writings count	372.6	587.2	424.9	542.5
Avg. words per writing	41.2	20.9	35.7	20.9

4.2   Embeddings Generation

The training corpus used to generate the embeddings, named *anorexia corpus*, consisted on the concatenation of all the writings from all the training users. A set of stop-words were removed. This resulted on a training corpus with a size of 1,267,208 tokens and a vocabulary size of 87,197 tokens. In order to consider the bigrams defined by our predictive pairs, the words belonging to a bigram were paired and formatted as if they were a single term.

For the predictive pairs generation with χ^2, each user is an instance represented by a document composed by all the user's posts concatenated. χ^2 is applied over the train set considering the users classes (anorexic or control) as the possible categories for the documents. The process described in Sect. 3.1 is followed in order to obtain a list of 854 positive (anorexia) and 15 negative (control) predictive terms. Some of these terms can be seen on Table 2, which displays the top 15 most predictive terms for both classes. *Anorexia* itself resulted to be the term with the highest χ^2 score, denoted as t_1 in Sect. 3.

The anorexia domain related terms from [1] were added as the topic related vocabulary, and the top 20 words with the highest similarity to *anorexia* coming from a set of pre-learned embeddings from *GloVe* [13] were also paired to it to define the predictive pairs sets. The GloVe's pre-learned vectors considered are the 100 dimensions representations learned over 2B tweets with 27B tokens, and with 1.2M vocabulary terms.

---

[1] eRisk task: https://early.irlab.org/2018/index.html.

Table 2. List of some of the most predictive terms for each class.

Positive Terms (Anorexia class)			Negative terms (Control class)		
anorexia	diagnosed	binges	war	sky	song
anorexic	macros	calories don't	bro	plot	master
meal plan	cal	relapsed	Trump	game	Russian
underweight	weight gain	restriction	players	Earth	video
eating disorder(s)	anorexia nervosa	caffeine	gold	America	trailer

The term *anorexia* was paired to 901 unique terms and, likewise, each of these terms was paired to *anorexia*. The same approach was followed for the negative predictive terms (15), which were also paired with *anorexia*. An instance of a positive predictive pair is *(anorexia, underweight)*, whereas an instance of a negative predictive pair is *(anorexia, game)*. For learning the embeddings through our approach, and as it extends *Word2vec*, we used as parameters a window size of 5, the number of random negative pairs chosen for negative sampling was 5, and we trained with one thread/worker and 5 epochs.

4.3 Evaluation Based on the Average Cosine Similarity

This evaluation is done over the embeddings generated through *Variation 0* over the anorexia corpus. It averages the cosine similarities (sim) between t_1 and all the terms that were defined either as its p positive predictive pairs, obtaining a positive score denoted as PS on Eq. 6a; or as its n negative predictive pairs, with a negative score denoted as NS on Eq. 6b. On these equations v_a represents the vector of the term *anorexia*; v_{PPT_i} represents the vector of the positive predictive term (PPT) i belonging to the set of positive predictive pairs of *anorexia* of size p; and v_{NPT_i} represents the vector of the negative predictive term (NPT) i belonging to the set of negative predictive pairs of *anorexia* of size n:

$$PS(a) = \frac{\sum_{i=1}^{p} sim(v_a, v_{PPT_i})}{p} \tag{6a}$$

$$NS(a) = \frac{\sum_{i=1}^{n} sim(v_a, v_{NPT_i})}{n} \tag{6b}$$

We designed our experiments using PS and NS in order to analyze three main aspects: (1) we verify that through the application of our method, the predictive terms for the positive class are closer to the pivot term representation, and that the predictive terms for the negative class were moved away from it; (2) we evaluate the impact of using different values of the parameters β_P and β_N to obtain the best representations where PS has the highest possible value, keeping NS as low as possible; and (3) we compare our generation method with *Word2vec* as baseline since this is the case for which our predictive pairs would not be considered ($\beta_P = 0$ and $\beta_N = 0$). We expect for our embeddings to obtain higher values for PS and lower values for NS in comparison to the baseline.

Results. Table 3 shows first the values for PS and NS obtained by what we consider our baseline, *Word2vec* ($\beta_P = 0$ and $\beta_N = 0$), and then the values obtained by embeddings models generated using our approach (*Variation 0*), with different yet equivalent values given to the parameters β_P and β_N, as they proved to provide the best results for PS and PN. We also evaluated individually the effects of varying exclusively the values for β_P, leaving $\beta_N = 0$, and then the effects of varying only the values of β_N, with $\beta_P = 0$. On the last row of the table we show a model corresponding to the combination of the parameters with the best individual performance ($\beta_P = 75$ and $\beta_N = 25$).

After applying our approach the value of PS becomes greater than NS for most of our generated models, meaning that we were able to obtain a representation where the positive predictive terms are closer to the pivot term *anorexia*, and the negative predictive terms are more apart from it. Then, we can also observe that the averages change significantly depending on the values of the parameters β_P and β_N, and for this case the best results according to PS are obtained when $\beta_P = 50$ and $\beta_N = 50$. Finally, when we compare our scores with *Word2vec*, we can observe that after applying our method, we can obtain representations where the values of PS and NS are respectively higher and lower than the ones obtained by the baseline model.

Table 3. Positive Scores (PS) and Negative Scores (NS) for *Variation 0*. Different values for β_P and β_N are tested.

Values for β_P and β_N	Positive score (PS)	Negative score (NS)
$\beta_P = 0$, $\beta_N = 0$ (baseline)	0.8861	0.8956
$\beta_P = 0.25$, $\beta_N = 0.25$	0.7878	0.7424
$\beta_P = 0.5$, $\beta_N = 0.5$	0.7916	0.5158
$\beta_P = 1$, $\beta_N = 1$	0.7996	0.5879
$\beta_P = 10$, $\beta_N = 10$	0.8495	**0.4733**
$\beta_P = 50$, $\beta_N = 50$	**0.9479**	0.6009
$\beta_P = 100$, $\beta_N = 100$	0.9325	0.6440

4.4   Evaluation Based on Visualization

We focus on the comparison of embeddings generated using *word2vec* (baseline), *Variation 0* of our enhanced embeddings, and *Variation 4*. In order to plot over the space the vectors of the embeddings generated (see Fig. 1), we performed dimensionality reduction, from the original 200 dimensions to 2, through Principal Component Analysis (PCA) over the vectors of the terms in Table 2 for the embeddings generated with these three representations. We focused over the embeddings representing the positive and negative predictive terms. For the resulting embeddings of our method (*Variation 0*), we selected $\beta_P = 50$ and $\beta_N = 50$ as parameter values.

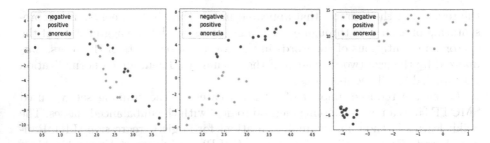

Fig. 1. Predictive terms sample represented on two dimensions after PCA was applied on their embeddings as dimensionality reduction method. From left to right each plot shows the vectorial representation of the predictive terms according to the embeddings obtained through (1) *Word2vec* (baseline), (2) *Variation 0*, and (3) *Variation 4*.

The positive predictive terms representations are closer after applying our method (*Variation 0*), and the negative predictive terms are displayed farther, in comparison to the baseline. The last plot displays the terms for the embeddings generated through Variation 4. For this case, given the input format for the retrofitting method, *anorexia* was linked with all the remaining predictive terms of the anorexia class (901), and likewise, each of these predictive terms was linked to the term *anorexia*. Notice that the retrofitting approach converges to changes in Euclidean distance of adjacent vertices, whereas the closeness between terms for our approach is given by the cosine distance.

4.5   Evaluation Based on the Predictive Task

In order to test our generated embeddings for the classification task dedicated to AN screening, we conduct a series of experiments to compare our method with related approaches. We define 5 baselines for our task: the first one is a BoW model based on word level unigrams and bigrams (*Baseline 1*), this model is kept mainly as a reference since our main focus is to evaluate our approach compared to other word embedding based models. We create a second model using *GloVe*'s pre-learned embeddings (*Baseline 2*), and a third model that uses word embeddings learned on the training set with the *Word2vec* approach (*Baseline 3*). We evaluate a fourth approach (*Baseline 4*) given by the enhancement of the *Baseline 3* embeddings, with Faruqui's *et al.* [7] retrofitting method. *Baseline 5* uses the same retrofitting method over GloVe's pre-learned embeddings, as we expected that a domain adaptation of the embeddings learned on a external source could be achieved this way.

Predictive Models Generation. To create our predictive models, again, each user is an instance represented by their writings (see Sect. 4.2). For *Baseline 1* we did a $tf \cdot idf$ vectorization of the users' documents, by using the *TfIdfVectorizer* provided by the *Scikit-learn* Python library, with a stop-words list and the removal of the n-grams that appeared in less than 5 documents. The representation of each user through embeddings was given by the aggregation of the vector representations of the words in the concatenated texts of the users, normalized by the size (words count) of the document. Then, an L_2 normalization was applied to all the instances.

Given the reduced amount of anorexia cases on the training set, we used SMOTE [4] as an over-sampling method to deal with the unbalanced classes. The Scikit learn's Python library implementations for Logistic regression (LR), Random Forest (RF), Multilayer Perceptron (MLP), and Support Vector Machines (SVM) were tested as classifiers over the training set with a 5-fold cross validation approach. A grid search over each method to find the best parameters for the models was done.

Results. The results of the baselines are compared to models with our variations. For *Variation 4* and baselines 4 and 5 we use the 901 predictive terms of Sect. 4.4. To define the parameters of *Variation 3*, we test different configurations, as on Sect. 4.3, and chose the ones with the best results according to PS.

Precision (P), Recall (R), F1-Score ($F1$) and Accuracy (A) are used as evaluation measures. The scores for P, R and F1 reported over the test set on Table 4 correspond to the Anorexia (positive) class, as this is the most relevant one, whereas A corresponds to the accuracy computed on both classes. Seeing that there are 6 times more control cases than AN and that false negative (FN) cases are a bigger concern compared to false positives, we prioritize R and F1 over P and A. This is done because as with most medical screening tasks, classifying a user at risk as a control case (FN) is worst than the opposite (FP), in particular on a classifier that is intended to be a first filter to detect users at risk and eventually alert clinicians, who are the ones that do an specialized screening of the user profile. Table 4 shows the results for the best classifiers. The best scores are highlighted for each measure.

Comparing the baselines, we can notice that the embeddings based approaches provide an improvement on R compared to the BoW model, however this is given with a significant loss on P.

Regarding the embeddings based models, our variations outperform the results obtained by the baselines. The model with the embeddings generated with our method (*Variation 0*) provides significantly better results compared to the *Word2vec* model (*Baseline 3*), and even the model with pre-learned embeddings (*Baseline 2*), with a wider vocabulary.

The combination of pre-learned embeddings and embeddings learned on our training set, provide the best results in terms of F1 and R. They also provide a good accuracy considering that most of the test cases are controls. We can also observe that using the weights of pre-learned embeddings (*Variation 3*) to start our learning process over our corpus improves significantly the R score in comparison to *Word2vec*'s generated embeddings (*Baseline 3*).

The worst results for our variations are given by *Variation 1* that obtains equivalent results to *Baseline 2*. The best model in terms of F1 corresponds to

Variation 2. Also, better results are obtained for P when the embeddings are enhanced by the retrofitting approach (*Variation 4*).

Table 4. Baselines and enhanced embeddings evaluated in terms of Precision (*P*), Recall (*R*), F1-Score (*F*1) and Accuracy (*A*).

Model	Description	P	R	$F1$	A	Classifier
Baseline 1	BoW Model	**90.00%**	65.85%	76.06%	**94.69%**	MLP
Baseline 2	GloVe's pre-learned embeddings	69.57%	78.05%	73.56%	92.81%	MLP
Baseline 3	Word2vec embeddings	70.73%	70.73%	70.73%	92.50%	SVM
Baseline 4	Word2vec retrofitted embeddings	71.79%	68.29%	70.00%	92.50%	SVM
Baseline 5	GloVe's pre-learned embeddings retrofitted	67.35%	**80.49%**	73.33%	92.50%	MLP
Variation 0	Predictive pairs embeddings ($\beta_P = 50\ \beta_N = 50$)	77.50%	75.61%	76.54%	94.03%	MLP
Variation 1	Predictive pairs embeddings + GloVe embeddings	69.57%	78.05%	73.56%	92.81%	MLP
Variation 2	Predictive pairs embeddings ($\beta_P = 50\ \beta_N = 50$) + GloVe embeddings	75.00%	**80.49%**	**77.65%**	94.06%	MLP
Variation 3	Predictive pairs embeddings + GloVe embeddings starting weights ($\beta_P = 0.25\ \beta_N = 50$)	72.73%	78.05%	75.29%	93.44%	MLP
Variation 4	Predictive pairs ($\beta_P = 50\ \beta_N = 50$) retrofitted embeddings	82.86%	70.73%	76.32%	94.37%	SVM

5   Conclusions and Future Work

We presented an approach for enhancing word embeddings towards a classification task on the detection of AN. Our method extends *Word2vec* considering positive and negative costs for the objective function of a target term. The costs are added by defining predictive terms for each of the target classes. The combination of the generated embeddings with pre-learned embeddings is also evaluated. Our results show that the usage of our enhanced embeddings outperforms the results obtained by pre-learned embeddings and embeddings learned through *Word2vec* regardless of the small size of the corpus. These results are promising as they might lead to new research paths to explore.

Future work involves the evaluation of the method on similar tasks, which can be formalized as document categorization problems, addressing small corpora. Also, ablation studies will be performed to assess the impact of each component into the results obtained.

References

1. Arseniev, A., Lee, H., McCormick, T., Moreno, M.: Proana: pro-eating disorder socialization on twitter. J. Adolesc. Health **58**, 659–664 (2016)
2. Bojanowski, P., Grave, E., Joulin, A., Mikolov, T.: Enriching word vectors with subword information. Trans. Assoc. Comput. Linguist. **5**, 135–146 (2017)
3. Çano, E., Morisio, M.: Word embeddings for sentiment analysis: a comprehensive empirical survey. CoRR abs/1902.00753 (2019)
4. Chawla, N.V., Bowyer, K.W., Hall, L.O., Kegelmeyer, W.P.: Smote: synthetic minority over-sampling technique. J. Artif. Intell. Res. **16**, 321–357 (2002)
5. Coppersmith, G., Leary, R., Crutchley, P., Fine, A.: Natural language processing of social media as screening for suicide risk. Biomed. Inform. Insights **10** (2018)
6. Devlin, J., Chang, M., Lee, K., Toutanova, K.: BERT: pre-training of deep bidirectional transformers for language understanding. CoRR abs/1810.04805 (2018)
7. Faruqui, M., Dodge, J., Jauhar, S.K., Dyer, C., Hovy, E., Smith, N.A.: Retrofitting word vectors to semantic lexicons. In: Proceedings of the 2015 Conference of the North American Chapter of the Association for Computational Linguistics, pp. 1606–1615. Association for Computational Linguistics (2015)
8. Guntuku, S.C., Yaden, D.B., Kern, M.L., Ungar, L.H., Eichstaedt, J.C.: Detecting depression and mental illness on social media: an integrative review. Curr. Opin. Behav. Sci. **18**, 43–49 (2017)
9. Kuang, S., Davison, B.D.: Learning word embeddings with chi-square weights for healthcare tweet classification. Appl. Sci. **7**(8), 846 (2017)
10. Losada, D.E., Crestani, F., Parapar, J.: Overview of eRisk: early risk prediction on the internet. In: Bellot, P., et al. (eds.) CLEF 2018. LNCS, vol. 11018, pp. 343–361. Springer, Cham (2018). https://doi.org/10.1007/978-3-319-98932-7_30
11. Mikolov, T., Chen, K., Corrado, G., Dean, J.: Efficient estimation of word representations in vector space. CoRR abs/1301.3781 (2013)
12. Mowafy, M., Rezk, A., El-Bakry, H.: An efficient classification model for unstructured text document. Am. J. Comput. Sci. Inf. Technol. **06**, 16 (2018)
13. Pennington, J., Socher, R., Manning, C.: Glove: global vectors for word representation. In: Proceedings of the 2014 Conference on Empirical Methods in Natural Language Processing (EMNLP), pp. 1532–1543. Association for Computational Linguistics (2014)
14. Peters, M.E., et al.: Deep contextualized word representations. In: Proceedings of NAACL (2018)
15. Tissier, J., Gravier, C., Habrard, A.: Dict2vec : learning word embeddings using lexical dictionaries. In: Proceedings of the 2017 Conference on Empirical Methods in Natural Language Processing, pp. 254–263. Association for Computational Linguistics, Copenhagen, September 2017
16. Yin, W., Schütze, H.: Learning word meta-embeddings. In: Proceedings of the 54th Annual Meeting of the Association for Computational Linguistics, pp. 1351–1360. Association for Computational Linguistics, Berlin, August 2016
17. Zhang, Y., Li, H.J., Wang, J., Cohen, T., Roberts, K., Xu, H.: Adapting word embeddings from multiple domains to symptom recognition from psychiatric notes. In: AMIA Joint Summits on Translational Science Proceedings. AMIA Joint Summits on Translational Science (2018)

Event Recognition Based on Classification of Generated Image Captions

Andrey V. Savchenko[1,2(✉)] and Evgeniy V. Miasnikov[1]

[1] Samsung-PDMI Joint AI Center, St. Petersburg Department of Steklov Institute
of Mathematics, Fontanka Street, St. Petersburg, Russia
[2] National Research University Higher School of Economics,
Laboratory of Algorithms and Technologies for Network Analysis,
Nizhny Novgorod, Russia
avsavchenko@hse.ru

Abstract. In this paper, we consider the problem of event recognition on single images. In contrast to conventional fine-tuning of convolutional neural networks (CNN), we proposed to use image captioning, i.e., a generative model that converts images to textual descriptions. The motivation here is the possibility to combine conventional CNNs with a completely different approach in an ensemble with high diversity. As event recognition task has nothing serial or temporal, obtained captions are one-hot encoded and summarized into a sparse feature vector suitable for the learning of an arbitrary classifier. We provide the experimental study of several feature extractors for Photo Event Collection, Web Image Dataset for Event Recognition and Multi-Label Curation of Flickr Events Dataset. It is shown that the image captions trained on the Conceptual Captions dataset can be classified more accurately than the features from an object detector, though they both are obviously not as rich as the CNN-based features. However, an ensemble of CNN and our approach provides state-of-the-art results for several event datasets.

Keywords: Image captioning · Event recognition · Ensemble of classifiers · Convolutional neural network (CNN)

1 Introduction

Nowadays, social networks and mobile devices create a vast stream of multimedia data because people are taking more photos in recent years than ever before [1]. To organize a large gallery of personal photos, they may be assigned to albums according to some events. Social events are happenings that are attended and shared by the people [2,3] and take place in a specific environment [4], e.g., holidays, sports events, weddings, various activities, etc. The album labels are usually assigned either manually or by using locations from EXIF data if the GPS tags in a camera are switched on. However, content-based image analysis has been recently introduced in photo organizing systems. Such analysis can be

© The Author(s) 2020
M. R. Berthold et al. (Eds.): IDA 2020, LNCS 12080, pp. 418–430, 2020.
https://doi.org/10.1007/978-3-030-44584-3_33

used to selectively look for photos for a particular event in order to keep nice memories of some episodes of our lives [4] or to gather our specific interests for personalized recommender systems.

There exist two different event recognition tasks [2]. In the first task, the event categories are recognized for the whole album (a sequence of photos). However, the assignments of images of the same event into albums may be unknown in practice. Hence, in this paper, we focus on the second task, namely, event recognition in single images from social media. As an event here is a complex scene with large variations in visual appearance [4], deep learning techniques [5] are widely used. It is typical to fine-tune existing convolutional neural networks (CNNs) on event datasets [4]. Sometimes CNN-based object detection is applied [6] for discovering particular categories, e.g., interior objects, food, transport, sports equipment, animals, etc. [7,8].

However, in this paper, a slightly different approach is considered. Despite the conventional usage of a CNN as a discriminative model in a classifier design [9], we propose to borrow generative models to represent an input image in the other domain. In particular, we use existing methods of image captioning [10] that generate textual descriptions of images. Our main contribution is a demonstration that the generated descriptions can be fed to the input of a classifier in an ensemble in order to improve the event recognition accuracy of traditional methods. Though the proposed visual representation is not as rich as features extracted by fine-tuned CNNs, they are better than the outputs of object detectors [8]. As our approach is completely different than traditional CNNs, it can be combined with them into an ensemble that possesses high diversity and, as a consequence, high accuracy.

The rest of the paper is organized as follows. In Sect. 2, the survey of image captioning models is given. In Sect. 3, we introduce the proposed pipeline for event recognition based on generated captions. Experimental results for several event datasets are presented in Sect. 4. Finally, concluding comments and future works are discussed in Sect. 5.

2   Literature Survey

Most existing methods of event recognition on single photos tend to applications of the CNN-based architectures [2]. Four layers of fine-tuned CNN were used to extract features for LDA (Linear Discriminant Analysis) classifier in the ChaLearn LAP 2015 cultural event recognition challenge [11]. The iterative selection method [4] identifies the most relevant subset of classes for transferring representations from CNN learned from the object (ImageNet) and scene (Places2) datasets. The bounding boxes of detected objects are projected onto multi-scale spatial maps in the paper [6]. An ensemble of scene classifiers and object detectors provided the high accuracy [12] for the Photo Event Collection (PEC) [13]. Unfortunately, there is a significant gap in the accuracies of event classification in still photos [4] and albums [14], so that there is a huge demand in all-the-more accurate methods of single image processing.

That is why in this paper, we proposed to concentrate on other suitable visual features extracted with the generative models and, in particular, image captioning techniques. There is a wide range of applications of image captioning: from the automatic generation of descriptions for photos posted in social networks to image retrieval from databases using generated text descriptions [15]. The image captioning methods are usually based on an encoder-decoder neural network, which first encodes an image into a fixed-length vector representation using pre-trained CNN, and then decodes the representation into captions (a natural language description). During the training of a decoder (generator), the input image and its ground-truth textual description are fed as inputs to the neural network, while one hot encoded description presents the desired network output. The description is encoded using text embeddings in the Embedding (look-up) layer [5]. The generated image and text embeddings are merged using concatenation or summation and form the input to the decoder part of the network. It is typical to include the recurrent neural network (RNN) layer followed by a fully connected layer with the Softmax output layer.

One of the first successful models, "Show and Tell" [16], won the first MS COCO Image Captioning Challenge in 2015. It uses RNN with long short-term memory (LSTM) units in a decoder part. Its enhancement "Show, Attend and Tell" [17] incorporates a soft attention mechanism to improve the quality of the caption generation. The "Neural Baby Talk" image captioning model [18] is based on generating the template with slot locations explicitly tied to specific image regions. These slots are then filled in by visual concepts identified in the object detectors. The foreground regions are obtained using the Faster-RCNN network [19], and LSTM with attention mechanism serves as a decoder. The "Multimodal Recurrent Neural Network" (mRNN) [20] is based on the Inception network for image features extraction and deep RNN for sentence generation. One of the best models nowadays is the Auto-Reconstructor Network (ARNet) [21], which uses the Inception-V4 network [22] in an encoder, and the decoder is based on LSTM. There exist two pre-trained models with greedy search (ARNet-g) and beam search (ARNet-b) with size 3 to generate the final caption for each input image.

3   Proposed Approach

Our task can be formulated as a typical image recognition problem [9]. It is required to assign an input photo X from a gallery to one of $C > 1$ event categories (classes). The training set of $N \geq 1$ images $\mathbf{X} = \{X_n | n \in \{1, ..., N\}\}$ with known event labels $c_n \in \{1, ..., C\}$ is available for classifier learning. Sometimes the training photos of the same event are associated with an album [13,14]. In such a case, the training albums are unfolded into a set $\mathbf{X}$ so that the collection-level label of the album is assigned to labels of each photo from this album. This task possesses several characteristics that makes it extremely challenging compared to album-based event recognition. One of these characteristics is the presence of irrelevant images or unimportant photos that can be associated with any event [2]. These images can be detected by attention-based models when the whole album is available [1] but may have a significant negative impact on the quality of event recognition in single images.

As N is usually rather small, transfer learning may be applied [5]. A deep CNN is firstly pre-trained on a large dataset, e.g., ImageNet or Places [23]. Secondly, this CNN is fine-tuned on $\mathbf{X}$, i.e., the last layer is replaced to the new layer with Softmax activations and C outputs. An input image X is classified by feeding it to the fine-tuned CNN to compute C scores from the output layer, i.e., the estimates of posterior probabilities for all event categories. This procedure can be modified by the extraction of deep image features (embeddings) using the outputs of one of the last layers of the pre-trained CNN [5,24]. The input image X and each training image $X_n, n \in \{1, ..., N\}$ are fed to the input of the CNN, and the outputs of the last-but-one layer are used as the D-dimensional feature vectors $\mathbf{x} = [x_1, ..., x_D]$ and $\mathbf{x}_n = [x_{n;1}, ..., x_{n;D}]$, respectively. Such deep learning-based feature extractors allow training of a general classifier $\mathcal{C}_{emb}$, e.g., k-nearest neighbor, random forest (RF), support vector machine (SVM) or gradient boosting [9,25]. The C-dimensional vector of $\mathbf{p}_{emb} = \mathcal{C}_{emb}(\mathbf{x})$ confidence scores is predicted given the input image in both cases of fine-tuning with the last Softmax layer in a role of classifier $\mathcal{C}_{emb}$ and feature extraction with general classifier. The final decision is made in favor of a class with maximal confidence.

In this paper, we use another approach to event recognition based on generative models and image captioning. The proposed pipeline is presented in Fig. 1. At first, the conventional extraction of embeddings $\mathbf{x}$ is implemented using pre-trained CNN. Next, these visual features and a vocabulary V are fed to a special RNN-based neural network (generator) that produces the caption, which describes the input image. Caption is represented as a sequence of $L > 0$ tokens

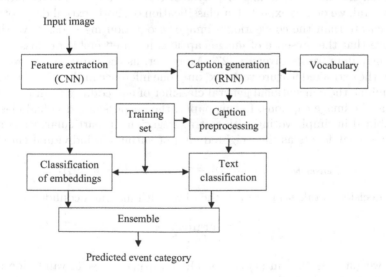

Fig. 1. Proposed event recognition pipeline based on image captioning

$\mathbf{t} = \{t_0, t_1 ..., t_{L+1}\}$ from the vocabulary ($t_l \in V, l \in \{0, ..., L\}$). It is generated sequentially, word-by-word starting from $t_0 = <START>$ token until a special $t_{L+1} = <END>$ word is produced [21].

The generated caption $\mathbf{t}$ is fed into an event classifier. In order to learn its parameters, every n-th image from the training set is fed to the same image captioning network to produce the caption $\mathbf{t}_n = \{t_{n;0}, t_{n;1} ..., t_{n;L_n+1}\}$. Since the number of tokens L_n is not the same for all images, it is necessary to either train a sequential RNN-based classifier or transform all captions into feature vectors with the same dimensionality. As the number of training instances N is not very large, we experimentally noticed that the latter approach is as accurate as the former, though the training time is significantly lower. This fact can be explained by the absence of anything temporal or serial in the initial task of event recognition in single images. Hence, we decided to use one-hot encoding and convert the sequences $\mathbf{t}$ and $\{\mathbf{t}_n\}$ into vectors of 0s and 1s as described in [26]. In particular, we select a subset of vocabulary $\tilde{V} \subset V$ by choosing the top most frequently occurring words in the training data $\{\mathbf{t}_n\}$ with the optional exclusion of stop words. Next, the input image is represented as the $|\tilde{V}|$-dimensional sparse vector $\tilde{\mathbf{t}} \subset \{0, 1\}^{|\tilde{V}|}$, where $|\tilde{V}|$ is the size of reduced vocabulary $\tilde{V}$ and the v-th component of vector $\tilde{\mathbf{t}}$ is equal to 1 only if at least one of L words in the caption $\mathbf{t}$ is equal to the v-th word from vocabulary $\tilde{V}$. This would mean, for instance, turning the sequence $\{1, 5, 10, 2\}$ into a $\tilde{V}$-dimensional sparse vector that would be all 0s except for indices 1, 2, 5 and 10, which would be 1s [26]. The same procedure is used to describe each n-th training image with $\tilde{V}$-dimensional sparse vector $\tilde{\mathbf{t}}_n$. After that an arbitrary classifier C_{txt} of such textual representations suitable for sparse data can be used to predict C confidence scores $\mathbf{p}_{txt} = C_{txt}(\tilde{\mathbf{t}})$. It is known [26] that such an approach is even more accurate than conventional RNN-based classifiers (including one layer of LSTMs) for the IMDB dataset.

In general, we do not expect that classification of short textual descriptions is more accurate than the conventional image recognition methods. Nevertheless, we believe that the presence of image captions in an ensemble of classifiers can significantly improve its diversity [27]. Moreover, as the captions are generated based on the extracted feature vector $\mathbf{x}$, only one inference in the CNN is required if we combine the conventional general classifier of embeddings from pre-trained CNN and the image captions. In this paper, the outputs of individual classifiers are combined in simple voting with soft aggregation. In particular, we compute aggregated confidences as the weighted sum of outputs of individual classifier:

$$\mathbf{p}_{ensemble} = [p_1, ..., p_C] = w \cdot \mathbf{p}_{emb} + (1 - w)\mathbf{p}_{txt}. \tag{1}$$

The decision is taken in favor of the class with maximal confidence:

$$c^* = \underset{c \in \{1, ..., C\}}{\operatorname{argmax}} p_c. \tag{2}$$

The weight $w \in [0, 1]$ in (1) can be chosen using a special validation subset in order to obtain the highest accuracy of criterion (2).

Let us provide qualitative examples for the usage of our pipeline (Fig. 1). The results of (correct) event recognition using our ensemble are presented in Fig. 2. Here the first line of the title contains the generated image caption. In addition, the title displays the result of event recognition using captions t (second line), embeddings x_{emb} (third line), and the whole ensemble (last line). As one can notice, the single classification of captions is not always correct. However, our ensemble is able to obtain a reliable solution even when individual classifiers make wrong decisions.

Fig. 2. Sample results of event recognition

4   Experimental Results

In the experimental study, we examined the following event datasets:

1. PEC [13] with 61,000 images from 807 collections of $C = 14$ social event classes (birthday, wedding, graduation, etc.).
2. WIDER (Web Image Dataset for Event Recognition) [6] with 50,574 images and $C = 61$ events (parade, dancing, meeting, press conference, etc.).
3. ML-CUFED (Multi-Label Curation of Flickr Events Dataset) [14] contains $C = 23$ common event types. Each album is associated with several events, i.e., it is a multi-label classification task.

We used standard train/test split for all datasets proposed by their creators. In PEC and ML-CUFED, the collection-level label is directly assigned to each image contained in this collection. Moreover, we completely ignore any metadata, e.g., temporal information, except the image itself similarly to the paper [4]. As a result, the training and validation sets are not ideally balanced. The majority classes in each dataset contains 5-times higher number of training images when compared to the minority classes. However, the class distribution in the training and validation sets remains more or less identical, so that the number of validation images for majority classes is also 5-times higher than the number of testing examples for minority classes.

As we mainly focus on the possibility of implementing offline event recognition on mobile devices [12], in order to compare the proposed approach with conventional classifiers, we used MobileNet v2 with $\alpha = 1$ [28] and Inception v4 [22] CNNs. At first, we pre-trained them on the Places2 dataset [23] for feature extraction. The linear SVM classifier from the scikit-learn library was used because it has higher accuracy than other classifiers from this library (RF, k-NN, and RBF SVM) on the considered datasets. Moreover, we fine-tuned these CNNs using the given training set as follows. At first, the weights in the base part of the CNN were frozen, and the new head (fully connected layer with C outputs and Softmax activation) was learned using the ADAM optimizer (learning rate 0.001) for 10 epochs with an early stop in the Keras 2.2 framework with the TensorFlow 1.15 backend. Next, the weights in the whole CNN were learned during 5 epochs using the ADAM. Finally, the CNN was trained using SGD during 3 epochs with 10-times lower learning rate.

In addition, we used features from object detection models that are typical for event recognition [6,12]. As many photos from the same event sometimes contain identical objects (e.g., ball in the football), they can be detected by contemporary CNN-based methods, i.e., SSDLite [28] or Faster R-CNN [19]. These methods detect the positions of several objects in the input image and predict the scores of each class from the predefined set of $K > 1$ types. We extract the sparse K-dimensional vector of scores for each type of object. If there are several objects of the same type, the maximal score is stored in this feature vector [8]. This feature vector is either classified by the linear SVM or used to train a feed-forward neural network with two hidden layers containing 32 units. Both classifiers were learned using the training set from each event dataset. In this study, we examined SSD with the MobileNet backbone and Faster R-CNN with the InceptionResNet backbone. The models pre-trained on the Open Images Dataset v4 ($K = 601$ objects) were taken from the TensorFlow Object Detection Model Zoo.

Our preliminarily experimental study with the pre-trained image captioning models discussed in Sect. 2 demonstrated that the best quality for MS COCO captioning dataset is achieved by the ARNet model [21]. Thus, in this experiment, we used ARNet's encoder-decoder model. However, it can be replaced with any other image captioning technique without modification of our event recognition algorithm.

Unfortunately, event datasets do not contain captions (textual descriptions), which are required to train or fine-tune the image captioning model. Due to this reason, the image captioning model was trained on the Conceptual Captions dataset. Today this dataset is the largest dataset used for image captioning. It contains more than 3.3M image-URL and caption pairs in the training set, and about 15 thousand pairs in the validation set. While there exist other smaller datasets, such as MS COCO and Flickr, in our preliminary experiments, the image captioning model, which were trained on the Conceptual Captions Dataset, provided better worse-case performance in the cross-dataset evaluation.

The feature extraction in the encoder is implemented not only with the same CNNs (Inception and MobileNet v2). We extracted $|\tilde{V}| = 5000$ most frequent words except special tokens $< START >$ and $< END >$. They are classified by either linear SVM or a feed-forward neural network with the same architecture as for the object detection case. Again, these classifiers are trained from scratch, given each event training set. The weight w in our ensemble (Eq. 1) was estimated using the same set.

The results of the lightweight mobile (MobileNet and SSD object detector) and deep models (Inception and Faster R-CNN) for PEC, WIDER and ML-CUFED are presented in Tables 1, 2, 3, respectively. Here we added the best-known results for the same experimental setups.

Certainly, the proposed recognition of image captions is not as accurate as conventional CNN-based features. However, the classification of textual descriptions is much better than the random guess with accuracy $100\%/14 \approx 7.14\%$, $100\%/61 \approx 1.64\%$ and $100\%/23 \approx 4.35\%$ for PEC, WIDER and ML-CUFED, respectively. It is important to emphasize that our approach has a lower error rate than the classification of the features based on object detection in most cases. This gain is especially noticeable for lightweight SSD models, which are 1.5–13% less accurate than the proposed classification of image captions due to the limitations of SSD-based models to detect small objects (food, pets, fashion accessories, etc.). The Faster R-CNN-based detection features can be classified more accurately, but the inference in Faster R-CNN with the InceptionResNet backbone is several times slower than the decoding in the image captioning model (6–10 s vs. 0.5–2 s on MacBook Pro 2015).

Table 1. Event recognition accuracy (%), PEC

Classifier	Features	Lightweight models	Deep models
SVM	Embeddings	59.72	61.82
	Objects	42.18	47.83
	Texts	43.77	47.24
	Proposed ensemble (1), (2)	60.56	62.87
Fine-tuned CNN	Embeddings	62.33	63.56
	Objects	40.17	47.42
	Texts	43.52	46.89
	Proposed ensemble (1), (2)	63.38	65.12
Aggregated SVM [13]			41.4
Bag of Sub-events [13]			51.4
SHMM [13]			55.7
Initialization-based transfer learning [4]			60.6
Transfer learning of data and knowledge [4]			62.2

Table 2. Event recognition accuracy (%), WIDER

Classifier	Features	Lightweight models	Deep models
SVM	Embeddings	48.31	50.48
	Objects	19.91	28.66
	Texts	26.38	31.89
	Proposed ensemble (1), (2)	48.91	51.59
Fine-tuned CNN	Embeddings	49.11	50.97
	Objects	12.91	21.27
	Texts	25.93	30.91
	Proposed ensemble (1), (2)	49.80	51.84
Baseline CNN [6]			39.7
Deep channel fusion [6]			42.4
Initialization-based transfer learning [4]			50.8
Transfer learning of data and knowledge [4]			53.0

Finally, the most appropriate way to use image captioning in event classification is its fusion with conventional CNNs. In such case, we improved the previous state-of-the-art for PEC from 62.2% [4] even for the lightweight models (63.38%) if the fine-tuned CNNs are used in an ensemble. Our Inception-based model is even better (accuracy 65.12%). We have not still reached the state-of-the-art accuracy 53% [4] for the WIDER dataset, though our best accuracy (51.84%) is up to 9% higher when compared to the best results (42.4%) from original paper [6]. Our experimental setup for the ML-CUFED dataset is studied for the

Table 3. Event recognition accuracy (%), ML-CUFED

Classifier	Features	Lightweight models	Deep models
SVM	Embeddings	53.54	57.27
	Objects	34.21	40.94
	Texts	37.24	41.52
	Proposed ensemble (1), (2)	55.26	58.86
Fine-tuned CNN	Embeddings	56.01	57.12
	Objects	32.05	40.12
	Texts	36.74	41.35
	Proposed ensemble (1), (2)	57.94	60.01

first time here because this dataset is developed mostly for album-based event recognition. We should highlight that our preliminary experiments in the latter task with this dataset and simple averaging of MobileNet features extracted from all images from an album slightly improved the state-of-the-art accuracy for this dataset, though it is necessary to study more complex feature aggregation techniques [1].

In practice, it is preferable to use pre-trained CNN as a feature extractor in order to prevent additional inference in fine-tuned CNN when it differs from the encoder in the image captioning model. Unfortunately, the accuracies of SVM for pre-trained CNN features are 1.5–3% lower when compared to the fine-tuned models for PEC and ML-CUFED. In this case, an additional inference may be acceptable. However, the difference in error rates between pre-trained and fine-tuned models for the WIDER dataset is not significant, so that the pre-trained CNNs are definitely worth being used here.

5   Conclusion

In this paper, we have proposed to apply generative models in the classical discriminative task [9]; namely, image captioning in event recognition in still images. We have presented the novel pipeline of visual preference prediction using image captioning with the classification of generated captions and retrieval of images based on their textual descriptions (Fig. 1). It has been experimentally demonstrated that our approach is more accurate than the widely-used image representations obtained by object detectors [6,8]. Moreover, our approach is much faster than Faster R-CNNs, which do not implement one-shot detection. What is especially useful for ensemble models [27] generated caption provides additional diversity to conventional CNN-based recognition.

The motivation behind the study of image captioning techniques in this paper is connected not only with generating compact informative descriptions of images, but also with the wide possibilities to ensure the privacy of user data if further processing at remote servers is necessary. Moreover, as the vocabulary of generated captions is restricted, such techniques are considered as effective anonymization methods. Since the textual descriptions can be easily perceived and understood by the user (as opposed to a vector of numeric features), his or her attitude to the use of such methods will be more trustworthy.

Unfortunately, short conceptual textual descriptions are obviously not enough to classify event categories with high accuracy even for a human due to errors and lack of specificity (see an example of generated captions in Fig. 2). Another disadvantage of the proposed approach is the need to repeat inference if fine-tuned CNN is applied in an ensemble. Hence, the decision-making time will be significantly increased, though the overall accuracy also becomes higher in most cases (Tables 1 and 3).

In the future, it is necessary to make the classification of generated captions more accurate. At first, though our preliminary experiments of LSTMs did not decrease the error rate of our simple approach with linear SVM and one-hot encoded words, we strongly believe that a thorough study of the RNN-based classifiers of generated textual descriptors is required. Second, the comparison of image captioning models trained on the Conceptual Captions dataset is needed to choose the best model for caption generation. Here the impact on event recognition accuracy arising from erroneous captions being generated should be examined. Third, additional research is needed to check if we can fine-tune a CNN on an event dataset and use it as an encoder for the caption generation without loss of quality. In this case, a more compact and fast solution can be achieved. Finally, the proposed pipeline should be extended for the album-based event recognition [2,13] with, e.g., attention models [12].

Acknowledgements. This research is based on the work supported by Samsung Research, Samsung Electronics. The work of A.V. Savchenko was conducted within the framework of the Basic Research Program at the National Research University Higher School of Economics (HSE).

References

1. Guo, C., Tian, X., Mei, T.: Multigranular event recognition of personal photo albums. IEEE Trans. Multimedia **20**(7), 1837–1847 (2017)
2. Ahmad, K., Conci, N.: How deep features have improved event recognition in multimedia: a survey. ACM Trans. Multimedia Comput. Commun. Appl. **15**(2), 39 (2019)
3. Papadopoulos, S., Troncy, R., Mezaris, V., Huet, B., Kompatsiaris, I.: Social event detection at MediaEval 2011: challenges, dataset and evaluation. In: MediaEval (2011)
4. Wang, L., Wang, Z., Qiao, Y., Van Gool, L.: Transferring deep object and scene representations for event recognition in still images. Int. J. Comput. Vis. **126**(2–4), 390–409 (2018)
5. Goodfellow, I., Bengio, Y., Courville, A.: Deep Learning. MIT Press, Cambridge (2016)
6. Xiong, Y., Zhu, K., Lin, D., Tang, X.: Recognize complex events from static images by fusing deep channels. In: Proceedings of the IEEE Conference on Computer Vision and Pattern Recognition (CVPR), pp. 1600–1609 (2015)

7. Grechikhin, I., Savchenko, A.V.: User modeling on mobile device based on facial clustering and object detection in photos and videos. In: Morales, A., Fierrez, J., Sánchez, J.S., Ribeiro, B. (eds.) IbPRIA 2019. LNCS, vol. 11868, pp. 429–440. Springer, Cham (2019). https://doi.org/10.1007/978-3-030-31321-0_37

8. Savchenko, A.V., Rassadin, A.G.: Scene recognition in user preference prediction based on classification of deep embeddings and object detection. In: Lu, H., Tang, H., Wang, Z. (eds.) ISNN 2019. LNCS, vol. 11555, pp. 422–430. Springer, Cham (2019). https://doi.org/10.1007/978-3-030-22808-8_41

9. Prince, S.J.: Computer Vision: Models, Learning and Inference. Cambridge University Press, Cambridge (2012)

10. Hossain, M., Sohel, F., Shiratuddin, M.F., Laga, H.: A comprehensive survey of deep learning for image captioning. ACM Comput. Surv. **51**(6), 1–36 (2019)

11. Escalera, S., et al.: ChaLearn looking at people 2015: apparent age and cultural event recognition datasets and results. In: Proceedings of the IEEE International Conference on Computer Vision Workshops (ICCVW), pp. 1–9 (2015)

12. Savchenko, A.V., Demochkin, K.V., Grechikhin, I.S.: User preference prediction in visual data on mobile devices. arXiv preprint arXiv:1907.04519 (2019)

13. Bossard, L., Guillaumin, M., Van Gool, L.: Event recognition in photo collections with a stopwatch HMM. In: Proceedings of the IEEE International Conference on Computer Vision, pp. 1193–1200 (2013)

14. Wang, Y., Lin, Z., Shen, X., Mech, R., Miller, G., Cottrell, G.W.: Recognizing and curating photo albums via event-specific image importance. In: Proceedings of British Conference on Machine Vision (BMVC) (2017)

15. Vijayaraju, N.: Image retrieval using image captioning. Master's Projects, p. 687 (2019). https://doi.org/10.31979/etd.vm9n-39ed

16. Vinyals, O., Toshev, A., Bengio, S., Erhan, D.: Show and tell: lessons learned from the 2015 MSCOCO image captioning challenge. IEEE Trans. Pattern Anal. Mach. Intell. **39**(4), 652–663 (2017)

17. Xu, K., et al.: Show, attend and tell: neural image caption generation with visual attention. In: Proceedings of the International Conference on Machine Learning (ICML), pp. 2048–2057 (2015)

18. Lu, J., Yang, J., Batra, D., Parikh, D.: Neural baby talk. In: Proceedings of the IEEE Conference on Computer Vision and Pattern Recognition (CVPR) (2018)

19. Ren, S., He, K., Girshick, R., Sun, J.: Faster R-CNN: towards real-time object detection with region proposal networks. In: Advances in Neural Information Processing Systems (NIPS), pp. 91–99 (2015)

20. Mao, J., Xu, W., Yang, Y., Wang, J., Yuille, A.L.: Deep captioning with multimodal recurrent neural networks (m-RNN). In: Proceedings of the International Conference on Learning Representations (ICLR) (2015)

21. Chen, X., Ma, L., Jiang, W., Yao, J., Liu, W.: Regularizing RNNs for caption generation by reconstructing the past with the present. In: Proceedings of the IEEE Conference on Computer Vision and Pattern Recognition (CVPR) (2018)

22. Szegedy, C., Ioffe, S., Vanhoucke, V., Alemi, A.A.: Inception-v4, inception-ResNet and the impact of residual connections on learning. In: Proceedings of the International Conference on Learning Representations (ICLR) Workshop (2016)

23. Zhou, B., Lapedriza, A., Khosla, A., Oliva, A., Torralba, A.: Places: a 10 million image database for scene recognition. IEEE Trans. Pattern Anal. Mach. Intell. **40**(6), 1452–1464 (2018)

24. Savchenko, A.V.: Sequential three-way decisions in multi-category image recognition with deep features based on distance factor. Inf. Sci. **489**, 18–36 (2019)

25. Savchenko, A.V.: Probabilistic neural network with complex exponential activation functions in image recognition. IEEE Trans. Neural Netw. Learn. Syst. **31**(2), 651–660 (2020)
26. Chollet, F.: Deep Learning with Python. Manning Publications Company, Shelter Island (2017)
27. Zhou, Z.H.: Ensemble Methods: Foundations and Algorithms. Chapman and Hall/CRC, London (2012)
28. Sandler, M., Howard, A.G., Zhu, M., Zhmoginov, A., Chen, L.: MobileNetV2: inverted residuals and linear bottlenecks. In: Proceedings of the Conference on Computer Vision and Pattern Recognition (CVPR), pp. 4510–4520. IEEE (2018)

Human-to-AI Coach:
Improving Human Inputs to AI Systems

Johannes Schneider[✉]

Institute of Information Systems, University of Liechtenstein,
Vaduz, Liechtenstein
johannes.schneider@uni.li

Abstract. Humans increasingly interact with Artificial intelligence (AI) systems. AI systems are optimized for objectives such as minimum computation or minimum error rate in recognizing and interpreting inputs from humans. In contrast, inputs created by humans are often treated as a given. We investigate how inputs of humans can be altered to reduce misinterpretation by the AI system and to improve efficiency of input generation for the human while altered inputs should remain as similar as possible to the original inputs. These objectives result in trade-offs that are analyzed for a deep learning system classifying handwritten digits. To create examples that serve as demonstrations for humans to improve, we develop a model based on a conditional convolutional autoencoder (CCAE). Our quantitative and qualitative evaluation shows that in many occasions the generated proposals lead to lower error rates, require less effort to create and differ only modestly from the original samples.

1 Introduction

Human-to-AI information flow is increasing rapidly in importance and extent across multiple modalities. For example, voice-machine interaction is becoming more and more popular with deep learning networks recognizing text from speech. Similar, the progress in image recognition has lowered error rates in gesture and optical character recognition. Still, key technologies in AI such as deep learning are not perfect. They might also error given ambiguous inputs created by humans. Errors might be more likely by humans being in a hurry, being unaware of the AI's recognition mechanism, sloppiness or lack of skill. Safety critical application areas such as autonomous driving or medical applications, where an AI might depend on inputs from humans in one way or another, are becoming more and more prominent. Thus, mistakes in recognizing and processing inputs should be avoided. Apart from avoiding errors, humans might also have an incentive to provide inputs with less effort, e.g. "Why try to speak clearly and loudly in the presence of noise, if mumbling works just as well? Why doing that extra stroke in writing a character, if detection works just as well without it?" In this work, we do not focus on how to improve AI systems that recognize and interpret human information. We aim at strategies how humans

© The Author(s) 2020
M. R. Berthold et al. (Eds.): IDA 2020, LNCS 12080, pp. 431–443, 2020.
https://doi.org/10.1007/978-3-030-44584-3_34

can convey information better to such a system by adjusting their behavior. Identifying potential improvements becomes more difficult when deep learning is involved. Improvements are often based on a deep understanding of mechanisms of the task at hand, i.e. how an AI system processes inputs. Deep learning is said to follow a black-box behavior. Even worse, deep learning is well-known to reason very differently from humans: Deep learning models might astonish due to their high accuracy rates, but disappoint at the same time by failing on simple examples that were just slightly modified as well-documented by so called "adversarial examples". As such, humans might depend even more on being shown opportunities for generating better data that serves as input to an AI. In this work, we formalize the aforementioned partially conflicting goals such as minimizing wrongly recognized human inputs and reducing effort for humans – both in terms of need to adjust their behavior as well as to interact effortlessly. We focus on the classification problem of digits, where we aim to provide suggestions to humans by altering their generated inputs as illustrated in Fig. 1. We express the problem in terms of a multi-objective optimization problem, i.e. as a linear weighted sum. As model we use a conditional convolutional autoencoder. Our qualitative and quantitative evaluation highlights that the generated samples are visually appealing, easy to interpret and also lead to a lower error rate in recognition.

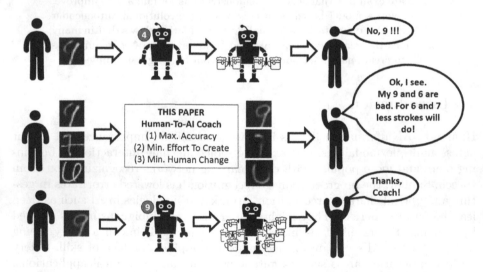

Fig. 1. "Human-to-AI" (H2AI) coach: From misunderstandings to understanding

2   Challenges of Human-to-AI Communication

We consider the problem of improving human generated inputs to an AI illustrated in Fig. 1. A human wants to convey information to an AI using some mode, e.g. speech, writing, or gestures. The processing of the received signals by the AI often involves two steps: (i) recognition, i.e. identifying and extracting relevant information in the input signal, and (ii) interpretation, i.e. deriving actions by utilizing the information in a specific context. For recognition, the information has to be extracted from a physical (analog) signal, e.g. using speech recognition, image recognition, etc. In case information is communicated in a digital manner using structured data, recognition is commonly obsolete. Often the extracted information has to be further processed by the AI using some form of sense-making or interpretation. The AI requires potentially semantic understanding capabilities and might rely on the use of context such as prior discourse or surrounding. We assume that the human interacts frequently with such a system, so that it is reasonable for the human to improve on objectives such as errors and efficiency in communication. In this paper, we consider the challenge of discovering variations of the original inputs that might help a human to improve.

More formally, we consider a classification problem, where a user provides data $D = (X, Y)$. Each sample X should be recognized as class Y by a classifier C_H. We denote by X_i the i-th feature of sample X. For illustration, for the case of handwritten digits a sample X is a gray-tone scan of a digit and $Y \in [0 - 9]$ the digitized number. $X_i \in [0, 1]$ gives the brightness of the i-th pixel in the scan. The classification model C_H was trained to optimize classification performance of human samples, i.e. maximize $P_{C_H}(Y|X)$. We regard the model C_H as a given, i.e. we do not alter it in any way, but use it in our optimization process. The Human-to-AI coach "H2AI" takes as input one sample X with its label Y. It returns at least one proposal $\hat{X}$, i.e. $\hat{X} := H2AI(X, Y)$. The suggestion $\hat{X}$ should be superior to X according to some objective, e.g. we might demand higher certainty in recognition $P_{C_H}(Y|X) < P_{C_H}(Y|\hat{X})$. In a handwriting scenario a human might use a proposal $\hat{X}$ based on an input X to adjust her strokes.

3 Model and Objectives

An essential requirement is that the modified samples are similar to the given input, otherwise a trivial solution is to always return "the perfect sample" that is the same for any input. This motivates utilizing an auto-encoder (Sect. 3.1) and adding multiple loss terms to handle various objectives (Sect. 3.2).

3.1 Architecture

Two approaches that allow to create (modified) samples are generative adversarial networks (GANs) and autoencoders (AEs). There are also combinations thereof, e.g. the pix2pix architecture [10] or conditional variational autoencoder [2]. [10] and [2] contain an AE which has a decoder serving as a generator based on a latent representation from the encoder and, additionally, a discriminator. AE tend to generate outcomes that are closer to the inputs. But they are often smoother and less realistic looking. In our application staying close to the input is a key requirement, since we only want to show how a sample can be modified rather than generating completely new samples. Thus, we decided to focus on an AE-based architecture. We also investigate including a discriminator to improve generated samples. More precisely, we utilize conditional AE with extra loss terms for regularization covering not only a discriminator loss but also losses for efficiency and classification of modified samples as shown in Fig. 3. Conditional AE are given as input the class of a sample in addition to the sample itself. This often improves generated samples, in particular for samples that are ambiguous, i.e. samples that seem to match multiple classes well.

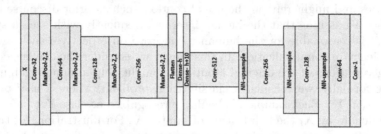

Fig. 2. H2AI implementation using a convolutional conditional autoencoder (CCAE)

Convolutional AE are known to work well on image data. Therefore, we propose convolutional conditional AE (CCAE) as shown in Fig. 2, where the NN-upsample layers in the decoder denote nearest-neighbor upsampling. After each convolutional layer, there is a ReLU layer that is not shown in Fig. 2. Compared to transposed convolutional layers, NN-upsampling with convolutional layers prevents checkerboard artifacts in the resulting images.

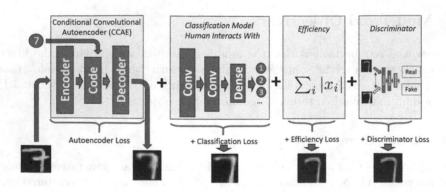

Fig. 3. Human-to-AI (H2AI) model with its components and regularizers

3.2 Objectives and Loss Terms

The generated input samples should meet multiple criteria, each of which is implemented as a loss term. The loss terms and their weighted sum (with parameters α.) are given in Eq. 1 and illustrated in Fig. 3. The total loss $L_{Tot}(X, Y)$ contains four parameters α_{RE}, α_{CL}, α_{EF} and α_D. It is possible to keep α_{RE} and use the other three to control the relative importance of the following objectives:

$$\hat{X} := CCAE(X, Y) \qquad \text{Sample proposed by H2AI-coach}$$

$$L_{RE}(X, \hat{X}) := \sum_i |X_i - \hat{X}_i| \qquad \text{Reconstruction or Change Loss}$$

$$L_{CL}(\hat{X}, Y) \qquad \text{Classification Loss}$$

$$L_{EF}(\hat{X}) := \sum_i |\hat{X}_i| \qquad \text{Efficiency Loss} \tag{1}$$

$$L_D(\hat{X}) := \log(1 - D(\hat{X})) \qquad \text{Discriminator Loss}$$

$$L_{Tot}(X, Y) := \alpha_{RE} L_{RE}(X, \hat{X}) + \alpha_{CL} L_{CL}(\hat{X}, Y) + \alpha_{EF} L_{EF}(\hat{X}) + \alpha_D L_D(\hat{X})$$

Minimal Effort to Change: Change might be difficult and tedious for humans. Thus, the effort for humans to adjust their behavior should be minimized. This implies that the original samples X created by humans and the newly generated variations $\hat{X}$ should be similar. This is covered by the reconstruction loss $L_{RE}(X, \hat{X})$ of the AE (see Eq. (1)). It enforces the output and the input to be similar. But parts of the input might be changed fairly drastically, i.e. for handwritten digits pixels might change from 0(black) to 1(white) and vice versa. For that reason, we do not employ an $L2$-metric, which heavily penalizes such differences, but rather opt for an $L1$-metric.

Reduce Mis-understanding: The amount of wrongly extracted or interpreted information by the AI should be reduced. AEs are known to have a denoising, averaging effect. They are also known to improve performance in some cases in conjunction with classification tasks [11]. To further foster a reduction in mis-understandings we minimize the classification loss $L_{C_H}(\hat{X}, Y)$ for generated examples $\hat{X}$ for the model C_H the human communicates with.

Realistic Samples: The generated samples $\hat{X}$ should still be comprehensible for humans or other systems, i.e. look realistic. It can happen that a generated proposal $\hat{X}$ is so optimized for the given AI model C_H that it is not meaningful in general. That is, the proposal $\hat{X}$ might appear not only very different from prototypical examples of its class but very different from any example occurring in reality. While AEs partially counteract this, AEs do not enforce that samples look real, but tend to create smooth (averaged) samples. Thus, we add a discriminator D resulting in a GAN architecture that should distinguish between real and generated samples and make them look crispier. The added discriminator loss $L_D(\hat{X})$ is $\log(1 - D(\hat{X}))$, where $\hat{X}$ is the generated sample $\hat{X} := CCAE(X, Y)$ for an input sample X of a human of class Y.

Minimal Effort to Create Samples: Interaction should be effortless for the human (and AI). To quantify effort of a human to create a sample, time might be a good option if available. If not, application specific measures might be more appropriate. For measuring effort in handwriting, the amount (and length) of strokes can be used. A good approximation can be the total amount of needed "ink", which corresponds to the $L1$-loss of the proposal $\hat{X}$, i.e. $L_{EF}(\hat{X}) := \sum_i |\hat{X}_i|$. We chose the $L1$ over the $L2$-metric, since having many low intensity pixels (as fostered by $L2$) is generally discouraged.

4   Evaluation

We conducted both a qualitative and quantitative evaluation on the MNIST dataset, since it has been used by recent work in similar contexts [6,8]. It consists of 50000 handwritten digits from 0 to 9 for training and 10000 digits for testing. The classification model C_H, i.e. the system a user is supposed to communicate well with, is a simple convolutional neural network (CNN) consisting of two convolutional layers (8 and 16 channels) that are both followed by a ReLU and 2×2 Max-Pooling Layer. The last layer is a fully connected layer. The network achieved a test accuracy of 95.97%. While this could be improved, it is not of prime relevance for our problem, since the classifier C_H is treated as a given. The architecture of the H2AI coach is shown in Fig. 3 with details of the AE in Fig. 2 and loss terms in Eq. 1. We did not employ any data augmentation. We used the AdamOptimizer with learning rate 1e−4 for all models. Training lasted for 10 epochs with a batchsize of 8. We trained 5 networks for each hyperparameter setting. We perform statistical analysis of our results using t-tests.

For the ablation study we consider adding each of the losses in isolation to the baseline with just the AE by varying parameters $\alpha_{CL}, \alpha_{EF}, \alpha_D$ that control their impact. For the AE we used $\alpha_{RE} = 32$ for all experiments.[1] Finally, we consider a model, where we add all losses. There are no fixed ranges for the parameters α, but they should be chosen so that all loss terms have a noticeable impact on the total loss – at least in the early phases of training.[2]

Our qualitative analysis is a visual assessment of the generated images. We investigate images that were improved (in terms of each of the metrics), worsened and remained roughly the same. As quantitative measures we used the losses as defined in Eq. 1 except for classification, where we used the more common accuracy metric.

4.1   Qualitative Analysis

Figure 4 shows unmodified samples (left most column) and various configurations of loss weights α. We use $R.x$ to denote "row x". The AE (2nd column, $\alpha_{RE} = 32$) on its own already has overall a positive impact yielding smoother images than the original ones. It tends to improve efficiency by removing "exotic" strokes,

---

[1] α_{RE} is not needed (could be set to 1). But, in practice, it is easier to vary α_{RE} than changing $\alpha_{CL}, \alpha_{EF}, \alpha_D$ since they behave non-linearly.

[2] We found that altering α during training requires much more tuning, but yields only modest improvements.

e.g. for the *2* in R.6 and the *5* in the last row, and sometimes helps also in improving readability (e.g. ease of classification), e.g. the *8* in R.1 row and the *6* in the 2nd last row both become more readable. Other digits might seem more readable but are actually worsened, e.g. the *6* in R.6 appears to become a *0* (it is actually a *6*) and the *7* in R.7 appears to become more of a *9*. When optimizing in addition for efficiency (3rd column), some parts of digits get deleted, which is sometimes positive and sometimes negative. Some benefits of the AE seem to get undone, e.g. the *6* in the 2nd last row now looks again more like the original with missing parts. The same holds for the *8* in R.1, though for both some improvement in shape remains. More interestingly, the digits in R.6 both get changed to *0*, which is incorrect. On the positive side, several figures become more readable through subtle changes, e.g. removals of parts like the *5* in the last row, the *2* in the 2nd last row or the *3* in R.3. When using the AE and the discriminator (4th column in Fig. 4), we can observe that the samples become slightly more realistic, i.e. crispier. We can see clear improvements for the *7* in R.7 and the *6* in R.9. Many digits remain the same. When using the AE and the classification loss (last column) smoothness increases and digits appear blurry. Readability worsens for a few digits, i.e. the left *4* in R.2 can now be easily confused with a *9*, the *6* in R.9 is no better than the original and worse than the one using a discriminator. Overall, the classification loss helps to improve many other samples. Some only now become readable, e.g. the *5* and *3* in R.8. Also some digits become simpler, e.g. the *1* R.1 and the *7*s in R.3, R.4 and R.7.

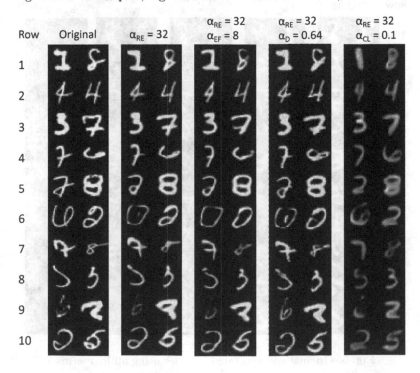

Fig. 4. Original and generated samples using a subset of all loss terms

When combining all losses (Fig. 5) it can be observed that for some parameters α larger values are possible to get reasonable results, since the objectives might counteract each other. For example, the discriminator loss pushes pixels to become brighter, whereas the efficiency loss pushes them to be darker. We noticed that the strong smoothing effect due to the classification loss is essentially removed mainly due to the discriminator loss but also partially due to the efficiency loss. The benefits of the classification loss, however, mainly remain and are also improved: The *4* in the R.2 and the *6* in R.9 become more readable. There are also differences in quality among the three configurations. Interestingly, the original images show somewhat more contrast, in particular compared to the second column. A careful observer will notice a few bright points in the upper part of both *4* in R.2. These seem to be artifacts of the optimization. It is well-known that training GANs might lead to non-convergence or mode-collapse. The former was observed for (too) large discriminator loss α_D. We also noticed mode collapse for large values of α_{CL} (not shown) and bad outcomes for large values of α_{EF} as shown in the last column. Degenerated examples still score high in some of the metrics, but are very poor in others, e.g. in the last column accuracy and efficiency loss are good, but reconstruction loss is large. Still, overall combining all losses leads to best results.

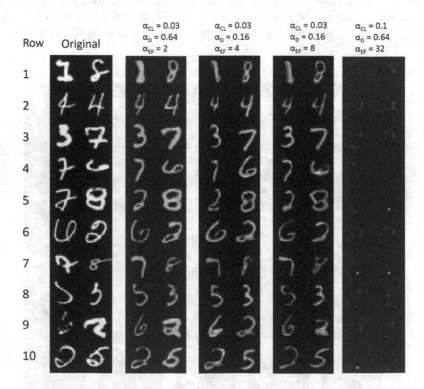

Fig. 5. Original and generated samples using all loss terms

4.2   Quantitative Analysis

Table 1 shows the loss terms (with accuracy instead of classification loss) for all loss configurations also shown in Fig. 4 for our ablation study with the reconstruction loss (AE only) as baseline. We first discuss accuracy. The AE on its own leads to a small gain in accuracy compared to the baseline classifier C_H of 95.97%. Not surprisingly, optimizing accuracy directly (using a classification loss, i.e. $\alpha_{CL} > 0$) leads to best results: even for a seemingly small α_{CL} accuracy exceeds .999%. While it appears that differences in accuracy between various values of α_{CL} are not significant, from a statistical perspective (using a t-test) they are (p-value $< .001$). For any α_{CL}, the network tends to always fail to learn the same samples, leading to very low variance in accuracy. The large accuracy values are no surprise, since also for the test set, the network is fed the correct label and therefore could in principle always return a "prototypical" class sample, ignoring all other information. When varying the efficiency loss weight α_{EF}, accuracy decreases, but the decrease was only statistically significant for $\alpha_{EF} \geq 8$ (p-value $< .001$). Adding a discriminator also negatively impacts accuracy with $\alpha_D \geq 0.64$ showing statistically significant worse results (p-value $< .01$).

Table 1. Results varying one loss term weight α_{CL}, α_{EF}, α_D

Loss	α_{CL}	α_{EF}	α_D	Accuracy	L_{RE}	L_{EF}
Baseline (AE only)	0.0	0.0	0.0	0.9609	0.00018	0.00097
Classific. loss	0.03	0.0	0.0	0.9994	0.00027	0.00096
	0.08	0.0	0.0	0.9997	0.00041	0.00096
	0.1	0.0	0.0	0.9998	0.00042	0.00092
	0.24	0.0	0.0	1.0	0.00062	0.00085
Efficiency loss	0.0	1.0	0.0	0.9587	0.00019	0.00095
	0.0	4.0	0.0	0.9607	0.00018	0.00093
	0.0	8.0	0.0	0.9578	0.00019	0.00091
	0.0	16.0	0.0	0.9458	0.00023	0.00081
	0.0	32.0	0.0	0.1135	0.00098	$<1e-5$
Discrim. loss	0.0	0.0	0.03	0.9608	0.00019	0.00099
	0.0	0.0	0.16	0.96	0.0002	0.00096
	0.0	0.0	0.64	0.9318	0.00032	0.00096

The reconstruction loss L_{RE} is most tightly correlated with the visual quality of the outcomes. In particular, large AE loss is likely to imply poor visual outcomes, despite the fact that other metrics such as accuracy are indicating good results. This can be observed in Table 1 for $\alpha_{CL} = 0.24$. Generally, the reconstruction loss worsens when optimizing for accuracy $\alpha_{CL} > 0$ or adding a discriminator $\alpha_D \geq 0$. Differences to the baseline are significant (p-value $< .01$). For adding an efficiency loss differences are only significant for values $\alpha_{EF} \geq 8$ (p-value $< .01$).

The efficiency loss decreases when adding other losses. For the discriminator differences are not significant compared to the baseline, while for all other losses they are for any value α_{EF} and $\alpha_{CL} \geq 0.1$ (p-value $< .01$).

5   Related Work

There are numerous types of AE. Related to our applications are denoising AE that are typically used through intentional noise injection with the goal of weight regularization. In contrast, we assume that noise is part of the input data and its removal is thus not motivated by regularization. The idea to combine AEs and GANs for image generation has been explored previously, e.g. [2] uses a conditional variational AE and applies it for image inpainting and attribute morphing. In this work, we consider a novel application of this architecture type. Our work is a form of image-to-image translation [10]. Typically, input and outputs are fairly different, e.g. the input could be a colored segmentation of an image not showing any details and the output could be a photo like image with many details. In contrast, in our scenario in- and outputs are fairly similar. For image in-painting or completion [9,16] a network learns to fill in blank spaces of an image. In contrast, we might both in-paint and erase. Image manipulation based on user edits has been studied in [18]. They learn the natural image manifold using a generative adversarial network and express manipulations as constraint optimization problem. They apply both spatial and channel, i.e. color, flow regularization. Their primary goal is to obtain realistically looking images after manipulations. Thus, their problem and approach is fairly different. Furthermore, in contrast to the mentioned prior works [2,9,10,16,18] our work can be classified as unsupervised learning. That is, we do not know the final outputs, i.e. the images that should be proposed to the human. Prior work trains by comparing their outcome to a target. In our case, we do not have pairs of human input (images) and improved input (images) in our training data.

The field of human-AI interaction is fairly broad. The effect of various user and system characteristics has been extensively studied [13]. There has been little work on how to improve communication and prevent misunderstandings. [12] discusses high level, non-technical strategies to deal with errors in communication using speech that originate either from humans or from machines. [4] lists some errors that occur when interacting with a robot using natural language, such as grammatical, geometrical misunderstandings as well as ambiguities. [5] highlighted the impact of nonverbal communication on efficiency and robustness in communication. It is shown that nonverbal communication can reduce errors. Our work also relates to the field of personalized explanations [15]. It aims to explain to a user how she might improve interaction with an AI. Explainability in the context of machine learning is generally more focused on interpreting decisions and models (see [1,15] for recent surveys). Counterfactual explanations also seek to identify some form of modification of the input. [6] explains by answering "How to modify an input to get classification Y?" and "What is minimally needed?". The former focuses on mis-classified examples with the goal of changing them with minimal effort to the correct class. For the latter all objectives except efficiency are ignored and there is only the constraint of maintaining classification confidence above a threshold. Thus, [6] discusses special cases of our work. Technically, [6] generates a perturbation added to the sample such that the perturbation is minimal given a threshold confidence of the prediction (either as the correct class or as an alternative class) has been achieved. They use an ordinary AE as an optional element on the perturbation, which does only slightly alter results. In contrast, we use a CCAE on the inputs, which is essential. We optimize for multiple linear weighted objectives without thresholds. [8] aims at explaining counterfactuals, i.e. showing how to change a class to another by combining images of both classes. That is, given a query image and a distractor image they generate a composite image that essentially uses parts of each input. For instance, in the right part of Fig. 6 the "7" in the second row serves as query image, the "2" in the middle as distractor and the right most column shows the outcome. The implementation relies on a gating mechanism to select image parts. Differences are also noticeable in the outcomes as shown in Fig. 6. The highlighted differences appear noisy in [6] and are not necessarily intuitive, e.g. for column CEM-PP for digit "3" a stroke on top is missing, but [6] finds a miniature "3" within the given digit. The generated images in [8] appear more natural, but do have artifacts, e.g. the "2" being a composition of a "7" and a "2" has a "dot" in the bottom originating from the "7". In conclusion, while counterfactual explanations [6,8] are related to our work, the objectives differ, e.g. we include efficiency, as well as methodology and outcomes. While we also make recommendations to a user, there are only weak ties to recommender systems. Even for interpretable recommendation systems [7] users typically primarily seek to understand decisions but do not commonly aim to alter their behavior to obtain better recommendations.

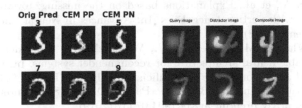

Fig. 6. Left digits are taken from [6]. Right digits stem from [8].

6   Discussion and Conclusions

Input from human to AI is likely to gain further in importance. This paper investigated improving information flow from human to AI by proposing adjustments to human generated examples based on optimizing multiple objectives. Our evaluation highlights that such an automatic approach is indeed feasible for handwriting. While we believe that our approach is suitable for other domains such as speech recognition, details of the network architecture, definition of loss terms and the loss weights likely need to be adjusted. Furthermore, our work focused on generating altered input samples fulfilling specific metrics, but it leaves many questions unanswered when applying it. For instance, it did not investigate how these samples are best shown or explained to users, e.g. by highlighting differences or, maybe, even in textual form. These points and more advanced multi-objective optimization, i.e. exploring the set of (Pareto) optimal solutions rather than manually adjusting parameters α, are subject to future work. Furthermore, one might include more objectives, e.g. generating proposals that require little energy to process by the AI [14] or taking into account behavioral norms expected by people as common for social robots [3,17]. We hope that in the future human-to-AI coaches will help non-experts to better interact with AI systems.

References

1. Adadi, A., Berrada, M.: Peeking inside the black-box: a survey on explainable artificial intelligence (XAI). IEEE Access **6**, 52138–52160 (2018)
2. Bao, J., Chen, D., Wen, F., Li, H., Hua, G.: CVAE-GAN: fine-grained image generation through asymmetric training. In: Proceeding of the International Conference on Computer Vision (2017)
3. Bartneck, C., Forlizzi, J.: A design-centred framework for social human-robot interaction. In: Workshop on Robot and Human Interactive Communication (2004)
4. Bisk, Y., Yuret, D., Marcu, D.: Natural language communication with robots. In: Proceedings of Conference of the North American Chapter of the Association for Computational Linguistics: Human Language Technologies (2016)
5. Breazeal, C., Kidd, C.D., Thomaz, A.L., Hoffman, G., Berlin, M.: Effects of nonverbal communication on efficiency and robustness in human-robot teamwork. In: International Conference on Intelligent Robots and Systems (2005)
6. Dhurandhar, A., et al.: Explanations based on the missing: towards contrastive explanations with pertinent negatives. In: Advances in Neural Information Processing Systems (2018)
7. Fusco, F., Vlachos, M., Vasileiadis, V., Wardatzky, K., Schneider, J.: RecoNet: an interpretable neural architecture for recommender systems. In: Proceedings of International Joint Conference on Artificial Intelligence (IJCAI) (2019)
8. Goyal, Y., Wu, Z., Ernst, J., Batra, D., Parikh, D., Lee, S.: Counterfactual visual explanations. arXiv preprint arXiv:1904.07451 (2019)
9. Iizuka, S., Simo-Serra, E., Ishikawa, H.: Globally and locally consistent image completion. ACM Trans. Graph. (ToG) **36**(4), 107 (2017)
10. Isola, P., Zhu, J.Y., Zhou, T., Efros, A.: Image-to-image translation with conditional adversarial networks. In: Proceedings of CVPR (2017)
11. Makhzani, A., Frey, B.: K-sparse autoencoders. arXiv:1312.5663 (2013)
12. Niculescu, A.I., Banchs, R.E.: Strategies to cope with errors in human-machine spoken interactions: using chatbots as back-off mechanism for task-oriented dialogues. In: Proceedings of the Errors by Humans and Machines in Multimedia, Multimodal and Multilingual Data Processing (ERRARE) (2015)
13. Rzepka, C., Berger, B.: User interaction with AI-enabled systems: a systematic review of IS research. In: International Conference on Information Systems (ICIS) (2018)

14. Schneider, J., Basalla, M., Seidel, S.: Principles of green data mining. In: Proceedings of Hawaii International Conference on System Sciences (HICSS) (2019)
15. Schneider, J., Handali, J.: Personalized explanation in machine learning. In: European Conference on Information Systems (ECIS) (2019)
16. Yu, J., Lin, Z., Yang, J., Shen, X., Lu, X., Huang, T.S.: Generative image inpainting with contextual attention. In: Computer Vision and Pattern Recognition (2018)
17. Zhao, S.: Humanoid social robots as a medium of communication. New Media Soc. **8**, 401–419 (2006)
18. Zhu, J.-Y., Krähenbühl, P., Shechtman, E., Efros, A.A.: Generative visual manipulation on the natural image manifold. In: Leibe, B., Matas, J., Sebe, N., Welling, M. (eds.) ECCV 2016. LNCS, vol. 9909, pp. 597–613. Springer, Cham (2016). https://doi.org/10.1007/978-3-319-46454-1_36

Aleatoric and Epistemic Uncertainty
with Random Forests

Mohammad Hossein Shaker$^{(\boxtimes)}$ and Eyke Hüllermeier

Heinz Nixdorf Institute and Department of Computer Science,
Paderborn University, Paderborn, Germany
{mhshaker,eyke}@upb.de

Abstract. Due to the steadily increasing relevance of machine learning for practical applications, many of which are coming with safety requirements, the notion of uncertainty has received increasing attention in machine learning research in the last couple of years. In particular, the idea of distinguishing between two important types of uncertainty, often refereed to as *aleatoric* and *epistemic*, has recently been studied in the setting of supervised learning. In this paper, we propose to quantify these uncertainties, referring, respectively, to inherent randomness and a lack of knowledge, with random forests. More specifically, we show how two general approaches for measuring the learner's aleatoric and epistemic uncertainty in a prediction can be instantiated with decision trees and random forests as learning algorithms in a classification setting. In this regard, we also compare random forests with deep neural networks, which have been used for a similar purpose.

Keywords: Machine learning · Uncertainty · Random forest

1 Introduction

The notion of uncertainty has received increasing attention in machine learning research in the last couple of years, especially due to the steadily increasing relevance of machine learning for practical applications. In fact, a trustworthy representation of uncertainty should be considered as a key feature of any machine learning method, all the more in safety-critical application domains such as medicine [9,22] or socio-technical systems [19,20].

In the general literature on uncertainty, a distinction is made between two inherently different sources of uncertainty, which are often referred to as *aleatoric* and *epistemic* [4]. Roughly speaking, aleatoric (*aka* statistical) uncertainty refers to the notion of randomness, that is, the variability in the outcome of an experiment which is due to inherently random effects. The prototypical example of aleatoric uncertainty is coin flipping. As opposed to this, epistemic (*aka* systematic) uncertainty refers to uncertainty caused by a lack of knowledge, i.e., it relates to the epistemic state of an agent or decision maker. This uncertainty can in principle be reduced on the basis of additional information. In other

M. R. Berthold et al. (Eds.): IDA 2020, LNCS 12080, pp. 444–456, 2020.
https://doi.org/10.1007/978-3-030-44584-3_35

words, epistemic uncertainty refers to the *reducible* part of the (total) uncertainty, whereas aleatoric uncertainty refers to the *non-reducible* part.

More recently, this distinction has also received attention in machine learning, where the "agent" is a learning algorithm [18]. In particular, a distinction between aleatoric and epistemic uncertainty has been advocated in the literature on deep learning [6], where the limited awareness of neural networks of their own competence has been demonstrated quite nicely. For example, experiments on image classification have shown that a trained model does often fail on specific instances, despite being very confident in its prediction. Moreover, such models are often lacking robustness and can easily be fooled by "adversarial examples" [14]: Drastic changes of a prediction may already be provoked by minor, actually unimportant changes of an object. This problem has not only been observed for images but also for other types of data, such as natural language text [17].

In this paper, we advocate the use of decision trees and random forests, not only as a powerful machine learning method with state-of-the-art predictive performance, but also for measuring and quantifying predictive uncertainty. More specifically, we show how two general approaches for measuring the learner's aleatoric and epistemic uncertainty in a prediction (recalled in Sect. 2) can be instantiated with decision trees and random forests as learning algorithms in a classification setting (Sect. 3). In an experimental study on uncertainty-based abstention (Sect. 4), we compare random forests with deep neural networks, which have been used for a similar purpose.

2   Epistemic and Aleatoric Uncertainty

We consider a standard setting of supervised learning, in which a learner is given access to a set of (i.i.d.) training data $\mathcal{D} := \{(\boldsymbol{x}_i, y_i)\}_{i=1}^N \subset \mathcal{X} \times \mathcal{Y}$, where $\mathcal{X}$ is an instance space and $\mathcal{Y}$ the set of outcomes that can be associated with an instance. In particular, we focus on the classification scenario, where $\mathcal{Y} = \{y_1, \ldots, y_K\}$ consists of a finite set of class labels, with binary classification ($\mathcal{Y} = \{0, 1\}$) as an important special case.

Suppose a *hypothesis space* $\mathcal{H}$ to be given, where a hypothesis $h \in \mathcal{H}$ is a mapping $\mathcal{X} \longrightarrow \mathbb{P}(\mathcal{Y})$, i.e., a hypothesis maps instances $\boldsymbol{x} \in \mathcal{X}$ to probability distributions on outcomes. The goal of the learner is to induce a hypothesis $h^* \in \mathcal{H}$ with low risk (expected loss)

$$R(h) := \int_{\mathcal{X} \times \mathcal{Y}} \ell(h(\boldsymbol{x}), y) \, d P(\boldsymbol{x}, y), \tag{1}$$

where P is the (unknown) data-generating process (a probability distribution on $\mathcal{X} \times \mathcal{Y}$), and $\ell : \mathcal{Y} \times \mathcal{Y} \longrightarrow \mathbb{R}$ a loss function. This choice of a hypothesis is commonly guided by the empirical risk

$$R_{emp}(h) := \frac{1}{N} \sum_{i=1}^N \ell(h(\boldsymbol{x}), y), \tag{2}$$

i.e., the performance of a hypothesis on the training data. However, since $R_{emp}(h)$ is only an estimation of the true risk $R(h)$, the empirical risk minimizer (or any other predictor)

$$\widehat{h} := \operatorname*{argmin}_{h \in \mathcal{H}} R_{emp}(h) \tag{3}$$

favored by the learner will normally not coincide with the true risk minimizer (Bayes predictor)

$$h^* := \operatorname*{argmin}_{h \in \mathcal{H}} R(h). \tag{4}$$

Correspondingly, there remains uncertainty regarding h^* as well as the approximation quality of $\widehat{h}$ (in the sense of its proximity to h^*) and its true risk $R(\widehat{h})$.

Eventually, one is often interested in the *predictive uncertainty*, i.e., the uncertainty related to the prediction $\widehat{y}_q$ for a concrete query instance $x_q \in \mathcal{X}$. In other words, given a partial observation $(x_q, \cdot)$, we are wondering what can be said about the missing outcome, especially about the uncertainty related to a prediction of that outcome. Indeed, estimating and quantifying uncertainty in a transductive way, in the sense of tailoring it to individual instances, is arguably important and practically more relevant than a kind of average accuracy or confidence, which is often reported in machine learning.

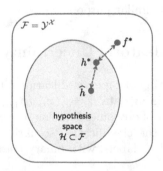

Fig. 1. Different types of uncertainties related to different types of discrepancies and approximation errors: f^* is the pointwise Bayes predictor, h^* is the best predictor within the hypothesis space, and $\widehat{h}$ the predictor produced by the learning algorithm.

As the prediction $\widehat{y}_q$ constitutes the end of a process that consists of different learning and approximation steps, all errors and uncertainties related to these steps may also contribute to the uncertainty about $\widehat{y}_q$ (cf. Fig. 1):

– Since the dependency between $\mathcal{X}$ and $\mathcal{Y}$ is typically non-deterministic, the description of a new prediction problem in the form of an instance x_q gives rise to a conditional probability distribution

$$p(y \,|\, x_q) = \frac{p(x_q, y)}{p(x_q)} \tag{5}$$

on $\mathcal{Y}$, but it does normally not identify a single outcome y in a unique way. Thus, even given full information in the form of the measure P (and its density p), uncertainty about the actual outcome y remains. This uncertainty is of an *aleatoric* nature. In some cases, the distribution (5) itself (called the predictive posterior distribution in Bayesian inference) might be delivered as a prediction. Yet, when having to commit to a point estimate, the best prediction (in the sense of minimizing the expected loss) is prescribed by the pointwise Bayes predictor f^*, which is defined by

$$f^*(\boldsymbol{x}) := \underset{\widehat{y} \in \mathcal{Y}}{\text{argmin}} \int_{\mathcal{Y}} \ell(y, \widehat{y}) \, dP(y \mid \boldsymbol{x}) \tag{6}$$

for each $\boldsymbol{x} \in \mathcal{X}$.

- The Bayes predictor (4) does not necessarily coincide with the pointwise Bayes predictor (6). This discrepancy between h^* and f^* is connected to the uncertainty regarding the right type of model to be fit, and hence the choice of the hypothesis space $\mathcal{H}$. We refer to this uncertainty as *model uncertainty*. Thus, due to this uncertainty, one can not guarantee that $h^*(\boldsymbol{x}) = f^*(\boldsymbol{x})$, or, in case the hypothesis h^* delivers probabilistic predictions $p(y \mid h^*, \boldsymbol{x})$ instead of point predictions, that $p(\cdot \mid h^*, \boldsymbol{x}) = p(\cdot \mid \boldsymbol{x})$.
- The hypothesis $\widehat{h}$ produced by the learning algorithm, for example the empirical risk minimizer (3), is only an estimate of h^*, and the quality of this estimate strongly depends on the quality and the amount of training data. We refer to the discrepancy between $\widehat{h}$ and h^*, i.e., the uncertainty about how well the former approximates the latter, as *approximation uncertainty*.

As already said, aleatoric uncertainty is typically understood as uncertainty that is due to influences on the data-generating process that are inherently random, that is, due to the non-deterministic nature of the sought input/output dependency. This part of the uncertainty is irreducible, in the sense that the learner cannot get rid of it. Model uncertainty and approximation uncertainty, on the other hand, are subsumed under the notion of epistemic uncertainty, that is, uncertainty due to a lack of knowledge about the perfect predictor (6). Obviously, this lack of knowledge will strongly depend on the underlying hypothesis space $\mathcal{H}$ as well as the amount of data seen so far: The larger the number $N = |\mathcal{D}|$ of observations, the less ignorant the learner will be when having to make a new prediction. In the limit, when $N \to \infty$, a consistent learner will be able to identify h^*. Moreover, the "larger" the hypothesis pace $\mathcal{H}$, i.e., the weaker the prior knowledge about the sought dependency, the higher the epistemic uncertainty will be, and the more data will be needed to resolve this uncertainty.

How to capture these intuitive notions of aleatoric and epistemic uncertainty in terms of quantitative measures? In the following, we briefly recall two proposals that have recently been made in the literature.

2.1 Entropy Measures

An attempt at measuring and separating aleatoric and epistemic uncertainty on the basis of classical information-theoretic measures of entropy is made in [2].

This approach is developed in the context of neural networks for regression, but the idea as such is more general and can also be applied to other settings. A similar approach was recently adopted in [10].

Given a query instance $\boldsymbol{x}$, the idea is to measure the total uncertainty in a prediction in terms of the (Shannon) entropy of the predictive posterior distribution, which, in the case of discrete $\mathcal{Y}$, is given as

$$H\big[p(y\,|\,\boldsymbol{x})\big] = \mathbf{E}_{p(y\,|\,\boldsymbol{x})}\big\{-\log_2 p(y\,|\,\boldsymbol{x})\big\} = -\sum_{y\in\mathcal{Y}} p(y\,|\,\boldsymbol{x})\log_2 p(y\,|\,\boldsymbol{x}). \qquad (7)$$

Moreover, the epistemic uncertainty is measured in terms of the mutual information between hypotheses and outcomes (i.e., the Kullback-Leibler divergence between the joint distribution of outcomes and hypotheses and the product of their marginals):

$$I(y,h) = \mathbf{E}_{p(y,h)}\left\{\log_2\left(\frac{p(y,h)}{p(y)p(h)}\right)\right\}, \qquad (8)$$

Finally, the aleatoric uncertainty is specified in terms of the difference between (7) and (8), which is given by

$$\mathbf{E}_{p(h\,|\,\mathcal{D})}H\big[p(y\,|\,h,\boldsymbol{x})\big] = -\int_{\mathcal{H}} p(h\,|\,\mathcal{D})\left(\sum_{y\in\mathcal{Y}} p(y\,|\,h,\boldsymbol{x})\log_2 p(y\,|\,h,\boldsymbol{x})\right) dh \qquad (9)$$

The idea underlying (9) is as follows: By fixing a hypothesis $h \in \mathcal{H}$, the epistemic uncertainty is essentially removed. Thus, the entropy $H[p(y\,|\,h,\boldsymbol{x})]$, i.e., the entropy of the conditional distribution on $\mathcal{Y}$ predicted by h for the query instance $\boldsymbol{x}$, is a natural measure of the aleatoric uncertainty. However, since h is not precisely known, aleatoric uncertainty is measured in terms of the expectation of this entropy with regard to the posterior probability $p(h\,|\,\mathcal{D})$.

The epistemic uncertainty (8) captures the dependency between the probability distribution on $\mathcal{Y}$ and the hypothesis h. Roughly speaking, (8) is high if the distribution $p(y\,|\,h,\boldsymbol{x})$ varies a lot for different hypotheses h with high probability. This is plausible, because the existence of different hypotheses, all considered (more or less) probable but leading to quite different predictions, can indeed be seen as a sign for high epistemic uncertainty.

Obviously, (8) and (9) cannot be computed efficiently, because they involve an integration over the hypothesis space $\mathcal{H}$. One idea, therefore, is to approximate these measures by means of ensemble techniques [10], that is, to represent the posterior distribution $p(h\,|\,\mathcal{D})$ by a finite ensemble of hypotheses $H = \{h_1,\ldots,h_M\}$. An approximation of (9) can then be obtained by

$$u_a(\boldsymbol{x}) := -\frac{1}{M}\sum_{i=1}^{M}\sum_{y\in\mathcal{Y}} p(y\,|\,h_i,\boldsymbol{x})\log_2 p(y\,|\,h_i,\boldsymbol{x}), \qquad (10)$$

an approximation of (7) by

$$u_t(\boldsymbol{x}) := -\sum_{y \in \mathcal{Y}} \left(\frac{1}{M} \sum_{i=1}^{M} p(y \mid h_i, \boldsymbol{x}) \right) \log_2 \left(\frac{1}{M} \sum_{i=1}^{M} p(y \mid h_i, \boldsymbol{x}) \right), \qquad (11)$$

and finally and approximation of (8) by $u_e(\boldsymbol{x}) := u_t(\boldsymbol{x}) - u_a(\boldsymbol{x})$.

2.2   Measures Based on Relative Likelihood

Another approach, put forward in [18], is based on the use of relative likelihoods, historically proposed by [1] and then justified in other settings such as possibility theory [21]. Here, we briefly recall this approach for the case of binary classification, i.e., where $\mathcal{Y} = \{0, 1\}$; see [13] for an extension to the case of multinomial classification.

Given training data $\mathcal{D} = \{(\boldsymbol{x}_i, y_i)\}_{i=1}^N \subset \mathcal{X} \times \mathcal{Y}$, the normalized likelihood of $h \in \mathcal{H}$ is defined as

$$\pi_{\mathcal{H}}(h) := \frac{L(h)}{L(h^{ml})} = \frac{L(h)}{\max_{h' \in \mathcal{H}} L(h')}, \qquad (12)$$

where $L(h) = \prod_{i=1}^N p(y_i \mid h, \boldsymbol{x}_i)$ is the likelihood of h, and $h^{ml} \in \mathcal{H}$ the maximum likelihood estimation. For a given instance $\boldsymbol{x}$, the degrees of support (plausibility) of the two classes are defined as follows:

$$\pi(1 \mid \boldsymbol{x}) = \sup_{h \in \mathcal{H}} \min \left[\pi_{\mathcal{H}}(h), p(1 \mid h, \boldsymbol{x}) - p(0 \mid h, \boldsymbol{x}) \right], \qquad (13)$$

$$\pi(0 \mid \boldsymbol{x}) = \sup_{h \in \mathcal{H}} \min \left[\pi_{\mathcal{H}}(h), p(0 \mid h, \boldsymbol{x}) - p(1 \mid h, \boldsymbol{x}) \right]. \qquad (14)$$

So, $\pi(1 \mid \boldsymbol{x})$ is high if and only if a highly plausible hypothesis supports the positive class much stronger (in terms of the assigned probability) than the negative class (and $\pi(0 \mid \boldsymbol{x})$ can be interpreted analogously). Given the above degrees of support, the degrees of epistemic and aleatoric uncertainty are defined as follows:

$$u_e(\boldsymbol{x}) = \min \left[\pi(1 \mid \boldsymbol{x}), \pi(0 \mid \boldsymbol{x}) \right], \qquad (15)$$

$$u_a(\boldsymbol{x}) = 1 - \max \left[\pi(1 \mid \boldsymbol{x}), \pi(0 \mid \boldsymbol{x}) \right]. \qquad (16)$$

Thus, epistemic uncertainty refers to the case where both the positive and the negative class appear to be plausible, while the degree of aleatoric uncertainty (16) is the degree to which none of the classes is supported. More specifically, the above measures have the following properties:

– $u_e(\boldsymbol{x})$ will be high if class probabilities strongly vary within the set of plausible hypotheses, i.e., if we are unsure how to compare these probabilities. In particular, it will be 1 if and only if we have $h(\boldsymbol{x}) = 1$ and $h'(\boldsymbol{x}) = 0$ for two totally plausible hypotheses h and h';

– $u_a(\boldsymbol{x})$ will be high if class probabilities are similar for all plausible hypotheses, i.e., if there is strong evidence that $h(\boldsymbol{x}) \approx 0.5$. In particular, it will be close to 1 if all plausible hypotheses allocate their probability mass around $h(\boldsymbol{x}) = 0.5$.

As can be seen, the measures (15) and (16) are actually quite similar in spirit to the measures (8) and (9).

3   Random Forests

Our basic idea is to instantiate the (generic) uncertainty measures presented in the previous section by means of decision trees [15,16], that is, with decision trees as an underlying hypothesis space $\mathcal{H}$. This idea is motivated by the fact that, firstly, decision trees can naturally be seen as probabilistic predictors [7], and secondly, they can easily be used as an ensemble in the form of a random forest—recall that ensembling is needed for the (approximate) computation of the entropy-based measures in Sect. 2.1.

3.1   Entropy Measures

The approach in Sect. 2.1 can be realized with decision forests in a quite straight-forward way. Let $H = \{h_1, \ldots, h_M\}$ be a classifier ensemble in the form of a random forest consisting of decision trees h_i. Moreover, recall that a decision tree h_i partitions the instance space $\mathcal{X}$ into (rectangular) regions $R_{i,1}, \ldots, R_{i,L_i}$ (i.e., $\bigcup_{l=1}^{L_i} R_{i,l} = \mathcal{X}$ and $R_{i,k} \cap R_{i,l} = \emptyset$ for $k \neq l$) associated with corresponding leafs of the tree (each leaf node defines a region R). Given a query instance $\boldsymbol{x}$, the probabilistic prediction produced by the tree h_i is specified by the Laplace-corrected relative frequencies of the classes $y \in \mathcal{Y}$ in the region $R_{i,j} \ni \boldsymbol{x}$:

$$p(y \mid h_i, \boldsymbol{x}) = \frac{n_{i,j}(y) + 1}{n_{i,j} + |\mathcal{Y}|},$$

where $n_{i,j}$ is the number of training instances in the leaf node $R_{i,j}$, and $n_{i,j}(y)$ the number of instances with class y. With probabilities estimated in this way, the uncertainty degrees (10) and (11) can directly be derived.

3.2   Measures Based on Relative Likelihood

Instantiating the approach in Sect. 2.2 essentially means computing the degrees of support (13–14), from which everything else can easily be derived.

As already said, a decision tree partitions the instance space into several regions, each of which can be associated with a constant predictor. More specifically, in the case of binary classification, the predictor is of the form h_θ, $\theta \in \Theta = [0,1]$, where $h_\theta(\boldsymbol{x}) \equiv \theta$ is the (predicted) probability $p(1 \mid \boldsymbol{x} \in R)$ of the positive class in the region. If we restrict inference to a local region, the underlying hypothesis space is hence given by $\mathcal{H} = \{h_\theta \mid 0 \leq \theta \leq 1\}$.

With p and n the number of positive and negative instances, respectively, within a region R, the likelihood and the maximum likelihood estimate of θ are respectively given by

$$L(\theta) = \binom{n+p}{n} \theta^n (1-\theta)^p \text{ and } \theta^{ml} = \frac{n}{n+p}. \tag{17}$$

Therefore, the degrees of support for the positive and negative classes are

$$\pi(1\,|\,\boldsymbol{x}) = \sup_{\theta \in [0,1]} \min \left(\frac{\theta^p (1-\theta)^n}{\left(\frac{p}{n+p}\right)^p \left(\frac{n}{n+p}\right)^n}, 2\theta - 1 \right), \tag{18}$$

$$\pi(0\,|\,\boldsymbol{x}) = \sup_{\theta \in [0,1]} \min \left(\frac{\theta^p (1-\theta)^n}{\left(\frac{p}{n+p}\right)^p \left(\frac{n}{n+p}\right)^n}, 1 - 2\theta \right). \tag{19}$$

Solving (18) and (19) comes down to maximizing a scalar function over a bounded domain, for which standard solvers can be used. From (18–19), the epistemic and aleatoric uncertainty associated with the region R can be derived according to (15) and (16), respectively. For different combinations of n and p, these uncertainty degrees can be pre-computed.

Note that, for this approach, the uncertainty degrees (15) and (16) can be obtained for a single tree. To leverage the ensemble H, we average both uncertainties over all trees in the random forest.

4   Experiments

The empirical evaluation of methods for quantifying uncertainty is a non-trivial problem. In fact, unlike for the prediction of a target variable, the data does normally not contain information about any sort of "ground truth" uncertainty. What is often done, therefore, is to evaluate predicted uncertainties *indirectly*, that is, by assessing their usefulness for improved prediction and decision making. Adopting an approach of that kind, we produced *accuracy-rejection curves*, which depict the accuracy of a predictor as a function of the percentage of rejections [5]: A classifier, which is allowed to abstain on a certain percentage p of predictions, will predict on those $(1-p)\%$ on which it feels most certain. Being able to quantify its own uncertainty well, it should improve its accuracy with increasing p, hence the accuracy-rejection curve should be monotone increasing (unlike a flat curve obtained for random abstention).

4.1   Implementation Details

For this work, we used the Random Forest Classifier from SKlearn. The number of trees within the forest is set to 50, with the maximum level of tree grows set to 10. We use bootstrapping to create diversity between the trees of the forest.

As a baseline to compare with, we used the DropConnect model for deep neural networks as introduced in [10]. The idea of DropConnect is similar to

Dropout, but here, instead of randomly deleting neurons, we randomly delete the connections between neurons. In this model, the act of dropping the connections is also active in the test phase. In this way, the data passes through a different network on each iteration, and therefore we can compute Monte Carlo samples for each query instance. The DropConnect model is a feed forward neural network consisting of two DropConnect layers with 32 neurons and a final softmax layer for the output. The model is trained for 20 epochs with mini batch size of 32. After the training is done, we take 50 Monte Carlo samples to create an ensemble, from which the uncertainty values can be calculated.

4.2   Results

Due to space limitations, we show results in the form of accuracy-rejection curves for only two exemplary data sets from the UCI repository[1], spect and diabetes—yet, very similar results were obtained for other data sets. The data is randomly split into 70% for training and 30% for testing, and accuracy-rejection curves are computed on the latter (the curves shown are averages over 100 repetitions). In the following, we abbreviate the aleatoric and epistemic uncertainty degrees produced by the entropy-based approach (Sect. 2.1) and the approach based on relative likelihood (Sect. 2.2) by AU-ent, EU-ent, AU-rl, and EU-rl, respectively.

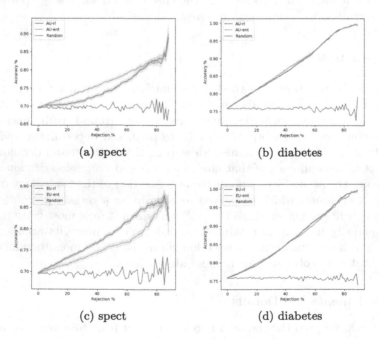

(a) spect   (b) diabetes

(c) spect   (d) diabetes

Fig. 2. Accuracy-rejection curves for aleatoric (above) and epistemic (below) uncertainty using random forests. The curve for random rejection is included as a baseline.

---

[1] https://archive.ics.uci.edu/ml/datasets/.

As can be seen from Figs. 1, 2, 3 and 4, both approaches to measuring uncertainty are effective in the sense of producing monotone increasing accuracy-rejection curves, and on the data sets we analyzed so far, we could not detect any systematic differences in performance. Besides, rejection seems to work well on the basis of both criteria, aleatoric as well as epistemic uncertainty. This is plausible, since both provide reasonable reasons for a learner to abstain from a prediction. Likewise, there are no big differences between random forests and neural networks, showing that the former are indeed a viable alternative to the latter—this was actually a major concern of our study.

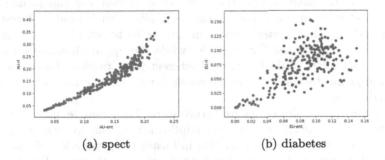

(a) spect   (b) diabetes

Fig. 3. Scatter plot for test set on diabetes data, showing the relationship between the uncertainty degrees (aleatoric left, epistemic right) estimated by the two approaches.

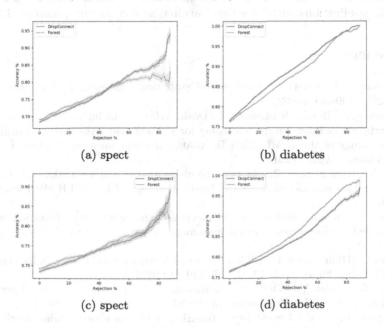

(a) spect   (b) diabetes

(c) spect   (d) diabetes

Fig. 4. Comparison between random forests and neural networks (DropConnect) for aleatoric (above) and epistemic (below) in the entropy-based uncertainty approach.

5   Conclusion

The distinction between aleatoric and epistemic uncertainty has recently received a lot of attention in machine learning, especially in the deep learning community [6]. Roughly speaking, the approaches in deep learning are either based on the idea of equipping networks with a probabilistic component, like in Bayesian deep learning [11], or on using ensemble techniques [8], which can be implemented (indirectly) through techniques such as Dropout [3] or DropConnect. The main purpose of this paper was to show that the use of decision trees and random forests is an interesting alternative to neural networks.

Indeed, as we have shown, the basic ideas underlying the estimation of aleatoric and epistemic uncertainty can be realized with random forests in a very natural way. In a sense, they even appear to be simpler and more flexible than neural networks. For example, while the approach based on relative likelihood (Sect. 2.2) could be realized efficiently for random forests, a neural network implementation is far from obvious (and was therefore not included in the experiments).

There are various directions for future work. For example, since the hyperparameters of random forests have an influence on the hypothesis space we are (indirectly) working with, they also influence the estimation of uncertainty degrees. This relationship calls for a thorough investigation. Besides, going beyond a proof of principle with statistics such as accuracy-rejection curves, it would be interesting to make use of uncertainty quantification with random forests in applications such as active learning, as recently proposed in [12].

References

1. Birnbaum, A.: On the foundations of statistical inference. J. Am. Stat. Assoc. **57**(298), 269–306 (1962)
2. Depeweg, S., Hernandez-Lobato, J., Doshi-Velez, F., Udluft, S.: Decomposition of uncertainty in Bayesian deep learning for efficient and risk-sensitive learning. In: Proceedings of the ICML, 35th International Conference on Machine Learning, Stockholm, Sweden (2018)
3. Gal, Y., Ghahramani, Z.: Bayesian convolutional neural networks with Bernoulli approximate variational inference. In: Proceedings of the ICLR Workshop Track (2016)
4. Hora, S.: Aleatory and epistemic uncertainty in probability elicitation with an example from hazardous waste management. Reliab. Eng. Syst. Saf. **54**(2–3), 217–223 (1996)
5. Hühn, J., Hüllermeier, E.: FR3: a fuzzy rule learner for inducing reliable classifiers. IEEE Trans. Fuzzy Syst. **17**(1), 138–149 (2009)
6. Kendall, A., Gal, Y.: What uncertainties do we need in Bayesian deep learning for computer vision? In: Proceedings of the NIPS, pp. 5574–5584 (2017)
7. Kruppa, J., et al.: Probability estimation with machine learning methods for dichotomous and multi-category outcome: theory. Biometrical J. **56**(4), 534–563 (2014)

8. Lakshminarayanan, B., Pritzel, A., Blundell, C.: Simple and scalable predictive uncertainty estimation using deep ensembles. In: Proceedings of the NeurIPS, 31st Conference on Neural Information Processing Systems, Long Beach, California, USA (2017)
9. Lambrou, A., Papadopoulos, H., Gammerman, A.: Reliable confidence measures for medical diagnosis with evolutionary algorithms. IEEE Trans. Inf. Technol. Biomed. **15**(1), 93–99 (2011)
10. Mobiny, A., Nguyen, H., Moulik, S., Garg, N., Wu, C.: DropConnect is effective in modeling uncertainty of Bayesian networks. CoRR abs/1906.04569 (2017). http://arxiv.org/abs/1906.04569
11. Neal, R.: Bayesian Learning for Neural Networks, vol. 118. Springer, Heidelberg (2012). https://doi.org/10.1007/978-1-4612-0745-0
12. Nguyen, V., Destercke, S., Hüllermeier, E.: Epistemic uncertainty sampling. In: Proceedings of the DS 2019, 22nd International Conference on Discovery Science, Split, Croatia (2019)
13. Nguyen, V.L., Destercke, S., Masson, M.H., Hüllermeier, E.: Reliable multi-class classification based on pairwise epistemic and aleatoric uncertainty. In: Proceedings of the IJCAI, pp. 5089–5095. AAAI Press (2018)
14. Papernot, N., McDaniel, P.: Deep k-nearest neighbors: towards confident, interpretable and robust deep learning. CoRR abs/1803.04765v1 (2018). http://arxiv.org/abs/1803.04765
15. Quinlan, J.R.: Induction of decision trees. Mach. Learn. **1**(1), 81–106 (1986)
16. Safavian, S.R., Landgrebe, D.: A survey of decision tree classifier methodology. IEEE Trans. Syst. Man Cybern. **21**(3), 660–674 (1991)
17. Sato, M., Suzuki, J., Shindo, H., Matsumoto, Y.: Interpretable adversarial perturbation in input embedding space for text. In: Proceedings IJCAI 2018, Stockholm, Sweden, pp. 4323–4330 (2018)
18. Senge, R., et al.: Reliable classification: learning classifiers that distinguish aleatoric and epistemic uncertainty. Inf. Sci. **255**, 16–29 (2014)
19. Varshney, K.: Engineering safety in machine learning. In: Proceedings of the Information Theory and Applications Workshop, La Jolla, CA (2016)
20. Varshney, K., Alemzadeh, H.: On the safety of machine learning: cyber-physical systems, decision sciences, and data products. CoRR abs/1610.01256 (2016). http://arxiv.org/abs/1610.01256
21. Walley, P., Moral, S.: Upper probabilities based only on the likelihood function. J. R. Stat. Soc.: Ser. B (Stat. Methodol.) **61**(4), 831–847 (1999)
22. Yang, F., Wanga, H.Z., Mi, H., de Lin, C., Cai, W.W.: Using random forest for reliable classification and cost-sensitive learning for medical diagnosis. BMC Bioinform. **10**, S22 (2009)

Master Your Metrics with Calibration

Wissam Siblini$^{(\boxtimes)}$ (ID), Jordan Fréry, Liyun He-Guelton, Frédéric Oblé,
and Yi-Qing Wang

Worldline, 53 avenue Paul Krüger, 69100 Villeurbanne, France
{wissam.siblini,jordan.frery,liyun.he-guelton,frederic.oble,
yi-qing.wang}@worldline.com

Abstract. Machine learning models deployed in real-world applications
are often evaluated with precision-based metrics such as F1-score or
AUC-PR (Area Under the Curve of Precision Recall). Heavily dependent
on the class prior, such metrics make it difficult to interpret the variation
of a model's performance over different subpopulations/subperiods in a
dataset. In this paper, we propose a way to calibrate the metrics so that
they can be made invariant to the prior. We conduct a large number of
experiments on balanced and imbalanced data to assess the behavior of
calibrated metrics and show that they improve interpretability and provide a better control over what is really measured. We describe specific
real-world use-cases where calibration is beneficial such as, for instance,
model monitoring in production, reporting, or fairness evaluation.

Keywords: Performance metrics · Class imbalance · Precision-recall

1 Introduction

In real-world machine learning systems, the predictive performance of a model is
often evaluated on multiple datasets, and comparisons are made. These datasets
can correspond to sub-populations in the data, or different periods in time [15].
Choosing the best suited metrics is not a trivial task. Some metrics may prevent
a proper interpretation of the performance differences between the sets [8,14],
especially because different datasets generally not only have a different likelihood
$\mathbb{P}(x|y)$ but also a different class prior $\mathbb{P}(y)$. A metric dependent on the prior (e.g.
precision) will be affected by both differences indiscernibly [3] but a practitioner
could be interested in isolating the variation of performance due to likelihood
which reflects the intrinsic model's performance (see illustration in Fig. 1). Take
the example of comparing the performance of a model across time periods: At
time t, we receive data drawn from $\mathbb{P}_t(x, y) = \mathbb{P}_t(x|y)\mathbb{P}_t(y)$ where x are the
features and y the label. Hence the optimal scoring function (i.e. model) for this
dataset is the likelihood ratio [11]:

$$s_t(x) := \frac{\mathbb{P}_t(x|y = 1)}{\mathbb{P}_t(x|y = 0)} \tag{1}$$

In particular, if $\mathbb{P}_t(x|y)$ does not vary with time, neither will $s_t(x)$. In this case,
even if the prior $\mathbb{P}_t(y)$ varies, it is desirable to have a performance metric $M(\cdot)$

© The Author(s) 2020
M. R. Berthold et al. (Eds.): IDA 2020, LNCS 12080, pp. 457–469, 2020.
https://doi.org/10.1007/978-3-030-44584-3_36

satisfying $M(s_t, \mathbb{P}_t) = M(s_{t+1}, \mathbb{P}_{t+1}), \forall t$ so that the model maintains the same metric value over time. That being said, this does not mean that dependence to prior is an intrinsically bad behavior. Some applications seek this property as it reflects a part of the difficulty to classify on a given dataset (e.g. the performance of the random classifier evaluated with a prior-dependent metric is more or less high depending on the skew of the dataset).

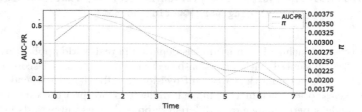

Fig. 1. Evolution of the AUC-PR of a fraud detection system and of the fraud ratio (π, i.e. the empirical $\mathbb{P}_t(y)$) over time. Both decrease, but, as the AUC-PR is dependent on the prior, it does not allow to tell if the performance variation is only due to the variation of π or if there was a drift in $\mathbb{P}_t(x|y)$

In binary classification, researchers often rely on the AUC-ROC (Area Under the Curve of Receiver Operating Characteristic) to measure a classifier's performance [6,9]. While this metric has the advantage of being invariant to the class prior, many real-world applications, especially when data are imbalanced, have recently begun to favor precision-based metrics such as AUC-PR and F-Score [12,13]. The reason is that AUC-ROC suffers from giving false positives too little importance [5] although the latter strongly deteriorate user experience and waste human efforts with false alerts. Indeed AUC-ROC considers a tradeoff between TPR and FPR whereas AUC-PR/F1-score consider a tradeoff between TPR (Recall) and Precision. With a closer look, the difference boils down to the fact that it normalizes the number of false positives with respect to the number of true negatives whereas precision-based metrics normalize it with respect to the number of true positives. In highly imbalanced scenarios (e.g. fraud/disease detection), the first is much more likely than the second because negative examples are in large majority.

Precision-based metrics give false positives more importance, but they are tied to the class prior [2,3]. A new definition of precision and recall into precision gain and recall gain has been recently proposed to correct several drawbacks of AUC-PR [7]. But, while the resulting AUC-PR Gain has some advantages of the AUC-ROC such as the validity of linear interpolation between points, it remains dependent on the class prior. Our study aims at providing metrics (i) that are precision-based to tackle problems where the class of interest is highly under-represented and (ii) that can be made independent of the prior for comparison purposes (e.g. monitoring the evolution of the performance of a classifier across several time periods). To reach this objective, this paper provides: (1) A formulation of calibration for precision-based metrics. It compute the value of precision as if the ratio π of the test set was equal to a reference class ratio π_0. We give theoretical arguments to explain why it allows invariance to the class prior. We also provide a calibrated version for precision gain and recall gain [7]. (2) An empirical analysis on both synthetic and real-world data to confirm our claims and show that new metrics are still able to assess the model's performance and are easier to interpret. (3) A large scale experiments on 614 datasets using openML [16] to (a) give more insights on correlations between popular metrics by analyzing how they rank models, (b) explore the links between the calibrated metrics and the regular ones.

Not only calibration solves the issue of dependence to the prior but also allows, with parameter π_0, anticipating a different ratio and controlling what the metric precisely reflects. This new property has several practical interests (e.g. for development, reporting, analysis) and we discuss them in realistic use-cases in Sect. 5.

2   Popular Metrics for Binary Classification: Advantages and Limits

We consider a usual binary classification setting where a model has been trained and its performance is evaluated on a test dataset of N instances. $y_i \in \{0, 1\}$ is the ground-truth label of the ith instance and is equal to 1 (resp. 0) if the instance belongs to the positive (resp. negative) class. The model provides $s_i \in \mathbb{R}$, a score for the ith instance to belong to the positive class. For a given threshold $\tau \in \mathbb{R}$, the predicted label is $\widehat{y}_i = 1$ if $s_i > \tau$ and 0 otherwise. Predictive performance is generally measured using the number of true positives (TP $= \sum_{i=1}^{N} \mathbb{1}(\widehat{y}_i = 1, y_i = 1)$), true negatives (TN $= \sum_{i=1}^{N} \mathbb{1}(\widehat{y}_i = 0, y_i = 0)$), false positives (FP $= \sum_{i=1}^{N} \mathbb{1}(\widehat{y}_i = 1, y_i = 0)$), false negatives (FN $= \sum_{i=1}^{N} \mathbb{1}(\widehat{y}_i = 0, y_i = 1)$). One can compute relevant ratios such as the True Positive Rate (TPR) also referred to as the Recall ($Rec = \frac{\text{TP}}{\text{TP} + \text{FN}}$), the False Positive Rate (FPR $= \frac{\text{FP}}{\text{TN} + \text{FP}}$) also referred to as the Fall-out and the Precision ($Prec = \frac{\text{TP}}{\text{TP} + \text{FP}}$). As these ratios are biased towards a specific type of error and can easily be manipulated with the threshold, more complex metrics have been proposed. In this paper, we discuss the most popular ones which have been widely adopted in binary classification: F1-Score, AUC-ROC, AUC-PR and AUC-PR Gain. F1-Score is the harmonic average between $Prec$ and Rec:

$$F_1 = \frac{2 * Prec * Rec}{Prec + Rec}. \tag{2}$$

The three other metrics consider every threshold τ from the highest s_i to the lowest. For each one, they compute TP, FP, TN and FN. Then, they plot one ratio against another and compute the Area Under the Curve (Fig. 2). AUC-ROC considers the Receiver Operating Characteristic curve where TPR is plotted against FPR. AUC-PR considers the Precision vs Recall curve. Finally, in AUC-PR Gain, the precision gain ($Prec_G$) is plotted against the recall gain (Rec_G).

They are defined in [7] as follows ($\pi = \frac{\sum_{i=1}^{N} y_i}{N}$ is the positive class ratio and we always consider that it is the minority class in this paper):

$$Prec_G = \frac{Prec - \pi}{(1 - \pi)Prec} \qquad (3)$$

$$Rec_G = \frac{Rec - \pi}{(1 - \pi)Rec} \qquad (4)$$

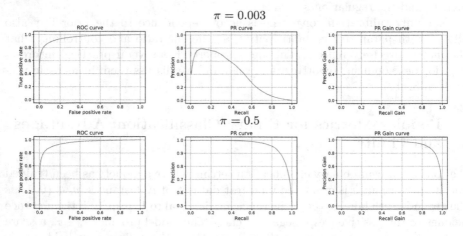

Fig. 2. ROC, PR and PR gain curves for the same model evaluated on an extremely imbalanced test set from a fraud detection application ($\pi = 0.003$, in the top row) and on a balanced sample ($\pi = 0.5$, in the bottom row).

PR Gain enjoys many properties of the ROC that the regular PR analysis does not (e.g. the validity of linear interpolations or the existence of universal baselines) [7]. However, AUC-PR Gain becomes hardly usable in extremely imbalanced settings. In particular, we can derive from (3) and (4) that $Prec_G/Rec_G$ will be mostly close to 1 if π is close to 0 (see top right chart in Fig. 2).

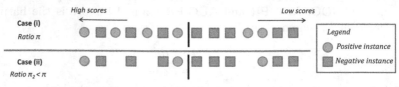

Fig. 3. Illustration of the impact of π on precision, recall, and the false positive rate. Instances are ordered from left to right according to their score given by the model. The threshold is illustrated as a vertical line between the instances: those on the left (resp. right) are classified as positive (resp. negative)

As explained in the introduction, precision-based metrics (F1, AUC-PR) are more adapted than AUC-ROC for problems with class imbalance. On the other hand, only AUC-ROC is invariant to the positive class ratio. Indeed, FPR and *Rec* are both unrelated to the class ratio because they only focus on one class but it is not the case for *Prec*. Its dependency on the positive class ratio π is illustrated in Fig. 3: when comparing a case (i) with a given ratio π and another case (ii) where a randomly selected half of the positive examples has been removed, one can visually understand that both recall and false positive rate are the same but the precision is lower in the second case.

3   Calibrated Metrics

We seek a metric that is based on *Prec* to tackle problems where data are imbalanced and the minority (positive) class is the one of interest but we want it to be invariant w.r.t. the class prior to be able to interpret its variation across different datasets (e.g. different time periods). To obtain such a metric, we will modify those based on *Prec* (AUC-PR, F1-Score and AUC-PR Gain) to make them independent of the positive class ratio π.

3.1   Calibration

The idea is to fix a reference ratio π_0 and to weigh the count of TP or FP in order to calibrate them to the value that they would have if π was equal to π_0. π_0 can be chosen arbitrarily (e.g. 0.5 for balanced) but it is preferable to fix it according to the task at hand (we analyze the impact of π_0 in Sect. 4 and describe simple guidelines to fix it in Sect. 5).

If the positive class ratio is π_0 instead of π, the ratio between negative examples and positive examples is multiplied by $\frac{\pi(1-\pi_0)}{\pi_0(1-\pi)}$. In this case, we expect the ratio between false positives and true positives to be multiplied by $\frac{\pi(1-\pi_0)}{\pi_0(1-\pi)}$. Therefore, we define the calibrated precision $Prec_c$ as follows:

$$Prec_c = \frac{TP}{TP + \frac{\pi(1-\pi_0)}{\pi_0(1-\pi)}FP} = \frac{1}{1 + \frac{\pi(1-\pi_0)}{\pi_0(1-\pi)}\frac{FP}{TP}} \tag{5}$$

Since $\frac{1-\pi}{\pi}$ is the imbalance ratio $\frac{N_-}{N_+}$ where N_+ (resp. N_-) is the number of positive (resp. negative) examples, we have: $\frac{\pi}{1-\pi}\frac{FP}{TP} = \frac{FP/N_-}{TP/N_+} = \frac{FPR}{TPR}$ which is independent of π.

Based on the calibrated precision, we can also define the calibrated F1-score, the calibrated $Prec_G$ and the calibrated Rec_G by replacing *Prec* by $Prec_c$ and π by π_0 in Eqs. (2), (3) and (4). Note that calibration does not change precision gain. Indeed, calibrated precision gain $\frac{Prec_c - \pi_0}{(1-\pi_0)Prec_c}$ can be rewritten as $\frac{Prec - \pi}{(1-\pi)Prec}$ which is equal to the regular precision gain. Also, the interesting properties of the recall gain were proved independently of the ratio π in [7] which means that calibration preserves them.

3.2   Robustness to Variations in π

In order to evaluate the robustness of the new metrics to variations in π, we create a synthetic dataset where the label is drawn from a Bernoulli distribution with parameter π and the feature is drawn from Normal distributions:

$$p(x|y = 1; \mu_1) = \mathcal{N}(x; \mu_1, 1), \qquad p(x|y = 0; \mu_0) = \mathcal{N}(x; \mu_0, 1) \qquad (6)$$

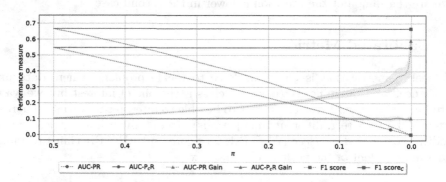

Fig. 4. Evolution of AUC-PR, AUC-PR Gain, F1-score and their calibrated version (AUC-P$_c$R, AUC-P$_c$R Gain, F1-score$_c$) as π decreases. We arbitrarily set $\pi_0 = 0.5$ for the calibrated metrics. The curves are obtained by averaging results over 30 runs and we show the confidence intervals.

For several values of π, data points are generated from (6) with $\mu_1 = 2$ and $\mu_0 = 1.8$. We consider a large number of points (10^6) so that the empirical class ratio π is approximately equal to the Bernouilli parameter π. We empirically study the evolution of several metrics (F_1-score, AUC-PR, AUC-PR Gain and their calibrated version) for the optimal model (as defined in (1)) as π decreases from $\pi = 0.5$ (balanced) to $\pi = 0.001$. We observe that the impact of the class prior on the regular metrics is important (Fig. 4). It can be a serious issue for applications where π sometimes vary by one order of magnitude from one day to another (see [4] for a real world example) as it leads to a significant variation of the measured performance (see the difference between AUC-PR when $\pi = 0.5$ and when $\pi = 0.05$) even if the optimal model remains the same. On the contrary, the calibrated versions remain very robust to changes in the class prior π even for extreme values. Note that we here experiment with synthetic data to have a full control over the distribution/prior and make the analysis easier but the conclusions are exactly the same on real world data.[1]

---

[1] See appendix in https://figshare.com/articles/Calibrated_metrics_IDA_Supplement-ary_material_pdf/11848146.

3.3   Assessment of the Model Quality

Besides the robustness of the calibrated metrics to changes in π, we also want them to be sensitive to the quality of the model. If this latter decreases regardless of the π value, we expect all metrics, calibrated ones included, to decrease in value. Let us consider an experiment where we use the same synthetic dataset as defined the previous section. However, instead of changing the value of π only, we change (μ_1, μ_0) to make the problem harder and harder and thus worsen the optimal model's performance. This can be done by reducing the distance between the two normal distributions in (6), because this would result in more overlapping between the classes and make it harder to discriminate between them. As a distance, we consider the KL-divergence that boils down to $\frac{1}{2}(\mu_1 - \mu_0)^2$.

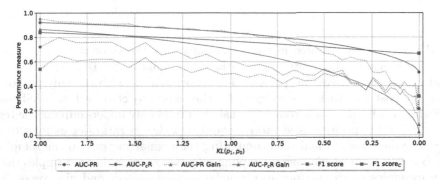

Fig. 5. Evolution of AUC-PR, AUC-PR Gain, F1-score and their calibrated version as $KL(p_1, p_0)$ tends to 0 and as π randomly varies. This curve was obtained by averaging results over 30 runs.

Figure 5 shows how the values of the metrics evolve as the KL-divergence gets closer to zero. For each run, we randomly chose the prior π in the interval $[0.001, 0.5]$. As expected, all metrics globally decrease as the problem gets harder. However, we can notice an important difference: the variation in the calibrated metrics are smooth and monotonic compared to those of the original metrics which are affected by the random changes in π. In that sense, variations of the calibrated metrics across the different generated datasets are much easier to interpret than the original metrics.

4   Link Between Calibrated and Original Metrics

4.1   Meaning of π_0

Let us first remark that for test datasets in which $\pi = \pi_0$, $Prec_c$ is equal to the regular precision $Prec$ since $\frac{\pi(1 - \pi_0)}{\pi_0(1 - \pi)} = 1$ (this is observable in Fig. 4 with the intersection of the metrics for $\pi = \pi_0 = 0.5$).

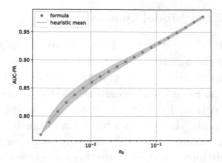

Fig. 6. Comparison between heuristic-based calibrated AUC-PR (red line) and our closed-form calibrated AUC-PR (blue dots). The red shadow represents the standard deviation of the heuristic-based calibrated AUC-PR over 1000 runs. (Color figure online)

If $\pi \neq \pi_0$, the calibrated metrics essentially have the value that the original ones would have if the positive class ratio π was equal to π_0. To further demonstrate that, we compare our proposal for calibration (5) with the only proposal from the past [10] that was designed for the same objective: a heuristic-based calibration. The approach from [10] consists in randomly undersampling the test set to make the positive class ratio π equal to a chosen ratio (let us refer to it as π_0 for the analogy) and then computing the regular metrics on the sampled set. Because of the randomness, sampling may remove more hard examples than easy examples so the performance can be over-estimated, and vice versa. To avoid that, the approach performs several runs and computes a mean estimation. In Fig. 6, we compare the results obtained with our formula and with their heuristic, for several reference ratio π_0, on a highly unbalanced ($\pi = 0.0017$) credit card fraud detection dataset available on Kaggle [4].

We can observe that our formula and the heuristic provide really close values. This can be theoretically explained (See Footnote 1) and confirms that our formula really computes the value that the original metric would have if the ratio π in the test set was π_0. Note that our closed-form calibration (5) can be seen as an improvement of the heuristic-based calibration from [10] as it directly provides the targeted value without running a costly Monte-Carlo simulation.

4.2 Do the Calibrated Metrics Rank Models in the Same Order as the Original Metrics?

Calibration results in evaluating the metric for a different prior. In this section, we analyze how this impacts the task of selectioning the best model for a given dataset. To do this, we empirically analyze the correlation of several metrics in terms of model ordering. We use OpenML [16] to select the 602 supervised binary classification datasets on which at least 30 models have been evaluated with a 10-fold cross-validation. For each one, we randomly choose 30 models, fetch their predictions, and evaluate their performance with the metrics. This leaves us with $614 \times 30 = 18,420$ different values for each metric. To analyze whether they rank the models in the same order, we compute the Spearman rank correlation coefficient between them for the 30 models for each of the 614 problems.[2] Most datasets roughly have balanced classes ($\pi > 0.2$ in more than 90% of the datasets). Therefore, to also specifically analyze the imbalance case, we run the same experiment with only the subset of 4 highly imbalanced datasets ($\pi < 0.01$). The compared metrics are AUC-ROC, AUC-PR, AUC-PR Gain and the best F1-score over all possible thresholds. We also add the calibrated version of the last three. In order to understand the impact of π_0, we use two different values: the arbitrary $\pi_0 = 0.5$ and another value $\pi_0 \approx \pi$ (for the first experiment with all datasets, $\pi_0 \approx \pi$ corresponds to $\pi_0 = 1.01\pi$ and for the second experiment where π is very small, we go further and $\pi_0 \approx \pi$ corresponds to $\pi_0 = 10\pi$ which remains closer to π than 0.5). The obtained correlation matrices are shown in Fig. 7. Each individual cell corresponds to the average Spearman correlation over all datasets between the row metric and the column metric.

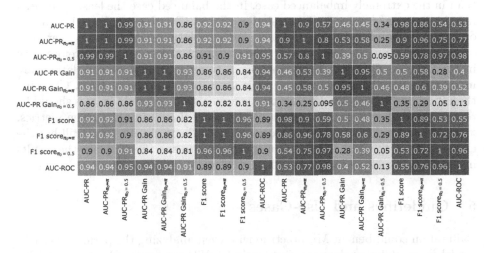

Fig. 7. Spearman rank correlation matrices between 10 metrics over 614 datasets for the left figure and the 4 highly imbalanced datasets for the right figure.

A general observation is that most metrics are less correlated with each other when classes are unbalanced (right matrix in Fig. 7). We also note that the best F1-score is more correlated to AUC-PR than to AUC-ROC or AUC-PR Gain. In the balanced case (left matrix in Fig. 7), we can see that metrics defined as area under curves are generally more correlated with each other than with the threshold sensitive classification metric F1-score. Let us now analyze the impact of calibration. As expected, in general, when $\pi_0 \approx \pi$, calibrated metrics have a behavior really close to that of the original metrics because $\frac{\pi(1-\pi_0)}{\pi_0(1-\pi)} \approx 1$ and therefore

---

[2] The implementation of the paper experiments can be found at https://github.com/wissam-sib/calibrated_metrics.

$Prec_c \approx Prec$. In the balanced case (left), since π is close to 0.5, calibrated metrics with $\pi_0 = 0.5$ are also highly correlated with the original metrics. In the imbalanced case (on the right matrix of Fig. 7), when π_0 is arbitrarily set to 0.5 the calibrated metrics seem to have a low correlation with the original ones. In fact, they are less correlated with them than with AUC-ROC. And this makes sense given the relative weights that each of the metric applies to FP and TP. The original precision gives the same weight to TP and FP, although false positives are $\frac{1-\pi}{\pi}$ times more likely to occur ($\frac{1-\pi}{\pi} > 100$ if $\pi < 0.01$). The calibrated precision with the arbitrary value $\pi_0 = 0.5$ boils down to $\frac{TP}{TP + \frac{\pi}{(1-\pi)}FP}$ and gives a weight $\frac{1-\pi}{\pi}$ times smaller to false positives which counterbalances their higher likelihood. ROC, like the calibrated metrics with $\pi_0 = 0.5$, gives $\frac{1-\pi}{\pi}$ less weight to FP because it is computed from FPR and TPR which are linked to TP and FP with the relationship $\frac{\pi}{1-\pi}\frac{FP}{TP} = \frac{FPR}{TPR}$.

To sum up the results, we first emphasize that the choice of the metrics to rank classifiers when datasets are rather balanced seems to be much less sensitive than in the extremely imbalanced case. In the balanced case the least correlated metrics have an average rank correlation of 0.81. For the imbalanced datasets, on the other hand, many metrics have low correlations which means that they often disagree on the best model. The choice of the metric is therefore very important here. Our experiment also seems to reflect that rank correlations are mainly a matter of how much weight is given to each type of error. Choosing these "weights" generally depends on the application at hand. An this should be remembered when using calibration. To preserve the nature of a given metrics, π_0 has to be fixed to a value close to π and not arbitrarily. The user still has the choice to fix it to another value if his purpose is to specifically place the results into a different reference with a different prior.

5   Guidelines and Use-Cases

Calibration could benefit ML practitioners when analyzing the performance of a model across different datasets/time periods. Without being exhaustive, we give four use-cases where it is beneficial (setting π_0 depends on the target use-case):

Comparing the Performance of a Model on Two Populations/Classes:
Consider a practitioner who wants to predict patients with a disease and evaluate the performance of his model on subpopulations of the dataset (e.g. children, adults and elderly people). If the prior is different from one population to another (e.g. elderly people are more likely to have the disease), precision will be affected, i.e. population with a higher disease ratio will be more likely to have a higher precision. In this case, the calibrated precision can be used to obtain the precision of each population set to the same reference prior (for instance, π_0 can be chosen as the average prior over all populations). This would provide an additional balanced point of view and make the analysis richer to draw more precise conclusions and perhaps study fairness [1].

Model Performance Monitoring in an Industrial Context: In systems where a model's performance is monitored over time with precision-based metrics like F1-score, using calibration in addition to the regular metrics makes it easier to understand the evolution especially when the class prior can evolve (cf. application in Fig. 1). For instance, it can be useful to analyze the drift (i.e. distinguish between variations linked to π or $P(X|y)$) and design adapted solutions; either updating the threshold or completely retraining the model. To avoid denaturing too much the F1-score, here π_0 has to be fixed based on realistic values (e.g. average π in historical data).

Establishing Agreements with Clients: As shown in previous sections, π_0 can be interpreted as the ratio to which we refer to compute the metric. This can be useful to establish a guarantee, in an agreement, that will be robust to uncontrollable events. Indeed, if we take the case of fraud detection, the real positive class ratio π can vary extremely from one day to another and on particular events (e.g. fraudster attacks, holidays) which significantly affects the measured metrics (see Fig. 4). Here, after having both parties to agree beforehand on a reasonable value for π_0 (based on their business knowledge), calibration will always compute the performance relative to this ratio and not the real π and thus be easier to guarantee.

Anticipating the Deployment of a Model in Production: Imagine one collects a sample of data to develop an algorithm and reaches an acceptable AUC-PR for production. If the prior in the collected data is different from reality, the non-calibrated metric might have given either a pessimistic or optimistic estimation of the post-deployment performance. This can be extremely harmful if the production has strict constraints. Here, if the practitioner uses calibration with π_0 equal to the minimal prior envisioned for the application at hand, he/she would be able to anticipate the worst case scenario.

6   Conclusion

In this paper, we provided a formula of calibration, empirical results, and guidelines to make the values of metrics across different datasets more interpretable. Calibrated metrics are a generalization of the original ones. They rely on a reference π_0 and compute the value that we would obtain if the positive class ratio π in the evaluated test set was equal to π_0. If the user chooses $\pi_0 = \pi$, this does not change anything and he retrieves the regular metrics. But, with different choices, the metrics can serve several purposes such as obtaining robustness to variation in the class prior across datasets, or anticipation. They are useful in both academic and industrial applications as explained in the previous section: they help drawing more accurate comparisons between subpopulations, or study incremental learning on streams by providing a point of view agnostic to virtual concept drift [17]. They can be used to provide more controllable performance indicators (easier to guarantee and report), help preparing deployment in production, and prevent false conclusions about the evolution of a deployed model. However, π_0 has to be chosen with caution as it controls the relative weights given to FP and TP and, consequently, can affect the selection of the best classifier.

References

1. Barocas, S., Hardt, M., Narayanan, A.: Fairness in machine learning. NIPS Tutorial (2017)
2. Branco, P., Torgo, L., Ribeiro, R.P.: A survey of predictive modeling on imbalanced domains. ACM Comput. Surv. (CSUR) **49**(2), 31 (2016)
3. Brzezinski, D., Stefanowski, J., Susmaga, R., Szczech, I.: On the dynamics of classification measures for imbalanced and streaming data. IEEE Trans. Neural Netw. Learn. Syst. (2019)
4. Dal Pozzolo, A., Boracchi, G., Caelen, O., Alippi, C., Bontempi, G.: Credit card fraud detection: a realistic modeling and a novel learning strategy. IEEE Trans. Neural Netw. Learn. Syst. **29**(8), 3784–3797 (2018)
5. Davis, J., Goadrich, M.: The relationship between precision-recall and ROC curves. In: Proceedings of the 23rd International Conference on Machine Learning, pp. 233–240. ACM (2006)
6. Fawcett, T.: An introduction to ROC analysis. Pattern Recogn. Lett. **27**(8), 861–874 (2006)
7. Flach, P., Kull, M.: Precision-recall-gain curves: PR analysis done right. In: Advances in Neural Information Processing Systems, pp. 838–846 (2015)
8. Garcia, V., Sánchez, J.S., Mollineda, R.A.: On the suitability of numerical performance measures for class imbalance problems. In: International Conference in Pattern Recognition Applications and Methods, pp. 310–313 (2012)
9. Hanley, J.A., McNeil, B.J.: The meaning and use of the area under a receiver operating characteristic (ROC) curve. Radiology **143**(1), 29–36 (1982)
10. Jeni, L.A., Cohn, J.F., De La Torre, F.: Facing imbalanced data-recommendations for the use of performance metrics. In: 2013 Humaine Association Conference on Affective Computing and Intelligent Interaction, pp. 245–251. IEEE (2013)
11. Neyman, J., Pearson, E.S.: IX. On the problem of the most efficient tests of statistical hypotheses. Philos. Trans. R. Soc. Lond. Ser. A Contain. Pap. Math. Phys. Character **231**(694–706), 289–337 (1933)
12. Saito, T., Rehmsmeier, M.: The precision-recall plot is more informative than the ROC plot when evaluating binary classifiers on imbalanced datasets. PLoS ONE **10**(3), e0118432 (2015)
13. Sajjadi, M.S., Bachem, O., Lucic, M., Bousquet, O., Gelly, S.: Assessing generative models via precision and recall. In: Advances in Neural Information Processing Systems, pp. 5228–5237 (2018)
14. Santafe, G., Inza, I., Lozano, J.A.: Dealing with the evaluation of supervised classification algorithms. Artif. Intell. Rev. **44**(4), 467–508 (2015). https://doi.org/10.1007/s10462-015-9433-y

15. Tatbul, N., Lee, T.J., Zdonik, S., Alam, M., Gottschlich, J.: Precision and recall for time series. In: Advances in Neural Information Processing Systems, pp. 1920–1930 (2018)
16. Vanschoren, J., Van Rijn, J.N., Bischl, B., Torgo, L.: OpenML: networked science in machine learning. ACM SIGKDD Explor. Newslett. **15**(2), 49–60 (2014)
17. Widmer, G., Kubat, M.: Effective learning in dynamic environments by explicit context tracking. In: Brazdil, P.B. (ed.) ECML 1993. LNCS, vol. 667, pp. 227–243. Springer, Heidelberg (1993). https://doi.org/10.1007/3-540-56602-3_139

Supervised Phrase-Boundary Embeddings

Manni Singh$^{(\boxtimes)}$, David Weston, and Mark Levene

Department of Computer Science and Information Systems,
Birkbeck, University of London, London WC1E 7HX, UK
{manni,dweston,mark}@dcs.bbk.ac.uk

Abstract. We propose a new word embedding model, called SPhrase, that incorporates supervised phrase information. Our method modifies traditional word embeddings by ensuring that all target words in a phrase have exactly the same context. We demonstrate that including this information within a context window produces superior embeddings for both intrinsic evaluation tasks and downstream extrinsic tasks.

Keywords: Phrase embeddings · Named entity recognition · Natural language processing

1 Introduction

Word embeddings represent words with multidimensional vectors that are used in various models for applications such as, named entity recognition [9], query expansion [13], and sentiment analysis [21]. These embeddings are usually generated from a huge corpus with unsupervised learning models [3,16,18,23,24]. These models are based on describing target words by their neighbouring words which are also considered as contexts. The selection of these context words is generally linear (i.e. n words surrounding the target). Alternatively, arbitrary context words were used in [16] where context selection is based on the syntactic dependencies to the target word.

These models treat words as lexical units and create a context window surrounding a target word. This approach can be problematic when the context window for a target word contains only part of a phrase. For example, consider a scenario where a target word is close to (and to the right of) the named entity "George W. Bush" but the context window only retains the word "George". Clearly this will generate ambiguity as the independent word "George" may refer another person (George Washington), location (George Street, Oxford) or a music band (George). To deal with the issue described above, [19] used a data-driven approach to identify and treat these phrases as individual tokens. While this technique may learn a phrase representation it cannot learn a representation of the individual words that comprise the phrase.

In our approach we obtain phrase information directly from Wikipedia. Terms from Wikipedia articles are formatted as hyperlinks to relevant articles. In a related method [22] these terms are extracted as named entities. This paper

© The Author(s) 2020
M. R. Berthold et al. (Eds.): IDA 2020, LNCS 12080, pp. 470–482, 2020.
https://doi.org/10.1007/978-3-030-44584-3_37

interprets these terms as phrases. By using Wikipedia for phrase information (unlike [16]) we avoid needing additional grammatical information. This also gives us the potential to generate multi-lingual embeddings, although we do not pursue this here.

In this work, we are using phrase boundary information to generate word embedding in a non-compositional manner rather than a phrase embedding. We consider each of the words in the phrase as a part of the unit, where a unit can either be single word (i.e. not a link in the Wikipedia) or otherwise a bag of words. The embeddings are then learned for each of the unit members by considering surrounding units in the context.

In the following section we present related work in this domain, Sect. 3 presents our model and in Sects. 4 to 6 we give details of the implementation and the experiments.

2   Related Work

Word representations can be obtained from a language model where the goal is to predict a future word based on some previously observed information such as, a sentence, a sequence, or a phrase. For this task, various models can be utilised including: joint probabilities of observation that may include the Markov assumption. Under this assumption, we may say that the immediate future is independent of the entire past given the present. N-gram language models [4] use this assumption to predict token(s) using the previous $N - 1$ tokens [17]. This can be constructed efficiently for very large datasets using neural network based language modelling (NNLM) [2].

The NNLM of [2] used a non-linear hidden layer between the input and output layers. A simpler network named the log bi-linear model was introduced in [20] by dropping the hidden layer between input and output layer. Instead of the hidden layer, context vectors were summed and projected to the output layer. This model was later used by [18] and named CBOW (Continuous Bag-of-words model), with a symmetric context (i.e. context words on both sides of the target word).

In addition, the Skip-gram model, was introduced in this work by reversing CBOW to predict context from the target word. Given a context range c and target word w_t the objective is to maximise the average log probability,

$$\sum_{-c \leq j \leq c} \log p(w_{t+j} | w_t)$$

The model defines $p(w_{t+j} | w_t)$ using the softmax function,

$$p(w_O | w_I) = \frac{\exp\left(v'_{w_O}{}^\top v_{w_I}\right)}{\sum_{w=1}^{W} \exp\left(v'_w{}^\top v_{w_I}\right)}$$

where v_w and v'_w are the "input" and "output" vector representations of w, and W is the number of words in the vocabulary. However, due to the large vocabulary, the computation becomes impractical. Thus, Noise Contrastive Estimation (NCE) [7] was used that performs the same operation by sampling a very small amount of words k from the vocabulary as noise.

A similar technique is called Candidate Sampling [10] that combines noise samples with the true class, denoted as the set $\mathcal{S}$, with the objective to predict the true class from it, where Y is a set of true classes. Embeddings are scored as,

$$\hat{Y}_s = (X_s * W_s + b_s) - \log(E(s)).$$

Where X_s is a vector (embedding) corresponding to a word $s \in \mathcal{S}$, W_s is the corresponding weight, b_s is the bias, and $\mathbb{E}(s)$ is the expectation for s. Each score is approximated to a probability using the softmax function,

$$Softmax(\hat{Y}_s) = \frac{\exp \hat{Y}_s}{\sum_{s' \in S} \exp \hat{Y}_{s'}}.$$

In addition to words, phrases may also be considered. In [18], the words comprising a phrase were joined using the delimiter '_' between them, and their joint embedding was learned. This scheme is called non-compositional embedding [8,26]. Alternatively, compositional embeddings [8] are generated by merging word embeddings of phrase components using a composition function. The main difference in these schemes is that the previous learns the phrase embeddings while the latter just merges already learned word embeddings to make the phrase embeddings. Similarly, [3] introduced an extension of the Skip-gram model [18] that composes sub-word embeddings to make word embeddings with summation as the composition function.

3   The SPhrase Model

The proposed model uses information about which words belong to which phrases. This information can be conveniently represented as simply the locations for where phrases start and end, hence the name, *Supervised Phrase Boundary Representations model* (SPhrase).

The key assumption is that each word that comprises a phrase has the same context. This will produce an embedding where words that occur in the same phrase are likely to be close in the vector space. For example consider the sentence:

British Airways to New York has Departed
This sentence includes the (noun) phrase 'New York'. Following the procedure for Word2vec we focus on the target word 'New' using a context window of size 1. The target, context pairs are (New, to) and (New, York). Repeating this procedure for the target word 'York', yields the target, context pairs (York, New) and (York, has).

For SPhrase, the context differs from Word2vec, both target words in 'New York' will have the same context based on the words immediately surrounding the phrase, hence the SPhrase target context pairs are (New, to), (New, has), (York, to), (York, has). Figure 1 highlights the context words for the word 'New' for both Word2vec and SPhrase.

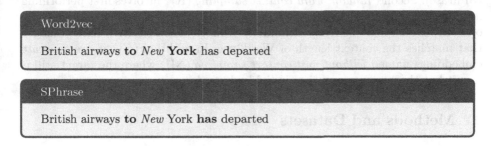

Fig. 1. Context words for the target word *New* using Word2vec and SPhrase. The context words are in bold. The context size is 1.

In the above, we demonstrated the target context pairs induced by a target word that is a member of a phrase, where its context are individual words. In the following, we generalise the approach to handle the situation where phrases are part of a context. We do this by introducing the concept of a *unit*, where a unit consist of a sequence of words. A unit of length 1 represents individual words, a unit of length 2 represents two word phrases and so on for larger phrases.

Thus we measure the context simply in terms of units. Figure 2 provides an example of a context of size 2 each side. Note that the left context for SPhrase contains 3 words. Thus the context size measured in words will be larger for SPhrase than Word2vec if there is a phrase within the context window.

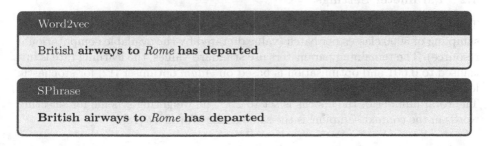

Fig. 2. Context words for the target word *Rome* using Word2vec and SPhrase. The context words are in bold. The context size is 2.

3.1 SPhrase Context Sampling

A standard approach to reduce the computation involved in generating embeddings is to shorten the effective context length by using only a sample of words from a context [18]. For SPhrase this can be achieved in several ways. First it can be done at the level of units not words, this is denoted *unit context sampling* (SPhrase). Second *random word context sampling* (R)[1] involves first performing unit context sampling, then for each unit that has a length greater than one only one word is sampled uniformly at random. This yields an effective context length that matches the context length of Word2vec. In addition to that, we generate embeddings named *without unit context sampling* (NU) where the target still is a unit but the context comprises individual words.

4 Methods and Datasets

4.1 Dataset

In order to generate an embedding using our approach, we require a corpus that has phrases annotated. Unfortunately this is not readily available, so we use a proxy for phrase annotation. In datasets that include hyperlinks we assume that the *hyperlink displayed text* is a phrase. One such data set is Wikipedia; we use the English Wikipedia dump version 20180920 that contains over 3 billion tokens. The proportion of tokens in phrases of length 2 is 2.5%; of length 3, 4, 5, and greater is respectively 0.8%, 0.3%, 0.2%, and less than 0.1%. Obviously not all phrases are represented as hyperlink text and not all hyperlink texts are phrases. Indeed the longest hyperlink text in our data set is of length 16,382 (it included internal formatting of Wikipedia). For our study we restricted maximum length to 10. The embedding vocabulary contained tokens with a frequency of at least 100 which gave us a total of 400,919 distinct tokens.

4.2 Parameter Settings

Training is performed in mini-batches of 60,000 tokens per batch with candidate sampling of 5000 classes per batch (value dictated by the available computational resource). The remaining parameters use standard values, the learning rate is initialised to 0.001 and optimisation is based on *Adam* optimiser [12] for stochastic learning. The learning decay is set to 10% (i.e. learning rate * 0.9) after each epoch. The total number of the epochs is set to 20. The weighting scheme for selecting words in the context sampling is the same as for Word2vec [18].

5 Evaluation

There are two types of evaluation tasks commonly accepted: intrinsic and extrinsic. Intrinsic evaluation tasks determine the quality of embeddings. Under this

---

[1] Pretrained embeddings are available at: https://github.com/ManniSingh/SPhrase

class, word similarity/relatedness tasks are generally based on cosine distance as a metric to find similarity between two word vectors. Extrinsic evaluation tasks, on the other hand, are based on specific downstream tasks such as, named entity recognition (NER), sentiment classification, topic detection. In this work, we are doing similarity based intrinsic evaluation and NER based extrinsic evaluation.

6   Experimental Design

6.1   Intrinsic Evaluation

The following experiments fit into the so-called *intrinsic* category of embedding evaluation. We aim to demonstrate that although the total number of phrases in our dataset is small compared to the number of words, they do have a positive impact on the resulting embeddings. In order to determine an optimal configuration of the method, intrinsic evaluation is done on embeddings trained on the first 10% of the corpus; see Fig. 3, As a result, the extrinsic evaluation described Sect. 6.2, the performance of the optimal configuration in this evaluations is: SPhrase (R) with window size 5. For the extrinsic evaluation only the optimal configuration is used and the embeddings are trained on the full corpus.

In the following experiments we compare SPhrase embeddings with the ones generated by Word2vec. It is known that increasing the context window size generally improves the quality of the embedding. Recall that the expected context size for each target word is the same for Word2vec and SPhrase due to word context sampling.

We expect that words in phrases should be mapped to similar locations in the embedding, i.e. words within a phrase should be closer together than words that are not in the same phrase. In the following we first perform experiment on pairwise similarity and then we investigate further structure with an analogy task.

Pairwise Similarity. For pairwise similarity experiments we use phrases from three datasets.

- CoNLL-2003 English dataset [25]. From this dataset multi-word named entities were extracted. These are used as phrases, in total there are 12,999. The maximum phrase length is 7 in this dataset, so we restricted the following two datasets to this as well.
- From our Wikipedia training corpus we obtained 16,470 phrases from the first 1,000000 tokens. This dataset comes from our training data, so we assume we should obtain good results in this case.
- Bristol [15] - from this dataset we selectively used the entity list and found 87,209 phrases.

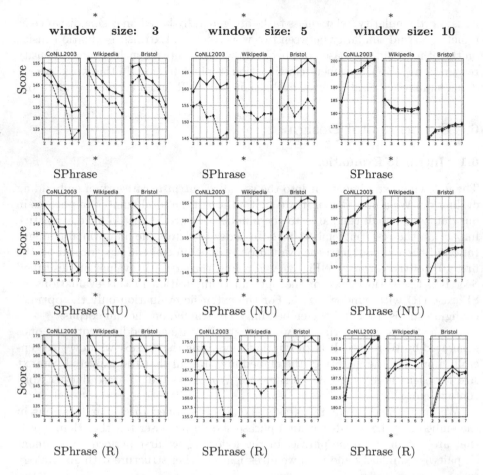

Fig. 3. Similarity scores comparison for the phrases relative to 100 random words representing: *unit context sampling* (SPhrase), Without *unit context sampling* (NU) and, with *random word context sampling* (R). Where SPhrase (in bold) and Word2vec (dashed) are compared on phrase lengths 2–7 (in horizontal axis) with higher the score the better it performed.

In order to investigate how the distances of words within a phrase compare to distances of words with random words in the datasets we use the following,

$$\text{Similarity Score} = \frac{1}{N_l(l-1)} \sum_{i=1}^{l-1} b(w_i, w_{i+1}, r)$$

where,

$$b(w_i, w_{i+1}, r) = \begin{cases} 1 & s(w_i, w_{i+1}) > s(w_i, r), \\ 0 & \text{otherwise}, \end{cases}$$

where r is a word selected at random from another phrase. A new word is drawn for each phrase pair comparison. The similarity score is calculated 100 times and the overall average is taken in order to reduce the noise generated by selecting only one word for each comparison. The interpretation of this is similar to the cosine score in that the larger the value the better.

We computed scores for phrase lengths up to and including length 7. We have used context window sizes 3, 5 and 10. Figure 3 shows these scores for the context sampling regimes: with *unit context sampling*, without *unit context sampling*, and *word context sampling*.

We can see that regardless of the embedding, the scores in general reduce as the phrase gets longer. However, the larger the window size the more Word2vec and SPhrase agree. This is what we should expect, since there will be greater overlap in the context words between SPhrase and Word2vec. Nevertheless we see that, overall, SPhrase performs better.

Google Analogy Test Set. Analogy based tasks are widely used, e.g. [5, 6, 11] to evaluate the quality of word embeddings. One well known test set is the Google analogy test set [18]. This dataset comprises rows of four words, such as known unknown informed uninformed. The analogy task is to predict the final word using the first three using simple vector addition/subtraction of their vector representations. Informally the task attempts to show how well words follow the vector relationship

unknown - known = uninformed - informed

Table 1. Scores on Google analogy dataset with *unit context sampling* (SPhrase), here accuracy is the total correct count on the total count of instances.

	Accuracy - displayed to 3 decimal places						Count
	Window size 3		Window size 5		Window size 10		
	SPhrase	Word2vec	SPhrase	Word2vec	SPhrase	Word2vec	
capital-world	0.727	0.628	0.746	0.658	0.815	0.782	4524
capital-common-countries	0.872	0.848	0.941	0.856	0.976	0.941	506
city-in-state	0.660	0.480	0.715	0.583	0.647	0.677	2467
gram3-comparative	0.848	0.806	0.758	0.813	0.643	0.670	1332
gram2-opposite	0.223	0.220	0.220	0.222	0.206	0.204	812
gram8-plural	0.755	0.736	0.715	0.744	0.641	0.727	1332
gram4-superlative	0.379	0.396	0.345	0.366	0.279	0.262	1122
gram9-plural-verbs	0.639	0.559	0.536	0.546	0.453	0.521	870
gram6-nationality-adjective	0.846	0.784	0.838	0.815	0.854	0.853	1599
family	0.603	0.595	0.595	0.638	0.581	0.543	506
gram7-past-tense	0.472	0.515	0.474	0.492	0.441	0.470	1560
currency	0.047	0.042	0.021	0.021	0.018	0.016	866
gram1-adjective-to-adverb	0.104	0.087	0.119	0.121	0.132	0.148	992
gram5-present-participle	0.517	0.520	0.509	0.486	0.479	0.455	1056
all	0.601	0.545	0.597	0.565	0.581	0.587	19544

Table 2. Scores on Google analogy dataset without *unit context sampling* (NU), here accuracy is the total correct count on the total count of instances.

	Accuracy - displayed to 3 decimal places						Count
	Window size 3		Window size 5		Window size 10		
	SPhrase	Word2vec	SPhrase	Word2vec	SPhrase	Word2vec	
capital-world	0.671	0.628	0.725	0.658	0.744	0.782	4524
capital-common-countries	0.881	0.848	0.935	0.856	0.929	0.941	506
city-in-state	0.653	0.480	0.645	0.583	0.652	0.677	2467
gram3-comparative	0.706	0.806	0.696	0.813	0.519	0.670	1332
gram2-opposite	0.217	0.220	0.197	0.222	0.172	0.204	812
gram8-plural	0.726	0.736	0.712	0.744	0.661	0.727	1332
gram4-superlative	0.273	0.396	0.298	0.366	0.269	0.262	1122
gram9-plural-verbs	0.577	0.559	0.548	0.546	0.477	0.521	870
gram6-nationality-adjective	0.855	0.784	0.821	0.815	0.827	0.853	1599
family	0.569	0.595	0.553	0.638	0.502	0.543	506
gram7-past-tense	0.453	0.515	0.483	0.492	0.414	0.470	1560
currency	0.039	0.042	0.024	0.021	0.028	0.016	866
gram1-adjective-to-adverb	0.130	0.087	0.173	0.121	0.168	0.148	992
gram5-present-participle	0.511	0.520	0.509	0.486	0.492	0.455	1056
all	0.565	0.545	0.576	0.565	0.553	0.587	19544

The dataset is divided into categories, some of which are inherently phrase-based. In the category `capital-common-countries` a typical line is:
`Athens Greece Baghdad Iraq`
Both *Athens Greece* and *Baghdad Iraq* can be reasonably construed to be phrases, unlike in the first example above. Two other categories have this same character, namely `capital-world` and `city-in-state`.
Example rows are: `Athens Greece Canberra Australia` and
`Chicago Illinois Houston Texas` respectively.

With this in mind we show the accuracy of SPhrase and Word2vec stratified by category, in addition to the overall accuracy that is usually reported. The categories that have a phrasal quality are italicised in Tables 1, 2 and 3. We see that, overall, SPhrase performs better in these categories.

6.2   Extrinsic Evaluation

We use Conll2003 English [25] and Wikigold [1] to evaluate the performance of the embeddings generated. The Conll dataset is widely used to evaluate various NER based models. It contains 203,621 tokens in the training set, while validation and test set contains 51,362 and 46,435 tokens respectively. On the other hand, Wikigold provides a single data file of 39,007 tokens that we used for testing while the NER models were trained with Conll train and validation data. We used SPhrase (R) model with window size 5 since this configuration demonstrated significant improvements over Word2vec as shown in Fig. 3. We recreated the BLSTMs and CRF based model [14] but without any feature engineering.

Table 3. Scores on Google analogy dataset with *random word context sampling* (R), here accuracy is the total correct count on the total count of instances.

	Accuracy - displayed to 3 decimal places						Count
	Window size 3		Window size 5		Window size 10		
	SPhrase	Word2vec	SPhrase	Word2vec	SPhrase	Word2vec	
capital-world	0.637	0.628	0.718	0.658	0.766	0.782	4524
capital-common-countries	0.858	0.848	0.903	0.856	0.953	0.941	506
city-in-state	0.664	0.480	0.623	0.583	0.663	0.677	2467
gram3-comparative	0.845	0.806	0.803	0.813	0.682	0.670	1332
gram2-opposite	0.224	0.220	0.245	0.222	0.196	0.204	812
gram8-plural	0.772	0.736	0.731	0.744	0.655	0.727	1332
gram4-superlative	0.373	0.396	0.392	0.366	0.257	0.262	1122
gram9-plural-verbs	0.575	0.559	0.586	0.546	0.474	0.521	870
gram6-nationality-adjective	0.818	0.784	0.824	0.815	0.831	0.853	1599
family	0.615	0.595	0.581	0.638	0.595	0.543	506
gram7-past-tense	0.479	0.515	0.520	0.492	0.460	0.470	1560
currency	0.040	0.042	0.024	0.021	0.023	0.016	866
gram1-adjective-to-adverb	0.090	0.087	0.127	0.121	0.172	0.148	992
gram5-present-participle	0.526	0.520	0.455	0.486	0.479	0.455	1056
all	0.576	0.545	0.588	0.565	0.576	0.587	19544

Table 4. Comparison of Word2vec with SPhrase(NU) on Conll2003 English and Wikigold dataset

Model	Conll2003Eng	Wikigold
Word2Vec	83.82 ± 0.3831	55.49 ± 0.4708
SPhrase	**88.93 ± 0.1115**	**66.01 ± 0.4172**

We trained this in 20 epochs with evaluating on validation data each time. We performed 10 instances for each of these models and presented the range of F1 scores (using Conll2003 evaluation script). Table 4 displays the results that show a significant improvement over the Word2vec model trained on the same corpus.

7 Concluding Remarks

This investigation demonstrates that using phrasal information can directly enrich word embeddings. In this work, we presented an alternative context sampling technique to that used in skip-gram Word2vec. We note that the SPhrase approach is not limited to augmenting Word2Vec, it can also be applied to morphological extensions such as Fasttext [3].

We used the displayed text from hyperlinks as a proxy for phrases, and in this sense SPhrase is supervised. We are, however, planning to generalise the methodology by investigating whether we can identify useful phrase boundaries in a completely unsupervised fashion.

References

1. Balasuriya, D., Ringland, N., Nothman, J., Murphy, T., Curran, J.R.: Named entity recognition in Wikipedia. In: Proceedings of the 2009 Workshop on the People's Web Meets NLP: Collaboratively Constructed Semantic Resources. People's Web 2009, pp. 10–18. Association for Computational Linguistics, Stroudsburg, PA, USA (2009). http://dl.acm.org/citation.cfm?id=1699765.1699767
2. Bengio, Y., Ducharme, R., Vincent, P., Jauvin, C.: A neural probabilistic language model. J. Mach. Learn. Res. **3**(Feb), 1137–1155 (2003)
3. Bojanowski, P., Grave, E., Joulin, A., Mikolov, T.: Enriching word vectors with subword information. Trans. Assoc. Comput. Linguist. (TACL) **5**(1), 135–146 (2017). http://www.aclweb.org/anthology/Q17-1010
4. Brants, T., Popat, A.C., Xu, P., Och, F.J., Dean, J.: Large language models in machine translation. In: Proceedings of the Joint Conference on Empirical Methods in Natural Language Processing and Computational Natural Language Learning, pp. 858–867 (2007)
5. Bruni, E., Tran, N.K., Baroni, M.: Multimodal distributional semantics. J. Arti. Intell. Res. **49**(1), 1–47 (2014). http://dl.acm.org/citation.cfm?id=2655713.2655714
6. Finkelstein, L., et al.: Placing search in context: the concept revisited. In: Proceedings of the 10th International Conference on World Wide Web, WWW 2001, pp. 406–414. ACM, New York (2001). https://doi.org/10.1145/371920.372094
7. Gutmann, M.U., Hyvärinen, A.: Noise-contrastive estimation of unnormalized statistical models, with applications to natural image statistics. J. Mach. Learn. Res. **13**(Feb), 307–361 (2012)
8. Hashimoto, K., Tsuruoka, Y.: Adaptive joint learning of compositional and non-compositional phrase embeddings. In: Proceedings of the 54th Annual Meeting of the Association for Computational Linguistics (Volume 1: Long Papers), pp. 205–215. Association for Computational Linguistics, Berlin, Germany, August 2016 (2016). https://doi.org/10.18653/v1/P16-1020, http://www.aclweb.org/anthology/P16-1020
9. Huang, Z., Xu, W., Yu, K.: Bidirectional LSTM-CRF models for sequence tagging. arXiv preprint arXiv:1508.01991 abs/1508.01991 (2015)
10. Jean, S., Cho, K., Memisevic, R., Bengio, Y.: On using very large target vocabulary for neural machine translation. In: Proceedings of the 53rd Annual Meeting of the Association for Computational Linguistics and the 7th International Joint Conference on Natural Language Processing (Volume 1: Long Papers), pp. 1–10. Association for Computational Linguistics (2015)
11. Jurgens, D.A., Turney, P.D., Mohammad, S.M., Holyoak, K.J.: Semeval-2012 task 2: measuring degrees of relational similarity. In: Proceedings of the First Joint Conference on Lexical and Computational Semantics - Volume 1: Proceedings of the Main Conference and the Shared Task, and Volume 2: Proceedings of the Sixth International Workshop on Semantic Evaluation, SemEval 2012, pp. 356–364. Association for Computational Linguistics, Stroudsburg, PA, USA (2012), http://dl.acm.org/citation.cfm?id=2387636.2387693
12. Kingma, D.P., Ba, J.: Adam: a method for stochastic optimization. arXiv abs/1412.6980 (2014)
13. Kuzi, S., Shtok, A., Kurland, O.: Query expansion using word embeddings. In: Proceedings of the 25th ACM International on Conference on Information and Knowledge Management, CIKM 2016, pp. 1929–1932. ACM, New York (2016)

14. Lample, G., Ballesteros, M., Subramanian, S., Kawakami, K., Dyer, C.: Neural architectures for named entity recognition. In: NAACL HLT 2016, the 2016 Conference of the North American Chapter of the Association for Computational Linguistics: Human Language Technologies, San Diego California, USA, 12–17 June 2016, pp. 260–270 (2016)
15. Lansdall-Welfare, T., Sudhahar, S., Thompson, J., Lewis, J., Team, F.N., Cristianini, N.: Content analysis of 150 years of british periodicals. Proc. Nat. Acad. Sci. **114**(4), E457–E465 (2017)
16. Levy, O., Goldberg, Y.: Dependency-based word embeddings. In: Proceedings of the 52nd Annual Meeting of the Association for Computational Linguistics (Volume 2: Short Papers), vol. 2, pp. 302–308 (2014)
17. Martin, J.H., Jurafsky, D.: Speech and Language Processing: an Introduction to Natural Language Processing, Computational Linguistics, and Speech Recognition. Pearson/Prentice Hall, Upper Saddle River (2009)
18. Mikolov, T., Chen, K., Corrado, G., Dean, J.: Efficient estimation of word representations in vector space. arXiv abs/1301.3781 (2013). http://arxiv.org/abs/1301.3781
19. Mikolov, T., Sutskever, I., Chen, K., Corrado, G., Dean, J.: Distributed representations of words and phrases and their compositionality. In: Proceedings of the 26th International Conference on Neural Information Processing Systems (NIPS), NIPS 2013, vol. 2, pp. 3111–3119. Curran Associates Inc., USA (2013)
20. Mnih, A., Hinton, G.: Three new graphical models for statistical language modelling. In: Proceedings of the 24th International Conference on Machine Learning. ICML 2007, pp. 641–648. ACM, New York (2007)
21. Nakov, P., Ritter, A., Rosenthal, S., Sebastiani, F., Stoyanov, V.: SemEval-2016 task 4: sentiment analysis in Twitter. In: Proceedings of the 10th International Workshop on Semantic Evaluation (SemEval-2016), pp. 1–18, USA (2016)
22. Nothman, J., Curran, J.R., Murphy, T.: Transforming Wikipedia into named entity training data. In: Proceedings of the Australasian Language Technology Association Workshop 2008, pp. 124–132, Hobart, Australia, December 2008. http://www.aclweb.org/anthology/U08-1016
23. Pennington, J., Socher, R., Manning, C.D.: Glove: global vectors for word representation. In: Empirical Methods in Natural Language Processing (EMNLP), pp. 1532–1543 (2014). http://www.aclweb.org/anthology/D14-1162
24. Salle, A., Villavicencio, A., Idiart, M.: Matrix factorization using window sampling and negative sampling for improved word representations. In: Proceedings of the 54th Annual Meeting of the Association for Computational Linguistics, ACL 2016, Berlin, Germany, 7–12 August 2016, vol. 2, Short Papers (2016)
25. Tjong Kim Sang, E.F., De Meulder, F.: Introduction to the CoNLL-2003 shared task: language-independent named entity recognition. In: Proceedings of the Seventh Conference on Natural Language Learning at HLT-NAACL 2003, vol. 4, pp. 142–147. Association for Computational Linguistics, Japan (2003)
26. Yu, M., Dredze, M.: Learning composition models for phrase embeddings. Trans. Assoc. Comput. Linguist. (TACL) **3**(1), 227–242 (2015). http://www.aclweb.org/anthology/Q15-1017

Predicting Remaining Useful Life with Similarity-Based Priors

Youri Soons[1], Remco Dijkman[2(✉)], Maurice Jilderda[3],
and Wouter Duivesteijn[2]

[1] Sitech Services B.V., Geleen, The Netherlands
`youri.soons@sitech.nl`
[2] Technische Universiteit Eindhoven, Eindhoven, The Netherlands
`{r.m.dijkman,w.duivesteijn}@tue.nl`
[3] Perfact Group, Munstergeleen, The Netherlands
`mauricejilderda@perfact-group.com`

Abstract. Prognostics is the area of research that is concerned with predicting the remaining useful life of machines and machine parts. The remaining useful life is the time during which a machine or part can be used, before it must be replaced or repaired. To create accurate predictions, predictive techniques must take external data into account on the operating conditions of the part and events that occurred during its lifetime. However, such data is often not available. Similarity-based techniques can help in such cases. They are based on the hypothesis that if a curve developed similarly to other curves up to a point, it will probably continue to do so. This paper presents a novel technique for similarity-based remaining useful life prediction. In particular, it combines Bayesian updating with priors that are based on similarity estimation. The paper shows that this technique outperforms other techniques on long-term predictions by a large margin, although other techniques still perform better on short-term predictions.

Keywords: Remaining useful life · Trajectory based similarity prediction · Bayesian updating · Similarity estimation · Prognostics · Prediction

1 Introduction

Prognostics is the area of research that concerns the prediction of the remaining useful life (RUL) of machines or machine parts. A RUL prediction is a prediction of the time until a machine or machine part must be replaced or repaired. It is important that such predictions are accurate: early predictions lead to unnecessarily frequent maintenance with associated costs, while late predictions increase the risk of a machine break down with associated loss of production time and possibly sales.

Data-driven RUL prediction is based on run to failure data, i.e., observations on what happened to a part or machine in a run from the last maintenance

© The Author(s) 2020
M. R. Berthold et al. (Eds.): IDA 2020, LNCS 12080, pp. 483–495, 2020.
https://doi.org/10.1007/978-3-030-44584-3_38

activity to the next. Figure 1 shows a typical example of run to failure data, in this case data of a filter in a chemical plant. The figure shows condition measurements on the filter over time, in terms of the difference in pressure before and after the filter. It shows that this difference is close to zero for some time. Then, the filter starts to clog up and the pressure builds up, until the filter is replaced and the pressure difference returns to normal. The resulting 'sawtooth' shape is frequently observed in run to failure data.

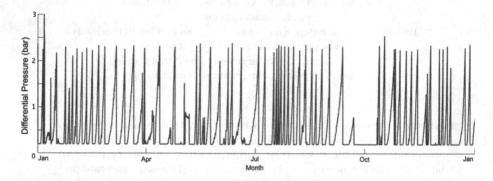

Fig. 1. Example run to failure data.

RUL prediction on run to failure data can be done by fitting a model, such as a regression model or a probability distribution, on the data. Many different techniques exist for those purposes [1]. However, as is evident from Fig. 1, different runs may have very different durations or shapes, and RUL prediction techniques rely on additional data to accurately predict the duration and shape of a particular run. Unfortunately, additional data is often unavailable or hard to relate to the run to failure data [2]. If additional data is unavailable, it is unclear which condition measurements are reliable and of course what their influence is on the RUL. One way to overcome these problems is to use similarity-based techniques, which work based on the hypothesis that, if a curve has developed similarly to some collection of other curves until now, it will likely continue to develop like that, and have a similar remaining useful life.

This paper explores the performance of two similarity-based techniques: trajectory-based similarity prediction, and Bayesian updating. It then adds its own: Bayesian updating with similarity-based priors. The contribution of this paper consists of this technique, described in Sect. 3.4, as well as a detailed evaluation of all three techniques in a case study from practice, described in Sect. 4.

Against this background, the remainder of this paper is structured as follows. Section 2 presents related work on remaining useful life prediction. Section 3 presents similarity-based remaining useful life prediction techniques, including the new technique. Section 4 compares the performance of the various techniques in a case study and Sect. 5 presents the conclusions.

2   Related Work

RUL prediction can be considered a specialized form of survival analysis [10]. Essentially, two types of techniques exist for predicting RUL: model-based and data-driven techniques. Model-based techniques use physical models to accurately represent the wear and tear of a component over time [5]. Data-driven techniques do not presume any knowledge about how a component wears out over time, but merely predicts the RUL based on past observations. Hybrid models, which are a combination of physical and data-driven techniques, also exist [9]. This paper focuses on data-driven models, which are most suited when the physical mechanisms that cause a component to fail are too complex to model cost-effectively, or if they are not sufficiently understood.

A large number of data-driven techniques is available that fall into two classes depending on whether or not a probability distribution of the RUL must be obtained or a point-estimate is sufficient [1]. A probability distribution of the RUL has several benefits [16,17,20]. For example, it facilitates stochastic decision making, where maintenance is done when the probability that a part will fail exceeds a certain threshold, which is in line with the way in which maintenance decisions are made. When it is not necessary to produce a probability density function, several models can be used. The most obvious choices include regression models that use time as the primary independent variable and time-series models. However, regression models require that the behavior of the curve is predictable over time [4,13] and time-series [12] models are only suitable for short-term predictions [3,16] or when the behavior of the curve is predictable over time. Regression models that take other variables into account can also be used [6]. Such models have the benefit that they do not only consider the dependency of the RUL on the time that the part has been in operation, but also on other relevant factors, such as the operational temperature or vibration of the part.

When the RUL depends on other factors beyond time, but data on such factors is not available, one can include them as a black box. While we may not know the values of relevant factors, we can still find historical runs that are similar to the current run. If we assume that the factors that influenced historically similar runs are also similar to the current run, then the future behavior of the current run will also be similar to the behavior of the historically similar runs. This is called Trajectory Based Similarity Prediction (TBSP) [11,18,19]. Bayesian updating techniques use a similar principle [7,8]. Such techniques create a prior probability distribution of the RUL (based on data from historical runs to failure), which updates as more data of the current run is revealed.

3   Prediction Techniques

This section presents similarity-based techniques that can be used for RUL prediction: TBSP and Bayesian updating, which are defined in related work as explained in Sect. 2. Subsequently, Sect. 3.4 presents a novel technique, Bayesian updating with similarity-based prior estimation, which is a combination of TBSP and Bayesian updating.

3.1   Preliminaries

The remaining useful life of a part is defined as follows.

Definition 1 (Remaining Useful Life (RUL)). *Let t be a moment in a run and t_E be the moment in the run at which the part fails. The Remaining Useful Life (RUL) at time t, $r(t)$, is defined as $r(t) = t_E - t$.*

Note that 'failure' can be interpreted broadly. It does not have to be the point at which the part breaks, but can also be the point at which the part reaches a condition in which it is not considered suitable for operation anymore, or a condition in which maintenance is considered necessary. Over time, multiple runs to failure will be observed, such as the runs to failure shown in Fig. 1.

Definition 2 (Run to failure library). *L is the library of past runs to failure. For each $l \in L$, t_E^l is the moment in the run at which the part fails, and $g^l(t)$ is the function that returns the condition of the part at time t of the run.*

The function $g^l(t)$ is created by fitting a curve on the condition measurements of the run. We consider the one-dimensional case here (i.e., the case in which we only measure the condition of the part), but this can easily be extended to a multi-dimensional case (i.e., the case in which we not only measure the condition of the part, but also external factors (i.e., other variables than the condition variable itself), such as the operating temperature or pressure) by considering the observations as vectors over multiple variables. We will also omit the superscript l if there can be no confusion about the run to which we refer.

3.2   Trajectory-Based Similarity Prediction

Figure 2 shows a different (cf. Fig. 1) representation of a run to failure library. It shows all runs in the library, starting from the moment at which the condition variable starts to increase from the base condition. It also shows a 'current' run as a thicker, unfinished curve. The idea of trajectory-based similarity prediction is to find some number k of runs that are most similar to the current run. For each of these k similar runs, we know the time it took until the part failed.

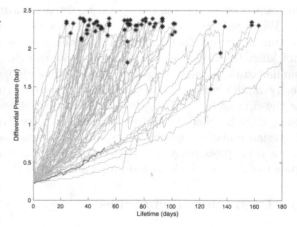

Fig. 2. Example library of runs.

Trajectory-based Similarity Prediction (TBSP) estimates the time until failure as the mean failure time of the similar runs.

Definition 3 (Distance of current run to library run). *At a moment in time t, let I be the number of observations made in the current run, with values $z_1, \ldots, z_I$ observed at times $t_1, \ldots, t_I$, and let $l \in L$ be a library run. We denote by $d^l(t)$ any distance measure contrasting $z_1, \ldots, z_I$ with $g^l(t_1), \ldots, g^l(t_I)$. Let $E^l(t)$ and $M^l(t)$ denote Euclidean and Manhattan distance, respectively.*

Clearly other distance functions can and indeed have been used as well in the context of remaining useful life prediction [21]. An in-depth analysis of the distance function that performs best for TBSP is beyond the scope of this work.

Definition 4 (Fit of current run to library run). *For each library run $l \in L$, let $d^l(t)$ be defined as in Definition 3. The fit of the current run to l is:*

$$S^l(t) = e^{-|d^l(t)|}$$

When, at time t of the current run, the library run l is found that fits the current run best, the remaining useful life of the current run can be predicted as the remaining useful life of that run l: $r(t) = t_E^l - t$. It is also possible to base the prediction of the remaining useful life on the best k runs; sensitivity to k is part of our experiments. If $k > 1$, we can also aggregate RUL predictions by weighted average, where the weights are the goodness of fit of the library runs to the current run.

Definition 5 (Trajectory-based Similarity Prediction). *For each library run $l \in L$, let $S^l(t)$ be the fit of the run to the current run as per Definition 4 and let $r^l(t)$ be the RUL of the run. Let $L' \subseteq L$ be the subset of past runs on which we want to base our RUL prediction. The predicted RUL of the current run, $\hat{r}(t)$, is:*

$$\hat{r}(t) = \frac{\sum_{l \in L'} S^l(t) \cdot r^l(t)}{\sum_{l \in L'} S^l(t)}$$

3.3 Bayesian Updating

A Bayesian updating method has also been proposed to create a probability distribution of the remaining useful life [7,8]. The probability distribution can be updated with each observation of the condition variable that is obtained. The method works by fitting an exponential model to the library runs and subsequently updating that model with observations of the current run.

Intuitively, looking at Fig. 2, Bayesian updating works by fitting a curve to each of the library runs or to a selection of library runs. Based on the resulting collection of curves, a prior probability distribution of the time until the part fails can be created, which represents the 'probable' curve that the current run —or in fact any run—will follow. The prior probability distribution can be updated each time a condition value is observed in the current run. This update leads to a posterior probability distribution that represents the curve that the current run will follow with a higher precision (smaller confidence interval).

Definition 6 (RUL probability density). *For each library run $l \in L$, let $g^l(t)$ be the function that returns the condition of the part at time t of the run. The condition function can be fitted as an exponential model that has the form:*

$$g^l(t) = \phi + \theta e^{\beta t + \epsilon(t) - \frac{1}{2}\sigma^2}$$

Here, ϕ is the intercept, $\epsilon(t)$ is the error term with mean 0 and variance σ^2, and θ and β are random variables.

If we set $\phi = 0$ and take the natural logarithm of both sides, we get:

$$\ln(g^l(t)) = \theta' + \beta t + \epsilon(t)$$

where $\theta' = \ln(\theta) + \frac{1}{2}\sigma^2$. Considering that we have multiple runs $l \in L$, it is possible to fit this equation multiple times to those runs and calculate values for θ', β and σ for each run. With these values, we can compute the prior probability distributions of θ' and β. We assume these distributions are normal distributions with means μ_0' and μ_1 and variances σ_0^2 and σ_1^2. While the prior distributions are created based on observations from library runs, the distribution can be updated as more observations become available in the current run.

Proposition 1 (RUL probability density updating). *Let $\pi(\theta')$ and $\pi(\beta)$ be the prior distributions of the random variables from Definition 6 with means μ_0' and μ_1 and variances σ_0^2 and σ_1^2, where $\theta' = \ln(\theta) + \frac{1}{2}\sigma^2$ and σ^2 is the variance of the error term. Furthermore, let there be I observed values, $z_1, \ldots, z_I$, in the current run, made at times $t_1, \ldots, t_I$, and for $i \in I$, let $L_i = \ln(z_i)$ the natural logarithm of each observation. The posterior distribution is a bivariate normal distribution with θ' and β, whose means $\mu_{\theta'}$ and μ_β, variances $\sigma_{\theta'}^2$ and σ_β^2, and correlation coefficient ρ can be calculated as follows:*

$$\mu_{\theta'} = \frac{\left(\sum_{i \in I} L_i \sigma_0^2 + \mu_0' \sigma^2\right)\left(\sum_{i \in I} t_i^2 \sigma_1^2 + \sigma^2\right) - \left(\sum_{i \in I} t_i \sigma_0^2\right)\left(\sum_{i \in I} L_i t_i \sigma_1^2 + \mu_1 \sigma^2\right)}{(|I|\sigma_0^2 + \sigma^2)\left(\sum_{i \in I} t_i^2 \sigma_1^2 + \sigma^2\right) - \left(\sum_{i \in I} t_i \sigma_1^2\right)\left(\sum_{i \in I} t_i \sigma_0^2\right)}$$

$$\mu_\beta = \frac{(|I|\sigma_0^2 + \sigma^2)\left(\sum_{i \in I} L_i t_i \sigma_1^2 + \mu_1 \sigma^2\right) - \left(\sum_{i \in I} t_i \sigma_1^2\right)\left(\sum_{i \in I} L_i \sigma_0^2 + \mu_0' \sigma^2\right)}{(|I|\sigma_0^2 + \sigma^2)\left(\sum_{i \in I} t_i^2 \sigma_1^2 + \sigma^2\right) - \left(\sum_{i \in I} t_i \sigma_1^2\right)\left(\sum_{i \in I} t_i \sigma_0^2\right)}$$

$$\sigma_{\theta'}^2 = \sigma^2 \sigma_0^2 \frac{\sum_{i \in I} t_i^2 \sigma_1^2 + \sigma^2}{(|I|\sigma_0^2 + \sigma^2)\left(\sum_{i \in I} t_i^2 \sigma_1^2 + \sigma^2\right) - \left(\sum_{i \in I} t_i\right)^2 \sigma_0^2 \sigma_1^2}$$

$$\sigma_\beta^2 = \sigma^2 \sigma_1^2 \frac{|I|\sigma_0^2 + \sigma^2}{(|I|\sigma_0^2 + \sigma^2)\left(\sum_{i \in I} t_i^2 \sigma_1^2 + \sigma^2\right) - \left(\sum_{i \in I} t_i\right)^2 \sigma_0^2 \sigma_1^2}$$

$$\rho = \frac{-\sigma_0 \sigma_1 \sum_{i \in I} t_i}{\sqrt{|I|\sigma_0^2 + \sigma^2}\sqrt{\sigma_1^2 \sum_{i \in I} t_i^2 + \sigma^2}}$$

The proof of this proposition is given in [8]. Consequently, $\ln(g^l(t))$ for the current run to failure l is normally distributed with mean and variance:

$$\mu(t) \cong \mu_{\theta'} + \mu_{\beta}t - \frac{1}{2}\sigma^2 \qquad\qquad \sigma(t) \cong \sigma_{\theta'}^2 + \sigma_{\beta}^2 t^2 + \sigma^2 + 2\rho t \sigma_{\theta'}\sigma_{\beta}$$

With this information, the probability that future values of $\ln(g^l(t))$ exceed the maximum acceptable condition at some time t can be computed.

3.4   Bayesian Updating with Similarity-Based Prior Estimation

The RUL probability density function in Definition 6 depends on estimated prior distributions of θ and β. These priors can be set through analyzing previous runs to failure, either based on the complete library of runs, or on a subset of the runs. More precisely, we can determine prior distributions as follows.

Definition 7 (Prior distributions). *For each library run $l \in L$, let $g^l(t)$ be the exponential curve that is fitted to the observations in that run with parameters θ'^l and β^l as in Definition 6. For a subset $M \subseteq L$ of runs, we can determine the mean and standard deviation of θ' and β over all θ'^m and β^m.*

Consequently, our priors depend on the subset $M \subseteq L$ of runs that we use. For example, we can determine our priors based on $M = L$, the complete set of runs. Here, we consider a variant of the Bayesian updating method in which the priors are set based on the runs that are most similar to the current run, using Definition 4 for similarity and thresholds to select the most similar runs. More precisely, we select our priors as follows.

Definition 8 (Similarity-based prior distributions). *Let t be the moment in time at which we determine our prior distributions and k be the number of similar runs on which we base them. Furthermore, let $S^l(t)$ be the similarity of a run l to the observations in the current run until time t as per Definition 4. The set of k most similar runs $M \subseteq L$ at moment t is then defined as the set in which, for all runs $m \in M$, there is no run $l \in L - M$, such that $S^l(t) > S^m(t)$.*

Note that this definition depends on variables t and k, which can therefore be expected to influence the performance of the technique. In our evaluation, we will explore the performance of the technique for different values of t and k.

4   Evaluation

In this section, we put the RUL prediction techniques introduced in Sect. 3 to the test, in a case study with data from practice.

4.1   Case Study

Our data originates from a chemical plant on the Chemelot Industrial Site[1]. The plant we investigate produces a steady flow of various chemical products; whatever the product happens to be, an unwanted byproduct is always generated. Filters have been installed to obtain an untainted final product. These filters have a variable service life, ranging between two and eight days. When the filter performs its function, it withholds residue of the unwanted byproduct. This residue gradually builds up, forming a cake which increases the resistance of the filter. The additional resistance is measured through an increase in differential pressure (δP), as illustrated in Fig. 1. An unclogged filter has a δP of 0.2 bar. When δP reaches a threshold of 2.4 bar, a valve in front of the filter is switched to let the product run through a parallel, clean filter, which returns δP to 0.2 bar and enables engineers to maintain the clogged filter.

Sensor data, including δP, is stored in a NoSQL database as time series. Preprocessing is needed in several aspects. First, the data has many missing values, which we replace by the last observed value. Second, the sensors generate a data point every second. We established experimentally that resampling the data to the minute barely loses any information from the signal, while still substantially reducing the size of the dataset. Third, to avoid the amplification of clear outliers, they are removed with a Hampel filter [14]. Fourth, we focus on the 'exponential deterioration stage' of the filter's life cycle [5], because—according to the company—the start of that stage is early enough to be able to act on time, and because it provides us with a dataset that is suitable for similarity-based RUL prediction techniques. The start and end of the exponential deterioration stage must be derived from data. We do that by comparing the average pressure over the last hour with its preceding hour. To ensure that every run has only one start per stop, a detected start is ignored if another start was already detected in the same run.

4.2   Results

We quantify our results using an $\alpha - \lambda$ graph. Intuitively, this graph represents the probability that, at a certain moment in the run to failure, the RUL prediction (λ) is within a pre-defined level of precision (α) [15]. We will use a concise representation of the $\alpha - \lambda$ quality: rather than time into the run, we put the RUL on the x-axis, while the y-axis displays the probability. This representation allows us to visually compare different techniques. All analysis is done using 5-fold cross validation. The results presented in the graphs are the averages over the 5 folds.

Figures 3a, b, and c show the performance of the TBSP technique for various parameter settings. Figure 3a compares the performance of TBSP when fitting various types of curves (second ('poly2') and third ('poly3') order polynomials,

---

[1] An anonymized version of the data is made available at: https://surfdrive.surf.nl/files/index.php/s/1dTFFXfZ7woeSUA.

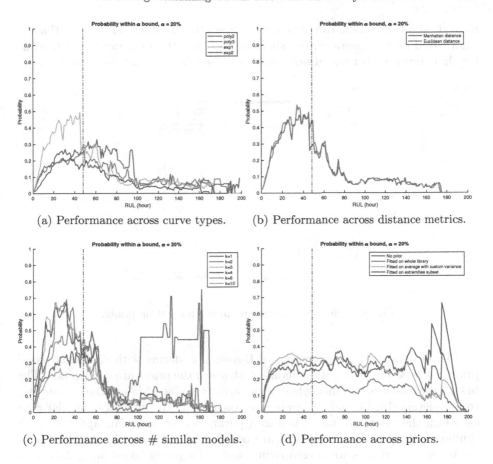

(a) Performance across curve types.   (b) Performance across distance metrics.

(c) Performance across # similar models.   (d) Performance across priors.

Fig. 3. Comparison of hyperparameter settings.

exponential curves ('exp1'), and the sum of two exponential curves ('exp2')),
Fig. 3b compares Manhattan and Euclidean distance, and Fig. 3c shows the sensitivity to the number of similar curves k. The graphs show that TBSP performs
best for an exponential curve in short term (<48 h) predictions, and for $k = 2, 3$,
or 4, while there is little to no performance difference between Manhattan and
Euclidean distance and between $k = 2, 3$, or 4. For those reasons, we parameterize TBSP with exponential curves, using Euclidean distance as a distance
metric, and using 3 similar curves to make the prediction.

Figure 3d shows the performance of the Bayesian updating technique for various prior sets of runs on which the prior is based. We consider four alternatives.
In the first alternative, no prior is defined and the prediction is only computed
based on the current run. In the second alternative, the prior distribution is
based on all runs in the library. In the third alternative, we create a prior distribution by fitting the run with the (closest to) average run to failure time. In
the fourth alternative, we create a prior distribution by fitting the shortest, the

longest, and the average run. The figure shows that for long term predictions, a prior fitted on the 'average', the shortest and longest run performs best, while for short term predictions, a prior fitted on the whole library performs best.

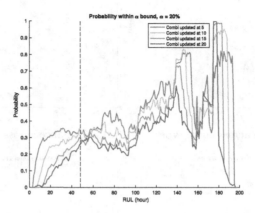

Fig. 4. Performance across moments for setting priors.

Figure 4 shows the performance of Bayesian updating with similarity-based priors for various settings of the moment at which the priors are determined. The best performance is obtained when priors are determined 5 h into the current run to completion; 10, 15, and 20 h were also considered. The number of similar runs on which the priors are based is also a parameter for Bayesian updating with similarity-based priors. The priors are based on the 3 most similar runs. This led to the best results when comparing results for priors based on 1, 2, 3, 4, 5, and 10 similar runs.

Figure 5 shows the results for the various prediction techniques: TBSP, Bayesian Updating, and Bayesian updating with similarity-based priors. The results show a clear distinction in the performance of the different techniques. TBSP performs best for short-term (<48 h before failure) predictions, while Bayesian updating with similarity-based priors performs best in the long term (150–200 h before failure). This is expected, because for long-term prediction, Bayesian updating with similarity-based priors benefits from being based both on similar runs and on general Bayesian behavior, while after some updates the impact of the priors is reduced and the behavior approaches that of normal Bayesian updating. TBSP on the other hand benefits from having a better estimate of the runs to which it is close as time progresses.

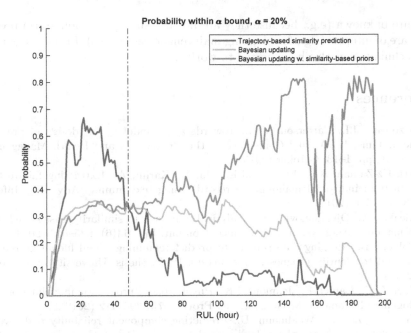

Fig. 5. Overall comparison of techniques.

5 Conclusions

In a case study, we show how techniques from literature can be combined and parameterized to accurately predict the Remaining Useful Life (RUL) of a machine or part. While curves of the degradation of a machine or part over time typically have a similar shape, the challenge is that operational constraints, which may be unknown, influence the exact parameterization of that curve, as evidenced by the real-life runs displayed in Figs. 1 and 2. Therefore, we propose a similarity-based prediction technique: while it makes no sense to compare the current run with all previously observed runs, it is quite likely that there are *some* historical runs that are similar to the current run, because they have similar operational constraints, hence providing us with powerful predictive information.

This paper proposes a new similarity-based prediction technique, in which we obtain a probability distribution of the RUL through Bayesian updating, where the priors of the Bayesian distribution are calculated based on a careful selection of previously seen runs. As evidenced by Fig. 5, our technique outperforms alternative techniques in a case study by a large margin within the long-term region. If we strive to predict the RUL shorter in advance, Fig. 5 clearly indicates that other methods work better.

While we studied the performance of RUL prediction techniques in the context of a particular case study, in many other domains degradation patterns have similar properties. In particular, in many other domains: run to failure data has a 'sawtooth' shape as in Fig. 1, degradation depends on operational conditions that are unknown (e.g., because they are not measured), and long-term predictions are of interest (e.g., for planning maintenance activities). In such situations our technique can also be expected to work well.

References

1. Aizpurua, J.I., Catterson, V.M.: Towards a methodology for design of prognostic systems. In: Annual Conference of the Prognostics and Health Management Society, pp. 1–13, October (2015)
2. Arif-Uz-Zaman, K., Cholette, M.E., Ma, L., Karim, A.: Extracting failure time data from industrial maintenance records using text mining. Adv. Eng. Inform. **33**, 388–396 (2017)
3. Bleakie, A., Djurdjanovic, D.: Analytical approach to similarity-based prediction of manufacturing system performance. Comput. Ind. **64**(6), 625–633 (2013)
4. Coble, J.B.: Merging data sources to predict remaining useful life-an automated method to identify prognostic parameters. Ph.D. thesis, University of Tennessee (2010)
5. Eker, O.F., Camci, F., Jennions, I.K.: Physics-based prognostic modelling of filter clogging phenomena. Mech. Syst. Sig. Process. **75**, 395–412 (2016)
6. Fink, O., Zio, E., Weidmann, U.: Predicting component reliability and level of degradation with complex-valued neural networks. Reliab. Eng. Syst. Saf. **121**, 198–206 (2014)
7. Gebraeel, N.: Sensory-updated residual life distributions for components with exponential degradation patterns. IEEE Trans. Autom. Sci. Eng. **3**(4), 382–393 (2006)
8. Gebraeel, N., et al.: Residual life distributions from component degradation signals: a Bayesian approach residual-life distributions from component degradation signals: A Bayesian approach. IIE Trans. **37**(6), 543–557 (2005). Research Collection Lee Kong Chian School of Business
9. Goebel, K., Eklund, N.: Prognostic fusion for uncertainty reduction. Soft Comput. (2007)
10. Kleinbaum, D.G., Klein, M.: Survival Analysis, 3rd edn. Springer, New York (2012). https://doi.org/10.1007/978-1-4419-6646-9
11. Lam, J., Sankararaman, S., Stewart, B.: Enhanced trajectory based similarity prediction with uncertainty quantification. In: Proceedings of the PHM, pp. 623–634 (2013)
12. Ling, Y.: Uncertainty quantification in time-dependent reliability analysis. Ph.D. thesis, Vanderbilt University (2013)
13. Liu, J., Djurdjanovic, D., Ni, J., Casoetto, N., Lee, J.: Similarity based method for manufacturing process performance prediction and diagnosis. Comput. Ind. **58**(6), 558–566 (2007)
14. Pearson, R.K., Neuvo, Y., Astola, J., Gabbouj, M.: Generalized hampel filters. EURASIP J. Adv. Sig. Process. **2016**(1), 1–18 (2016). https://doi.org/10.1186/s13634-016-0383-6
15. Saxena, A., Celaya, J., Saha, B., Saha, S., Goebel, K.: Metrics for offline evaluation of prognostic performance. Int. J. Progn. Health Manag. **1**, 1–20 (2010)
16. Si, X.S., Wang, W., Hu, C.H., Chen, M.Y., Zhou, D.H.: A Wiener-process-based degradation model with a recursive filter algorithm for remaining useful life estimation. Mech. Syst. Sig. Process. **35**(1–2), 219–237 (2013)

17. Tobon-Mejia, D.A., Medjaher, K., Zerhouni, N., Tripot, G.: A data-driven failure prognostics method based on mixture of Gaussian hidden Markov models. IEEE Trans. Reliab. **61**(2), 491–503 (2012)
18. Wang, T.: Trajectory similarity based prediction for remaining useful life estimation. Ph.D. thesis, University of Cincinnati (2010)
19. Wang, T., Yu, J., Siegel, D., Lee, J.: A similarity-based prognostics approach for remaining useful life estimation of engineered systems. In: 2008 International Conference on Prognostics and Health Management, pp. 1–6. IEEE, October 2008
20. Yildirim, M., Sun, X.A., Gebraeel, N.Z.: Sensor-driven condition-based generator maintenance scheduling - part I: maintenance problem. IEEE Trans. Power Syst. **31**(6), 4253–4262 (2016)
21. You, M.Y.: A predictive maintenance system for hybrid degradation processes. Int. J. Qual. Reliab. Manag. **34**(7), 1123–1135 (2017)

Orometric Methods in Bounded Metric Data

Maximilian Stubbemann[1,2]([✉])([ID]), Tom Hanika[2]([ID]), and Gerd Stumme[1,2]([ID])

[1] L3S Research Center, Leibniz University of Hannover, Hannover, Germany
{stubbemann,stumme}@l3s.de
[2] Knowledge and Data Engineering Group, University of Kassel, Kassel, Germany
{stubbemann,hanika,stumme}@cs.uni-kassel.de

Abstract. A large amount of data accommodated in knowledge graphs (KG) is metric. For example, the Wikidata KG contains a plenitude of metric facts about geographic entities like cities or celestial objects. In this paper, we propose a novel approach that transfers orometric (topographic) measures to bounded metric spaces. While these methods were originally designed to identify relevant mountain peaks on the surface of the earth, we demonstrate a notion to use them for metric data sets in general. Notably, metric sets of items enclosed in knowledge graphs. Based on this we present a method for identifying outstanding items using the transferred valuations functions isolation and prominence. Building up on this we imagine an item recommendation process. To demonstrate the relevance of the valuations for such processes, we evaluate the usefulness of isolation and prominence empirically in a machine learning setting. In particular, we find structurally relevant items in the geographic population distributions of Germany and France.

Keywords: Metric spaces · Orometry · Knowledge graphs · Classification

1 Introduction

Knowledge graphs (KG), such as DBpedia [15] or Wikidata [24], are the state of the art for storing information and to draw knowledge from. They represent knowledge through graphs and consist essentially of *items* which are related through *properties* and *values*. This enables them to fulfill the task of giving exact answers to exact questions. However, their ability to present a concise overview over collections of items with metric distances is limited. The number of such data sets in Wikidata is tremendous, e.g., the set of all cities of the world, including their geographic coordinates. Further examples are celestial bodies and their trajectories or, more general, feature spaces of data mining tasks.

One approach to understand such metric data is to identify outstanding elements, i.e., outstanding items. Based on such elements it is possible to compose or enhance item recommendations to users. For example, such recommendations could provide a set of the most relevant cities in the world with respect

M. R. Berthold et al. (Eds.): IDA 2020, LNCS 12080, pp. 496–508, 2020.
https://doi.org/10.1007/978-3-030-44584-3_39

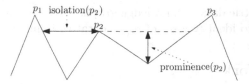

Fig. 1. Isolation: minimal horizontal distance to another point of at least equal height. Prominence: minimal vertical descent to reach a point of at least equal height.

to being outstanding in their local surroundings. However, it is a challenging task to identify outstanding items in metric data sets. In cases where the metric space is equipped with an additional valuation function, this task becomes more feasible. Such functions, often called *scores* or *height* functions, are often naturally provided: cities may be ranked by population; the importance of scientific authors by the *h*-index [12]. A naïve approach for recommending relevant items in such settings would be: items with higher scores are more relevant items. As this method seems reasonable for many applications, some obstacles arise if the "highest" items concentrate into a specific region of the underlying metric space. For example, representing the cities of the world by the twenty most populated ones would include no western European city.[1] Recommending the 100 highest mountains would not lead to knowledge about the mountains outside of Asia.[2]

Our novel approach shall overcome this problem: we combine the valuation measure (e.g., "height") and distances, to provide new valuation functions on the set of items, called *prominence* and *isolation*. These functions do rate items based on their height in relation to the valuations of the surrounding items. This results in valuation functions on the set of items that reflect the extend to which an item is locally outstanding. The basic idea is the following: the prominence values an item based on the minimal descent (w.r.t. the height function) that is needed to get to another point of at least same height. The isolation, sometimes also called *dominance radius*, values the distance to the next higher point w.r.t. the metric (Fig. 1). These measures are adapted from the field of topography where isolation and prominence are used in order to identify outstanding mountain peaks. We base our approach on [22], where the authors proposed prominence and dominance for networks. We generalize these to the realm of bounded metric space.

We provide insights to the novel valuation functions and demonstrate their ability to identify relevant items for a given topic in metric knowledge graph applications. The contributions of this paper are as follows: • We propose prominence and isolation for bounded metric spaces. For this we generalize the results in [22] and overcome the limitations to finite, undirected graphs. • We demonstrate an artificial machine learning task for evaluating our novel valuation functions in metric data. • We introduce an approach for using prominence and iso-

---

[1] https://en.wikipedia.org/wiki/List_of_largest_cities on 2019-06-16.
[2] https://en.wikipedia.org/wiki/List_of_highest_mountains_on_Earth on 2019-06-16.

lation to enrich metric data in knowledge graphs. We show empirically that this information helps to identify a set of representative items.

2   Related Work

Item recommendations for knowledge graphs is a contemporary topic of high interest in research. Investigations cover for example music recommendation using content and collaborative information [17] or movie recommendations using PageRank like methods [5]. The former is based on the common notion of embedding, i.e., embedding of the graph structure into d-dimensional real vector spaces. The latter operates on the relational structure itself. Our approach differs from those as it is based on combining a valuation measure with the metric of the data space. Nonetheless, given an embedding into an finite dimensional real vector space, one could apply isolation and prominence in those as well.

The novel valuation functions prominence and isolation are inspired by topographic measures, which have their origin in the classification of mountain peaks. The idea of ranking peaks solely by their absolute height was already deprecated in 1978 by Fry in his work [8]. The author introduced prominence for geographic mountains, a function still investigated in this realm, e.g., in Torres et al. [23], where the authors used deep learning methods to identify prominent mountain peaks. Another recent step for this was made in [14], where the authors investigated methods for discovering new ultra-prominent mountains. Isolation and more valuations functions motivated in the orometric realm are collected in [11]. A well-known procedure for identifying peaks and saddles in 3D terrain data is described in [6]. However, these approaches rely on data that approximates a continuous terrain surface via a regular square grid or a triangulation. Our data cannot fulfill this requirement. Recently the idea of transferring orometric functions to different realms of research gained attention: The authors of [16] used topographic prominence to identify population areas in several U.S. States. In [22] the authors Schmidt and Stumme transferred prominence and dominance, i.e., isolation, to co-author graphs in order to evaluate their potential of identifying ACM Fellows. We build on this for proposing our valuation functions on bounded metric data. This generalization results in a wide range of applications.

3   Mathematical Modeling

While the Wikidata knowledge graph itself could be analyzed with the prominence and isolation measures for networks, this paper focuses on bounded metric data sets. To analyze such data sets is more sufficient, since real world networks often suffer from a small average shortest path length [26]. This leads to a low amount of outstanding items: an item is outstanding if it is "higher" than the items that have a low distance to it. This leads to a strict measure for many real-world network data when the shortest path length is used as the metric function. Hence, we model our functions for bounded metric data instead of networks.

We consider the following scenario: We have a data set M, consisting of a set of items, in the following called *points*, equipped with a metric d and a valuation function h, in the following called *height function*. The goal of the orometric (topographic) measures prominence and isolation is, to provide measures that reflect the extent to which a point is locally outstanding in its neighborhood.

More precisely, let M be a non-empty set and $d : M \times M \to \mathbb{R}_{\geq 0}$. We call d a *metric* on the set M iff • $\forall x, y \in M : d(x,y) = 0 \iff x = y$, and • $d(x,y) = d(y,x)$ for all $x, y \in M$, called symmetry, and • $\forall x, y, z \in M :$ $d(x, z) \leq d(x, y) + d(y, z)$, called triangle inequality. If d is a metric on M, we call (M, d) a *metric space* and if M is finite we call (M, d) a *finite metric space*. If there exists a $C \in \mathbb{R}_{\geq 0}$ such that we have $d(m, n) \leq C$ for all $m, n \in M$, we call (M, d) *bounded*. For the rest of our work we assume that $|M| > 1$ and (M, d) is a bounded metric space. Additionally, we have that M is equipped with a height function (valuation/score function) $h : M \to \mathbb{R}_{\geq 0}, m \mapsto h(m)$.

Definition 1 (Isolation). *Let (M, d) be a bounded metric space and let $h :$ $M \to \mathbb{R}_{\geq 0}$ be a height function on M. The* isolation *of a point $x \in M$ is then defined as follows:*

- *If there is no point with at least equal height to m, than* $\mathrm{iso}(m) :=$ $\sup\{d(m, n) \mid n \in M\}$. *The boundedness of M guarantees the existence of this supremum.*
- *If there is at least one other point in M with at least equal height to m, we define its isolation by:*

$$\mathrm{iso}(m) := \inf\{d(m, n) \mid n \in M \setminus \{m\} \wedge h(n) \geq h(m)\}.$$

The isolation of a mountain peek is often called the *dominance radius* or sometimes the *dominance*. Since the term *orometric dominance* of a mountain sometimes refers to the quotient of prominence and height, we will stick to the term *isolation* to avoid confusion. While the isolation can be defined within the given setup, we have to equip our metric space with some more structure in order to transfer the notion of prominence. Informally, the prominence of a point is given by the minimal vertical distance one has to descend to get to a point of at least the same height. To adapt this measure to our given setup in metric spaces with a height function, we have to define what a path is. Structures that provide paths in a natural way are graph structures. For a given graph $G = (V, E)$ with vertex set V and edge set $E \subseteq \binom{V}{2}$, *walks* are defined as sequences of nodes $\{v_i\}_{i=0}^{n}$ which satisfy $\{v_{i-1}, v_i\} \in E$ for all $i \in \{1, ..., n\}$. If we also have $v_i \neq v_j$ for $i \neq j$, we call such a sequence a *path*. For $v, w \in V$ we say v and w are *connected* iff there exists a path connecting them. Furthermore, we denote by $G(v)$ the *connected component* of G containing v, i.e., $G(v) := \{w \in V \mid v \text{ is connected with } w\}$.

To use the prominence measure as introduced by Schmidt and Stumme in [22], which is indeed defined on graphs, we have to derive an appropriate graph structure from our metric space. The topic of graphs embedded in finite dimensional vector spaces, so called spatial networks [2], is a topic of current interest. These networks appear in real world scenarios frequently, for example in the modeling of urban street networks [13]. Note that our setting, in contrast to the afore mentioned, is not based on a priori given graph structure. In our scenario the graph structure must be derived from the structure of the given metric space.

Our approach is, to construct a *step size graph* or *threshold graph*, where we consider points in the metric space as nodes and connect two points through an edge, iff their distance is smaller then a given threshold δ.

Definition 2 (δ-Step Graph). *Let (M, d) be a metric space and $\delta > 0$. We define the δ-step graph or δ-threshold graph, denoted by G_δ, as the tuple (M, E_δ) via*

$$E_\delta := \{\{m, n\} \in \binom{M}{2} \mid d(m, n) \leq \delta\}. \tag{1}$$

This approach is similar to the one found in the realm of random geometric graphs, where it is common sense to define random graphs by placing points uniformly in the plane and connect them via edges if their distance is less than a given threshold [21]. Since we introduced a possibility to derive a graph that just depends on the metric space, we use a slight modification of the definition of prominence compared to [22] for networks.

Definition 3 (Prominence in Networks). *Let $G = (V, E)$ be a graph and let $h : V \rightarrow \mathbb{R}_{\geq 0}$ be a height function. The prominence $\text{prom}_G(v)$ of $v \in V$ is defined by*

$$\text{prom}_G(v) := \min\{h(v), \text{mindesc}_G(v)\} \tag{2}$$

where $\text{mindesc}_G(v) := \inf\{\max\{h(v) - h(u) \mid u \in p\} \mid p \in P_v\}$. The set P_v contains of all paths to vertices w with $h(w) \geq h(v)$, i.e., $P_v := \{\{v_i\}_{i=0}^n \in P \mid v_0 = v \wedge v_n \neq v \wedge h(v_n) \geq h(v)\}$, where P denotes the set of all paths of G.

Informally, $\text{mindesc}_G(v)$ reflects on the minimal descent in order to get to a vertex in G which has a height of at least $h(v)$. For this the definition makes use of the fact that $\inf \emptyset = \infty$. This case results in $\text{prom}_G(v)$ being the height of v. A distinction to the definition in [22] is, that we now consider all paths and not just shortest paths. This change better reflects the calculation of the prominence for mountains. Based on this we transfer the notions above to metric spaces.

Definition 4 (δ-Prominence). *Let (M, d) be a bounded metric space and $h : M \rightarrow \mathbb{R}_{\geq 0}$ be a height function. We define the δ-prominence $\text{prom}_\delta(m)$ of $m \in M$ as $\text{prom}_{G_\delta}(v)$, i.e., the prominence of m in G_δ from Definition 2.*

We now have a prominence term for all metric spaces that depends on a parameter δ to choose. For all knowledge procedures, choosing such a parameter is a demanding task. Hence, we want to provide in the following a natural choice for δ. We consider only those values for δ such that corresponding G_δ does not exhibit noise, i.e., there is no element without a neighbor.

Definition 5 (Minimal Threshold). *For a bounded metric space (M, d) with $|M| > 1$ we define the* minimal threshold δ_M *of M as*

$$\delta_M := \sup\{\inf\{d(m, n) \mid n \in M \setminus \{m\}\} \mid m \in M\}.$$

Based on this definition a natural notion of prominence for metric spaces (equipped with a height function) emerges via a limit process.

Lemma 1. *Let M be a bounded metric space and δ_M as in Definition 5. For $m \in M$ the following descending limit exists:*

$$\lim_{\delta \searrow \delta_M} \text{prom}_\delta(m). \tag{3}$$

Proof. Fix any $\hat{\delta} > \delta_M$ and consider on the open interval from δ_M to $\hat{\delta}$ the function that maps δ to $\text{prom}_\delta(m)$: $\text{prom}_{(.)}(m) :]\delta_M, \hat{\delta}[\to \mathbb{R}, \delta \mapsto \text{prom}_\delta(m)$. It is known that it is sufficient to show that $\text{prom}_{(.)}(m)$ is monotone decreasing and bounded from above. Since we have for any δ that $\text{prom}_\delta(m) \leq h(m)$ holds, we need to show the monotony. Let δ_1, δ_2 be in $]\delta_M, \hat{\delta}[$ with $\delta_1 \leq \delta_2$. If we consider the corresponding graphs (M, E_{δ_1}) and (M, E_{δ_2}), it easy to see $E_{\delta_1} \subseteq E_{\delta_2}$. Hence, we have to consider more paths in Eq. (2) for E_{δ_2}, resulting in a not larger value for the infimum. We obtain $\text{prom}_{\delta_1}(m) \geq \text{prom}_{\delta_2}(m)$, as required.

Definition 6 (Prominence in Metric Spaces). *If M is a bounded metric space with $|M| > 1$ and a height function h, the prominence $\text{prom}(m)$ of m is defined as:*

$$\text{prom}(m) := \lim_{\delta \searrow \delta_M} \text{prom}_\delta(m).$$

Note, if we want to compute prominence on a real world finite metric data set, it is possible to directly compute the prominence values: in that case the supremum in Definition 5 can be replaced by a maximum and the infimum by a minimum, which leads to $\text{prom}(m)$ being equal to $\text{prom}_{\delta_M}(m)$. There are results for efficiently creating such step graphs [3]. However, for our needs in this work, in particular in the experiment section, a quadratic brute force approach for generating all edges is sufficient. We want to show that our prominence definition for bounded metric spaces is a natural generalization of Definition 3.

Lemma 2. *Let $G = (V, E)$ be a finite, connected graph with $|V| \geq 2$. Consider V equipped with the shortest path metric as a metric space. Then the prominence $\text{prom}_G(\cdot)$ from Definition 3 and $\text{prom}(\cdot)$ from Definition 6 coincide.*

Proof. Let $M := V$ be equipped with the shortest path metric d on G. As G is connected and has more than one node, we have $\delta_M = 1$. Hence, (M, E_{δ_M}) from Definition 2 and G are equal. Therefore, the prominence terms coincide.

4   Application

Score Based Item Recommending. As an application we envisage a general approach for a score based item recommending process. The task of item recommending with knowledge graphs is a current research topic [17,18]. However,
most approaches are solely based on knowledge about preferences of the user
and graph structural properties, often accessed through KG embeddings [19].
The idea of the recommendation process we imagine differs from those. We stip-
ulate on a procedure that is based on the information entailed in the connection
of the metric aspects of the data together with some (often naturally present)
height function. We are aware that this limits our approach to metric data in
KGs. Nonetheless, given the large amounts of metric item sets in prominent KGs,
we claim the existence of a plenitude of applications. For example, while considering sets of cities, such a system could recommend a *relevant* subset, based on
a height function, like population, and a metric, like geographical distances. By
doing so, we introduce a source of information for recommending metric data in
relational structures, like KGs. A common approach for analyzing and learning
in KGs is embedding. There is an extensive amount of research about that, see
for example [4,25]. Since our novel methods rely solely on bounded metric spaces
and some valuation function, one may apply those after the embedding step as
well. In particular, one may use isolation and prominence for investigating or
completing KG embeddings. This constitutes our second envisioned application.
Finally, common item recommending scores/ranks can also be used as height
functions in our sense. Hence, computing prominence and isolation for already
setup recommendation systems is another possibility. Here, our valuation functions have the potential to enrich the recommendation process with additional
information. In such a way our measures can provide a novel additional aspect to
existing approaches. The realization and evaluation of our proposed recommendation approach is out of scope of this paper. Nonetheless, we want to provide
some first insights for the applicability of valuation functions for item sets based
on empirical experiments. As a first experiment, we will evaluate if isolation and
prominence help to separate important and unimportant items in specific item
sets in Wikidata. In detail, we evaluate if the valuation functions help to differentiate important and unimportant municipalities in France and Germany, solely
based on their geographic metric properties and their population as height.

4.1   Resulting Questions

Given a bounded metric space M which represents the data set and a given
height h. The following questions shall evaluate if our functions isolation and
prominence provide useful information about the relevance of given points in the
metric space. If (M, d, h) is a metric space equipped with an additional height
function, let $c : M \rightarrow \{0, 1\}$ be a binary function that classifies the points in the
data set as relevant (1) or not (0). We connect this to our running example using
a function that classifies municipalities having a university (1) and municipalities
that do not have an university (0). We admit that the underlying classification

is not meaningful in itself. It treats a real geographic case while our model could also handle more abstract scenarios. However, since this setup is essentially a benchmark framework (in which we assume cities with universities to be more relevant) we refrain from employing a more meaningful classification task in favor of a controllable classification scenario. Our research questions are now: **1. Are prominence and isolation alone characteristical for relevance?** We use isolation and/or prominence for a given set of data points as features. To which extend do these features improve learning a classification function for relevance? **2. Do prominence and isolation provide additional information, not catered by the absolute height?** Do prominence and isolation improve the prediction performance of relevance compared to just using the height? Does a classifier that uses prominence and isolation as additional features produce better results than a classifier that just uses the height? We will evaluate the proposed setup in the realm of a KG and take on the questions stated above in the following section and present some experimental evidence.

5   Experiments

We extract information about municipalities in the countries of Germany and France from the Wikidata KG. This KG is a structure that stores knowledge via *statements*, linking *entities* via *properties* to *values*. A detailed description can be found in [24], while [9] gives an explicit mathematical structure to the Wikidata graph and shows how to use the graph for extracting implicational knowledge from Wikidata subsets. We investigate if prominence and isolation of a given municipality can be used as features to predict university locations in a classification setup. We use the query service of Wikidata[3] to extract points in the country maps from Germany and France and to extract all their universities. We report all necessary SPAQRL queries employed on GitHub.[4]

- Wikidata provides different relations for extracting items that are instances of the notion city. The obvious choice is to employ the *instance of* (P31) property for the item *city* (Q515). Using this, including *subclass of* (P279), we find insufficient results. More specific, we find only 102 French cities and 2215 German cities.[5] For Germany, there exists a more commonly used item *urban municipality of Germany* (Q42744322) for extracting all cities, while to the best of our knowledge, a counterpart for France is not provided.
- The preliminary investigation leads us to use *municipality* (Q15284), again including the *subclass of* (P279) property, with more than 5000 inhabitants.
- Since there are multiple french municipalities that are not located in the mainland of France, we encounter problems for constructing the metric space. To cope with that we draw a basic approximating square around the mainland of France and consider only those municipalities inside.

---

[3] https://query.wikidata.org/.
[4] https://github.com/mstubbemann/Orometric-Methods-in-Bounded-Metric-Data.
[5] Queried on 2019-08-07.

- We find the class of every municipality, i.e, university location or non-university location as follows. We use the properties *located in the administrative territorial entity* (P131) and *headquarters location* (P159) on the set of all universities and checked if these are set in Germany or France. An example of a University that has not set P131 is *TU Dortmund* (Q685557).[6]
- We match the municipalities with the university properties. This is necessary because some universities are not related to municipalities through P131, e.g., *Hochschule Niederrhein* (Q1318081) is located in the administrative location *North Rhine-Westphalie* (Q1198) (See footnote 6), which is a federal state containing multiple municipalities. For these cases we check the university locations manually. This results in 2064 municipalities (89 university loc.) in France and 2986 municipalities (160 university loc.) in Germany.
- While constructing the data set we encounter twenty-two universities that are associated to a country having neither *located in the administrative territorial entity* (P131) nor *headquarters location* (P159). We check them manually and are able to discard them all for different reasons.

5.1   Binary Classification Task

Setup. We compute prominence and isolation for all data points and normalize them as well as the height. The data that is used for the classification task consists of the following information for each city: The height, the prominence, the isolation and the binary information whether the city has a university. Since our data set is highly imbalanced, common classifiers tend to simply predict the majority class. To overcome the imbalance, we use inverse penalty weights with respect to the class distribution. We want to stress out again that the goal for the to be introduced classification task is not to identify the best classifier. Rather we want to produce evidence for the applicability of employing isolation and prominence as features for learning a classification function. We decide to use logistic regression with L^2 regularization and Support Vector Machines [7] with a radial kernel. For our experiment we use Scikit-Learn [20]. As penalty factor for the SVC we set $C = 1$, and experiment with $C \in \{0.5, 1, 2, 5, 10, 100\}$. For γ we rely on previous work by [1] and set it to one. For all combinations of population, isolation and prominence we use 100 iterations of 5-fold-cross-validation.

Evaluation. We use the g-mean (i.e., geometric mean) as evaluation function. Consider for this denotations TN (True Negative), FP (False Positive), FN (False Negative), and TP (True Positive). Overall accuracy is highly misleading for heavily imbalanced data. Therefore, we evaluate the classification decisions by using the geometric mean of the accuracy on the positive instances, $acc_+ := \frac{TP}{TP+FN}$ and the accuracy on the negative instances $acc_- := \frac{TN}{TN+FP}$. Hence, the g-mean score is then defined by the formula $g_{mean} := \sqrt{acc_+ \cdot acc_-}$. The evaluation function g-mean is established in the topic of imbalanced data mining. It is mentioned in [10] and used for evaluation in [1]. We compare the values for

---

[6] Last checked on 2019-10-26.

Table 1. Results of the classification task. We do 100 rounds of 5-fold-cross-validation and shuffle the data between the rounds. For all rounds we compute the g-mean value and then compute the average over the 100 rounds.

Country	France				Germany			
Classifier	SVM		LR		SVM		LR	
	Mean	Std	Mean	Std	Mean	Std	Mean	Std
iso	0.7416	0.0059	0.7703	0.0034	0.7463	0.0028	0.7761	0.0035
pro	0.4861	0.0053	0.6362	0.0055	0.3998	0.0068	0.5750	0.0049
pop	0.6940	0.0031	0.7593	0.0086	0.5982	0.0038	0.7134	0.0043
iso+pro	0.7329	0.0067	0.7657	0.0066	0.7320	0.0042	0.7642	0.0041
iso+pop	**0.7668**	0.0086	**0.7812**	0.0039	**0.7971**	0.0041	**0.8068**	0.0038
pro+pop	0.7011	0.0040	0.7496	0.0051	0.6134	0.0050	0.7108	0.0065
iso+pro+pop	0.7653	0.0078	0.7778	0.0052	0.7947	0.0042	0.8006	0.0042

po = population, pr = prominence, is = isolation
SVM = Support Vector Machine, LR = Logistic Regression

g-mean for the following cases. First, we train a classifier function purely on the features population, prominence or isolation. Secondly, we try combinations of them for the training process. We consider the classifier trained using the population feature as baseline. An increase in g-mean while using prominence or isolation together with the population function is evidence for the utility of the introduced valuation functions. Even stronger evidence is a comparison of isolation/prominence trained classifiers versus baseline.

In our experiments, we are not expecting high g-mean values, since the placement of university locations depends on many additional features, including historical evolution of the country and political decisions. Still, the described evaluation setup is sufficient to demonstrate the potential of the novel features.

Results. The results of the computations are depicted in Table 1. • *Isolation is a good indicator for structural relevance.* For both countries and classifiers isolation outperforms population. • *Combining absolute height with our valuation functions leads to better results.* • *Prominence is not useful as a solo indicator.* We draw from our result that prominence solely is not a useful indicator. Prominence is a very strict valuation function: recall that we constructed the graphs by using distance margins as indicators for edges, leading to a dense graph structure in more dense parts of the metric space. Hence, a point in a more dense part has many neighbors and thus many potential paths that may lead to a very low prominence value. From Definition 3 we see that having a higher neighbor always leads to a prominence value of zero. This threshold is about 34 km for Germany and 54 km for France. Thus, a municipality has a not vanishing prominence if it is the most populated point in a radius of over 34 km, respectively 54 km. Only 75 municipalities of France have non zero prominence, with 40 of them being university locations. Germany has 104 municipalities with positive prominence with 72 of them being university locations. Thus, prominence alone as a feature is insufficient for the prediction of university locations. • *Support vector machine and logistic regression lead to similar results.* To the question, whether our valuation functions improve the classification compared with the population feature, support vector machines and logistic regressions provide the same answer: isolation always outperforms population, a combination of all features is always better then using just the plain population feature. • *Support vector machine penalty parameter.* Finally, for our last test we check the different results for support vector machines using the penalty parameters $C \in \{0.5, 1, 2, 5, 10, 100\}$. We observe that increasing the penalty results in better performance using the population feature. However, for lower values of C, i.e., less overfitting models, we see better performance in using the isolation feature. In short, the more the model overfits due to C, the less useful are the novel valuation functions we introduced in this paper.

6   Conclusion and Outlook

In this work, we presented a novel approach to identify outstanding elements in item sets. For this we employed orometric valuation functions, namely prominence and isolation. We investigated a computationally reasonable transfer to the realm of bounded metric spaces. In particular, we generalized previously known results that were researched in the field of finite networks.

The theoretical work was motivated by the observation that KGs, like Wikidata, do contain huge amounts of metric data. These are often equipped with some kind of height functions in a natural way. Based on this we proposed in this work the groundwork for a locally working item recommending scheme.

To evaluate the capabilities for identifying locally outstanding items we selected an artificial classification task. We identified all French and German municipalities from Wikidata and evaluated if a classifier can learn a meaningful connection between our valuation functions and the relevance of a municipality. To gain a binary classification task and to have a benchmark, we assumed that universities are primarily located at relevant municipalities. In consequence, we evaluated if a classifier can use prominence and isolation as features to predict university locations. Our results showed that isolation and prominence are indeed helpful for identifying relevant items.

For future work we propose to develop the conceptualized item recommender system and to investigate its practical usability in an empirical user study. Furthermore, we urge to research the transferability of other orometric based valuation functions.

Acknowledgement. The authors would like to express thanks to Dominik Dürrschnabel for fruitful discussions. This work was funded by the German Federal Ministry of Education and Research (BMBF) in its program "Quantitative Wissenschaftsforschung" as part of the REGIO project under grant 01PU17012.

References

1. Akbani, R., Kwek, S., Japkowicz, N.: Applying support vector machines to imbalanced datasets. In: Boulicaut, J.-F., Esposito, F., Giannotti, F., Pedreschi, D. (eds.) ECML 2004. LNCS (LNAI), vol. 3201, pp. 39–50. Springer, Heidelberg (2004). https://doi.org/10.1007/978-3-540-30115-8_7
2. Barthélemy, M.: Spatial networks. Phys. Rep. **499**(1), 1–101 (2011)
3. Bentley, J.L.: A survey of techniques for fixed radius near neighbor searching. Technical report, SLAC, SCIDOC, Stanford, CA, USA (1975). SLAC-R-0186, SLAC-0186
4. Bordes, A., Weston, J., Collobert, R., Bengio, Y.: Learning structured embeddings of knowledge bases. In: Burgard, W., Roth, D. (eds.) Proceedings of the 25th Conference on Artificial Intelligence, pp. 301–306. AAAI Press, Palo Alto (2011)
5. Catherine, R., Cohen, W.: Personalized recommendations using knowledge graphs: a probabilistic logic programming approach. In: Proceedings of the 10th ACM Conference on Recommender Systems, RecSys, pp. 325–332. ACM, New York (2016)
6. Čomić, L., De Floriani, L., Papaleo, L.: Morse-smale decompositions for modeling terrain knowledge. In: Cohn, A.G., Mark, D.M. (eds.) COSIT 2005. LNCS, vol. 3693, pp. 426–444. Springer, Heidelberg (2005). https://doi.org/10.1007/11556114_27
7. Cortes, C., Vapnik, V.: Support-vector networks. Mach. Learn. **20**(3), 273–297 (1995)
8. Fry, S.: Defining and sizing-up mountains. Summit, pp. 16–21, January-February 1987
9. Hanika, T., Marx, M., Stumme, G.: Discovering implicational knowledge in wikidata. In: Cristea, D., Le Ber, F., Sertkaya, B. (eds.) ICFCA 2019. LNCS (LNAI), vol. 11511, pp. 315–323. Springer, Cham (2019). https://doi.org/10.1007/978-3-030-21462-3_21
10. He, H., Garcia, E.A.: Learning from imbalanced data. IEEE Trans. Knowl. Data Eng. **21**(9), 1263–1284 (2009)
11. Helman, A.: The Finest Peaks-Prominence and Other Mountain Measures. Trafford, Victoria (2005)
12. Hirsch, J.E.: An index to quantify an individual's scientific research output. Proc. Nat. Acad. Sci. **102**(46), 16569–16572 (2005)
13. Jiang, B., Claramunt, C.: Topological analysis of urban street networks. Environ. Plan. B: Plan. Des. **31**(1), 151–162 (2004)
14. Kirmse, A., de Ferranti, J.: Calculating the prominence and isolation of every mountain in the world. Prog. Phys. Geogr.: Earth Environ. **41**(6), 788–802 (2017)
15. Lehmann, J., et al.: DBpedia - a large-scale, multilingual knowledge base extracted from wikipedia. Semant. Web **6**(2), 167–195 (2015)
16. Nelson, G.D., McKeon, R.: Peaks of people: using topographic prominence as a method for determining the ranked significance of population centers. Prof. Geogr. **71**(2), 342–354 (2019)
17. Oramas, S., Ostuni, V.C., Noia, T.D., Serra, X., Sciascio, E.D.: Sound and music recommendation with knowledge graphs. ACM Trans. Intell. Syst. Technol. **8**(2), 21:1–21:21 (2016)
18. Palumbo, E., Rizzo, G., Troncy, R.: Entity2rec: learning user-item relatedness from knowledge graphs for top-n item recommendation. In: Proceedings of the Eleventh ACM Conference on Recommender Systems, pp. 32–36. ACM (2017)

19. Palumbo, E., Rizzo, G., Troncy, R., Baralis, E., Osella, M., Ferro, E.: Knowledge graph embeddings with node2vec for item recommendation. In: Gangemi, A., et al. (eds.) ESWC 2018. LNCS, vol. 11155, pp. 117–120. Springer, Cham (2018). https://doi.org/10.1007/978-3-319-98192-5_22

20. Pedregosa, F., et al.: Scikit-learn: machine learning in Python. JMLR **12**, 2825–2830 (2011)

21. Penrose, M.: Random Geometric Graphs. Oxford Studies in Probability, vol. 5. Oxford University Press, Oxford (2003)

22. Schmidt, A., Stumme, G.: Prominence and dominance in networks. In: Faron Zucker, C., Ghidini, C., Napoli, A., Toussaint, Y. (eds.) EKAW 2018. LNCS (LNAI), vol. 11313, pp. 370–385. Springer, Cham (2018). https://doi.org/10.1007/978-3-030-03667-6_24

23. Torres, R.N., Fraternali, P., Milani, F., Frajberg, D.: A deep learning model for identifying mountain summits in digital elevation model data. In: First IEEE International Conference on Artificial Intelligence and Knowledge Engineering, AIKE 2018, Laguna Hills, CA, USA, 26–28 September 2018, pp. 212–217. IEEE Computer Society (2018)

24. Vrandečić, D., Krötzsch, M.: Wikidata: a free collaborative knowledge base. Commun. ACM **57**, 78–85 (2014)

25. Wang, Z., Zhang, J., Feng, J., Chen, Z.: Knowledge graph embedding by translating on hyperplanes. In: Brodley, C.E., Stone, P. (eds.) Proceedings of the 28th Conference on Artificial Intelligence, pp. 1112–1119. AAAI Press (2014)

26. Watts, D.J.: Six Degrees: The Science of a Connected Age. W. W. Norton, New York (2003)

Interpretable Neuron Structuring
with Graph Spectral Regularization

Alexander Tong[1], David van Dijk[2], Jay S. Stanley III[2], Matthew Amodio[1],
Kristina Yim[2], Rebecca Muhle[2], James Noonan[2], Guy Wolf[3],
and Smita Krishnaswamy[1,2(✉)]

[1] Yale Department of Computer Science, New Haven, USA
smita.krishnaswamy@yale.edu
[2] Yale Department of Genetics, New Haven, USA
[3] Department of Mathematics and Statistics,
Université de Montréal, Mila, Montreal, Canada

Abstract. While neural networks are powerful approximators used to classify or embed data into lower dimensional spaces, they are often regarded as black boxes with uninterpretable features. Here we propose *Graph Spectral Regularization* for making hidden layers more interpretable without significantly impacting performance on the primary task. Taking inspiration from spatial organization and localization of neuron activations in biological networks, we use a graph Laplacian penalty to structure the activations within a layer. This penalty encourages activations to be smooth either on a predetermined graph or on a feature-space graph learned from the data via co-activations of a hidden layer of the neural network. We show numerous uses for this additional structure including cluster indication and visualization in biological and image data sets.

Keywords: Neural Network Interpretability · Graph learning · Feature saliency

1 Introduction

Common intuitions and motivating explanations for the success of deep learning approaches rely on analogies between artificial and biological neural networks, and the mechanism they use for processing information. However, one aspect that is overlooked is the spatial organization of neurons in the brain. Indeed, the hierarchical spatial organization of neurons, determined via fMRI and other technologies [13,16], is often leveraged in neuroscience works to explore, understand, and interpret various neural processing mechanisms and high-level brain functions. In artificial neural networks (ANN), on the other hand, hidden layers offer no organization that can be regarded as equivalent to the biological one. This lack of organization poses great difficulties in exploring and interpreting

---

A. Tong, D. Dijk, G. Wolf and S. Krishnaswamy—Equal contribution.

© The Author(s) 2020
M. R. Berthold et al. (Eds.): IDA 2020, LNCS 12080, pp. 509–521, 2020.
https://doi.org/10.1007/978-3-030-44584-3_40

the internal data representations provided by hidden layers of ANNs and the information encoded by them. This challenge, in turn, gives rise to the common treatment of ANNs as black boxes whose operation and data processing mechanisms cannot be easily understood. To address this issue, we focus on the problem of modifying ANNs to learn more interpretable feature spaces without degrading their primary task performance.

While most neural networks are treated as black boxes, we note that there are methods in ANN literature for understanding the activations of filters in convolutional neural networks (CNNs) [11], either by examining trained networks [24], or by learning a better representation [12,17,18,22,25], but such methods rarely apply to other types of networks, in particular dense neural networks (DNNs) where a single activation is often not interpretable on its own. Furthermore, convolutions only apply to datatypes where we know the feature structure apriori, as in the case of images and natural language. In layers of a DNN, there is no enforced structure between neurons. The correspondence between neurons and concepts is only determined based on the random initialization of the network. In this work, we encourage *structure between neurons* in the same layer, creating more localized and interpretable layers in dense architectures.

More specifically we propose a *Graph Spectral Regularization* to encourage arbitrary graph structure between neurons within a layer. The internal layers of a neural network are constrained to take the structure of a graph, with graph neighbors activating on similar inputs. This allows us to map the activations of a given layer over the graph and interpret new input by examining the activations. We show that graph-structuring a hidden layer causes useful, interpretable features to emerge. For instance, we show that grid-structuring a layer of a classification network creates a structure over which convolution can be applied, and local receptive fields can be traced to understand classification decisions.

While a majority of the time imposing a known graph structure gives interpretable results, there are circumstances where we would like to learn the graph structure from data. In such cases we can learn and emphasize the natural graph structure of the feature space. We do this by an iterative process of encoding the data, and modifying the graph based on the feature co-activation patterns. This procedure reinforces existing patterns in the data. This allows us to learn an abstracted graph structure of features in high-dimensional domains such as single-cell RNA sequencing.

The main contributions of this work are as follows: (1) Demonstration of hierarchical, spatial, and smoothed feature maps for interpretability in dense networks. (2) A novel method for learning and reinforcing the natural graph structure for complex feature spaces. (3) Demonstration of graph learning and abstraction on single-cell RNA-sequencing data.

2   Related Work

Disentangled Representation Learning: While there is no precise definition of what makes for a disentangled representation, the aim is to learn a representation that axis aligns with the generative factors of the data [2,8]. [9] suggest a way to disentangle the representation of variational autoencoders [10] with β-VAE. Subsequent work has generalized this to discrete representations [5], and simple hierarchical representations [6]. These works focus on learning a single vector representation of the data, where each element represents a single concept. In contrast, our work learns a representation where groups of neurons may be involved in representing a single concept. Moreover, disentangled representation learning can only be applied to unsupervised models and only the most compressed level of either an autoencoder [9] or generative adversarial network as in [4], whereas graph spectral regularization (GSR) can be applied to any or all layers of the network.

Graph Structure in ANNs: Graph based penalties have been used in the graph signal processing literature [3,21,26], but are rarely used in an ANN setting. In the biological data setting, [14] used a graph penalty in sparse logistic regression on gene expression data. Another way of utilizing graph structure is through graph convolutional networks (GCN). GCNs are a related body of work introduced by [7], and expanded on by [19], but focus on a different set of problems (For an overview see [23]). GCNs require a known graph structure. We focus on learning a graph representation of general data. This learned graph representation could be used as the input to a GCN similar to our MNIST example.

3 Enforcing Graph Structure

We consider the intra-layer relationships between neurons or larger structures such as capsules. For a given layer of neurons we construct a graph $G = (V, E)$ with $V = \{v_1, \ldots, v_N\}$ the set of vertices and $E \subseteq V \times V$ the set of edges. Let W be the weighted symmetric adjacency matrix of size $N \times N$ with $W_{ij} = W_{ji} \geq 0$ representing the weight of the edge between v_i and v_j. The graph Laplacian L is then defined as $L = D - W$ where $D_{ii} = \sum_j W_{ij}$ and $D_{ij} = 0$ for $i \neq j$.

To enforce smoothing we use the Laplacian smoothing loss. On some activation vector z and fixed Laplacian L we formulate the graph spectral regularization function G as:

$$G(z, \mathbf{L}) = z^T \mathbf{L} z = \sum_{ij} W_{ij} ||z_i - z_j|| \tag{1}$$

where $|| \cdot ||$ denotes the Frobenius norm. We add it to the reconstruction or classification loss with a weighting term α. This adds an additional objective that activations should be smooth along the graph defined by L. This optimization procedure applies to any multi-layer model and valid graph Laplacian. We apply this algorithm to grid, and hierarchical graph structures on both autoencoder and classification dense architectures.

Algorithm 1. Graph Learning

Input batches x_i, model M with latent layer activations z_i, regularization weight α.
Pre-train M on x_i with $\alpha = 0$
for $i = 1$ **to** T **do**
    Create Graph Laplacian L_i from activations z_i
    for $j = 1$ **to** m **do**
        Train M on x_i with $\alpha = w$ and $L = L_i$ with MSE + loss in eq. 1
    end for
end for

3.1 Learning and Reinforcing an Abstracted Feature-Space Graph

Instead of enforcing smoothness over a fixed graph, we can learn a feature graph from the data (See Algorithm 1) using neural network activations themselves to bootstrap the process. Note, that most graph and kernel-based methods are applied over the space of observations but not over the space of features. One of the reasons is because it is even more difficult to define a distance between features than it is between observations. To circumvent this problem, we propose to learn a feature graph in the latent space of a neural network using feature co-activations as a measure of similarity.

We proceed by creating a graph using feature activation similarity, then applying this graph using Laplacian smoothing for a number of iterations. This converges to a graph of a latent feature space at the level of granularity of the number of dimensions in the corresponding layer.

Our algorithm for learning the graph consists of two phases. First, a pretraining phase where the model is learned with no graph regularization. Second, we alternate between constructing the graph from the similarities of the embedding layer features and further training the network for reconstruction and smoothness on the graph. There are many ways to create a graph from the feature $\times$ datapoint activation matrix. We use an adaptive Gaussian kernel,

$$K(z_i, z_j) = \frac{1}{2} exp\left(- \frac{||z_i - z_j||_2^2}{\sigma_i^2} \right) + \frac{1}{2} exp\left(- \frac{||z_i - z_j||_2^2}{\sigma_j^2} \right)$$

where σ_i is the adaptive bandwidth for node i which we set as the distance to the kth nearest neighbor of feature. An adaptive bandwidth Gaussian kernel is necessary for general architectures as the scale of the activations is not fixed. Batch normalization can also be used to limit the activation scale.

Since we are smoothing on the graph then constructing a new graph from the smoothed signal the learned graph converges to a steady state where the mean squared error acts as a repulsive force to stop the graph collapsing any further. We present the results of graph learning a biological dataset and show that the learned structure adds interpretability to the activations.

4 Experiments

Through examples, we show that visualizing the activations of data on the regularized layer highlights relationships in the data that are not easily visible without it. We establish this with two examples on fixed graphs, then move to graphs learned from the structure of the data with two examples of hierarchical structure and two with progression structure.

4.1 Fixed Structure

Enforcing fixed graph structure localizes activations for similar datapoints to a region of the graph. Here we show that enforcing a 8×8 grid graph on a layer of a dense MNIST classifier causes receptive fields to form, where each digit occupies a localized group of neurons on the grid. This can, in principle, be applied to any neural network layer to group neurons activating to similar features. Like in FMRI data or a convolutional neural network, we can examine the activation patterns for each localized group of neurons. For a second example, we show the usefulness in encouraging localized structure on a capsulenet architecture [18]. Where we are able to create globally consistent structure for better alignment of features between capsules.

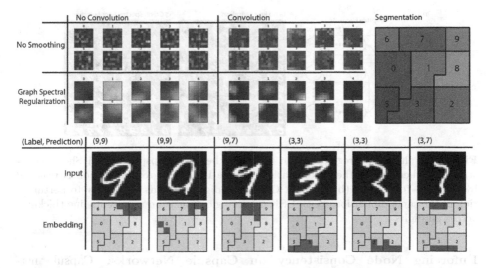

Fig. 1. Shows average activation by digit over an (8×8) 2D grid using graph spectral regularization and convolutions following the regularization layer. Next, we segment the embedding space by class to localize portions of the embedding associated with each class. Notice that the digit 4 here serves as the null case and does not show up in the segmentation. Finally, we show the top 10% activation on the embedding of some sample images. For two digits (9 and 3) we show a normal input, a correctly classified but transitional input, and a misclassified input. The highlighted regions of the embedding space correlate with the semantic description of the input.

Enforcing Grid Structure on Mnist. Without GSR, activations are unstructured and as a result are difficult to interpret, in that it is difficult to visually identify even which class a digit comes from based on the activation pattern (See Fig. 1). With GSR we can organize the activations making this representation more visually distinguishable. Since we can now take this embedding as an image, it is possible to use a standard convolutional architecture in subsequent layers in order to further filter the encodings. When we add 3 layers of 3×3 2D convolutions with 2×2 max pooling we see that representations for each digit are compressed into specific areas of the image. This leads to the formation of receptive fields over the network pertaining to similar datapoints. Using these receptive fields, we can now extract the features responsible for digit classification. For example, features that contribute to the activation of the top right of our grid we can associate with those features that contribute to being the digit 9.

The activation patterns on the embedding layer correspond well to a human perception of the digit type. The 9 that is misclassified as 7 both has significant activation in the 7 region of the embedding layer, and looks visually close to a 7. We can now interpret the embedding layer as a sort of brain map, where the map can map regions of activations, to types of inputs. This is not possible in a standard neural network, where activations are not spatially organized.

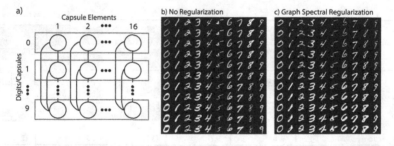

Fig. 2. (a) shows the regularization structure between capsules. (b–c) Show reconstruction when one of the 16 dimensions in the DigitCaps representation is tweaked by $0.05 \in [-0.25, 0.25]$. (b) Without GSR each digit responds differently to perturbation of the same dimension. With GSR (c) a single dimension represents line thickness across all digits.

Enforcing Node Consistency on Capsule Networks. Capsule networks [18] represent the input as a set of vectors where norm denotes activation and each component corresponds to some abstract feature. These elements are generally unordered. Here we use GSR to order these features consistently between digits. We train a capsule net on MNIST with GSR on 16 fully connected graphs between the 10 digit capsules. In the standard capsule network, each capsule orders features randomly based on initialization. However, with GSR we obtain a *consistent feature ordering*, e.g. node 1 corresponds to line thickness across all digits. GSR enforces a more ordered and interpretable encoding where localized regions are similarly organized, and the global line thickness feature is consistently learned between digits. More generally, GSR can be used to order nodes such that features common across capsules appear together. Finally, GSR does not degrade performance much, as can be seen by the digit reconstructions in Fig. 2.

In these examples the goal was to enforce a specified structure on unstructured features, but next we will examine the case where the goal is to learn the structure of the reduced feature space.

4.2   Learning Graph Structure

Using the procedure defined in Sect. 3.1, we can learn a graph structure. We first show that depending on the data, the learned graph exhibits either cluster or trajectory structure. We then show that our framework can learn structures that are hierarchical, i.e. subclusters within clusters or trajectories within clusters. Hierarchies are a difficult structure for other interpretability methods to learn [6]. However, our method naturally captures this by allowing for arbitrary graph structure among neurons in a layer.

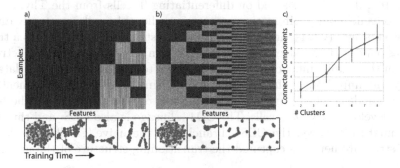

Fig. 3. We show the structure of the training data and snapshots of the learned graph for (a) three modules and (b) eight modules. (c) shows we have the mean and 95% CI of the number of connected components in the trained graph for over 50 trials.

Cluster Structure on Generated Data. We structure our n^{th} dataset to have exactly n feature clusters. We generate the data with n clusters by first creating 2^n data points representing the binary numbers from 0 to $2^n - 1$, then added gaussian noise $N(0, 0.1)$. This creates a dataset with a ground truth number of feature clusters. In the nth dataset the learned graph should have n connected components for n independent features. In Fig. 3 (a–b) we can see how this graph evolves over time for 3 and 8 modules. (c) shows how the learned graph learns the correct number of connected components for each ground truth number of clusters.

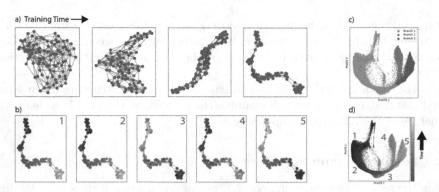

Fig. 4. Shows (a) graph structure over training iterations (b) feature activations of parts of the trajectory. PHATE [15] embedding plots colored by (c) branch number and (b) inferred trajectory location showing the branching structure of the data.

Trajectory Structure on T Cell Development Data. Next, we test graph learning on biological mass cytometry data, which is a high dimensional, single-cell protein dataset, measured on differentiating T cells from the Thymus [20]. The T cells lie along a bifurcating progression where the cells eventually diverge into two lineages (CD4+ and CD8+). Here, the structure of the data is a trajectory (as opposed to a pattern of clusters). We can see in Fig. 4 how the activated nodes in the graph embedding layer correspond to locations along the data trajectory, and importantly, the learned graph is a single connected component. The activated nodes (yellow) move from the bottom of the embedding to the top as T-cells develop into CD8+ cells. The CD4+ lineage is also CD8- and thus looks like a mixture between the CD8+ branch and the naive T cells. The learned graph structure here has captured the transitioning structure of the underlying data.

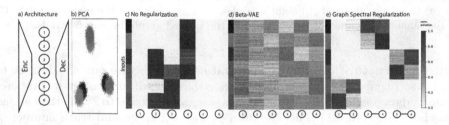

Fig. 5. Graph architecture, PCA plot, activation heatmaps of a standard autoencoder, β-VAE [9] and a graph regularized autoencoder. With relu activations normalized to $[0, 1]$ for comparison. In the model with graph spectral we are able to clearly decipher the hierarchical structure of the data, whereas with the standard autoencoder or the β-VAE the structure of the data is not clear.

Clusters Within Clusters on Generated Data. We demonstrate graph spectral regularization on data that is generated with a structure containing sub-clusters. Our data contains three large-scale structures, each comprising two Gaussian sub clusters generated in 15 dimensions (See Fig. 5). We use this dataset as it has both global and local structure. We demonstrate that our graph spectral regularized model is able to pick up on both the global and local structure of this dataset where disentangling methods such as β-VAE cannot. We use a graph-structure layer with six nodes with three connected node pairs and employ the graph spectral regularization. After training, we find that each node pair acts as a "super node" that detects each large-scale cluster. Within each super node, each of the two nodes encodes one of each of the two Gaussian substructures. Thus, this specific graph topology is able to extract the hierarchical topology of the data.

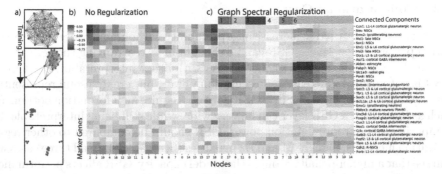

Fig. 6. Shows correlation between a set of marker genes for specific cell types and embedding layer activations. First with the standard autoencoder, then our autoencoder with graph spectral regularization. The left heatmap is biclustered, the right heatmap is grouped by connected components in the learned graph. We can see progression especially in the largest connected component where features on the right of the component correspond to less developed neurons.

Hierarchical Cluster and Trajectory Structure on Developing Mouse Cortex Data. In Fig. 6 we learn a graph on a single-cell RNA-sequencing dataset of over 4000 cells and over 8000 genes. The data contains a set of cells in the process of developing from neural stem cells to full neurons in the mouse brain. While there are many gene modules that contribute to the neuronal development, there are some states that have been studied. We use a list of cell type marker genes to validate our method. We use 1000 PCA components of the data in an autoencoder with a 20-dimensional embedding space. We learn the graph using an adaptive bandwidth gaussian kernel with the bandwidth for each feature set to the Euclidean distance to the nearest neighboring feature.

Our graph learns six components that represent meta features over the gene space. We can identify each with a specific type of cell or related types of cells.

For example, the light green component (cluster 2) represents the very early stage neural stem cells as it is highly correlated with increased Aldoc, Pax6 and Sox2 gene expression. Most interesting to examine is cluster 6, the largest component, which represents development into mature neurons. Within this component we can see a progression from just after intermediate progenitors on the left (showing Eomes expression) to more mature neurons with higher expression of Tbr1 and Sox5. With a standard autoencoder we cannot see progression structure of this dataset. While some of the more global structure is captured, we fail to see the data progression from intermediate progenitors to mature neurons. Learning a graph allows us to create receptive fields e.g. clusters of neurons that correspond to specific structures within the data, in this case cell types. Within these neighborhoods, we can pick up on the substructure within a single cell type, i.e. their developmental trajectory.

4.3   Computational Cost

Our method can be used to increase interpretability without much loss in representation power. At low levels, GSR can be thought of as rearranging the activations so that they become spatially coherent. As with other interpretability methods, GSR is not meant to increase representation power, but create useful representations with low cost in power. Since GSR does not require an information bottleneck such as in β-VAE, a GSR layer can be very wide, while still being interpretable. In comparing loss of representation power, GSR should be compared to other regularization methods, namely L1 and L2 penalties (See Table 1). In all three cases we can see that a higher penalty reduces the model capacity. GSR affects performance in approximately the same way as L1 and L2 regularizations do. To confirm this, we ran a MNIST classifier and measured train and test accuracy with 10 replicates. Graph spectral regularization adds a bit more overhead than elementwise activation penalties. However, the added cost can be seen as containing one matrix vector operation per pass. Empirically, GSR shows similar computational cost as other simple regularizations such as L1 and L2. To compare costs, we used a Keras model with Tensorflow backend [1] on a Nvidia Titan X GPU and a dual Intel(R) Xeon(R) CPU E5-2697 v4 @ 2.30 GHz, and with batchsize 256. we observed during training 233 milliseconds (ms) per step with no regularization, 266 ms for GSR, and 265 ms for L2 penalties.

Table 1. MNIST classification training and test accuracies for coefficient selected using cross validation over regularization weights in $[10^{-7}, 10^{-6}, \ldots, 10^{-2}]$ for various regularization methods with standard deviation over 10 replicates.

Regularization	Training accuracy	Test accuracy	Coefficient
None	99.1 ± 0.3	97.5 ± 0.3	N/A
L1	98.9 ± 0.3	97.4 ± 0.4	10^{-4}
L2	98.3 ± 0.3	98.0 ± 0.2	10^{-4}
GSR (ours)	99.3 ± 0.3	98.0 ± 0.3	10^{-3}

5 Conclusion

We have introduced a novel biologically inspired method for regularizing features of the internal layers of dense neural networks to take the shape of a graph. We show that coherent features emerge and can be used to interpret the underlying structure of the dataset. Furthermore, when the intended graph is not known apriori, we have presented a method for learning the graph structure, which learns a graph relevant to the data. This regularization framework takes a step towards more interpretable neural networks, and has applicability for future work seeking to reveal important structure in real-world biological datasets as we have demonstrated here.

Acknowledgements. This research was partially funded by IVADO (l'institut de valorisation des données) [*G.W.*]; Chan-Zuckerberg Initiative grants 182702 & CZF2019-002440 [*S.K.*]; and NIH grants R01GM135929 & R01GM130847 [*G.W.,S.K.*].

References

1. Abadi, M., et al.: TensorFlow: a system for large-scale machine learning. In: OSDI, p. 21 (2016)
2. Achille, A., Soatto, S.: Emergence of invariance and disentanglement in deep representations (2017). arXiv:1706.01350 [cs, stat]
3. Belkin, M., Matveeva, I., Niyogi, P.: Regularization and semi-supervised learning on large graphs. In: Shawe-Taylor, J., Singer, Y. (eds.) COLT 2004. LNCS (LNAI), vol. 3120, pp. 624–638. Springer, Heidelberg (2004). https://doi.org/10.1007/978-3-540-27819-1_43
4. Chen, X., Duan, Y., Houthooft, R., Schulman, J., Sutskever, I., Abbeel, P.: InfoGAN: interpretable representation learning by information maximizing generative adversarial nets (2016). arXiv:1606.03657 [cs, stat]
5. Dupont, E.: Learning disentangled joint continuous and discrete representations. In: Bengio, S., Wallach, H., Larochelle, H., Grauman, K., Cesa-Bianchi, N., Garnett, R. (eds.) Advances in Neural Information Processing Systems vol. 31, pp. 710–720. Curran Associates, Inc. (2018)
6. Esmaeili, B., et al.: Structured disentangled representations. In: AISTATS, p. 10 (2019)
7. Gori, M., Monfardini, G., Scarselli, F.: A new model for learning in graph domains. In: Proceedings. 2005 IEEE International Joint Conference on Neural Networks, 2005. vol. 2, pp. 729–734. IEEE, Montreal (2005). https://doi.org/10.1109/IJCNN.2005.1555942
8. Higgins, I., et al.: Towards a definition of disentangled representations (2018). arXiv:1812.02230 [cs, stat]
9. Higgins, I., et al.: β-VAE: learning basic visual concepts with a constrained variational framework. In: ICLR, p. 22 (2017)
10. Kingma, D.P., Welling, M.: Auto-encoding variational bayes (2013). arXiv:1312.6114 [Cs, Stat]
11. LeCun, Y., et al.: Backpropagation applied to handwritten zip code recognition. In: Neural Computation (1989)

12. Liao, R., Schwing, A., Zemel, R.S., Urtasun, R.: Learning deep parsimonious representations. In: NeurIPS (2016)
13. Logothetis, N.K., Pauls, J., Augath, M., Trinath, T., Oeltermann, A.: Neurophysiological investigation of the basis of the fMRI signal. Nature **412**(6843), 150–157 (2001). https://doi.org/10.1038/35084005
14. Min, W., Liu, J., Zhang, S.: Network-regularized sparse logistic regression models for clinical risk prediction and biomarker discovery. IEEE/ACM Trans. Comput. Biol. Bioinf. **15**(3), 944–953 (2018). https://doi.org/10.1109/TCBB.2016.2640303
15. Moon, K.R., et al.: Visualizing transitions and structure for high dimensional data exploration. bioRxiv (2017). https://doi.org/10.1101/120378, https://www.biorxiv.org/content/early/2017/12/01/120378
16. Ogawa, S., Lee, T.M.: Magnetic resonance imaging of blood vessels at high fields: in vivo and in vitro measurements and image simulation. Mag. Reson. Med. **16**(1), 9–18 (1990). https://doi.org/10.1002/mrm.1910160103
17. Ross, A.S., Hughes, M.C., Doshi-Velez, F.: Right for the right reasons: training differentiable models by constraining their explanations (2017). arXiv:1703.03717 [cs, stat]
18. Sabour, S., Frosst, N., Hinton, G.E.: Dynamic routing between capsules. In: 31st Conference on Neural Information Processing Systems (2017)
19. Scarselli, F., Gori, M., Tsoi, A.C., Hagenbuchner, M., Monfardini, G.: The graph neural network model. IEEE Trans. Neural Netw. **20**(1), 61–80 (2009). https://doi.org/10.1109/TNN.2008.2005605
20. Setty, M., et al.: Wishbone identifies bifurcating developmental trajectories from single-cell data. Nat. Biotechnol. **34**(6), 637 (2016)
21. Shuman, D.I., Narang, S.K., Frossard, P., Ortega, A., Vandergheynst, P.: The emerging field of signal processing on graphs: extending high-dimensional data analysis to networks and other irregular domains. IEEE Sign. Process. Mag. **30**(3), 83–98 (2013)
22. Stone, A., Wang, H., Stark, M., Liu, Y., Phoenix, D.S., George, D.: Teaching compositionality to CNNs. In: 2017 IEEE Conference on Computer Vision and Pattern Recognition (CVPR), pp. 732–741. IEEE, Honolulu (2017). https://doi.org/10.1109/CVPR.2017.85
23. Wu, Z., Pan, S., Chen, F., Long, G., Zhang, C., Yu, P.S.: A comprehensive survey on graph neural networks (2019). arXiv:1901.00596 [cs, stat]
24. Zeiler, M.D., Fergus, R.: Visualizing and understanding convolutional networks (2013). arXiv:1311.2901 [cs]
25. Zhang, Q., Wu, Y.N., Zhu, S.C.: Interpretable convolutional neural networks. In: 2018 IEEE/CVF Conference on Computer Vision and Pattern Recognition, pp. 8827–8836. IEEE, Salt Lake City (2018). https://doi.org/10.1109/CVPR.2018.00920
26. Zhou, D., Schölkopf, B.: A regularization framework for learning from graph data. In: ICML Workshop on Statistical Relational Learning and Its Connections to Other Fields, vol. 15, pp. 67–78 (2004)

Comparing the Preservation of Network Properties by Graph Embeddings

Rémi Vaudaine[1]([✉]), Rémy Cazabet[2], and Christine Largeron[1]

[1] Univ Lyon, UJM-Saint-Etienne, CNRS, Institut d Optique Graduate School
Laboratoire Hubert Curien UMR 5516, 42023 Saint-Etienne, France
{remi.vaudaine,christine.largeron}@univ-st-etienne.fr
[2] Univ Lyon, UCBL, CNRS, LIRIS UMR 5205, 69621 Lyon, France
remy.cazabet@gmail.com

Abstract. Graph embedding is a technique which consists in finding a new representation for a graph usually by representing the nodes as vectors in a low-dimensional real space. In this paper, we compare some of the best known algorithms proposed over the last few years, according to four structural properties of graphs: first-order and second-order proximities, isomorphic equivalence and community membership. To study the embedding algorithms, we introduced several measures. We show that most of the algorithms are able to recover at most one of the properties and that some algorithms are more sensitive to the embedding space dimension than some others.

Keywords: Graph embedding · Network properties

1 Introduction

Graphs are useful to model complex systems in a broad range of domains. Among the approaches designed to study them, graph embedding has attracted a lot of interest in the scientific community. It consists in encoding parts of the graph (node, edge, substructure) or a whole graph into a low dimensional space while preserving structural properties. Because it allows all the range of data mining and machine learning techniques that require vectors as input to be applied to relational data, it can benefit a lot of applications.

Several surveys have been recently published [5,6,8,20,21], some of them including a comparative study of the performance of the methods to solve specific tasks. Among them, Cui *et al.* [6] propose a typology of network embedding methods into three families: matrix factorization, random walk and deep learning methods. Following the same typology, Goyal et al. [8] compare state of the art methods on few tasks such as link prediction, graph reconstruction or node classification and analyze the robustness of the algorithms with respect

Electronic supplementary material The online version of this chapter (https://doi.org/10.1007/978-3-030-44584-3_41) contains supplementary material, which is available to authorized users.

M. R. Berthold et al. (Eds.): IDA 2020, LNCS 12080, pp. 522–534, 2020.
https://doi.org/10.1007/978-3-030-44584-3_41

to hyper-parameters. Recently, Cai et al. [5] extended the typology by adding deep learning based methods without random walks but also two other families: graph kernel based methods notably helpful to represent the whole graph as a low-dimensional vector and generative models which provide a latent space as embedding space. For their part, Zhang et al. [21] classify embedding techniques into two types: unsupervised network representation learning or semi-supervised and they list a number of embedding methods depending on the information sources they use to learn. Like Goyal et al. [8], they compare the methods on different tasks. Finally, Hamilton et al. [10] introduce an encoder-decoder framework to describe representative embedding algorithms from a methodological perspective. In this framework, the encoder corresponds to the function which maps the elements of a graph as vectors. The decoder is a function which associates a specific graph statistic to the obtained vectors, for instance for a pair of node embeddings the decoder can give their similarity in the vector space, allowing the similarity of the nodes in the original graph to be quantified.

From this last work, we retained the encoder-decoder framework and we propose to use it for evaluating the different embedding methods. To that end, we compare, using metrics that we introduce, the value computed by the decoder with the value associated to the corresponding nodes in the graph for the equivalent function. Thus, in this paper, we adopt a different point of view from the previous task-oriented evaluations. Indeed, all of them consider embeddings as a *black box*, i.e., using obtained features without considering their properties. They ignore the fact that embedding algorithms are designed, explicitly or implicitly, to preserve some particular structural properties and their usefulness for a given task depends on how they succeed to capture it. Thus, in this paper, through an experimental comparative study, we compare the ability of embedding algorithms to capture specific properties, i.e., first-order proximity of nodes, structural equivalence (second-order proximity), isomorphic equivalence and community structure.

In Sect. 2, these topological properties are formally defined and measures are introduced to evaluate to what extent embedding methods encode them. Section 3 presents the studied embedding methods. Section 4 describes the datasets used for the experiments, while Sect. 5 presents the results.

2   Structural Properties and Metrics

There is a wide range of graph properties that are of interest. We propose to study several of them which are at the basis of network analysis and are directly linked with usual learning and mining tasks on graphs [13]. First, we measure the ability of an embedding method to recover the set of neighbors of the nodes which is the first-order proximity (P1). This property is important for several downstream tasks: clustering where vectors of the same cluster represent nodes of the same community, graph reconstruction where two similar vectors represent two nodes that are neighbors in the graph, and node classification based for instance on majority vote of the neighbors. Secondly, we evaluate the ability of embedding methods to capture the second-order proximity (P2) which is the fact that two nodes have the same set of neighbors. This property is especially interesting when dealing with link prediction since, in social graphs, it is assumed that two nodes that share the same friends are likely to become friends too. Thirdly, we measure how much an embedding method is able to capture the roles of nodes in a graph which is the isomorphic equivalence (P3). This property is interesting when looking for specific nodes like leaders or outsiders. Finally, we evaluate the ability of an embedding method to detect communities (P4) in a graph which has been an on going field of research for the last 20 years. Next, we define both properties and measures we use in order to quantify how much an embedding method is able to capture those properties.

Let $G(V, E)$ be an unweighted and undirected graph where $V = \{v_0, ..., v_{n-1}\}$ is the set of n vertices, $E = \{e_{ij}\}_{i,j=0}^{n-1}$ the set of m edges and A is its binary adjacency matrix. Graph embedding consists in encoding the graph into a low-dimensional space R^d, where d is the dimension of the real space, with a function $f : V \longmapsto Y$ which maps vertices to vector embeddings while preserving some properties of the graph. We note $Y \in \mathbb{R}^{n \times d}$ the embedding matrix and Y_i its i-th row representing the node v_i.

Neighborhood or first-order proximity (P1): capturing the neighborhood for an embedding method means that it aims at keeping any two nodes v_i and v_j that are linked in the original graph ($A_{ij} = 1$) close in the embedding space. The measure S designed for this property is based on the comparison between the set $N(v_i)$ of neighbors in the graph of every node v_i and the set $N_E(v_i)$ of its $|N(v_i)|$ nearest neighbors in the embedding space where $|N(v_i)|$ is its degree. Finally, by averaging over all nodes, S quantifies the ability of an embedding to respect the neighborhood. The higher S, the more P1 is preserved.

$$S(v_i) = \frac{|N(v_i) \bigcap N_E(v_i)|}{|N(v_i)|}, \quad S = \frac{1}{n} \sum_i S(v_i) \tag{1}$$

Structural equivalence or second-order proximity (P2): two vertices are structurally equivalent if they share many of the same neighbors [13]. To measure the efficiency of an embedding method to recover the structural equivalence, we define the distance $dist_A(A_i, A_j)$ between the lines of the adjacency matrix corresponding to each pair of nodes (v_i, v_j), and $dist_E(Y_i, Y_j)$ the distance between their representative vectors in the embedding space. The metric for P2 is defined by the correlation coefficient (Spearman or Pearson) $Struct\_eq$ between those values for all pairs of nodes. The higher $Struct\_eq$ (close to 1), the better P2 is preserved by the algorithm.

$$L_A(v_i, v_j) = dist_A(A_i, A_j), \quad L_E(v_i, v_j) = dist_E(Y_i, Y_j) \tag{2}$$

with $dist_A$ the distance in the adjacency matrix (cosine or euclidean) and $dist_E$, the embedding similarity which is indicated in Table 1. Finally,

$$Struct\_eq = pearson(L_A, L_E) \tag{3}$$

Isomorphic equivalence (P3): two nodes are isomorphically equivalent, i.e they share the same role in the graph, if their ego-networks are isomorphic [4]. The ego-network of node v_i is defined as the subgraph EN_i made up of its neighbors and the edges between them (without v_i itself). To go beyond a binary evaluation, for each pair of nodes (v_i, v_j), we compute the Graph Edit Distance $GED(EN_i, EN_j)$ between their ego-networks EN_i and EN_j thanks to the Graph Matching Toolkit [16] and the distance between their representative vectors in the embedding space $dist_E(Y_i, Y_j)$. $dist_E$ is indicated in Table 1. Finally, the Pearson and Spearman correlation coefficients between those values computed on all pairs of nodes are used to have an indicator for the whole graph. A negative correlation means that if the distance in the embedding space is large then exp(-GED), as in [15], is small. So, to ease one's reading, we take the opposite of the correlation coefficient such that, for all measures, the best result is 1. Thus, the higher $Isom\_eq$, the better P3 is preserved by the algorithm.

$$L_{Egonet}(v_i, v_j) = exp(-GED(EN_i, EN_j)), \; L_E(v_i, v_j) = dist_E(Y_i, Y_j) \quad (4)$$

$$Isom\_eq = -pearson(L_{Egonet}, L_E) \quad (5)$$

Community/cluster membership (P4): communities can be defined as "groups of vertices having higher probability of being connected to each other than to members of other groups" [7]. On the other hand, clusters can be defined as sets of elements such that elements in the same cluster are more similar to each other than to those in other clusters. We propose to study the ability of an embedding method to transfer a community structure to a cluster structure. Given a graph with k ground-truth communities, we cluster, using KMeans (since k, the number of communities, is known), the node embeddings into k clusters. Finally, we compare this partition with the ground-truth partition using the adjusted mutual information (AMI). We also used the normalized mutual information (NMI) but both measures showed similar results. Let $L_{Community}$ be the ground-truth labeling and $L_{Clusters}$ the one found by KMeans.

$$Score = AMI(L_{Community}, L_{Clusters}) \quad (6)$$

3 Embeddings

There are many different graph embedding algorithms. We present a non-exhaustive list of recent methods, representative of the different families proposed in the state-of-the-art. We refer the reader to the full papers for more information. In Table 1 we mention all the embedding methods we used in our comparative study with the graph similarity they are supposed to preserve and the distance that is used in the embedding space to relate any pair of nodes of the graph. Two versions of N2V are used (A: $p = 0.5, q = 4$ for local random walks, B: $p = 4, q = 0.5$ for deeper random walks).

Table 1. Studied methods with complexity, their graph similarity (encoder) and their distance in the embedding space (decoder)

Name of the method	Graph sim.	Embedding sim.
Laplacian Eigenmaps (LE) [1] - $O(N^2)$	1st-order prox	Euclidean
Locally Linear Emb. (LLE) [17] - $O(N^2)$	1st-order prox	Euclidean
HOPE [14] - $O(N^2)$	Katz-Index	Dot-product
SVD of the adjacency matrix - $O(N^2)$	2nd-order prox	Dot-product
struc2vec (S2V) [15] - $O(Nlog(N))$	Co-occurence proba	Dot-product
node2vec (N2V) [9] - $O(N)$	Co-occurence proba	Dot-product
Verse [18] - $O(N)$	Perso. Page-Rank	Dot-product
Kamada-Kawai layout (KKL) [11] - $O(N^2)$		Euclidean
Multi-dim Scaling (MDS) [12]	1st-order prox	Euclidean
SDNE [19] - $O(N)$	1st & 2nd-order prox	Euclidean

4   Graphs

To evaluate embedding algorithms, we choose real graphs and generated graphs having different sizes and types: random (R), with preferential attachment (PA), social (S), social with community structure (SC) as shown in Table 2. While real graphs correspond to common datasets, generators allow to control the characteristics of the graphs. Thus, we have prior knowledge which makes evaluation easier and more precise. Table 2 gives the characteristics of these graphs divided in three groups: small, medium and large graphs.

Table 2. Dataset characteristics. All graphs are provided in our GitHub

Name of the graph	Number of nodes	Number of edges	Type
Zachary Karate Club (ZKC)	34	77	SC
Erdos-Renyi (Gnp100)	100	474	R
Barabasi-Albert (BA100)	100	900	PA
Dancer (Dancer_100)	100	243	SC
Email network (Email)	1133	5452	S
Erdos-Renyi (Gnp1000)	1 000	4985	R
Barabasi-Albert (BA1000)	1000	9900	PA
Dancer (Dancer_1k)	1 000	3627	SC
PGP	10 680	24316	S
Erdos-Renyi (Gnp10000)	10 000	49722	R
Barabasi-Albert (BA10k)	10 000	99900	PA
Dancer (Dancer_10k)	10 000	189886	SC

5   Results

We used the metrics presented in Sect. 2 to quantify the ability of the embedding algorithms described in Sect. 3 to recover four properties of the graphs: first order proximity (P1), structural and isomorphic equivalences (P2 and P3), community membership (P4). Due to lack of space, we show only the most representative results and provide the others as additional materials[1]. For the same reason, to evaluate P2 and P3, both Pearson and Spearman correlation coefficients have been computed but we only show results for Pearson as they are similar with Spearman. For readability, every algorithm successfully captures a property when its corresponding score is at 1 and 0 means unsuccessful. Moreover, a dash (-) in a Table indicates that a method has not been able to provide a result. Note that due to high complexity, KKL and MDS are not computed for every graph. Finally, the code and datasets are available online on our GitHub (see footnote 1).

5.1   Neighborhood (P1)

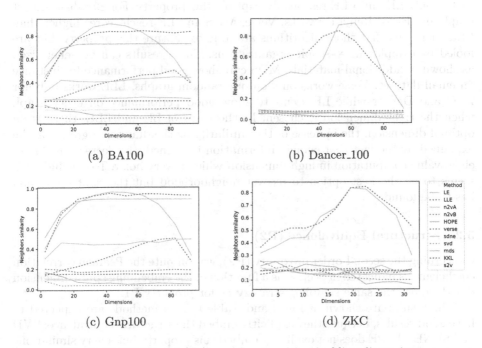

(a) BA100

(b) Dancer_100

(c) Gnp100

(d) ZKC

Fig. 1. Neighborhood (P1) as a function of embedding dimension.

---

[1] https://github.com/vaudaine/Comparing_embeddings.

Table 3. Neighborhood (P1) *Italic*: Best in row. **Bold**: best.

Dimensions	2	10	100	1128
LE	0.086	0.196	*0.371*	0.007
LLE	0.193	0.352	*0.589*	0.021
HOPE	0.022	0.104	*0.177*	0.018
S2V	0.02	0.022	0.021	*0.022*
N2VA	0.044	0.245	0.37	*0.437*
N2VB	0.04	0.29	0.414	*0.45*
SDNE	0.024	0.047	*0.055*	0.041
SVD	0.054	*0.138*	0.134	0.026
Verse	0.019	0.021	*0.021*	0.021
MDS	0.104	0.287	0.793	**0.919**

(a) Email

Dimensions	2	10	100	1000
LE	0.004	0.097	0.72	*0.933*
LLE	-	-	0.045	*0.117*
HOPE	0.002	0.01	*0.226*	0.094
S2V	0.001	0.001	0.001	*0.001*
N2VA	0.002	0.032	0.914	*0.945*
N2VB	0.002	0.045	0.935	*0.935*
SDNE	0.001	0.001	*0.001*	0.001
SVD	*0.001*	0.001	0.001	0.0
Verse	0.002	0.052	**0.961**	0.854

(b) Gnp10000

For the first order proximity (P1), we measure the similarity S as a function of the dimension d for all the embedding methods. For computational reasons, for large graphs, the measure is computed on 10% of the nodes. Results are shown in Fig. 1 and Table 3, for d varying from 2 until approximately the number of nodes. We can make several observations: for networks with communities (Dancer and ZKC), only LE and LLE reasonably capture this property. For Barabasi Albert graph and Erdos-Renyi networks, Verse, MDS and LE reach scores higher than LLE. It means that those algorithms are able to capture this property, but are fooled by complex meso-scopic organizations. These results can be generalized as shown in additional materials. MDS can show good performance for instance on email dataset, Verse works only on our random graphs, LLE works only for ZKC and Dancer while LE seems to show good performance on every graph when the right dimension is chosen. In the cases of LE and LLE, there is an optimal dimension: the increase of the similarity as the dimension grows can be explained by the fact that enough information is learned; the decrease is due to eigen-value computation in high-dimension which is very noisy. To conclude, LE seems to be the best option to recover neighborhood but the right dimension has to be found.

5.2   Structural Equivalence (P2)

Concerning the second-order proximity (P2), we compute the Pearson correlation coefficient, as indicated in Sect. 2, as a function of the embedding space dimension d and we use the same sampling strategy as for property P1.

The results are shown in Fig. 2 and Table 4. Two methods are expected to have good results, because they explicitly embed the structural equivalence: SVD and SDNE. HOPE does not explicitly embed this property but a very similar one which is Katz-Index. On every small graph, SVD effectively performs the best and with the lowest dimension. HOPE still has very good results. The Pearson coefficient grows as the dimension of the embedding grows which implies that the best results are obtained when the dimension of the space is high enough. The other algorithms fail to recover the structural equivalence. For medium

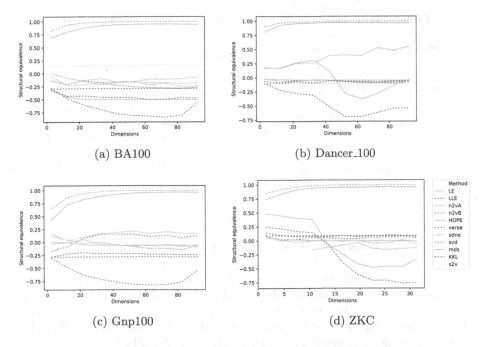

(a) BA100

(b) Dancer_100

(c) Gnp100

(d) ZKC

Fig. 2. Structural equivalence (P2) as a function of embedding dimension.

and large graphs as presented in Table 4, SVD and HOPE still show very good performance and the higher the dimension of the embedding space, the higher the correlation. For large graphs, SDNE shows also very good results but it seems to need more data to be able to learn properly. In the end, SVD seems to be the best algorithm to capture the second order proximity. It computes a singular value decomposition which is fast and scalable but SDNE performs also very well on the largest graphs and, in that case, it can outperform SVD.

5.3   Isomorphic Equivalence (P3)

With the property P3, we investigate the ability of an embedding algorithm to capture roles in a graph. To do so, we compute the graph edit distance (GED) between every pair of nodes in the graph and the distance between the vectors of the embedding. Moreover, we sample nodes at random and compute the GED only between every pair of the sampled nodes thus reducing the computing time drastically. We sample 10% of the nodes for medium graphs and 1% of the nodes for large graphs. Experiments have demonstrated that results are robust to sampling. We present, in Fig. 3 and Table 5, the evolution of the correlation coefficient according to the dimension of the embedding space. The only algorithm that is supposed to perform well for this property is Struc2vec. Note also that algorithms which capture the structural equivalence can also give results since two nodes that are structurally equivalent are also isomorphically equivalent

Table 4. Structural equivalence (P2). *Italic*: Best in row. **Bold**: best.

Dimensions	2	10	100	995
LE	*0.593*	0.281	0.052	0.044
LLE	0.079	−0.069	−0.244	*−0.441*
HOPE	0.726	0.909	*0.967*	0.947
S2V	0.041	0.134	*0.137*	0.131
N2VA	*0.043*	−0.038	−0.018	−0.033
N2VB	0.05	*−0.055*	−0.042	−0.036
SDNE	0.174	0.037	0.034	*0.626*
SVD	0.823	0.933	0.987	**1.0**
Verse	0.036	−0.038	0.023	*0.141*
MDS	−0.053	−0.015	−0.048	−0.079

(a) Dancer_1k

Dimensions	2	10	100	1000
LE	0.06	0.077	0.189	*0.192*
LLE	-	-	-0.724	*-0.785*
HOPE	0.844	0.723	0.799	*0.967*
S2V	0.003	0.457	*0.744*	0.717
N2VA	*0.438*	0.144	−0.289	0.297
N2VB	*0.445*	−0.175	−0.342	0.402
SDNE	0.678	0.787	0.952	*0.954*
SVD	0.795	0.621	0.873	**0.983**
Verse	−0.036	−0.386	−0.186	*0.642*

(b) BA10k

but the converse is not true. For small graphs, as illustrated in Fig. 3, Struc2vec (S2V) is nearly always the best. It performs well on medium and large graphs too as shown in Table 5. However results obtained on other graphs (available in supplementary material) indicate that Stru2vec is not always much better than the other algorithms. As a matter of fact, Struc2vec remains the best algorithm for this measure but it is not totally accurate since the correlation coefficient is not close to 1 on every graph e.g on Dancer10k in Table 5(b).

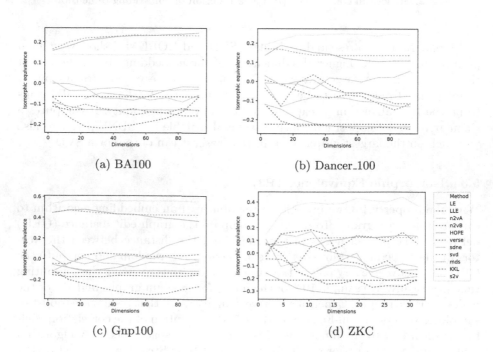

(a) BA100   (b) Dancer_100

(c) Gnp100   (d) ZKC

Fig. 3. Isomorphic equivalence (P3) as a function of embedding dimension.

Table 5. Isomorphic equivalence (P3). *Italic*: Best in row. **Bold**: best.

(a) Gnp1000

Dimensions	2	10	100	995
LE	*0.058*	0.053	0.023	0.023
LLE	0.004	−0.055	−0.05	−0.111
HOPE	*0.687*	0.295	0.299	0.126
S2V	0.468	**0.761**	0.759	0.753
N2VA	*0.18*	0.08	−0.119	−0.107
N2VB	*0.327*	0.041	−0.053	−0.03
SDNE	nan	0.088	−0.057	0.004
SVD	*0.39*	0.295	0.284	0.165
Verse	0.077	−0.017	0.006	*0.101*
MDS	*0.018*	−0.011	0.001	0.01

(b) Dancer_10k

Dimensions	2	10	100	1000
LE	−0.068	*0.072*	0.05	−0.052
LLE	−0.088	0.009	−0.008	*−0.102*
HOPE	0.086	0.075	*0.108*	0.103
S2V	0.11	0.258	**0.431**	0.401
N2VA	0.123	0.166	*0.38*	0.203
N2VB	0.123	0.161	*0.204*	0.081
SDNE	0.057	0.083	0.035	*0.086*
SVD	0.053	0.076	0.1	*0.102*
Verse	0.036	−0.032	−0.071	*−0.148*

5.4 Community Membership (P4)

To study the ability of an embedding to recover the community structure of a graph (P4), we compare, using Adjusted Mututal Information (AMI) and Normalized (NMI), the partition given by KMeans on the node embeddings and the ground-truth partition. The results are given only for PPG (averaged over 3 instances) and Dancer graphs (for 20 different graphs) for which the community structure (ground truth) is provided by the generators. To obtain them, we generated planted partition graphs (PPG) with 10 communities and 100 nodes

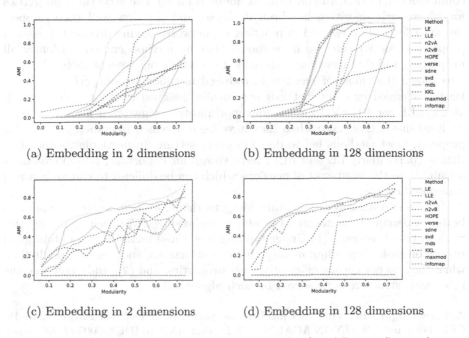

(a) Embedding in 2 dimensions   (b) Embedding in 128 dimensions

(c) Embedding in 2 dimensions   (d) Embedding in 128 dimensions

Fig. 4. AMI for community detection on PPG (top) and Dancer (bottom)

per community. We set the probability of an edge existing between communities $p_{out} = 0.01$ and vary the probability of an edge existing within a community p_{in} from 0.01 (no communities) to 1 (clearly defined communities), thus varying the modularity of the graph from 0 to 0.7. For Dancer, we generate 20 graphs with varying community structure by adding between-community edges and removing within-community edges. Moreover, we apply also usual community detection algorithms such as Louvain's modularity maximisation (maxmod) [2] and Infomap [3] on the graphs. Results are shown in Fig. 4. In low dimension ($d = 2$, left of the Figure), every embedding is less efficient than the usual community detection algorithms. In higher dimension ($d = 128$, right of the Figure), many embedding techniques, Verse, MDS, N2V (both versions) and HOPE (on PPG), are able to have the same results as the best community detection algorithm: Louvain and obvioulsly for all the methods, AMI increases with the modularity.

6   Conclusion

In this paper, we studied how a wide range of graph embedding techniques preserve essential structural properties of graphs. Most of recent works on graph embeddings focused on the introduction of new methods and on task-oriented evaluation but they ignore the rationale of the methods, and only focus on their performance on a specific task in a particular setting. As a consequence, methods that have been designed to embed local structures are compared with methods that should embed global structures on tasks as diverse as link prediction or community detection. In contrast, we focused on (i) The structural properties for which each algorithm has been *designed*, and (ii) How well these properties are effectively preserved in practice, on networks having diverse topological properties. As a result, we have shown that no method embed efficiently all properties, and that most methods embed effectively only one of them. We have also shown that most of recently introduced methods are outperformed or at least challenged by older methods specifically designed for that purpose, such as LE/LLE for P1, SVD for P2, and modularity optimization for P4. Finally, we have shown that, even when they have been designed to embed a particular property, most methods fail to do so in every setting. In particular, some algorithms (particularly LE and LLE) have shown an important, non-monotonous sensibility to the number of dimensions which can be difficult to choose in a non supervised context.

In order to improve graph embedding methods, we believe that we need to better understand the nature of produced embeddings. We wish to pursue this work in two directions, (1) Understanding how those methods can obtain good results on tasks depending mainly on local structures, such as link prediction, when they do not encode efficiently local properties, and (2) study how well the meso-scale structure is preserved by such algorithms.

Acknowledgement. This work has been supported by BITUNAM Project ANR-18-CE23-0004 and IDEXLYON ACADEMICS Project ANR-16-IDEX-0005 of the French National Research Agency.

References

1. Belkin, M., Niyogi, P.: Laplacian eigenmaps for dimensionality reduction and data representation. Neural Comput. **15**(6), 1373–1396 (2003)
2. Blondel, V.D., Guillaume, J.L., Lambiotte, R., Lefebvre, E.: Fast unfolding of communities in large networks. J. Stat. Mech: Theory Exp. **2008**(10), P10008 (2008)
3. Bohlin, L., Edler, D., Lancichinetti, A., Rosvall, M.: Community detection and visualization of networks with the map equation framework. In: Ding, Y., Rousseau, R., Wolfram, D. (eds.) Measuring Scholarly Impact, pp. 3–34. Springer, Cham (2014). https://doi.org/10.1007/978-3-319-10377-8_1
4. Borgatti, S., Everett, M., Freeman, L.: Software for social network analysis. Ucinet for windows (2002)
5. Cai, Z., Chang, K.: A comprehensive survey of graph embedding: problems, techniques, and applications. TKDE **30**(9), 1616–1637 (2018)
6. Cui, P., Wang, X., Pei, J., Zhu, W.: A survey on network embedding. CoRR, abs/1711.08752 (2017)
7. Fortunato, S., Hric, D.: Community detection in networks: a user guide. CoRR, abs/1608.00163 (2016)
8. Goyal, P., Ferrara, E.: Graph embedding techniques, applications, and performance: a survey. Knowl.-Based Syst. **151**, 78–94 (2018)
9. Grover, A., Leskovec, J.: Node2vec: acalable feature learning for networks. In: Proceedings of the 22nd ACM SIGKDD Conference. ACM (2016)
10. Hamilton, W.L., Ying, R., Leskovec, J.: Representation learning on graphs: methods and applications. CoRR, abs/1709.05584 (2017)
11. Kamada, T., Kawai, S.: An algorithm for drawing general undirected graphs. Inf. Process. Lett. **31**(7), 7–15 (1989)
12. Kruskal, J.B.: Multidimensional scaling by optimizing goodness of fit to a nonmetric hypothesis. Psychometrika (1964)
13. Lorrain, F., White, H.C.: Structural equivalence of individuals in social networks. J. Math. Sociol. **1**(1), 49–80 (1971)
14. Ou, M., Cui, P., Pei, J., Zhang, Z., Zhu, W.: Asymmetric transitivity preserving graph embedding. In: Proceedings of the 22Nd ACM SIGKDD Conference on Knowledge Discovery and Data Mining (2016)
15. Ribeiro, L.F., Saverese, P.H., Figueiredo, D.R.: Struc2vec: learning node representations from structural identity. In: ACM SIGKDD, New York, NY, USA (2017)
16. Riesen, K., Emmenegger, S., Bunke, H.: A novel software toolkit for graph edit distance computation. In: Graph-Based Representations in Pattern Recognition (2013)
17. Roweis, S.T., Saul, L.K.: Nonlinear dimensionality reduction by locally linear embedding. Science **290**(5500), 2323–2326 (2000)
18. Tsitsulin, A., Mottin, D., Karras, P., Müller, E.: Verse: versatile graph embeddings from similarity measures. In: WWW 2018 (2018)
19. Wang, D., Cui, P., Zhu, W.: Structural deep network embedding. In: SIGKDD (2016)
20. Wu, Z., Pan, S., Chen, F., Long, G., Zhang, C., Yu, P.S.: A comprehensive survey on graph neural networks. CoRR, abs/1901.00596 (2019)
21. Zhang, D., Yin, J., Zhu, X., Zhang, C.: Network representation learning: a survey. CoRR, abs/1801.05852 (2018)

534   R. Vaudaine et al.

Making Learners (More) Monotone

Tom Julian Viering[1]([✉]) [ID], Alexander Mey[1] [ID], and Marco Loog[1,2] [ID]

[1] Delft University of Technology, Delft, The Netherlands
{t.j.viering,a.mey,m.loog}@tudelft.nl
[2] University of Copenhagen, Copenhagen, Denmark

Abstract. Learning performance can show non-monotonic behavior. That is, more data does not necessarily lead to better models, even on average. We propose three algorithms that take a supervised learning model and make it perform more monotone. We prove consistency and monotonicity with high probability, and evaluate the algorithms on scenarios where non-monotone behaviour occurs. Our proposed algorithm MT$_{HT}$ makes less than 1% non-monotone decisions on MNIST while staying competitive in terms of error rate compared to several baselines. Our code is available at https://github.com/tomviering/monotone.

Keywords: Learning curve · Model selection · Learning theory

1 Introduction

It is a widely held belief that more training data usually results in better generalizing machine learning models—cf. [11,17] for instance. Several learning problems have illustrated, however, that more training data can lead to worse generalization performance [3,9,12]. For the peaking phenomenon [3], this occurs exactly at the transition from the underparametrized to the overparametrized regime. This double-descent behavior has found regained interest in the context of deep neural networks [1,18], since these models are typically overparametrized. Recently, also several new examples have been found, where in quite simple settings more data results in worse generalization performance [10,19].

It can be difficult to explain to a user that machine learning models can actually perform worse when more, possibly expensive to collect data has been used for training. Besides, it seems generally desirable to have algorithms that guarantee increased performance with more data. How to get such a guarantee? That is the question we investigate in this work and for which we use learning curves. Such curves plot the expected performance of a learning algorithm versus the amount of training data.[1] In other words, we wonder how we can make learning curves monotonic.

The core approach to make learners monotone is that, when more data is gathered and a new model is trained, this newly trained model is compared to

---

[1] Not to be confused with training curves, where the loss versus epochs (optimization iterations) is plotted.

© The Author(s) 2020
M. R. Berthold et al. (Eds.): IDA 2020, LNCS 12080, pp. 535–547, 2020.
https://doi.org/10.1007/978-3-030-44584-3_42

the currently adopted model that was trained on less data. Only if the new model performs better should it be used. We introduce several wrapper algorithms for supervised classification techniques that use the holdout set or cross-validation to make this comparison. Our proposed algorithm MT_{HT} uses a hypothesis test to switch if the new model improves significantly upon the old model. Using guarantees from the hypothesis test we can prove that the resulting learning curve is monotone with high probability. We empirically study the effect of the parameters of the algorithms and benchmark them on several datasets including MNIST [8] to check to what degree the learning curves become monotone.

This work is organized as follows. The notion of monotonicity of learning curves is reviewed in Sect. 2. We introduce our approaches and algorithms in Sect. 3, and prove consistency and monotonicity with high probability in Sect. 4. Section 5 provides the empirical evaluation. We discuss the main findings of our results in Sect. 6 and end with the most important conclusions.

2   The Setting and the Definition of Monotonicity

We consider the setting where we have a learner that now and then receives data and that is evaluated over time. The question is then, how to make sure that the performance of this learner over time is monotone—or with other words, how can we guarantee that this learner over time improves its performance?

We analyze this question in a (frequentist) classification framework. We assume there exists an (unknown) distribution P over $\mathcal{X} \times \mathcal{Y}$, where $\mathcal{X}$ is the input space (features) and $\mathcal{Y}$ is the output space (classification labels). To simplify the setup we operate in rounds indicated by i, where $i \in \{1, \ldots, n\}$. In each round, we receive a batch of samples S^i that is sampled i.i.d. from P. The learner L can use this data in combination with data from previous rounds to come up with a hypothesis h_i in round i. The hypothesis comes from a hypothesis space $\mathcal{H}$. We consider learners L that, as subroutine, use a supervised learner $A : \mathcal{S} \to \mathcal{H}$, where $\mathcal{S}$ is the space of all possible training sets.

We measure performance by the error rate. The true error rate on P equals

$$\epsilon(h_i) = \int_{x \in \mathcal{X}} \sum_{y \in \mathcal{Y}} l_{0\text{-}1}(h_i(x), y) dP(x, y) \tag{1}$$

where $l_{0\text{-}1}$ is the zero-one loss. We indicate the empirical error rate of h on a sample S as $\hat{\epsilon}(h, S)$. We call n rounds a run. The true error of the returned h_i by the learner L in round i is indicated by ϵ_i, all the ϵ_i's of a run form a learning curve. By averaging multiple runs one obtains the expected learning curve, $\bar{\epsilon}_i$.

The goal for the learner L is twofold. The error rates of the returned models ϵ_i's should (1) be as small as possible, and (2) be monotonically decreasing. These goals can be at odds with another. For example, always returning a fixed model ensures monotonicity but incurs large error rates. To measure (1), we summarize performance of a learning curve using the Area Under the Learning Curve (AULC) [6,13,16]. The AULC averages all ϵ_i's of a run. Low AULC indicates that a learner manages to quickly reduce the error rate.

Monotone in round i means that $\epsilon_{i+1} \leq \epsilon_i$. We may care about monotonicity of the expected learning curve *or* individual learning curves. In practice, however, we typically get one chance to gather data and submit models. In that case, we rather want to make sure that then any additional data also leads to better performance. Therefore, we are mainly concerned with monotonicity of *individual* learning curves. We quantify monotonicity of a run by the fraction of non-monotone transitions in an individual curve.

3 Approaches and Algorithms

We introduce three algorithms (learners L) that wrap around supervised learners with the aim of making them monotone. First, we provide some intuition how to achieve this: ideally, during the generation of the learning curve, we would check whether $\epsilon(h_{i+1}) \leq \epsilon(h_i)$. A fix to make a learner monotone would be to output h_i instead of h_{i+1} if the error rate of h_{i+1} is larger. Since learners do not have access to $\epsilon(h_i)$, we have to estimate it using the incoming data. The first two algorithms, $\mathrm{MT_{SIMPLE}}$ and $\mathrm{MT_{HT}}$, use the holdout method to this end; newly arriving data is partitioned into training and validation sets. The third algorithm, $\mathrm{MT_{CV}}$, makes use of cross validation.

$\mathrm{MT_{SIMPLE}}$: Monotone Simple. The pseudo-code for $\mathrm{MT_{SIMPLE}}$ is given by Algorithm 1 in combination with the function UpdateSimple. Batches S^i are split into training (S_t^i) and validation (S_v^i). The training set S_t is enlarged each round with S_t^i and a new model h_i is trained. S_v^i is used to estimate the performance of h_i and h_{best}. We store the previously best performing model, h_{best}, and compare its performance to that of h_i. If the new model h_i is better, it is returned and h_{best} is updated, otherwise h_{best} is returned.

Because h_i and h_{best} are both compared on S_v^i the comparison is more accurate because the comparison is paired. After the comparison S_v^i can safely be added to the training set (line 7 of Algorithm 1).

We call this algorithm $\mathrm{MT_{SIMPLE}}$ because the model selection is a bit naive: for small validation sets, the variance in the performance measure could be quite large, leading to many non-monotone decisions. In the limit of infinitely large S_v^i, however, this algorithm should always be monotone (and very data hungry).

$\mathrm{MT_{HT}}$: Monotone Hypothesis Test. The second algorithm, $\mathrm{MT_{HT}}$, aims to resolve the issues of $\mathrm{MT_{SIMPLE}}$ with small validation set sizes. In addition, for this algorithm, we prove that individual learning curves are monotone with high probability. The same pseudo-code is used as for $\mathrm{MT_{SIMPLE}}$ (Algorithm 1), but with a different update function UpdateHT. Now a hypothesis test *HT* determines if the newly trained model is significantly better than the previous model. The hypothesis test makes sure that the newly trained model is not better due to chance (such as an unlucky sample). The hypothesis test is conservative, and only switches to a new model if we are reasonably sure it is significantly better, to avoid non-monotone decisions. Japkowicz and Shah [7] provide an accessible introduction to understand the frequentist hypothesis testing.

Algorithm 1. M_{SIMPLE} and M_{HT}

---

input: supervised learner A, rounds n, batches S^i
$\qquad u \in \{\text{updateSimple, updateHT}\}$
$\qquad$ if $u = \text{updateHT}$: confidence level α, hypothesis test HT

1   $S_t = \{\}$
2   **for** $i = 1, \ldots, n$ **do**
3   Split S^i in S_t^i and S_v^i
4   Append to $S_t : S_t = [S_t; S_t^i]$
5     $h_i \leftarrow A(S_t)$
6     $Update_i \leftarrow \text{u}(S_v^i, h_i, h_{\text{best}}, \alpha, HT)$ `// see below`
7   Append to $S_t : S_t = [S_t; S_v^i]$
8   **if** $Update_i$ or $i = 1$ **then**
9 | $h_{\text{best}} \leftarrow h_i$
10   **end**
11   Return h_{best} in round i
12 **end**

Function UpdateSimple	**Function** UpdateHT
input: S_v^i, h_i, h_{best}	**input:** S_v^i, h_i, h_{best}, confidence level α,
	$\qquad\qquad$ hypothesis test HT
1 $P_{current} \leftarrow \hat{\epsilon}(h_i, S_v^i)$	1 $p = HT(S_v^i, h_i, h_{\text{best}})$ `// p-value`
2 $P_{best} \leftarrow \hat{\epsilon}(h_{\text{best}}, S_v^i)$	2 return $(p \leq alpha)$
3 return $(P_{current} \leq P_{best})$	

The choice of hypothesis test depends on the performance measure. For the error rate the McNemar test can be used [7,14]. The hypothesis test should use paired data, since we evaluate two models on one sample, and it should be one-tailed. One-tailed, since we only want to know whether h_i is better than h_{best} (a two tailed test would switch to h_i if its performance is significantly different). The test compares two hypotheses: $H_0 : \epsilon(h_i) = \epsilon(h_{\text{best}})$ and $H_1 : \epsilon(h_i) < \epsilon(h_{\text{best}})$.

Several versions of the McNemar test can be used [4,7,14]. We use the McNemar exact conditional test which we briefly review. Let b be the random variable indicating the number of samples classified correctly by h_{best} and incorrectly by h_i of the sample S_v^i, and let N_d be the number of samples where they disagree. The test conditions on N_d. Assuming H_0 is true, $P(b = x | H_0, N_d) = \binom{N_d}{x}(\frac{1}{2})^{N_d}$. Given x b's, the p-value for our one tailed test is $p = \sum_{i=0}^{x} P(b = i | H_0, N_d)$.

The one tailed p-value is the probability of observing a more extreme sample given hypothesis H_0 considering the tail direction of H_1. The smaller the p-value, the more evidence we have for H_1. If the p-value is smaller than α, we accept H_1, and thus we update the model h_{best}. The smaller α, the more conservative the hypothesis test, and thus the smaller the chance that a wrong decision is made due to unlucky sampling. For the McNemar exact conditional test [4] the False Positive Rate (FPR, or the probability to make a Type I error) is bounded by α: $P(p \leq \alpha | H_0) \leq \alpha$. We need this to prove monotonicity with high probability.

MT$_{CV}$: Monotone Cross Validation. In practice, often K-fold cross validation (CV) is used to estimate model performance instead of the holdout. This is what MT$_{CV}$ does, and is similar to MT$_{SIMPLE}$. As described in Algorithm 2, for each incoming sample an index I maintains to which fold it belongs. These indices are used to generate the folds for the K-fold cross validation.

During CV, K models are trained and evaluated on the validation sets. We now have to memorize K previously best models, one for each fold. We average the performance of the newly trained models over the K-folds, and compare that to the average of the best previous K models. This averaging over folds is essential, as this reduces the variance of the model selection step as compared to selecting the best model overall (like MT$_{SIMPLE}$ does).

In our framework we return a single model in each iteration. We return the model with the optimal training set size that performed best during CV. This can further improve performance.

Algorithm 2. M$_{CV}$

---

input: K folds, learner A, rounds n, batches S^i

1   $b \leftarrow 1$ // keeps track of best round
2   $S = \{\}, I = \{\}$
3   **for** $i = 1, \ldots, n$ **do**
4   Generate stratified CV indices for S^i and put in I^i. Each index i indicates to which validation fold the corresponding sample belongs.
5   Append to S: $S \leftarrow [S; S^i]$
6   Append to I: $I \leftarrow [I; I^i]$
7   **for** $k = 1, \ldots, K$ **do**
8          $h_i^k \leftarrow A(S[I \neq k])$ // training set of kth fold
9          $P_i^k \leftarrow \hat{e}(h_i^k, S[I = k])$ // validation set of kth fold
10         $P_b^k \leftarrow \hat{e}(h_b^k, S[I = k])$ // update performance of prev. models
11   **end**
12      $Update_i \leftarrow (mean(P_i^k) \leq mean(P_b^k))$ // mean w.r.t. k
13   **if** $Update_i$ or $i = 1$ **then**
14          $b \leftarrow i$
15   **end**
16      $k \leftarrow \arg\min_k P_b^k$ // break ties
17   Return h_b^k in round i
18 **end**

---

4   Theoretical Analysis

We derive the probability of a monotone learning curve for MT$_{SIMPLE}$ and MT$_{HT}$, and we prove our algorithms are consistent if the model updates enough.

Theorem 1. *Assume we use the McNemar exact conditional test (see Sect. 3) with $\alpha \in (0, \frac{1}{2}]$, then the individual learning curve generated by Algorithm MT$_{HT}$ with n rounds is monotone with probability at least $(1 - \alpha)^n$.*

Proof. First we argue that the probability of making a non-monotone decision in round i is at most α. If $H_1 : \epsilon(h_i) < \epsilon(h_{\text{best}})$ or $H_0 : \epsilon(h_i) = \epsilon(h_{\text{best}})$ is true, we are monotone in round i, so we only need to consider a new alternative hypothesis $H_2 : \epsilon(h_i) > \epsilon(h_{\text{best}})$. Under H_0 we have [4]: $P(p \leq \alpha | H_0) \leq \alpha$. Conditioned on H_2, b is binomial with larger mean than in the case of H_0, thus we observe larger p-values if $\alpha \in (0, \frac{1}{2}]$, thus $P(p \leq \alpha | H_2) \leq P(p \leq \alpha | H_0) \leq \alpha$. Therefore the probability of being non-monotone in round i is at most α. This holds for any model h_i, h_{best} and anything that happened before round i. Since S_v^i are independent samples, being non-monotone in each round can be seen as independent events, resulting in $(1 - \alpha)^n$.   □

If the probability of being non-monotone in all rounds is at most β, we can set $\alpha = 1 - \beta^{\frac{1}{n}}$ to fulfill this condition. Note that this analysis also holds for $\text{MT}_{\text{SIMPLE}}$, since running MT_{HT} with $\alpha = \frac{1}{2}$ results in the same algorithm as $\text{MT}_{\text{SIMPLE}}$ for the McNemar exact conditional test.

We now argue that all proposed algorithms are consistent under some conditions. First, let us revisit the definition of consistency [17].

Definition 1 (Consistency [17]). *Let L be a learner that returns a hypothesis $L(S) \in \mathcal{H}$ when evaluated on S. For all $\epsilon_{excess} \in (0, 1)$, for all distributions D over $X \times Y$, for all $\delta \in (0, 1)$, if there exists a $n(\epsilon_{excess}, D, \delta)$, such that for all $m \geq n(\epsilon_{excess}, D, \delta)$, if L uses a sample S of size m, and the following holds with probability (over the choice of S) at least $1 - \delta$,*

$$\epsilon(L(S)) \leq \min_{h \in \mathcal{H}} \epsilon(h) + \epsilon_{excess}, \qquad (2)$$

then L is said to be consistent.

Before we can state the main result, we have to introduce a bit of notation. U_i indicates the event that the algorithm updates h_{best} (or in case of M_{CV} it updates the variable b). H_i^{i+z} to indicates the event that $\neg U_i \cap \neg U_{i+1} \cap \ldots \cap \neg U_{i+z}$, or in words, that in round i to $i + z$ there has been no update. To fulfill consistency, we need that when the number of rounds grows to infinity, the probability of updating is large enough. Then consistency of A makes sure that h_{best} has sufficiently low error. For this analysis it is assumed that the number of rounds of the algorithms is not fixed.

Theorem 2. *$\text{MT}_{\text{SIMPLE}}$, MT_{HT} and MT_{CV} are consistent, if A is consistent and if for all i there exists a $z_i \in \mathbb{N} \setminus 0$ and $C_i > 0$ such that for all $k \in \mathbb{N} \setminus 0$ it holds that $P(H_i^{i+kz_i}) \leq (1 - C_i)^k$.*

Proof. Let A be consistent with $n_A(\epsilon_{excess}, D, \delta)$ samples. Let us analyze round i where i is big enough such that[2] $|S_t| > n_A(\epsilon_{excess}, D, \frac{\delta}{2})$. Assume that

$$\epsilon(h_{\text{best}}) > \min_{h \in \mathcal{H}} \epsilon(h) + \epsilon_{excess}, \qquad (3)$$

---

[2] In case of MT_{CV}, take $|S_t|$ to be the smallest training fold size in round i.

otherwise the proof is trivial. For any round $j \geq i$, since A produces hypothesis h_j with $|S_t| > n_A(\epsilon_{\text{excess}}, D, \frac{\delta}{2})$ samples,

$$\epsilon(h_j) \leq \min_{h \in \mathcal{H}} \epsilon(h) + \epsilon_{\text{excess}} \tag{4}$$

holds with probability of at least $1 - \frac{\delta}{2}$. Now L should update. The probability that in the next kz_i rounds we don't update is, by assumption, bounded by $(1 - C_i)^k$. Since $C_i > 0$, we can choose k big enough so that $(1 - C_i)^k \leq \frac{\delta}{2}$. Thus the probability of not updating after kz_i more rounds is at most $\frac{\delta}{2}$, and we have a probability of $\frac{\delta}{2}$ that the model after updating is not good enough. Applying the union bound we find the probability of failure is at most δ.   □

A few remarks about the assumption. It tells us, that an update is more and more likely if we have more consecutive rounds where there has been no update. It holds if each z_i rounds the update probability is nonzero. A weaker but also sufficient assumption is $\forall_i : \lim_{z \to \infty} P(H_i^{i+z}) \to 0$.

For $\text{MT}_{\text{SIMPLE}}$ and MT_{CV} the assumption is always satisfied, because these algorithms look directly at the mean error rate—and due to fluctuations in the sampling there is always a non-zero probability that $\hat{\epsilon}(h_i) \leq \hat{\epsilon}(h_{\text{best}})$. However, for MT_{HT} this may not always be satisfied. Especially if the validation batches N_v are small, the hypothesis test may not be able to detect small differences in error—the test then has zero power. If N_v stays small, even in future rounds the power may stay zero, in which case the learner is not consistent.

5   Experiments

We evaluate $\text{MT}_{\text{SIMPLE}}$ and MT_{HT} on artificial datasets to understand the influence of their parameters. Afterward we perform a benchmark where we also include MT_{CV} and a baseline that uses validation data to tune the regularization strength. This last experiment is also performed on the MNIST dataset to get an impression of the practicality of the proposed algorithms. First we describe the experimental setup in more detail.

Experimental Setup. The peaking dataset [3] and dipping dataset [9] are artificial datasets that cause non-monotone behaviour. We use stratified sampling to obtain batches S^i for the peaking and dipping dataset, for MNIST we use random sampling. For simplicity all batches have the same size. N indicates batch size, and N_v and N_t indicate the sizes of the validation and training sets.

As model we use least squares classification [5,15]. This is ordinary linear least squares regression on the classification labels $\{-1, +1\}$ with intercept. For MNIST one-versus-all is used to train a multi-class model. In case there are less samples for training than dimensions, the required inverse of the covariance matrix is ill-defined and we resort to the Moore-Penrose Pseudo-Inverse.

Monotonicity is calculated by the fraction of non-monotone iterations per run. AULC is also calculated per run. We do 100 runs with different batches and average to reduce variation from the randomness in the batches. Each run uses a newly sampled test set consisting of 10000 samples. The test set is used to estimate the true error rate and is not accessible by any of the algorithms.

We evaluate M_{SIMPLE}, M_{HT} and M_{CV} and several baselines. The standard learner just trains on all received data. A second baseline, λ_S, splits the data in train and validation like M_{SIMPLE} and uses the validation data to select the optimal L_2 regularization parameter λ for the least square classifier. Regularization is implemented by adding λI to the estimate of the covariance matrix.

In the first experiment we investigate the influence of N_v and α for MT_{SIMPLE} and MT_{HT} on the decisions. A complicating factor is that if N_v changes, not only decisions change, but also training set sizes because S_v is appended to the training set (see line 7 of Algorithm 1). This makes interpretation of the results difficult because decisions are then made in a different context. Therefore, for the first set of experiments, we do not add S_v to the training sets, also not for the standard learner. For this set of experiment We use $N_t = 4$, $n = 150$, $d = 200$ for the peaking dataset, and we vary α and N_v.

For the benchmark, we set $N_t = 10$, $N_v = 40$, $n = 150$ for peaking and dipping, and we set $N_t = 5$, $N_v = 20$, $n = 40$ for MNIST. We fix $\alpha = 0.05$ and use $d = 500$ for the peaking dataset. For MNIST, as preprocessing step we extract 500 random Fourier-features as also done by Belkin et al. [1]. For MT_{CV} we use $K = 5$ folds. For λ_S we try $\lambda \in \{10^{-5}, 10^{-4.5}, \ldots, 10^{4.5}, 10^5\}$ for peaking and dipping, and we try $\lambda \in \{10^{-3}, 10^{-2}, \ldots, 10^3\}$ for MNIST.

Results. We perform a preliminary investigation of the algorithms M_{SIMPLE} and M_{HT} and the influence of the parameters N_v and α. We show several learning curves in Fig. 1a and d. For small N_v and α we observe MT_{HT} gets stuck: it does not switch models anymore, indicating that consistency could be violated.

In Fig. 1b and e we give a more complete picture of all tried hyperparameters in terms of the AULC. In Fig. 1c and f we plot the fraction of non-monotone decisions during a run (note that the legends for the subfigures are different). Observe that the axes are scaled differently (some are logarithmic). In some cases zero non-monotone decisions were observed, resulting in a missing value due to $\log(0)$. This occurs for example if MT_{HT} always sticks to the same model, then no non-monotone decisions are made. The results of the benchmark are shown in Fig. 2. The AULC and fraction of monotone decisions are given in Table 1.

6   Discussion

First Experiment: Tuning α and N_v. As predicted MT_{SIMPLE} typically performs worse than MT_{HT} in terms of AULC and monotonicity unless N_v is very large. The variance in the estimate of the error rates on S_v^i is so large that in most cases the algorithm doesn't switch to the correct model. However, MT_{SIMPLE} seems to be consistently better than the standard learner in terms of monotonicity and AULC, while MT_{HT} can perform worse if badly tuned.

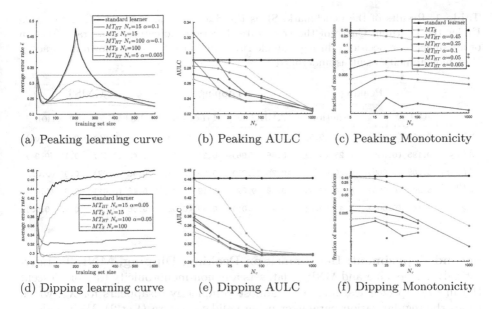

(a) Peaking learning curve   (b) Peaking AULC   (c) Peaking Monotonicity

(d) Dipping learning curve   (e) Dipping AULC   (f) Dipping Monotonicity

Fig. 1. Influence of N_v and α for MT_{SIMPLE} and MT_{HT} on the Peaking and Dipping dataset. Note that some axes are logarithmic and b, c, e, f have the same legend.

Larger N_v leads typically to improved AULC for both. $\alpha \in [0.05, 0.1]$ seems to work best in terms of AULC for most values of N_v. If α is too small, MT_{HT} can get stuck, if α is too large, it switches models too often and non-monotone behaviour occurs. If $\alpha \to \frac{1}{2}$, MT_{HT} becomes increasingly similar to MT_{SIMPLE} as predicted by the theory.

The fraction of non-monotone decisions of MT_{HT} is much lower than α. This is in agreement with Theorem 1, but could indicate in addition that the hypothesis test is rather pessimistic. The standard learner and MT_{SIMPLE} often make non-monotone decisions. In some cases almost 50% of the decisions are not-monotone.

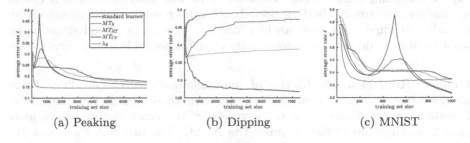

(a) Peaking   (b) Dipping   (c) MNIST

Fig. 2. Expected learning curves on the benchmark datasets.

Table 1. Results of the benchmark. SL is the Standard Learner. AULC is the Area Under the Learning Curve of the error rate. Fraction indicates the average fraction of non-monotone decisions during a single run. Standard deviation shown in (braces). Best monotonicity result is <u>underlined</u>.

	Peaking		Dipping		MNIST	
	AULC	Fraction	AULC	Fraction	AULC	Fraction
SL	0.198 (0.003)	0.31 (0.02)	0.49 (0.01)	0.50 (0.03)	0.44 (0.01)	0.27 (0.04)
MT_S	0.195 (0.005)	0.23 (0.03)	0.45 (0.06)	0.37 (0.15)	0.42 (0.02)	0.11 (0.04)
MT_{HT}	0.208 (0.009)	<u>0.00</u> (0.00)	0.38 (0.08)	<u>0.00</u> (0.00)	0.45 (0.02)	<u>0.00</u> (0.00)
MT_{CV}	0.208 (0.005)	0.34 (0.03)	0.28 (0.02)	0.19 (0.08)	0.45 (0.01)	0.30 (0.06)
λ_S	0.147 (0.003)	0.43 (0.03)	0.49 (0.01)	0.50 (0.03)	0.36 (0.02)	0.46 (0.05)

Second Experiment: Benchmark on Peaking, Dipping, MNIST. Interestingly, for peaking and MNIST datasets any non-monotonicity (double descent [1]) in the *expected* learning curve almost completely disappears for λ_S, which tunes the regularization parameter using validation data (Fig. 2). We wonder if regularization can also help reducing the severity of double descent in other settings. For the dipping dataset, regularization doesn't help, showing that it cannot prevent non-monotone behaviour. Furthermore, the fraction of non-monotone decisions *per run* is largest for this learner (Table 1).

For the dipping dataset M_{CV} has a large advantage in terms of AULC. We hypothesize that this is largely due to tie breaking and small training set sizes due to the 5-folds. Surprisingly on the peaking dataset it seems to learn quite slowly. The expected learning curves of MT_{HT} look better than that of MT_{SIMPLE}, however, in terms of AULC the difference is quite small.

The fraction of non-monotone decisions for MT_{HT} per run is very small as guaranteed. However, it is interesting to note that this does not always translate to monotonicity in the expected learning curve. For example, for peaking and dipping the expected curve doesn't seem entirely monotone. But MT_{CV}, which makes many non-monotone decisions per run, still seems to have a monotone expected learning curve. While monotonicity of each individual learning curves guarantees monotonicity in the expected curve, this result indicates monotonicity of each individual curve may not be necessary. This raises the question: under what conditions do we have monotonicity of the expected learning curve?

General Remarks. The fraction of non-monotone decisions of MT_{HT} being so much smaller than α could indicate the hypothesis test is too pessimistic. Fagerland et al. [4] note that the asymptotic McNemar test can have more power, which could further improve the AULC. For this test the guarantee $P(p \leq \alpha | H_0) \leq \alpha$ can be violated, but in light of the monotonicity results obtained, practically this may not be an issue.

MT$_{HT}$ is inconsistent at times, but this does not have to be problematic. If one knows the desired error rate, a minimum N_v can be determined that ensures the hypothesis test will not get stuck before reaching that error rate. Another possibility is to make the size N_v dependent on i: if N_v is monotonically increasing this directly leads to consistency of MT$_{HT}$. It would be ideal if somehow N_v could be automatically tuned to trade off sample size requirements, consistency and monotonicity. Since for CV N_v automatically grows and thus also directly implies consistency, a combination of MT$_{HT}$ and MT$_{CV}$ is another option.

Devroye et al. [2] conjectured that it is impossible to construct a consistent learner that is monotone in terms of the expected learning curve. Since we look at individual curves, our work does not disprove this conjecture, but some of the authors on this paper believe that the conjecture can be disproved. One step to make is to get to an essentially better understanding of the relation between individual learning curves and the expected one.

Currently, our definition judges any decision that increases the error rate, by however small amount, as non-monotone. It would be desirable to have a broader definition of non-monotonicity that allows for small and negligible increases of the error rate. Using a hypothesis test satisfying such a less strict condition could allow us to use less data for validation.

Finally, the user of the learning system should be notified that non-monotonicity has occurred. Then the cause can be investigated and mitigated by regularization, model selection, etc. However, in automated systems our algorithm can prevent any known and unknown causes of non-monotonicity (as long as data is i.i.d.), and thus can be used as a failsafe that requires no human intervention.

7   Conclusion

We have introduced three algorithms to make learners more monotone. We proved under which conditions the algorithms are consistent and we have shown for MT$_{HT}$ that the learning curve is monotone with high probability. If one cares only about monotonicity of the expected learning curve, MT$_{SIMPLE}$ with very large N_v or MT$_{CV}$ may prove sufficient as shown by our experiments. If N_v is small, or one desires that individual learning curves are monotone with high probability (as practically most relevant), MT$_{HT}$ is the right choice. Our algorithms are a first step towards developing learners that, given more data, improve their performance in expectation.

Acknowledgments. We would like to thank the reviewers for their useful feedback for preparing the camera ready version of this paper.

References

1. Belkin, M., Hsu, D., Ma, S., Mandal, S.: Reconciling modern machine-learning practice and the classical bias-variance trade-off. Proc. Nat. Acad. Sci. **116**(32), 15849–15854 (2019)
2. Devroye, L., Györfi, L., Lugosi, G.: A Probabilistic Theory of Pattern Recognition. Stochastic Modelling and Applied Probability. Springer, Heidelberg (1996). https://doi.org/10.1007/978-1-4612-0711-5
3. Duin, R.: Small sample size generalization. In: Proceedings of the Scandinavian Conference on Image Analysis, vol. 2, pp. 957–964 (1995)
4. Fagerland, M.W., Lydersen, S., Laake, P.: The McNemar test for binary matched-pairs data: mid-p and asymptotic are better than exact conditional. BMC Med. Res. Methodol. **13**, 91 (2013). https://doi.org/10.1186/1471-2288-13-91
5. Hastie, T., Tibshirani, R., Friedman, J.: The Elements of Statistical Learning. SSS. Springer, New York (2009). https://doi.org/10.1007/978-0-387-84858-7
6. Huijser, M., van Gemert, J.C.: Active decision boundary annotation with deep generative models. In: ICCV, pp. 5286–5295 (2017)
7. Japkowicz, N., Shah, M.: Evaluating Learning Algorithms: A Classification Perspective. Cambridge University Press, Cambridge (2011)
8. LeCun, Y., Bottou, L., Bengio, Y., Haffner, P.: Gradient-based learning applied to document recognition. Proc. IEEE **86**(11), 2278–2324 (1998)
9. Loog, M., Duin, R.: The dipping phenomenon. In: S+SSPR, Hiroshima, Japan, pp. 310–317 (2012)
10. Loog, M., Viering, T., Mey, A.: Minimizers of the empirical risk and risk monotonicity. In: NeuRIPS, vol. 32, pp. 7476–7485 (2019)
11. Mohri, M., Rostamizadeh, A., Talwalkar, A.: Foundations of Machine Learning. MIT Press, Cambridge (2012)
12. Opper, M., Kinzel, W., Kleinz, J., Nehl, R.: On the ability of the optimal perceptron to generalise. J. Phys. A: Math. General **23**(11), L581 (1990)
13. O'Neill, J., Jane Delany, S., MacNamee, B.: Model-free and model-based active learning for regression. In: Angelov, P., Gegov, A., Jayne, C., Shen, Q. (eds.) Advances in Computational Intelligence Systems. AISC, vol. 513, pp. 375–386. Springer, Cham (2017). https://doi.org/10.1007/978-3-319-46562-3_24
14. Raschka, S.: Model evaluation, model selection, and algorithm selection in machine learning (2018). arXiv preprint arXiv:1811.12808
15. Rifkin, R., Yeo, G., Poggio, T.: Regularized least-squares classification. Nato Sci. Ser. Sub Ser. III Comput. Syst. Sci. **190**, 131–154 (2003)
16. Settles, B., Craven, M.: An analysis of active learning strategies for sequence labeling tasks. In: EMNLP, pp. 1070–1079 (2008)
17. Shalev-Shwartz, S., Ben-David, S.: Understanding Machine Learning: From Theory to Algorithms. Cambridge University Press, Cambridge (2014)
18. Spigler, S., Geiger, M., D'Ascoli, S., Sagun, L., Biroli, G., Wyart, M.: A jamming transition from under- to over-parametrization affects loss landscape and generalization (2018). arXiv preprint arXiv:1810.09665
19. Viering, T., Mey, A., Loog, M.: Open problem: monotonicity of learning. In: Conference on Learning Theory, COLT, pp. 3198–3201 (2019)

Combining Machine Learning and Simulation to a Hybrid Modelling Approach: Current and Future Directions

Laura von Rueden[1,2](✉), Sebastian Mayer[1,3], Rafet Sifa[1,2],
Christian Bauckhage[1,2], and Jochen Garcke[1,3,4]

[1] Fraunhofer Center for Machine Learning, Sankt Augustin, Germany
[2] Fraunhofer IAIS, Sankt Augustin, Germany
[3] Fraunhofer SCAI, Sankt Augustin, Germany
[4] Institute for Numerical Simulation, University of Bonn, Bonn, Germany
laura.von.rueden@iais.fraunhofer.de

Abstract. In this paper, we describe the combination of machine learning and simulation towards a hybrid modelling approach. Such a combination of data-based and knowledge-based modelling is motivated by applications that are partly based on causal relationships, while other effects result from hidden dependencies that are represented in huge amounts of data. Our aim is to bridge the knowledge gap between the two individual communities from machine learning and simulation to promote the development of hybrid systems. We present a conceptual framework that helps to identify potential combined approaches and employ it to give a structured overview of different types of combinations using exemplary approaches of simulation-assisted machine learning and machine-learning assisted simulation. We also discuss an advanced pairing in the context of Industry 4.0 where we see particular further potential for hybrid systems.

Keywords: Machine learning · Simulation · Hybrid approaches

1 Introduction

Machine learning and *simulation* have a similar goal: To predict the behaviour of a system with data analysis and mathematical modelling. On the one side, machine learning has shown great successes in fields like image classification [21], language processing [24], or socio-economic analysis [7], where causal relationships are often only sparsely given but huge amounts of data are available. On the other side, simulation is traditionally rooted in natural sciences and engineering, e.g. in computational fluid dynamics [35], where the derivation of causal relationships plays an important role, or in structural mechanics for the performance evaluation of structures regarding reactions, stresses, and displacements [6].

However, some applications can benefit from combining machine learning and simulation. Such an hybrid approach can be useful when the processing

© The Author(s) 2020
M. R. Berthold et al. (Eds.): IDA 2020, LNCS 12080, pp. 548–560, 2020.
https://doi.org/10.1007/978-3-030-44584-3_43

capabilities of classical simulation computations can not handle the available dimensionality of the data, for example in earth system sciences [30], or when the behaviour of a system that is supposed to be predicted is based on both known, causal relationships and unknown, hidden dependencies, for example in risk management [25]. However, such challenges are in practice often still approached distinctly with either machine learning or simulation, apparently because they historically originate from distinct fields. This raises the question how these two modelling approaches can be combined into a hybrid approach in order to foster intelligent data analysis. Here, a key challenge in developing a hybrid modelling approach is to bridge the knowledge gap between the two individual communities, which are mostly either experts for machine learning or experts for simulation. Both groups have extremely deep knowledge about the methods used in their particular fields. However, the respectively used terminologies are different, so that an exchange of ideas between both communities can be impeded.

Related work that describes a combination of machine learning with simulation can roughly be divided in two groups, not surprisingly, either from a machine learning or a simulation point of view. The first group frequently describes the integration of simulation into machine learning as an additional source for training data, for example in autonomous driving [23], thermodynamics [19], or biomedicine [13]. A typical motivation is the augmentation of data for scenarios that are not sufficiently represented in the available data. The second group of related works describes the integration of machine learning techniques in simulation, often for a specific application, such as car crash simulation [6], fluid simulation [38], or molecular simulation [26]. A typical motivation is to identify surrogate models [16], which offer an approximate but cheaper to evaluate model to replace the full simulation. Another technique that is used to adapt a dynamical simulation model to new measurements is data assimilation, which is traditionally used in weather forecasting [22]. Related work that considers an equal combination of machine learning and simulation is quite rare. A work that is closest to describing such a hybrid, symbiotic modelling approach is [4].

More general, the integration of prior knowledge into machine learning can be described as *informed machine learning* [34] or *theory-guided data science* [18]. The paper [34] presents a survey with a taxonomy that structures approaches according to the knowledge type, representation, and integration stage. We reuse those categories in this paper. However, that survey considers a much broader spectrum of knowledge representations, from logic rules over simulation results to human interaction, while this paper puts an explicit focus on simulations.

Our goal is to make the key components of the two modelling approaches *machine learning* and *simulation* transparent and to show the versatile, potential combination possibilities in order to inspire and foster future developments of hybrid systems. We do not intend to go into technical details but rather give a high-level methodological overview. With our paper we want to outline a vision of a stronger, more automated interplay between data- and simulation-based analysis methods. We mainly aim our findings at the data analysis and machine

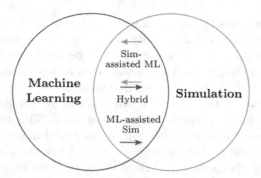

Fig. 1. Subfields of Combining Machine Learning and Simulation. The fields of machine learning and simulation have an intersecting area, which we partition into three subfields: 1. Simulation-assisted machine learning describes the integration of simulations into machine learning. 2. Machine-learning assisted simulation describes the integration of machine learning into simulation. 3. A hybrid combination describes a combination of machine learning and simulation with a strong mutual interplay.

learning community, but also those from the simulation community are welcome to read on. Generally, our target audience are researchers and users of one of the two modelling approaches who want to learn how they can use the other one.

The contributions of this paper are: 1. A conceptual framework serving as an orientation aid for comparing and combining machine learning and simulation, 2. a structured overview of combinations of both modelling approaches, 3. our vision of a hybrid approach with a stronger interplay of data- and simulation based analysis.

The paper is structured as follows: In Sect. 2 we give a brief overview of the subfields that result from combining machine learning and simulation. In Sect. 3 we present these two separate modelling approaches along our conceptual framework. In Sect. 4 we describe the versatile combinations by giving exemplary references and applications. In Sect. 5 we further discuss our observations in Industry 4.0 projects that lead us to a vision for the advanced pairing of machine learning and simulation. Finally we conclude in Sect. 6.

2   Overview

In this section, we give a short overview about the subfields that result from a combination of machine learning with simulation. We view the combination with equal focus on both fields, driving our vision of a hybrid modelling approach with a stronger and automated interplay. Figure 1 illustrates our view on the fields' overlap, which can be partitioned into the three subfields simulation-assisted machine learning, machine-learning assisted simulation, and a hybrid combination. Even though the first two can be regarded as one-sided approaches because they describe the integration with a point of view from one approach, the last one can be regarded as a two-sided approach. Although the term *hybrid*

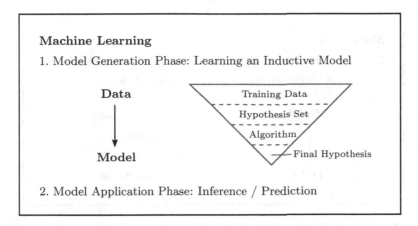

Fig. 2. Components of Machine Learning. Machine Learning consists of two phases 1. model generation, and 2. model application, where the focus is usually made on the first phase, in which an inductive model is learned from data. The components of this phase are the training data, a hypothesis set, a learning algorithm, and a final hypothesis [1,34]. It describes the finding of patterns in an initially large data space, which are finally represented in a condensed form by the final hypothesis. This is illustrated by the reversed triangle and can be described as a "bottom-up approach".

is in the literature often used for the above one-sided approaches, we prefer to use it only for the two-sided approach where machine learning and simulation have a strong mutual, symbiotic-like interplay.

3   Modelling Approaches

In this section, we describe the two modelling approaches by means of a conceptual framework that aims to make them and their components transparent and comparable.

3.1   Machine Learning

The main goal of machine learning is that a machine automatically learns a model that describes patterns in given data. The typical components of machine learning are illustrated in Fig. 2. In the first, main phase an inductive model is learned. Inductive means that the model is built by drawing conclusions from samples and is thus not guaranteed to depict causal relationships, but can instead identify hidden, previously unknown patterns, meaning that the model is usually not knowledge-based but rather data-based. This inductive model can finally be applied to new data in order to predict or infer a desired target variable.

The model generation phase can be roughly split into four sub-phases or respective components [1,34]. Firstly, training data is prepared that depicts historical records of the investigated process or system. Secondly, a hypothesis set

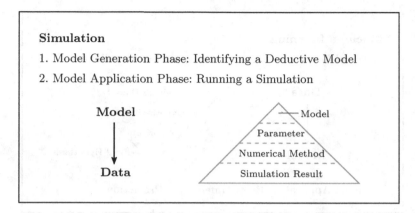

Fig. 3. Components of Simulation. Simulation comprises the two phases 1. model generation, and 2. model application, where the focus often is on the second phase, in which an earlier identified deductive model is used in order to create simulation results. The components of this phase are the simulation model, input parameters, a numerical method, and the simulation result. It describes the unfolding of local interactions from a compactly represented initial model into an expanded data space. This is supposed be illustrated by the triangle and can be described as a "top-down approach".

is defined in the form of a function class or network architecture that is assumed to map input features to the target variables. Thirdly, a learning algorithm tunes the parameters of the hypothesis set so that the performance of the mapping is maximized by using optimization algorithms like gradient descent and results in, fourthly, the final hypothesis, which is the desired inductive model. This model generation phase is often repeated in a loop-like manner by tuning hyper-parameters until a sufficient model performance is achieved.

3.2   Simulation

The goal of a simulation is to predict the behaviour of a system or process for a particular situation. There are different types of simulations, ranging from cellular automata, over agent-based simulations, to equation-based simulations [9,15,36]. In the following we concentrate on the last type, which is based on mathematical models and is especially used in science and engineering. The first, required stage preceding the actual simulation is the identification of a deductive model, often in the form of differential equations. Deductive in this context means that the model describes causal relationships and can thus be called knowledge-based. Such models are often developed through extensive research, starting with a derivation, for example in theoretical physics, and continuing with plentiful experimental validations. Some recent research exists of proof-of-concepts for identifying models directly from data [8,33].

The main phase of a simulation is the application of the identified model for a specific scenario, often called running a simulation. This phase can be described in four typical main components or sub-phases, which are, as illustrated in Fig. 3, the mathematical model, the input parameters, the numerical

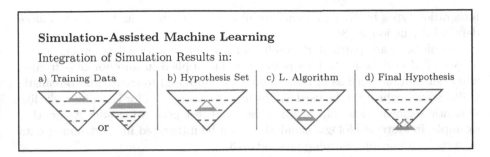

Fig. 4. Types of Simulation-Assisted Machine Learning. Simulations, in particular the simulation results, can be generally integrated into the four different components of machine learning. The triangles illustrate the machine learning (blue/dark gray) or the simulation (orange/light gray) approach and their components, which are themselves presented in Figs. 2 and 3. The simulation results can be used to (a) augment the training data, (b) define parts of the hypothesis set in the form of empirical functions, (c) steer the training algorithm in generative adversarial networks, or (d) verify the final hypothesis against scientific consistency. (Color figure online)

method, and finally the simulation result [36]. After the selection of a mathematical model, the input parameters that describe the specific scenario are defined in the second sub-phase. They can comprise general parameters such as the spatial domain or time of interest, as well as initial conditions quantifying the systems' or processes' initial status and boundary conditions defining the behaviour at domain borders. In the third sub-phase, a numerical method computes the solution of the given model observing the constraints resulting from the input parameters. Examples for numerical methods are finite differences, finite elements or finite volume methods for spatial discretization [36], or particle methods based on interaction forces [26]. These form the basis for an approximate solution, which is the final simulation result. This model application phase is often repeated in a loop-like manner, e.g., by tuning the discretization to achieve a desired approximation accuracy and stability of the solution.

4 Combining Machine Learning and Simulation

In this section, we describe combinations of machine learning and simulation by using our conceptual framework from Sect. 3. Here, we focus on simulation-assisted machine learning and machine-learning assisted simulation. For each of the methodical combination types, we give exemplary application references.

4.1 Simulation-Assisted Machine Learning

Simulation offers an additional source of information for machine learning that goes beyond typically available data and that is rich of knowledge. This additional information can be integrated into the four components of machine learning as illustrated in Fig. 4. In the following, we will give an overview about these integration types by giving for each an illustrative example and refer for a more detailed discussion to [34].

Simulations are particularly useful for creating additional training data in a controlled environment. This is for example applied in autonomous driving, where simulations such as physics engines are employed to create photo-realistic traffic scenes, which can be used as synthetic training data for learning tasks like semantic segmentation [14], or for adversarial test generation [40]. As another example, in systems biology, simulations can be integrated in the training data of kernelized machine learning methods [13].

Moreover, simulations can be integrated into the hypothesis set, either directly as the solvers or through deduced, empirical functions that compactly describe the simulations results. These functions can be built into the architecture of a neural network, as shown for the application of finding an optimal design strategy for a warm forming process [20].

The integration of simulations into the learning algorithm can for example be realized by generative adversarial networks (GANs), which learn a prediction function that obeys constraints, which might be unknown but are implicitly given through a simulation [31].

Another important integration type is in the validation of the final hypothesis by simulations. An example for this comes from material discovery, where first a machine learning model suggests new compounds based on patterns in a data basis, and second the physical properties are computed and thus checked by a density functional theory simulation [17].

An approach that uses simulations along the whole machine learning pipeline is reinforcement learning (RL), when the model is learned in a simulated environment [2]. Studies under the keyword "sim-to-real" are often concerned with robots learning to grip or move unknown objects in simulations and usually require retraining in reality. An application for controlling the temperature of plasma follows the analogous approach, i.e., a training based on a software-physics model, where the learned RL model is then further adapted for use in reality [41].

4.2   Machine-Learning Assisted Simulation

Machine learning is often used in simulation with the intention to support the solution process or to detect patterns in the simulation data. With respect to our conceptual framework presented in Sect. 3, machine learning techniques can be used for the initial model, the input parameters, the numerical method, and the final simulation results, as illustrated in Fig. 4. In the following we will give an overview about the integration types. Again, we do not intend to cover the full spectrum of machine-learning assisted simulation, we rather want to illustrate its diverse approaches through representative examples.

A prominent integration type of machine learning techniques into simulation is the identification of simpler models, such as surrogate models [11,12,16,26].

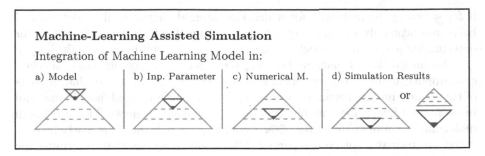

Fig. 5. Types of Machine-Learning Assisted Simulation. Machine learning techniques, in particular the final hypothesis, can be used in different simulation components. The triangles illustrate the machine learning (blue/dark gray) or the simulation (orange/light gray) approach and their components, which are themselves explained in Figs. 2 and 3. Exemplary use cases for machine learning models in simulation are (a) model order reduction and the development of surrogate models that offer approximate but simpler solutions, (b) the automated inference of an intelligent choice of input parameters for a next simulation run, (c) a partly trainable solver for differential equations, or d) the identification of patterns in simulation results for scientific discovery. (Color figure online)

These are approximate and cheap to evaluate models that are particularly of interest when the solution of the original, more precise model is very time- or resource-consuming. The surrogate model can then be used to analyse the overall behaviour of the system in order to reveal scenarios that should be further investigated with the detailed original simulation model. Such surrogate models can be developed with machine-learning techniques either with data from real-world experiments, or with data from high-fidelity simulations. One application example is the optimization of process parameters using deep neural networks as surrogate models [27]. Kernel-based approaches are also commonly used as surrogate models for simulations, an example to improve the energetic efficiency of a gas transport network is shown in [10]. A well-established approach for surrogate modelling is model order reduction, for example with proper orthogonal decomposition, which is closely related to principal component analysis [5,37].

Data assimilation, which includes the calibration of constitutive models and the estimation of system states, is another area where machine learning techniques enhance simulations. Data assimilation problems can be modelled using dynamic Bayesian networks with continuous physically interpretable state spaces where the evaluation of transition kernels and observation operators requires forward-simulation runs [29].

Machine learning techniques can also be used to study the parameter dependence of simulation results. For example, after an engineer executes a sequence of simulations, a machine learning model can detect different behavioral modes in the results and thus reduce the analysis effort during the engineering process [6]. This supports the selection of the parameter setting for the next simulation, for which active learning techniques can also be employed. For example, [39] studied it for selecting the molecules for which the internal energy shall be determined by computationally expensive quantum-mechanical calculations, as well as for determining a surrogate model for the fluid flow in a well-bore while drilling.

The integration of machine learning techniques into the numerical method can support to obtain the numerical solution. One approach is to exchange parts of the model that are resource-consuming to solve, with learned models that can be computed faster, for example with machine learning generated force fields in molecular dynamics simulations [26]. Another approach that is recently investigated are trainable solvers for partial differential equations that determine the complete solution through a neural network [28].

A further, very important integration type is the application of machine learning techniques on the simulation results in order to detect patterns, often motivated by the goal of scientific discovery. While there are plenty of application domains, two exemplary representatives are particle physics [3] and earth-sciences, for example with the use of convolutional neural networks for the detection of weather patterns on climate simulation data [30]. For further examples we refer to a survey about explainable machine learning for scientific discovery [32].

5   Advanced Pairing of Machine Learning and Simulation

Section 4 gave a brief overview of the versatile existing approaches that integrate aspects of machine learning into simulation and vice versa, or that combine simulation and machine learning sequentially. Yet, we think that the integration of these two established worlds is only at the beginning, both in terms of modelling approaches and in terms of available software solutions.

In the following, we describe a number of observations from our project experience in the development of cyber-physical systems for Industry 4.0 applications that support this assessment. Note that the key technical goal of Industry 4.0 is the flexibilization of production processes. In addition to the broad integration of digital equipment in the production machinery, a key provider of flexibilization is a decrease of process design and dimensioning times and ideally, a merging of planning and production phase that are today still strictly separated. This requires a new generation of computer-aided engineering (CAE) software systems that allow for very fast process optimization cycles with real time feedback loops to the production machinery. An advanced pairing of machine learning and simulation will be key to realize such systems by addressing the following issues:

- **Simulation results are not fully exploited:** Especially in the industrial practice, simulations are run with a very specific analysis goal based on expert-designed quantities of interest. This ignores that the simulation result might reveal more patterns and regularities, which might be irrelevant for the current analysis goal but useful in other contexts.
- **Selective surrogate modelling**: Even if modern machine learning approaches are used, surrogate models are built for very specific purposes and the decision when and where to use a surrogate model is left to domain

experts. In this way, it is exploited too little that similar underlying systems might lead to similar surrogate models and in consequence, too many costly high-fidelity simulations are run to generate the data basis, although parts of the learned surrogate models could be transferred.

- **Parameter studies and simulation engines:** Parameter and design studies are well-established tools in many fields of engineering. Surprisingly, the frameworks to conduct these studies and to build the surrogate models are third-party solutions that are separated from the core simulation engines. For the parameter study framework, the simulation engine is a black box, which does not know that it is currently used for a parameter study. In turn, the standard rules to generate sampling points in the parameter space are not aware about the internals of the simulation engine. This raises the question how much more efficient parameter studies could be conducted so that both software systems were stronger connected to each other.

These observations lead us to a research concept that we propose in this paper and call it **learning simulation engines**. A learning simulation engine is a hybrid system that combines machine learning and simulation in an optimal way. Such an engine can automatically decide when and where to apply learned surrogate models or high-fidelity simulations. Surrogate models are efficiently organized and re-used through the use of transfer learning. Parameter and design optimization is an integral component of the learning simulation engine and active learning methods allow the efficient re-use of costly high-fidelity computations.

Of course, the vision of a learning simulation engine raises numerous research questions. We describe some of them in view of Fig. 1. First of all, the question is how learning and simulation can be technically combined to such an advanced hybrid approach, especially, if they can only be integrated into each other by using the final simulation results and the final hypothesis (as shown in Figs. 4 and 5), or if they can also be combined at an earlier sub-phase. Moreover, the counterparts of the learning's model generation phase and the simulation's model application phase (see Figs. 2 and 3) should be investigated further in order to better understand the similarities and differences to the simulation's model generation phase and a learning's model application phase.

6   Conclusion

In this paper, we described the combination of machine learning and simulation motivated by fostering intelligent analysis of applications that can benefit from a combination of data- and knowledge-based solution approaches.

We categorized the overlap between the two fields into three sub-fields, namely, simulation-assisted machine learning, machine-learning assisted simulation, and a hybrid approach with a strong and mutual interplay. We presented a conceptual framework for the two separate approaches, in order to make them and their components transparent for the development of a potential combined approach. In summary, it describes machine learning as a bottom-up approach that generates an inductive, data-based model and simulation as a top-down approach that applies a deductive, knowledge-based model. Using this conceptual framework as an orientation aid for their integration into each other, we gave a structured overview about the combination of machine learning and simulation. We showed the versatility of the approaches through exemplary methods and use cases, ranging from simulation-based data augmentation and scientific consistency checking of machine learning models, to surrogate modelling and pattern detection in simulations for scientific discovery. Finally, we described the scenario of an advanced pairing of machine learning and simulation in the context of Industry 4.0 where we see particular further potential for hybrid systems.

References

1. Abu-Mostafa, Y.S., Magdon-Ismail, M., Lin, H.T.: Learning From Data (2012)
2. Akkaya, I., et al.: Solving rubik's cube with a robot hand (2019). arXiv:1910.07113
3. Albertsson, K., Altoe, P., Anderson, D., Andrews, M., Espinosa, J.P.A., Aurisano, A., Basara, L., Bevan, A., Bhimji, W., et al.: Machine learning in high energy physics community white paper. J. of Phys.: Conf. Ser. **1085**, 022008 (2018)
4. Baker, R.E., Pena, J.M., Jayamohan, J., Jérusalem, A.: Mechanistic models versus machine learning, a fight worth fighting for the biological community? Biol. Lett. **14**(5), 20170660 (2018)
5. Benner, P., Gugercin, S., Willcox, K.: A survey of projection-based model reduction methods for parametric dynamical systems. SIAM Rev. **57**(4), 483–531 (2015)
6. Bohn, B., Garcke, J., Iza-Teran, R., Paprotny, A., Peherstorfer, B., Schepsmeier, U., Thole, C.A.: Analysis of car crash simulation data with nonlinear machine learning methods. Proc. Comput. Sci. **18**, 621–630 (2013)
7. Bollen, J., Mao, H., Pepe, A.: Modeling public mood and emotion: Twitter sentiment and socio-economic phenomena. In: AAAI Conference Weblogs and Social Media (2011)
8. Brunton, S.L., Proctor, J.L., Kutz, J.N.: Discovering governing equations from data by sparse identification of nonlinear dynamical systems. Proc. Nat. Acad. Sci. **113**(15), 3932–3937 (2016)
9. Bungartz, H.J., Zimmer, S., Buchholz, M., Pflger, D.: Modeling and Simulation (2014)
10. Clees, T., Hornung, N., Nikitin, I., Nikitina, L., Steffes-lai, D.: RBF-metamodel driven multi-objective optimization and its applications. Int. J. Adv. Intell. Syst. **9**(1), 19–24 (2016)
11. Cozad, A., Sahinidis, N.V., Miller, D.C.: Learning surrogate models for simulation-based optimization. AIChE J. **60**(6), 2211–2227 (2014)
12. Cranmer, K., Brehmer, J., Louppe, G.: The frontier of simulation-based inference (2019). arXiv:1911.01429
13. Deist, T.M., Patti, A., Wang, Z., Krane, D., Sorenson, T., Craft, D.: Simulation-assisted machine learning. Bioinformatics **35**(20), 4072–4080 (2019)
14. Dosovitskiy, A., Ros, G., Codevilla, F., Lopez, A., Koltun, V.: Carla: An open urban driving simulator (2017). arXiv:1711.03938
15. Durán, J.M.: Computer Simulations in Science and Engineering. TFC. Springer, Heidelberg (2018). https://doi.org/10.1007/978-3-319-90882-3

16. Forrester, A., Sobester, A., Keane, A.: Engineering Design via Surrogate Modelling: A Practical Guide. John Wiley, Hoboken (2008)
17. Hautier, G., Fischer, C.C., Jain, A., Mueller, T., Ceder, G.: Finding natures missing ternary oxide compounds using machine learning and density functional theory. Chem. Mater. **22**(12), 3762–3767 (2010)
18. Karpatne, A., Atluri, G., Faghmous, J.H., Steinbach, M., Banerjee, A., Ganguly, A., Shekhar, S., Samatova, N., Kumar, V.: Theory-guided data science: a new paradigm for scientific discovery from data. IEEE Trans. Knowl. Data Eng. **29**(10), 2318–2331 (2017)
19. Karpatne, A., Watkins, W., Read, J., Kumar, V.: Physics-guided neural networks (pgnn): an application in lake temperature modeling (2017). arXiv:1710.11431
20. Kim, H.S., Koc, M., Ni, J.: A hybrid multi-fidelity approach to the optimal design of warm forming processes using a knowledge-based artificial neural network. Int. J. Mach. Tools Manuf. **47**(2), 211–222 (2007)
21. Krizhevsky, A., Sutskever, I., Hinton, G.E.: Imagenet classification with deep convolutional neural networks. In: NIPS (2012)
22. Lahoz, W., Khattatov, B., Menard, R. (eds.): Data Assimilation. Making Sense of Observations. Springer, Heidelberg (2010). https://doi.org/10.1007/978-3-540-74703-1
23. Lee, K.H., Li, J., Gaidon, A., Ros, G.: Spigan: Privileged adversarial learning from simulation. In: ICLR (2019)
24. Mikolov, T., Sutskever, I., Chen, K., Corrado, G.S., Dean, J.: Distributed representations of words and phrases and their compositionality. In: NIPS (2013)
25. Mitchell-Wallace, K., Foote, M., Hillier, J., Jones, M.: Natural Catastrophe Risk Management and Modelling: A practitioner's Guide. John Wiley, Hoboken (2017)
26. Noé, F., Tkatchenko, A., Müller, K.R., Clementi, C.: Machine learning for molecular simulation (2019). arXiv:1911.02792
27. Pfrommer, J., Zimmerling, C., Liu, J., Kärger, L., Henning, F., Beyerer, J.: Optimisation of manufacturing process parameters using deep neural networks as surrogate models. Proc. CIRP **72**(1), 426–431 (2018)
28. Raissi, M., Perdikaris, P., Karniadakis, G.E.: Physics informed deep learning (part i): Data-driven solutions of nonlinear partial differential equations (2017). arXiv:1711.10561
29. Reich, S., Cotter, C.: Probabilistic Forecasting and Bayesian Data Assimilation. Cambridge University Press, Cambridge (2015)
30. Reichstein, M., Camps-Valls, G., Stevens, B., Jung, M., Denzler, J., Carvalhais, N., et al.: Deep learning and process understanding for data-driven earth system science. Nature **566**(7743), 195–204 (2019)
31. Ren, H., Stewart, R., Song, J., Kuleshov, V., Ermon, S.: Adversarial constraint learning for structured prediction. In: IJCAI (2018)
32. Roscher, R., Bohn, B., Duarte, M.F., Garcke, J.: Explainable machine learning for scientific insights and discoveries (2020). IEEE Access
33. Rudy, S.H., Brunton, S.L., Proctor, J.L., Kutz, J.N.: Data-driven discovery of partial differential equations. Sci. Adv. **3**(4), e1602614 (2017)
34. von Rueden, L., Mayer, S., Beckh, K., Georgiev, B., Giesselbach, S., Heese, R., Kirsch, B., Pfrommer, J., Pick, A., Ramamurthy, R., Walczak, M., Garcke, J., Bauckhage, C., Schuecker, J.: Informed machine learning - a taxonomy and survey of integrating knowledge into learning systems (2020). arXiv:1903.12394v2
35. Shaw, C.T.: Using Computational Fluid Dynamics (1992)
36. Strang, G.: Computational Science and Engineering, vol. 791 (2007)

37. Swischuk, R., Mainini, L., Peherstorfer, B., Willcox, K.: Projection-based model reduction: formulations for physics-based machine learning. Comput. Fluids **179**, 704–717 (2019)
38. Tompson, J., Schlachter, K., Sprechmann, P., Perlin, K.: Accelerating Eulerian fluid simulation with convolutional networks. In: ICML (2017)
39. Tsymbalov, E., Makarychev, S., Shapeev, A., Panov, M.: Deeper connections between neural networks and gaussian processes speed-up active learning (2019). arXiv:1902.10350
40. Tuncali, C.E., Fainekos, G., Ito, H., Kapinski, J.: Simulation-based adversarial test generation for autonomous vehicles with machine learning components. In: IEEE Intelligent Vehicles Symposium (2018)
41. Witman, M., Gidon, D., Graves, D.B., Smit, B., Mesbah, A.: Sim-to-real transfer reinforcement learning for control of thermal effects of an atmospheric pressure plasma jet plasma sources. Sci. Technol. **28**(9), 095019 (2019)

LiBRe: Label-Wise Selection of Base Learners in Binary Relevance for Multi-label Classification

Marcel Wever[1]([✉]), Alexander Tornede[1], Felix Mohr[2], and Eyke Hüllermeier[1]

[1] Heinz Nixdorf Institut, Paderborn University, Paderborn, Germany
{marcel.wever,alexander.tornede,eyke}@upb.de
[2] Universidad de La Sabana, Chia, Cundinamarca, Colombia
felix.mohr@unisabana.edu.co

Abstract. In multi-label classification (MLC), each instance is associated with a set of class labels, in contrast to standard classification, where an instance is assigned a single label. Binary relevance (BR) learning, which reduces a multi-label to a set of binary classification problems, one per label, is arguably the most straight-forward approach to MLC. In spite of its simplicity, BR proved to be competitive to more sophisticated MLC methods, and still achieves state-of-the-art performance for many loss functions. Somewhat surprisingly, the optimal choice of the base learner for tackling the binary classification problems has received very little attention so far. Taking advantage of the label independence assumption inherent to BR, we propose a label-wise base learner selection method optimizing label-wise macro averaged performance measures. In an extensive experimental evaluation, we find that or approach, called LiBRe, can significantly improve generalization performance.

Keywords: Multi-label classification · Algorithm selection · Binary relevance

1 Introduction

By relaxing the assumption of mutual exclusiveness of classes, the setting of *multi-label classification* (MLC) generalizes standard (binary or multinomial) classification—subsequently also referred to as single-label classification (SLC). MLC has received a lot of attention in the recent machine learning literature [23, 29]. The motivation for allowing an instance to be associated with several classes simultaneously originated in the field of text categorization [19], but nowadays multi-label methods are used in applications as diverse as image processing [4, 26] and video annotation [14], music classification [18], and bioinformatics [2].

Common approaches to MLC either adapt existing algorithms (*algorithm adaptation*) to the MLC setting, e.g., the structure and the training procedure for neural networks, or reduce the original MLC problem to one or multiple SLC problems (*problem transformation*). The most intuitive and straight-forward

© The Author(s) 2020
M. R. Berthold et al. (Eds.): IDA 2020, LNCS 12080, pp. 561–573, 2020.
https://doi.org/10.1007/978-3-030-44584-3_44

problem transformation is to decompose the original task into several binary classification tasks, one per label. More specifically, each task consists of training a classifier that predicts whether or not a specific label is relevant for a query instance. This approach is called *binary relevance* (BR) learning [3]. Beyond BR, many more sophisticated strategies have been developed, most of them trying to exploit correlations and interdependencies between labels [28]. In fact, BR is often criticized for ignoring such dependencies, implicitly assuming that the relevance of one label is (statistically) independent of the relevance of another label. In spite of this, or perhaps just because of this simplification, BR proved to achieve state-of-the-art performance, especially for so-called decomposable loss functions, for which its optimality can even be corroborated theoretically [7,9].

Techniques for reducing MLC to SLC problems involve the choice of a base learner for solving the latter. Somewhat surprisingly, this choice is often neglected, despite having an important influence on generalization performance [10–12,15]. Even in more extensive studies [10,12], a base learner is fixed a priori in a more or less arbitrary way. Broader studies considering multiple base learners, such as [6,22], are relatively rare and rather limited in terms of the number of base learners considered. Only recently, greater attention to the choice of the base learner has been paid in the field of automated machine learning (AutoML) [17,24,25], where the base learner is considered as an important "hyper-parameter" to tune. Indeed, while optimizing the selection of base learners is laborious and computationally expensive in general, which could be one reason for why it has been tackled with reservation, AutoML now offers new possibilities in this direction.

Motivated by these opportunities, and building on recent AutoML methodology, we investigate the idea of base learner selection for BR in a more systematic way. Instead of only choosing a single base learner to be used for all labels simultaneously, we even allow for selecting an individual learner for each label (i.e., each binary classification task) separately. In an extensive experimental study, we find that customizing BR in a label-wise manner can significantly improve generalization performance.

2   Multi-label Classification

The setting of *multi-label classification* (MLC) allows an instance to belong to several classes simultaneously. Consequently, several class labels can be assigned to an instance at the same time. For example, a single image could be tagged with labels Sun and Beach and Sea and Yacht.

2.1   Problem Setting

To formalize this learning problem, let $\mathcal{X}$ denote an instance space and $\mathcal{L} = \{\lambda_1, \ldots, \lambda_m\}$ a finite set of m class labels. An instance $x \in \mathcal{X}$ is then (non-deterministically) associated with a subset of class labels $L \in 2^{\mathcal{L}}$. The subset L is often called the set of relevant labels, while its complement $\mathcal{L} \setminus L$ is considered irrelevant for $\boldsymbol{x}$. Furthermore, a set L of relevant labels can be identified by a binary vector $\boldsymbol{y} = (y_1, \ldots, y_m)$ where $y_i = 1$ if $\lambda_i \in L$ and $y_i = 0$ otherwise (i.e., if $\lambda_i \in \mathcal{L} \setminus L$). The set of all label combinations is denoted by $\mathcal{Y} = \{0,1\}^m$.

Generally speaking, a multi-label classifier $\boldsymbol{h}$ is a mapping $\boldsymbol{h} : \mathcal{X} \longrightarrow \mathcal{Y}$ returning, for a given instance $\boldsymbol{x} \in \mathcal{X}$, a prediction in the form of a vector

$$\boldsymbol{h}(\boldsymbol{x}) = \big(h_1(\boldsymbol{x}), h_2(\boldsymbol{x}), \ldots, h_m(\boldsymbol{x})\big).$$

The MLC task can be stated as follows: Given a finite set of observations as training data $\mathcal{D}_{\text{train}} := (X_{\text{train}}, Y_{\text{train}}) = \{(\boldsymbol{x}_i, \boldsymbol{y}_i)\}_{i=1}^{N} \subset \mathcal{X}^N \times \mathcal{Y}^N$, the goal is to learn a classifier $\boldsymbol{h} : \mathcal{X} \longrightarrow \mathcal{Y}$ that generalizes well beyond these observations in the sense of minimizing the risk with respect to a specific loss function.

2.2 Loss Functions

A wide spectrum of loss functions has been proposed for MLC, many of which are generalizations or adaptations of losses for single-label classification. In general, these loss functions can be divided into two major categories: instance-wise and label-wise. While the latter first compute a loss for each label and then aggregate the values obtained across the labels, e.g., by taking the mean, instance-wise loss functions first compute a loss for each instance and subsequently aggregate the losses over all instances in the test data. As an obvious advantage of label-wise loss functions, note that they can be optimized by optimizing a standard SLC loss for each label separately. In other words, label-wise losses naturally harmonize with label-wise decomposition techniques such as BR. Since this allows for a simpler selection of the base learner per label, we focus on two such loss functions in the following. For additional details on MLC and loss functions, especially instance-wise losses, we refer to [23,29].

Let $\mathcal{D}_{\text{test}} := (X_{\text{test}}, Y_{\text{test}}) = \{(\boldsymbol{x}_i, \boldsymbol{y}_i)\}_{i=1}^{S} \subset \mathcal{X}^S \times \mathcal{Y}^S$ be a test set of size S. Further, let $H = (\boldsymbol{h}(\boldsymbol{x}_1), \ldots, \boldsymbol{h}(\boldsymbol{x}_S)) \subset \mathcal{Y}^S$. Then, the Hamming loss, which can be seen as a generalized form of the error rate, is defined[1] as

$$\mathcal{L}_H(Y_{\text{test}}, H) := \frac{1}{m} \sum_{j=1}^{m} \frac{1}{S} \sum_{i=1}^{S} [\![y_{i,j} \neq h_j(\boldsymbol{x}_i)]\!] . \tag{1}$$

Moreover, the label-wise macro-averaged F-measure (which is actually a measure of accuracy, not a loss function, and thus to be maximized) is given by

$$F(Y_{\text{test}}, H) := \frac{1}{m} \sum_{j=1}^{m} \frac{2 \sum_{i=1}^{S} y_{i,j} h_j(\boldsymbol{x}_i)}{\sum_{i=1}^{S} y_{i,j} + \sum_{i=1}^{S} h_j(\boldsymbol{x}_i)} . \tag{2}$$

Obviously, to optimize the measures (1) and (2), it is sufficient to optimize each label individually, which corresponds to optimizing the inner term of the (first) sum.

---

[1] $[\![\cdot]\!]$ is the indicator function.

2.3   Binary Relevance

As already said, binary relevance learning decomposes the MLC task into several binary classification tasks, one for each label. For every such task, a single-label classifier, such as an SVM, random forest, or logistic regression, is trained. More specifically, a classifier for the jth label is trained on the dataset $\{(\boldsymbol{x}_i, y_{i,j})\}_{i=1}^N$. Formally, BR induces a multi-label predictor

$$\mathbf{BR}_b : \mathcal{X} \longrightarrow \mathcal{Y}, \quad \boldsymbol{x} \mapsto \big(b_1(\boldsymbol{x}), b_2(\boldsymbol{x}), \ldots, b_m(\boldsymbol{x})\big) \ ,$$

where $b_j : \mathcal{X} \longrightarrow \{0, 1\}$ represents the prediction of the base learner for the jth label.

3   Related Work

Binary relevance has been subject to modifications in various directions, an excellent overview of which is provided in a recent survey [28]. Extensions of BR mainly focus on its inability to exploit label correlations, due to treating all labels independently of each other. Three types of approaches have been proposed to overcome this problem. The first is to use *classifier chains* [15]. In this approach, one first defines a total order among the m labels and then trains binary classifiers in this order. The input of the classifier for the ith label is the original data plus the predictions of *all classifiers* for labels preceding this label in the chain. Similarly, in addition to the binary classifiers for the m labels, *stacking* uses a second layer of m meta-classifiers, one for each label, which take as input the original data augmented by the predictions of *all* base learners [11,21]. A third approach seeks to capture the dependencies in a Bayesian network, and to learn such a network from the data [1,20]. One can then use probabilistic inference to compute the probability for each possible prediction.

Another line of research looks at how the problem of imbalanced classes can be addressed using BR. Class imbalance constitutes an important challenge in multi-label classification in general, since most labels are usually irrelevant for an instance, i.e., the overwhelming majority of labels in a binary task is negative. Using BR, the imbalance can be "repaired" in a label-wise manner, using techniques for standard binary classification, such as sampling [5] or thresholding the decision boundary [13]. An approach taking dependencies among labels into account (and hence applied prior to splitting the problem) is presented in [27].

To the best of our knowledge, this is the first approach in which the base learner used for the different labels is subject to optimization itself. In fact, except for AutoML tools, we are not even aware of an approach optimizing a single base learner applied to all labels. In all the above approaches, the choice of the base learners is an external decision and not part of the learning problem itself.

4   Label-Wise Selection of Base Learners

As already stated before, while various attempts at improving binary relevance learning by capturing label dependencies have been made, the choice of the base learner for tackling the underlying binary problems—as another potential source of improvement—has attracted much less attention in the literature so far. If considered at all, this choice has been restricted to the selection of a *single* learner, which is applied to all m binary problems simultaneously.

We proceed from a portfolio of base learners

$$\mathcal{A} := \{a \mid a : (\mathcal{X}^n \times \{0,1\}^n) \longrightarrow (\mathcal{X} \longrightarrow \{0,1\})\} \, .$$

Then, given training data $\mathcal{D}_{\text{train}} = (X_{\text{train}}, Y_{\text{train}})$, the objective is to find the base learner a for which BR performs presumably best on test data $\mathcal{D}_{\text{test}} = (X_{\text{test}}, Y_{\text{test}})$ with respect to some loss function $\mathcal{L}$:

$$\arg\min_{a \in \mathcal{A}} \mathcal{L}\big(Y_{\text{test}}, \mathbf{BR}_b(X_{\text{test}})\big), \text{ with } b_j := a\left(X_{\text{train}}, Y_{\text{train}}^{(j)}\right) \, , \qquad (3)$$

where $Y_{\text{train}}^{(i)}$ denotes the jth column of the label matrix Y_{train}.

Moreover, we propose to leverage the independence assumption underlying BR to select a different base learner for each of the labels, and refer to this variant as LiBRe. We are thus interested in solving the following problem:

$$\arg\min_{a \in \mathcal{A}^m} \mathcal{L}\big(Y_{\text{test}}, \mathbf{BR}_b(X_{\text{test}})\big), \text{ with } b_j := a_j\left(X_{\text{train}}, Y_{\text{train}}^{(j)}\right) \, . \qquad (4)$$

Compared to (3), we thus significantly increase flexibility. In fact, by taking advantage of the different behavior of the respective base learners, and the ability to model the relationship between features and a class label differently for each binary problem, one may expect to improve the overall performance of BR. On the other side, the BR learner as a whole is now equipped with many degrees of freedom, namely the choice of the base learners, which can be seen as "hyperparameters" of LiBRe. Since this may easily lead to undesirable effects such as over-fitting of the training data, an improvement in terms of generalization performance (approximated by the performance on the test data) is by no means self-evident. From this point of view, the restriction to a single base learner in (3) can also be seen as a sort of regularization. Such kind of regulation can indeed be justified for various reasons. In most cases, for example, the binary problems are indeed not completely different but share important characteristics.

Computationally, (4) may appear more expensive than choosing a single base learner jointly for all the labels, at least at first sight. However, the complexity in terms of the number of base learners to be evaluated remains exactly the same. In fact, just like in (3), we need to fit a BR model for every base learner exactly once. The only difference is that, instead of picking one of the base learners for all labels in the end, LiBRe assembles the base learners performing best for the respective labels (recall that we head for label-wise decomposable performance measures).

5   Experimental Evaluation

This section presents an empirical evaluation of LiBRe, comparing it to the use of a single base learner as a baseline. We first describe the experimental setup (Sect. 5.1), specify the baseline with the single best base learner (Sect. 5.2), and define the oracle performance (Sect. 5.3) for an upper bound. Finally, the experimental results are presented in Sect. 5.4.

5.1   Experimental Setup

For the evaluation, we considered a total of 24 MLC datasets. These datasets stem from various domains, such as text, audio, image classification, and biology, and range from small datasets with only a few instances and labels to larger datasets with thousands of instances and hundreds of labels. A detailed overview is given in Table 1, where, in addition to the number of instances (#I) and number of labels (#L), statistics regarding the label-to-instance ratio (L2IR), the percentage of unique label combinations (ULC), and the average label cardinality (card.) are given.

The train and validation folds were derived by conducting a nested 2-fold cross validation, i.e., to assess the test performance we have an outer loop of 2-fold cross validation. To tune the thresholds and select the base learner, we again split the training fold of the outer loop into train and validation sets by 2-fold cross validation. The entire process is repeated 5 times with different random seeds for the cross validation. Throughout this study, we trained and evaluated a total of 14,400 instances of BR and 649,800 base learners accordingly.

Furthermore, we consider two performance measures, namely the Hamming loss $\mathcal{L}_H$ and the macro-averaged label-wise F-measure as defined in (1) and (2), respectively. A binary prediction is obtained by thresholding the prediction of an underlying scoring classifier, which produces values in the unit interval (the higher the value, the more likely a label is considered relevant). The thresholds $\tau = (\tau_1, \tau_2, \ldots, \tau_m)$ are optimized by a grid search considering values for $\tau_i \in [0, 1]$ and a step size of 0.01. When optimizing the thresholds, we either allow for label-wise optimization or constrain the threshold to be the same for all labels (uniform τ), i.e., $\tau_i = \tau_j$ for all $i, j \in \{1, \ldots, m\}$.

In order to determine significance of results, we apply a Wilcoxon signed rank test with a threshold for the p-value of 0.05. Significant improvements of LiBRe are marked by ● and significant degradations by ○.

We executed the single BR evaluation runs, i.e., training and evaluating either on the validation or test split, on up to 300 nodes in parallel, each of them equipped with 8 CPU cores and 32 GB of RAM, and a timeout of 6 h. Due to the limitation of the memory and the runtime, some of the evaluations failed due to memory overflows or timeouts.

The implementation is based on the Java machine learning library WEKA [8] and an extension for multi-label classification called MEKA [16]. In our study, we consider a total of 20 base learners from WEKA: BayesNet (BN), DecisionStump (DS), IBk, J48, JRip (JR), KStar (KS), LMT, Logistic (L), MultilayerPerceptron

Table 1. The datasets used in this study. Furthermore, the number of instances (#I), the number of labels (#L), the label-to-instance ratio (L2IR), the percentage of unique label combinations (ULC), and the label cardinality (card.) are given.

Dataset	#I	#L	L2IR	ULC	card.	Dataset	#I	#L	L2IR	ULC	card.
arts1	7484	26	0.0035	0.08	1.65	bibtex	7395	159	0.0215	0.39	2.40
birds	645	19	0.0295	0.21	1.01	bookmarks	87856	208	0.0024	0.21	2.03
business1	11214	30	0.0027	0.02	1.60	computers1	12444	33	0.0027	0.03	1.51
education1	12030	33	0.0027	0.04	1.46	emotions	593	6	0.0101	0.05	1.87
enron-f	1702	53	0.0311	0.44	3.38	entertainment1	12730	21	0.0016	0.03	1.41
flags	194	12	0.0619	0.53	4.12	genbase	662	27	0.0408	0.05	1.25
health1	9205	32	0.0035	0.04	1.64	llog-f	1460	75	0.0514	0.21	1.18
mediamill	43907	101	0.0023	0.15	4.38	medical	978	45	0.0460	0.10	1.25
recreation1	12828	22	0.0017	0.04	1.43	reference1	8027	33	0.0041	0.03	1.17
scene	2407	6	0.0025	0.01	1.07	science1	6428	40	0.0062	0.07	1.45
social1	12111	39	0.0032	0.03	1.28	society1	14512	27	0.0019	0.07	1.67
tmc2007	28596	22	0.0008	0.05	2.16	yeast	2417	14	0.0058	0.08	4.24

(MlP), NaiveBayes (NB), NaiveBayesMultinomial (NBM), OneR (1R), PART (P), REPTree (REP), RandomForest (RF), RandomTree (RT), SMO, SimpleLogistic (SL), VotedPerceptron (VP), ZeroR (0R). All the data and source code is made available via GitHub (https://github.com/mwever/LiBRe).

5.2   Single Best Base Learner

To figure out how much we can benefit from selecting a base learner for each label individually, and whether this flexibility is beneficial at all, we define the single best base learner, subsequently referred to as SBB, as a baseline. In principle, SBB is nothing but a grid search over the portfolio of base learners (3).

When considering a base learner a, it is chosen to be employed as a base learner for every label. After training and validating the performance, we pick the base learner that performs best overall. This baseline thus gives an upper bound on the performance of what can be achieved when the base learner is not chosen for each label individually. As simple and straight-forward as it is, this baseline represents what is currently possible in implementations of MLC libraries, and already goes beyond what is most commonly done in the literature.

5.3   Optimistic Versus Validated Optimization

In addition to the results obtained by selecting the base learner(s) according to the validation performance (obtained in the inner loop of the nested cross validation), we consider optimistic performance estimates, which are obtained as follows: After having trained the base learners on the training data, we select the presumably best one, not on the basis of their performance on validation data, but based on their actual test performance (as observed in the outer loop

Fig. 1. The heat map shows the average share of each base learner being employed for a label with respect to the optimized performance measure: Hamming ($\mathcal{L}_H$) or the label-wise macro averaged F-measure (F).

of the nested cross-validation). Intuitively, this can be understood as a kind of "oracle" performance: Given a set of candidate predictors to choose from, the oracle anticipates which of them will perform best on the test data.

Although these performances should be treated with caution, and will certainly tend to overestimate the true generalization performance of a classifier, they can give some information about the potential of the optimization. More specifically, these optimistic performance estimates suggest an upper bound on what can be obtained by the nested optimization routine.

5.4   Results

In Fig. 1, the average share of a base learner per label is shown. From this heatmap, it becomes obvious that for the SBB baseline only a subset of base learners plays a role. However, one can also notice that the distribution of the shares varies when different performance measures are optimized. Furthermore, although random forest (RF) achieves significant shares of 0.8 for the Hamming loss and around 0.6 for the F-measure, it is not best on all the datasets. To put it differently, one still needs to optimize the base learner per dataset. This is especially true, when different performance measures are of interest.

In the case of LiBRe, it is clearly recognizable how the shares are distributed over the base learners, in contrast to SBB. For example, the shares of RF decrease to 0.29 for F-measure and to 0.25 for Hamming, respectively. Moreover, base learners that did not even play any role in SBB are now gaining in importance and are selected quite often. Although there are significant differences in the frequency of base learners being picked, there is not a single base learner in the portfolio that was never selected.

In Table 2, the results for optimizing Hamming loss are presented. The optimistic performance estimates already indicate that there is not much room for improvement. This comes at no surprise, since the datasets are already pretty much saturated, i.e., the loss is already close to 0 for most of the datasets. While LiBRe performs competitively to SBB for the setting with uniform τ, SBB compares favourably to LiBRe in the case where the thresholds can be tuned in a label-wise manner. Apparently, the additional degrees of freedom make LiBRe more prone to over-fitting, especially on smaller datasets.

In contrast to the previous results, for the optimization of the F-measure, the optimistic performance estimates already give a promising outlook on the

Table 2. Results obtained for minimizing $\mathcal{L}_H$ optimistically resp. with validation performances. Thresholds are optimized either jointly for all the labels (uniform τ) or label-wise. Best performances per setting and dataset are highlighted in bold. Significant improvements of LiBRe are marked by a • and degradations by ○.

Dataset	Optimistic uniform τ		Validated uniform τ		Optimistic label-wise τ		Validated label-wise τ	
	LiBRe	SBB	LiBRe	SBB	LiBRe	SBB	LiBRe	SBB
arts1	**0.0515**	0.0536	**0.0531**	0.0538	**0.0504**	0.0513	0.0526	**0.0525**
bibtex	**0.0118**	0.0126	**0.0126**	0.0127	**0.0115**	0.0120	0.0151	**0.0139**
birds	**0.0357**	0.0397	0.0476	**0.0420** ○	**0.0329**	0.0352	0.0470	**0.0422** ○
bookmarks	**0.0085**	0.0087	**0.0086**	0.0087 •	**0.0085**	0.0086	**0.0105**	0.0114 •
business1	**0.0233**	0.0248	**0.0241**	0.0249 •	**0.0218**	0.0223	**0.0227**	0.0228
computers1	**0.0313**	0.0334	**0.0329**	0.0335	**0.0301**	0.0306	0.0323	**0.0312**
education1	**0.0352**	0.0365	**0.0359**	0.0369 •	**0.0340**	0.0344	0.0354	**0.0349** ○
emotions	**0.1762**	0.1800	0.1926	**0.1856** ○	**0.1684**	0.1712	0.1961	**0.1875** ○
enron-f	**0.0447**	0.0474	0.0481	**0.0477**	**0.0437**	0.0445	0.0485	**0.0469** ○
entertainment1	**0.0432**	0.0466	**0.0440**	0.0469 •	**0.0414**	0.0434	**0.0430**	0.0443 •
flags	**0.1732**	0.1979	0.2134	**0.2088**	**0.1635**	0.1799	**0.2105**	0.2158
genbase	**7.0E-4**	0.0014	0.0069	**0.0016** ○	**6.0E-4**	7.0E-4	**0.0070**	0.0023 ○
health1	**0.0305**	0.0344	**0.0313**	0.0347 •	**0.0282**	0.0297	0.0303	**0.0302**
llog-f	**0.0149**	0.0153	0.0202	**0.0157** ○	**0.0145**	0.0149	0.0230	**0.0178** ○
mediamill	**0.0268**	0.0270	0.0271	**0.0270**	**0.0261**	0.0262	**0.0265**	0.0265
medical	**0.0084**	0.0103	0.0115	**0.0109**	**0.0078**	0.0093	0.0136	**0.0116**
recreation1	**0.0459**	0.0472	**0.0472**	0.0473	**0.0446**	0.0453	0.0468	**0.0462**
reference1	**0.0244**	0.0264	**0.0267**	0.0268	**0.0230**	0.0245	0.0255	**0.0251**
scene	**0.0781**	0.0788	0.0817	**0.0794** ○	**0.0757**	0.0762	0.0816	**0.0800** ○
science1	**0.0281**	0.0311	**0.0311**	0.0317	**0.0269**	0.0291	0.0304	**0.0302**
social1	**0.0197**	0.0208	0.0227	**0.0210**	**0.0188**	0.0196	0.0223	**0.0200**
society1	**0.0474**	0.0495	**0.0479**	0.0496 •	**0.0444**	0.0455	**0.0455**	0.0461 •
tmc2007	**0.0601**	0.0611	**0.0600**	0.0611 •	**0.0590**	0.0611	0.0613	**0.0611**
yeast	**0.1914**	0.1926	0.2002	**0.1930** ○	**0.1886**	0.1890	0.1940	**0.1929** ○

potential for improving the generalization performance through the label-wise selection of the base learners. More precisely, they indicate that performance gains of up to 11% points are possible. Independent of the threshold optimization variant, LiBRe outperforms the SBB baseline, yielding the best performance on two third of the considered datasets, 13 improvements of which are significant in the case of uniform τ, and 11 in the case of label-wise τ. Significant degradations of LiBRe compared to SBB can only be observed for 2 respectively 3 datasets. Hence, for the F-measure, LiBRe compares favorably to the SBB baseline.

In summary, we conclude that LiBRe does indeed yield performance improvements. However, increasing the flexibility of BR also makes it more prone to over-fitting. Furthermore, these results were obtained by conducting a nested 2-fold cross validation. While keeping the computational costs of this evaluation reasonable, this implies that, for the purpose of validation, the base learners were trained on only one fourth of the original dataset. Therefore, considering nested 5-fold or 10-fold cross validation could help to reduce the observed over-fitting.

Table 3. Results for maximizing the F-measure optimistically resp. with validation performances. Thresholds are optimized either jointly for all the labels (uniform τ) or label-wise. Best performances per setting and dataset are highlighted in bold. Significant improvements of LiBRe are marked by a • and degradations by ○.

Dataset	Optimistic uniform τ		Validated uniform τ		Optimistic label-wise τ		Validated label-wise τ	
	LiBRe	SBB	LiBRe	SBB	LiBRe	SBB	LiBRe	SBB
arts1	**0.3445**	0.2749	**0.3018**	0.2684 •	**0.3680**	0.3211	**0.3184**	0.3001 •
bibtex	**0.4020**	0.3027	**0.3391**	0.2998 •	**0.4194**	0.3516	**0.3378**	0.3041 •
birds	**0.5404**	0.4424	0.3707	**0.3961** ○	**0.5832**	0.5310	0.3843	**0.3981** ○
bookmarks	**0.2495**	0.2244	**0.2347**	0.2239 •	**0.2646**	0.2516	**0.2435**	0.2416
business1	**0.3692**	0.2854	**0.2970**	0.2659 •	**0.3874**	0.3197	**0.3006**	0.2790 •
computers1	**0.3646**	0.2861	**0.3099**	0.2810 •	**0.3833**	0.3486	**0.3224**	0.3190
education1	**0.3346**	0.2468	**0.2594**	0.2437 •	**0.3591**	0.3022	**0.2652**	0.2612
emotions	**0.7068**	0.6946	0.6670	**0.6779**	**0.7186**	0.7135	0.6761	**0.6859** ○
enron-f	**0.2870**	0.2192	0.2056	**0.2096**	**0.3138**	0.2773	**0.2077**	0.2069
entertainment1	**0.4470**	0.3673	**0.3929**	0.3500 •	**0.4639**	0.4049	**0.3950**	0.3774 •
flags	**0.6280**	0.5634	**0.5230**	0.5098	**0.6474**	0.5981	**0.5150**	0.5145
genbase	**0.8126**	0.7798	0.6039	**0.7421** ○	**0.8141**	0.8119	0.6201	**0.6390**
health1	**0.4203**	0.3259	**0.3486**	0.3208 •	**0.4312**	0.3582	**0.3464**	0.3225 •
llog-f	**0.1569**	0.0808	**0.0730**	0.0689	**0.1834**	0.1264	**0.0744**	0.0741
mediamill	**0.3766**	0.3499	0.3481	**0.3483**	**0.4010**	0.3898	0.3543	**0.3600** ○
medical	**0.4960**	0.3852	0.3560	**0.3639**	**0.5251**	0.4523	**0.3547**	0.3208 •
recreation1	**0.4964**	0.4224	**0.4669**	0.4160 •	**0.5093**	0.4675	**0.4670**	0.4494 •
reference1	**0.3185**	0.2254	**0.2477**	0.2021 •	**0.3393**	0.2860	**0.2587**	0.2418 •
scene	**0.7831**	0.7816	0.7734	**0.7776**	**0.7909**	0.7897	0.7759	**0.7812**
science1	**0.3824**	0.2724	**0.2928**	0.2637 •	**0.4033**	0.3240	**0.3036**	0.2662 •
social1	**0.3629**	0.3073	0.3046	**0.3060**	**0.3737**	0.3119	**0.3103**	0.2769 •
society1	**0.3437**	0.2807	**0.3180**	0.2688 •	**0.3597**	0.3382	0.3215	**0.3238**
tmc2007	**0.5659**	0.5342	**0.5467**	0.5342	**0.5782**	0.5525	**0.5656**	0.5484 •
yeast	**0.4970**	0.4750	**0.4800**	0.4731 •	**0.5145**	0.5084	0.4922	**0.4947**

6   Conclusion

In this paper, we have not only demonstrated the potential of binary relevance to optimize label-wise macro averaged measures, but also the importance of the base learner as a hyper-parameter for each label. Especially for the case of optimizing for F1 macro-averaged over the labels, we could achieve significant performance improvements by choosing a proper base learner in a label-wise manner. Compared to selecting the best single base learner, choosing the base learner for each label individually comes at no additional cost in terms of base learner evaluations. Moreover, the label-wise selection of base learners can be realized by a straight-forward grid search.

As the label-wise choice of a base learner has already led to considerable performance gains, we plan to examine to what extent the optimization of the hyper-parameters of those base learners can lead to further improvements. Furthermore, we want to increase the efficiency of the tuning by replacing the grid search with a heuristic approach.

Another direction of future work concerns the avoidance of over-fitting effects due to an overly excessive flexibility of LiBRe. As already explained, the restriction to a single base learner can be seen as a kind of regularization, which, however, appears to be too strong, at least according to our results. On the other side, the full flexibility of LiBRe does not always pay off either. An interesting compromise could be to restrict the number of different base learners used by LiBRe to a suitable value $k \in \{1, \ldots, m\}$. Technically, this comes down to finding the arg min in (4), not over $a \in \mathcal{A}^m$, but over $\{a \in \mathcal{A}^m \mid \#\{a_1, \ldots, a_m\} \leq k\}$.

Acknowledgement. This work was supported by the German Research Foundation (DFG) within the Collaborative Research Center "On-The-Fly Computing" (SFB 901/3 project no. 160364472). The authors also gratefully acknowledge support of this project through computing time provided by the Paderborn Center for Parallel Computing (PC^2).

References

1. Antonucci, A., Corani, G., Mauá, D.D., Gabaglio, S.: An ensemble of Bayesian networks for multilabel classification. In: IJCAI 2013, Proceedings of the 23rd International Joint Conference on Artificial Intelligence, Beijing, China, 3–9 August 2013, pp. 1220–1225 (2013)
2. Barutcuoglu, Z., Schapire, R.E., Troyanskaya, O.G.: Hierarchical multi-label prediction of gene function. Bioinformatics **22**(7), 830–836 (2006). https://doi.org/10.1093/bioinformatics/btk048
3. Boutell, M.R., Luo, J., Shen, X., Brown, C.M.: Learning multi-label scene classification. Pattern Recogn. **37**(9), 1757–1771 (2004). https://doi.org/10.1016/j.patcog.2004.03.009
4. Cabral, R.S., la Torre, F.D., Costeira, J.P., Bernardino, A.: Matrix completion for multi-label image classification. In: 25th Annual Conference on Neural Information Processing Systems 2011, Advances in Neural Information Processing Systems, Granada, Spain, vol. 24, pp. 190–198 (2011)
5. Charte, F., Rivera, A.J., del Jesús, M.J., Herrera, F.: Addressing imbalance in multilabel classification: measures and random resampling algorithms. Neurocomputing **163**, 3–16 (2015). https://doi.org/10.1016/j.neucom.2014.08.091
6. Cherman, E.A., Metz, J., Monard, M.C.: Incorporating label dependency into the binary relevance framework for multi-label classification. Exp. Syst. Appl. **39**(2), 1647–1655 (2012). https://doi.org/10.1016/j.eswa.2011.06.056
7. Dembczynski, K., Waegeman, W., Cheng, W., Hüllermeier, E.: On label dependence and loss minimization in multi-label classification. Mach. Learn. **88**(1–2), 5–45 (2012). https://doi.org/10.1007/s10994-012-5285-8
8. Frank, E., Hall, M.A., Witten, I.H.: The Weka workbench. Online appendix. In: Frank, E., Hall, M.A., Witten, I.H. (eds.) Data Mining: Practical Machine Learning Tools and Techniques. Morgan Kaufmann, Cambridge (2016)
9. Luaces, O., Díez, J., Barranquero, J., del Coz, J.J., Bahamonde, A.: Binary relevance efficacy for multilabel classification. Prog. AI **1**(4), 303–313 (2012). https://doi.org/10.1007/s13748-012-0030-x
10. Madjarov, G., Kocev, D., Gjorgjevikj, D., Dzeroski, S.: An extensive experimental comparison of methods for multi-label learning. Pattern Recogn. **45**(9), 3084–3104 (2012). https://doi.org/10.1016/j.patcog.2012.03.004

11. Montañés, E., Senge, R., Barranquero, J., Quevedo, J.R., del Coz, J.J., Hüllermeier, E.: Dependent binary relevance models for multi-label classification. Pattern Recogn. **47**(3), 1494–1508 (2014). https://doi.org/10.1016/j.patcog.2013.09.029
12. Moyano, J.M., Galindo, E.L.G., Cios, K.J., Ventura, S.: Review of ensembles of multi-label classifiers: models, experimental study and prospects. Inf. Fusion **44**, 33–45 (2018). https://doi.org/10.1016/j.inffus.2017.12.001
13. Pillai, I., Fumera, G., Roli, F.: Threshold optimisation for multi-label classifiers. Pattern Recogn. **46**(7), 2055–2065 (2013). https://doi.org/10.1016/j.patcog.2013.01.012
14. Qi, G., Hua, X., Rui, Y., Tang, J., Mei, T., Zhang, H.: Correlative multi-label video annotation. In: Proceedings of the 15th International Conference on Multimedia 2007, Augsburg, Germany, 24–29 September 2007, pp. 17–26 (2007). https://doi.org/10.1145/1291233.1291245
15. Read, J., Pfahringer, B., Holmes, G., Frank, E.: Classifier chains for multi-label classification. Mach. Learn. **85**(3), 333–359 (2011). https://doi.org/10.1007/s10994-011-5256-5
16. Read, J., Reutemann, P., Pfahringer, B., Holmes, G.: MEKA: a multi-label/multi-target extension to Weka. J. Mach. Learn. Res. **17**(21), 667–671 (2016)
17. de Sá, A.G.C., Freitas, A.A., Pappa, G.L.: Automated selection and configuration of multi-label classification algorithms with grammar-based genetic programming. Parallel Prob. Solving Nat. - PPSN XV **2018**, 308–320 (2018). https://doi.org/10.1007/978-3-319-99259-4_25
18. Sanden, C., Zhang, J.Z.: Enhancing multi-label music genre classification through ensemble techniques. In: Proceeding of the 34th International ACM SIGIR Conference on Research and Development in Information Retrieval, Beijing, China, pp. 705–714 (2011). https://doi.org/10.1145/2009916.2010011
19. Schapire, R.E., Singer, Y.: BoosTexter: a boosting-based system for text categorization. Mach. Learn. **39**(2/3), 135–168 (2000). https://doi.org/10.1023/A:1007649029923
20. Sucar, L.E., Bielza, C., Morales, E.F., Hernandez-Leal, P., Zaragoza, J.H., Larrañaga, P.: Multi-label classification with bayesian network-based chain classifiers. Pattern Recogn. Lett. **41**, 14–22 (2014). https://doi.org/10.1016/j.patrec.2013.11.007
21. Tahir, M.A., Kittler, J., Bouridane, A.: Multi-label classification using stacked spectral kernel discriminant analysis. Neurocomputing **171**, 127–137 (2016). https://doi.org/10.1016/j.neucom.2015.06.023
22. Tsoumakas, G., Katakis, I.: Multi-label classification: an overview. IJDWM **3**(3), 1–13 (2007). https://doi.org/10.4018/jdwm.2007070101
23. Tsoumakas, G., Katakis, I., Vlahavas, I.P.: Mining multi-label data. In: Maimon, O., Rokach, L. (eds.) Data Mining and Knowledge Discovery Handbook, pp. 667–685. Springer, Boston (2010). https://doi.org/10.1007/978-0-387-09823-4_34
24. Wever, M., Mohr, F., Hüllermeier, E.: Automated multi-label classification based on ML-Plan. CoRR abs/1811.04060 (2018)
25. Wever, M.D., Mohr, F., Tornede, A., Hüllermeier, E.: Automating multi-label classification extending ML-Plan (2019)
26. Xue, X., Zhang, W., Zhang, J., Wu, B., Fan, J., Lu, Y.: Correlative multi-label multi-instance image annotation. In: IEEE International Conference on Computer Vision, pp. 651–658 (2011). https://doi.org/10.1109/ICCV.2011.6126300
27. Zhang, M., Li, Y., Liu, X.: Towards class-imbalance aware multi-label learning. In: Proceedings of the 24th International Joint Conference on Artificial Intelligence, IJCAI 2015, Buenos Aires, Argentina, 2015, pp. 4041–4047 (2015)

28. Zhang, M.-L., Li, Y.-K., Liu, X.-Y., Geng, X.: Binary relevance for multi-label learning: an overview. Frontiers Comput. Sci. **12**(2), 191–202 (2018). https://doi.org/10.1007/s11704-017-7031-7
29. Zhang, M., Zhou, Z.: A review on multi-label learning algorithms. IEEE Trans. Knowl. Data Eng. **26**(8), 1819–1837 (2014). https://doi.org/10.1109/TKDE.2013.39

Angle-Based Crowding Degree Estimation for Many-Objective Optimization

Yani Xue[1(✉)], Miqing Li[2], and Xiaohui Liu[1]

[1] Department of Computer Science, Brunel University London,
Uxbridge, Middlesex UB8 3PH, UK
ynxue6219@gmail.com
[2] CERCIA, School of Computer Science, University of Birmingham,
Edgbaston, Birmingham B15 2TT, UK

Abstract. Many-objective optimization, which deals with an optimization problem with more than three objectives, poses a big challenge to various search techniques, including evolutionary algorithms. Recently, a meta-objective optimization approach (called bi-goal evolution, BiGE) which maps solutions from the original high-dimensional objective space into a bi-goal space of proximity and crowding degree has received increasing attention in the area. However, it has been found that BiGE tends to struggle on a class of many-objective problems where the search process involves *dominance resistant solutions*, namely, those solutions with an extremely poor value in at least one of the objectives but with (near) optimal values in some of the others. It is difficult for BiGE to get rid of dominance resistant solutions as they are Pareto nondominated and far away from the main population, thus always having a good crowding degree. In this paper, we propose an angle-based crowding degree estimation method for BiGE (denoted as aBiGE) to replace distance-based crowding degree estimation in BiGE. Experimental studies show the effectiveness of this replacement.

Keywords: Many-objective optimization · Evolutionary algorithm · Bi-goal evolution · Angle-based crowding degree estimation

1 Introduction

Many-objective optimization problems (MaOPs) refer to the optimization of four or more conflicting criteria or objectives at the same time. MaOPs exist in many fields, such as environmental engineering, software engineering, control engineering, industry, and finance. For example, when assessing the performance of a machine learning algorithm, one may need to take into account not only accuracy but also some other criteria such as efficiency, misclassification cost, interpretability, and security.

There is often no one best solution for an MaOP since the performance increase in one objective will lead to a decrease in some other objectives.

M. R. Berthold et al. (Eds.): IDA 2020, LNCS 12080, pp. 574–586, 2020.
https://doi.org/10.1007/978-3-030-44584-3_45

In the past three decades, multi-objective evolutionary algorithms (MOEAs) have been successfully applied in many real-world optimization problems with low-dimensional search space (two or three conflicting objectives) to search for a set of trade-off solutions.

The major purpose of MOEAs is to provide a population (a set of optimal individuals or solutions) that balance proximity (converging a population to the Pareto front) and diversity(diversifying a population over the whole Pareto front). By considering the two goals above, traditional MOEAs, such as SPEA2 [13] and NSGA-II [1] mainly focus on the use of Pareto dominance relations between solutions and the design of diversity control mechanisms.

However, compared with a low-dimensional optimization problem, well-known Pareto-based evolutionary algorithms lose their efficiency in solving MaOPs. In MaOPs, most solutions in a population become equally good solutions, since the Pareto dominance selection criterion fails to distinguish between solutions and drive the population towards the Pareto front. Then the density criterion is activated to guide the search, resulting in a substantial reduction of the convergence of the population and the slowdown of the evolution process. This is termed the *active diversity promotion* (ADP) phenomenon in [11].

Some studies [6] observed that the main reason for ADP phenomenon is the preference of dominance resistant solutions (DRSs). DRSs refer to those solutions that are extremely inferior to others in at least one objective but have near-optimal values in some others. They are considered as Pareto-optimal solutions despite having very poor performance in terms of proximity. As a result, Pareto-based evolutionary algorithms could search a population that is widely covered but far away from the true Pareto front.

To address the difficulties of MOEAs in high-dimensional search space, one approach is to modify the Pareto dominance relation. Some powerful algorithms in this category include: ϵ-MOEA [2] and fuzzy Pareto dominance [5]. These methods work well under certain circumstances but they often involve extra parameters and the performance of these algorithms often depends on the setting of parameters. The other approach, without considering Pareto dominance relation, may be classified into two categories: aggregation-based algorithms [15] and indicator-based algorithms [14]. These algorithms have been successfully applied to some applications, however, the diversity performance of these aggregation-based algorithms depends on the distribution of weight vectors. The latter defines specific performance indicators to guide the search.

Recently, a meta-objective optimization algorithm, called Bi-Goal Evolution (BiGE) [8] for MaOPs is proposed and becomes the most cited paper published in the Artificial Intelligence journal over the past four years. BiGE was inspired by two observations in many-objective optimization: (1) the conflict between proximity and diversity requirement is aggravated when increasing the number of objectives and (2) the Pareto dominance relation is not effective in solving MaOPs. In BiGE, two indicators were used to estimate the proximity and crowding degree of solutions in the population, respectively. By doing so, BiGE maps solutions from the original objective space to a bi-goal objective space and deals with the two goals by the nondominated sorting. This is able to provide sufficient selection pressure towards the Pareto front, regardless of the number of objectives that the optimization problem has.

However, despite its attractive features, it has been found that BiGE tends to struggle on a class of many-objective problems where the search process involves DRSs. DRSs are far away from the main population and always ranked as good solutions by BiGE, thus hindering the evolutionary progress of the population. To address this issue, this paper proposes an angle-based crowding degree estimation method for BiGE (denoted as aBiGE). The rest of the paper is organized as follows. Section 2 gives some concepts and terminology about many-objective optimization. In Sect. 3, we present our angle-based crowding degree estimation method and its incorporation with BiGE. The experimental results are detailed in Sect. 4. Finally, the conclusions and future work are set out in Sect. 5.

2 Concepts and Terminology

When dealing with optimization problems in the real world, sometimes it may involve more than three performance criteria to determine how "good" a certain solution is. These criteria, termed as objectives (e.g., cost, safety, efficiency) need to be optimized simultaneously, but usually conflict with each other. This type of problem is called many-objective optimization problem (MaOP). A minimization MaOP can be mathematically defined as follows:

$$
\begin{aligned}
\text{minimize} \quad & F(x) = (f_1(x), f_2(x), ..., f_N(x)) \\
\text{subject to} \quad & g_j(x) \leq 0, \quad j = 1, 2, ..., J \\
& h_k(x) = 0, \quad k = 1, 2, ..., K \\
& x = (x_1, x_2, ..., x_M), \quad x \in \Omega
\end{aligned}
\tag{1}
$$

where x denotes an M-dimensional decision variable vector from the feasible region in the decision space Ω, $F(x)$ represents an N-dimensional objective vector (N is larger than three), $f_i(x)$ is the i-th objective to be minimized, objective functions $f_1, f_2, ..., f_N$ constitute N-dimensional space called the objective space, $g_j(x) \leq 0$ and $h_k(x) = 0$ define J inequality and K equality constraints, respectively.

Definition 1 (Pareto Dominance). *Given two decision vectors $x, y \in \Omega$ of a minimization problem, x is said to (Pareto) dominate y (denoted as $x \prec y$), or equivalently y is dominated by x, if and only if [4]*

$$
\forall i \in (1, 2, ..., N) : f_i(x) \leq f_i(y) \land \exists i \in (1, 2, ..., N) : f_i(x) < f_i(y).
\tag{2}
$$

Namely, given two solutions, one solution is said to dominate the other solution if it is at least as good as the other solution in any objective and is strictly better in at least one objective.

Definition 2 (Pareto Optimality). *A solution* $x \in \Omega$ *is said to be Pareto optimal if and only if there is no solution* $y \in \Omega$ *dominates it. Those solutions that are not dominated by any other solutions is said to be Pareto-optimal (or non-dominated).*

Definition 3 (Pareto Set). *All Pareto-optimal (or non-dominated) solutions in the decision space constitute the Pareto set (PS).*

Definition 4 (Pareto Front). *The Pareto front (PF) is referred to corresponding objective vectors to a Pareto set.*

Definition 5 (Dominance Resistant Solution). *Given a solution set, dominance resistant solution (DRS) is referred to the solution with an extremely poor value in at least one objective, but with near-optimal value in some other objective.*

3   The Proposed Algorithm: aBiGE

3.1   A Brief Review of BiGE

Algorithm 1 shows the basic framework of BiGE. First, a parent population with M solutions is randomly initialized. Second, proximity and crowding degree for each solution is estimated, respectively. Third, in the mating selection, individuals that have better quality with regards to the proximity and crowding degree tend to become parents of the next generation. Afterward, variation operators (e.g., crossover and mutation) are applied to these parents to produce an offspring population. Finally, the environmental selection is applied to reduce the expanded population of parents and offspring to M individuals as the new parent population of the next generation.

Algorithm 1. Basic Framework of BiGE

Require: P (current population), M (population size)
  1: $P = Initialization(P)$
  2: **while** termination criterion not fulfilled **do**
  3:      $Proximity\_Estimation(P)$
  4:      $Crowding\_Degree\_Estimation(P)$
  5:      $P' = Mating\_Selection(P)$
  6:      $P'' = Variation(P')$
  7:      $Q = P' \bigcup P''$
  8:      $P = Environmental\_Selection(Q)$
  9: **end while**
 10: **return** P

In particular, a simple aggregation function is adopted to estimate the proximity of an individual. For an individual x in a population, denoted as $f_p(x)$, its aggregation value is calculated by the sum of each normalized objective value in the range $[0, 1]$ (lines 3 in Algorithm 1), formulated as [8]:

$$f_p(x) = \sum_{j=1}^{N} \widetilde{f}_j(x). \tag{3}$$

where $\widetilde{f}_j(x)$ denotes the normalized objective value of individual x in the j-th objective, and N is the number of objectives. A smaller f_p value of an individual usually indicates a good performance on proximity. In particular, for a DRS, it is more likely to obtain a significantly large f_p value in comparison with other individuals in a population.

In addition, the crowding degree of an individual x (lines 4 in Algorithm 1) is defined as follows [8]:

$$f_c(x) = (\sum_{y \in \Omega, x \neq y} sh(x, y))^{1/2}. \tag{4}$$

where $sh(x, y))^{1/2}$ denotes a sharing function. It is a penalized Euclidean distance between two individuals x and y by using a weight parameter, defined as follows:

$$sh(x, y) = \begin{cases} (0.5(1 - \frac{d(x,y)}{r}))^2, & \text{if } d(x,y) < r, f_p(x) < f_p(y) \\ (1.5(1 - \frac{d(x,y)}{r}))^2, & \text{if } d(x,y) < r, f_p(x) > f_p(y) \\ rand(), & \text{if } d(x,y) < r, f_p(x) = f_p(y) \\ 0, & \text{otherwise} \end{cases} \tag{5}$$

where r is the radius of a niche, adaptively calculated by $r = 1/\sqrt[N]{M}$ (M is the population size and N is the number of objectives). The function $rand()$ means to assign $either$ $sh(x, y) = (0.5(1-[d(x, y)/r]))^2$ and $sh(y, x) = (1.5(1-[d(x, y)/r]))^2$ or $sh(x, y) = (1.5(1-[d(x, y)/r]))^2$ and $sh(y, x) = (0.5(1-[d(x, y)/r]))^2$ randomly. Individuals with lower crowding degree imply better performance on diversity.

It is observed that BiGE tends to struggle on a class of MaOPs where the search process involves DRSs, such as DTLZ1 and DTLZ3 (in a well-known benchmark test suite DTLZ [3]). Figure 1 shows the true Pareto front of the eight-objective DTLZ1 and the final solution set of BiGE in one typical run on the eight-objective DTLZ1 by parallel coordinates. The parallel coordinates map the original many-objective solution set to a 2D parallel coordinates plane. Particularly, Li et al. in [9] systematically explained how to read many-objective solution sets in parallel coordinates, and indicates that parallel coordinates can $partly$ reflect the quality of a solution set in terms of convergence, coverage, and uniformity.

Clearly, there are some solutions that are far away from the Pareto front in BiGE, with the solution set of eight-objective DTLZ1 ranging from 0 to around 450 compared to the Pareto front ranging from 0 to 0.5 on each objective. Such solutions always have a poor proximity degree and a good crowding degree (estimated by Euclidean distance)in bi-goal objective space (i.e., convergence and

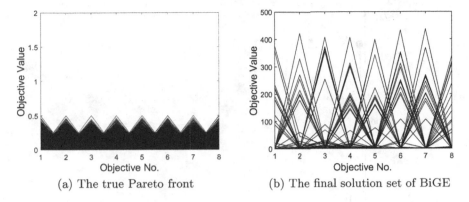

(a) The true Pareto front   (b) The final solution set of BiGE

Fig. 1. The true Pareto front and the final solution set of BiGE on the eight-objective DTLZ1, shown by parallel coordinates.

diversity), and will be preferred since there is no solution in the population that dominates them in BiGE. These solutions are detrimental for BiGE to converge the population to the Pareto front considering their poor performance in terms of convergence. A straightforward method to remove DRSs is to change the crowding degree estimation method.

3.2   Basic Idea

The basic idea of the proposed method is based on some observations of DRSs. Figure 2 shows one typical situation of a non-dominated set with five individuals including two DRSs (i.e, A and E) in a two-dimensional objective minimization scenario.

As seen, it is difficult to find a solution that could dominate DRSs by estimating the crowding degree using Euclidean distance. Take individual A as an example, it performs well on objective f_1 (slightly better than B with a near-optimal value 0) but inferior to all the other solutions on objective f_2. It is difficult to find a solution with better value than A on objective f_1, same as individual E on objective f_2. A and E (with poor proximity degree and good crowding degree) are considered as good solutions and have a high possibility to survive in the next generation in BiGE. However, the results would be different if the distance-based crowding degree estimation is replaced by a vector angle. It can be observed that (1) an individual in a crowded area would have a smaller vector angle to its neighbor compared to the individual in a sparse area, e.g., C and D, (2) a DRS would have an extremely small value of vector angle to its neighbor, e.g., the angle between A and B or the angle between E and D. Namely, these DRSs would be assigned both poor proximity and crowding degrees, and have a high possibility to be deleted during the evolutionary process. Therefore, vector angles have the advantage to distinguish DRSs in the population and could be considered into crowding degree estimation.

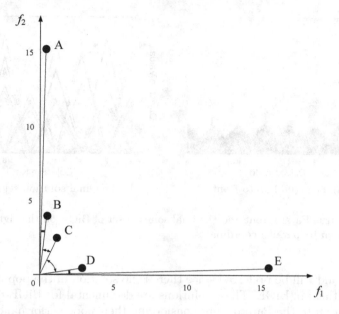

Fig. 2. An illustration of a population of five solutions with two DRSs - A and E. They have good crowding degrees estimated by the Euclidean distance, but poor crowding degrees calculated by the vector angle between two neighbors.

3.3   Angle-Based Crowding Degree Estimation

Inspired by the work in [12], we propose a novel angle-based crowding degree estimation method, and integrate it into the BiGE framework (line 4 in Algorithm 1), called aBiGE. Before estimating the diversity of an individual in a population in aBiGE, we first introduce some basic definitions.

Norm. For individual x_i, its norm, denoted as $norm(x_i)$ in the normalzied objective space defined as [12]:

$$norm(x_i) = \sqrt{\sum_{j=1}^{N} \widetilde{f}_j(x_i)^2}. \tag{6}$$

Vector Angles. The vector angle between two individuals x_i and x_k is defined as follows [12]:

$$angle_{x_i \to x_k} = arccos \left| \frac{F'(x_i) \bullet F'(x_k)}{norm(x_i) \cdot norm(x_k)} \right|. \tag{7}$$

where $F'(x_i) \bullet F'(x_k)$ is the inner product between $F'(x_i)$ and $F'(x_k)$ defined as:

$$F'(x_i) \bullet F'(x_k) = \sum_{j=1}^{N} \widetilde{f}_j(x_i) \cdot \widetilde{f}_j(x_k). \tag{8}$$

Note that angle $_{x_i \to x_k} \in [0, \pi/2]$.

The vector angle from an individual $x_i \in \Omega$ to the population is defined as the minimum vector angle between x_i and another individual in a population P: $\theta(x_i) = angle_{x_i \to P}$

When an individual x is selected into archive in the environmental selection, respectively, $\theta(x)$ value will be punished. There are several factors need to be considered in order to achieve a good balance between proximity and diversity.

- A severe penalty should be imposed on individuals that have more adjacent individuals in a niche. Inspired by the punishment method of crowding degree estimation, a punishment to an individual x is based on the number of individuals that have a lower proximity degree compared to x is counted (denote as c). The punishment is aggravated with an increase of c.
- In order to avoid the situation that some individuals have the same vector angle value to the population, individuals should be further punished. Therefore, the penalty is implemented according to the proportion value of $\theta(x)$ to all the individuals in the niche, denoted as p.

Keep the above factors in mind, in aBiGE, the diversity estimation of individual $x \in \Omega$ based on vector angles is defined as

$$f_a(x) = \frac{c+1}{\theta(x) \cdot (p+1) + \dfrac{\pi}{90}}. \tag{9}$$

By applying the angle-based crowding degree estimation method to BiGE framework in minimizing many-objective optimization problems, we aim to enhance the selection pressure on those non-dominated solutions in the population of each generation and avoid the negative influence of DRSs in the optimization process. Note that, a smaller value of $f_a(x)$ is preferred.

4   Experiments

4.1   Experimental Design

To test the performance of the proposed aBiGE on those MaOPs where the search process involves DRSs, the experiments are conducted on nine DTLZ test problems. For each test problem (i.e., DTLZ1, DTLZ3, and DTLZ7), five, eight, and ten objectives will be considered, respectively.

To make a fair comparison with the state-of-the-art BiGE for MaOPs, we kept the same settings as [8]. Settings for both BiGE and aBiGE are:

- The population size of both algorithms is set to 100 for all test problems.

- 30 runs for each algorithm per test problem to decrease the impact of their stochastic nature.
- The termination criterion of a run is a predefined maximum of 30,000 evaluations, namely 300 generations for test problems.
- For crossover and mutation operators, crossover and mutation probability are set to 1.0 and $1/M$ (where M represents the number of decision variables) respectively. In particular, uniform crossover and polynomial mutation are used.

Algorithms performance is assessed by performance indicators that consider both proximity and diversity. In this paper, a modified version of the original inverted generational distance indicator (IGD) [15], called (IGD+) [7] is chosen as the performance indicator. Although IGD has been widely used to evaluate the performance of MOEAs on MaOPs, it has been shown [10] that IGD needs to be replaced by IGD+ to make it compatible with Pareto dominance. IGD+ evaluates a solution set in terms of both convergence and diversity, and a smaller value indicates better quality.

4.2   Performance Comparison

Test Problems with DRSs. Table 1 shows the mean and standard deviation of IGD+ metric results on nine DTLZ test problems with DRSs. For each test problem, among different algorithms, the algorithm that has the best result based on the IGD+ metric is shown in bold. As can be seen from the table, for MaOPs with DRSs, the proposed aBiGE performs significantly better than BiGE on all test problems in terms of convergence and diversity.

Table 1. Mean and standard deviation of IGD+ metric on nine DTLZ test problems. The best result for each test problem is highlighted in boldface.

Problem	Obj.	BiGE	aBiGE
DTLZ1	5	8.4207E−01 (3.59E−01)	**1.1768E−01** (3.41E−02)
	8	1.9350E+00 (1.27E+00)	**1.9495E−01** (9.44E−02)
	10	1.9653E+00 (1.36E+00)	**2.2763E−01** (9.57E−02)
DTLZ3	5	1.5705E+01 (5.87E+00)	**6.0008E+00** (3.50E+00)
	8	3.3434E+01 (1.17E+01)	**9.6401E+00** (6.30E+00)
	10	3.5720E+01 (1.58E+01)	**1.2780E+01** (5.40E+00)
DTLZ7	5	4.6666E−01 (1.52E−01)	**3.1701E−01** (6.48E−02)
	8	3.0415E+00 (6.03E−01)	**2.6350E+00** (8.59E−01)
	10	5.6152E+00 (7.41E−01)	**4.0059E+00** (4.53E−01)

To visualize the experimental results, Figs. 3 and 4 plot, by parallel coordinate, the final solutions of one run with respect to five-objective DTLZ1 and five-objective DTLZ7, respectively. This run is associated with the particular run with the closest results to the mean value of IGD+. As shown in Fig. 3(a), the approximation set obtained by BiGE has an inferior convergence on the five-objective DTLZ1, with the range of its solution set is between 0 and about 400 in contrast to the Pareto front ranging from 0 to 0.5 on each objective. From Fig. 3 (b), it can be observed that the obtained solution set of the proposed aBiGE converge to the Pareto front and only a few individuals do not converge.

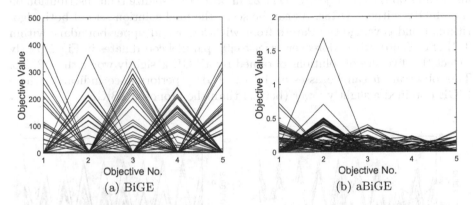

(a) BiGE   (b) aBiGE

Fig. 3. The final solution sets of the two algorithms on the five-objective DTLZ1, shown by parallel coordinates.

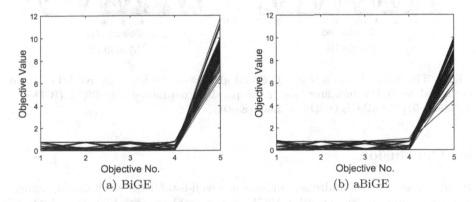

(a) BiGE   (b) aBiGE

Fig. 4. The final solution sets of the two algorithms on the five-objective DTLZ7, shown by parallel coordinates.

For the solutions of the five-objective DTLZ7, the boundary of the first four objectives is in the range [0, 1], and the boundary of the last objective is in the range [3.49, 10] according to the formula of DTLZ7. As can be seen from (Fig. 4), all solutions of the proposed aBiGE appear to converge into the Pareto front. In contrast, some solutions (with objective value beyond the upper boundary in 5th objective) of BiGE fail to reach the Pareto front. In addition, the solution set of the proposed aBiGE has better extensity than BiGE on the boundaries. In particular, the solution set of BiGE fails to cover the region from 3.49 to 6 of the last objective and the solution set of the proposed aBiGE does not cover the range of Pareto front below 4 on 5th objective.

Test Problem Without DRSs. Figure 5 gives the final solution set of both algorithms on the ten-objective DTLZ2 in order to visualize their distribution on the MaOPs without DRSs. As can be seen, the final solution sets of both algorithms could coverage the Pareto front with lower and upper boundary within [0,1] of each objective. Moreover, refer to [9], parallel coordinates in Fig. 5 partly reflect the diversity of solutions obtained by aBiGE is sightly worse than BiGE. This observation can be assessed by the IGD+ performance indicator where BiGE obtained a slightly lower (better) than the proposed aBiGE.

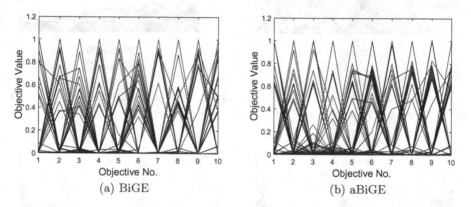

Fig. 5. The final solution sets of BiGE and aBiGE on the ten-objective DTLZ2 and evaluated by IGD+ indicator, shown by parallel coordinates. (a) BiGE (IGD+ = 2.4319E−01) (b) aBiGE (IGD+ = 2.5021E−01).

5   Conclusion

In this paper, we have addressed an issue of a well-established evolutionary many-objective optimization algorithm BiGE on the problems with high probability to produce dominance resistant solutions during the search process. We have proposed an angle-based crowding distance estimation method to replace distance-based estimation in BiGE, thus significantly reducing the effect of dominance resistant solutions to the algorithm. The effectiveness of the proposed method has been well evaluated on three representative problems with dominance resistant solutions. It is worth mentioning that for problems without dominance resistant solutions the proposed method performs slightly worse than the original BiGE.

In the near future, we would like to focus on the problems without dominance resistant solutions, aiming at a comprehensive improvement of the algorithm on both types of problems.

References

1. Deb, K., Pratap, A., Agarwal, S., Meyarivan, T.: A fast and elitist multiobjective genetic algorithm: NSGA-II. IEEE Trans. Evol. Comput. **6**(2), 182–197 (2002)
2. Deb, K., Mohan, M., Mishra, S.: Evaluating the ε-domination based multi-objective evolutionary algorithm for a quick computation of pareto-optimal solutions. Evol. Comput. **13**(4), 501–525 (2005)
3. Deb, K., Thiele, L., Laumanns, M., Zitzler, E.: Scalable test problems for evolutionary multiobjective optimization. In: Abraham, A., Jain, L., Goldberg, R. (eds.) Evolutionary Multiobjective Optimization. Advanced Information and Knowledge Processing. Springer, London (2005). https://doi.org/10.1007/1-84628-137-7_6
4. Fonseca, C.M., Fleming, P.J.: An overview of evolutionary algorithms in multiobjective optimization. Evol. Comput. **3**(1), 1–16 (1995)
5. He, Z., Yen, G.G., Zhang, J.: Fuzzy-based pareto optimality for many-objective evolutionary algorithms. IEEE Trans. Evol. Comput. **18**(2), 269–285 (2014)
6. Ishibuchi, H., Tsukamoto, N., Nojima, Y.: Evolutionary many-objective optimization: a short review. In: 2008 IEEE Congress on Evolutionary Computation. IEEE World Congress on Computational Intelligence, pp. 2419–2426. IEEE (2008)
7. Ishibuchi, H., Masuda, H., Tanigaki, Y., Nojima, Y.: Modified distance calculation in generational distance and inverted generational distance. In: Gaspar-Cunha, A., Henggeler Antunes, C., Coello, C.C. (eds.) EMO 2015. LNCS, vol. 9019, pp. 110–125. Springer, Cham (2015). https://doi.org/10.1007/978-3-319-15892-1_8
8. Li, M., Yang, S., Liu, X.: Bi-goal evolution for many-objective optimization problems. Artif. Intell. **228**, 45–65 (2015)
9. Li, M., Zhen, L., Yao, X.: How to read many-objective solution sets in parallel coordinates (educational forum). IEEE Comput. Intell. Mag. **12**(4), 88–100 (2017)
10. Li, M., Yao, X.: Quality evaluation of solution sets in multiobjective optimisation: a survey. ACM Comput. Surv. (CSUR) **52**(2), 1–38 (2019)
11. Purshouse, R.C., Fleming, P.J.: On the evolutionary optimization of many conflicting objectives. IEEE Trans. Evol. Comput. **11**(6), 770–784 (2007)
12. Xiang, Y., Zhou, Y., Li, M., Chen, Z.: A vector angle-based evolutionary algorithm for unconstrained many-objective optimization. IEEE Trans. Evol. Comput. **21**(1), 131–152 (2017)
13. Zitzler, E., Laumanns, M., Thiele, L.: SPEA2: improving the strength Pareto evolutionary algorithm. TIK-report **103** (2001)
14. Zitzler, E., Künzli, S.: Indicator-based selection in multiobjective search. In: Yao, X., et al. (eds.) PPSN 2004. LNCS, vol. 3242, pp. 832–842. Springer, Heidelberg (2004). https://doi.org/10.1007/978-3-540-30217-9_84
15. Zhang, Q., Li, H.: MOEA/D: a multiobjective evolutionary algorithm based on decomposition. IEEE Trans. Evol. Comput. **11**(6), 712–731 (2007)

Author Index

Printed in the United States
By Bookmasters